Lecture Notes in Computer Science 4271

Commenced Publication in 1973
Founding and Former Series Editors:
Gerhard Goos, Juris Hartmanis, and Jan van Leeuwen

Fedor V. Fomin (Ed.)

Graph-Theoretic Concepts in Computer Science

32nd International Workshop, WG 2006
Bergen, Norway, June 22-24, 2006
Revised Papers

 Springer

Volume Editor

Fedor V. Fomin
Institutt for Informatikk
Universitetet i Bergen
Postboks 7800
5020 Bergen
Norway
E-mail: fomin@ii.uib.no

Library of Congress Control Number: 2006934979

CR Subject Classification (1998): F.2, G.2, G.1.6, G.1.2, E.1, I.3.5

LNCS Sublibrary: SL 1 – Theoretical Computer Science and General Issues

ISSN   0302-9743
ISBN-10   3-540-48381-0 Springer Berlin Heidelberg New York
ISBN-13   978-3-540-48381-6 Springer Berlin Heidelberg New York

Springer is a part of Springer Science+Business Media

springer.com

© Springer-Verlag Berlin Heidelberg 2006
Printed in Germany

Typesetting: Camera-ready by author, data conversion by Scientific Publishing Services, Chennai, India
Printed on acid-free paper   SPIN: 11917496   06/3142   5 4 3 2 1 0

Preface

The 32nd International Workshop on Graph-Theoretic Concepts in Computer Science (WG 2006) was held on the island of Sotra close to the city of Bergen on the west coast of Norway. The workshop was organized by the Algorithms Research Group at the Department of Informatics, University of Bergen, and it took place from June 22 to June 24. The 78 participants of WG 2006 came from the universities and research institutes of 17 different countries.

The WG 2006 workshop continues the series of 31 previous WG workshops. Since 1975, WG has taken place 20 times in Germany, four times in The Netherlands, twice in Austria as well as once in France, in Italy, in Slovakia, in Switzerland and in the Czech Republic, and has now been held for the first time in Norway. The workshop aims at uniting theory and practice by demonstrating how graph-theoretic concepts can be applied to various areas in computer science, or by extracting new problems from applications. The goal is to present recent research results and to identify and explore directions of future research. The talks showed how recent research results from algorithmic graph theory can be used in computer science and which graph-theoretic questions arise from new developments in computer science. There were two fascinating invited lectures by Hans Bodlaender (Utrecht, The Netherlands) and Tandy Warnow (Austin, USA).

The number of submitted papers was 91. In a careful reviewing process with four reports per submission, the Program Committee selected 30 papers for presentation at the workshop and several high quality papers had to be rejected.

With much pleasure, I thank all those who contributed to the success of WG 2006: the authors who submitted their work to the workshop, the speakers, the Program Committee members and the referees. I am indebted to the members of the Local Organization Committee: Frederic Dorn, Federico Mancini, Fredrik Manne, Alexey Stepanov, Jan Arne Telle, and especially Pinar Heggernes. Without their engagement and the help of various students during the meeting, WG 2006 could not have been such a great event.

Special thanks go to the sponsoring organizations: the University of Bergen, the Research Council of Norway, and IBM.

August 2006   Fedor V. Fomin

Organization

The Tradition of WG

1975   U. Pape – Berlin, Germany

1976   H. Noltemeier – Göttingen, Germany

1977   J. Mühlbacher – Linz, Austria

1978   M. Nagl, H.J. Schneider – Castle Feuerstein, Germany

1979   U. Pape – Berlin, Germany

1980   H. Noltemeier – Bad Honnef, Germany

1981   J. Mühlbacher – Linz, Austria

1982   H.J. Schneider, H. Göttler – Neuenkirchen, Germany

1983   M. Nagl, J. Perl – Haus Ohrbeck, Germany

1984   U. Pape – Berlin, Germany

1985   H. Noltemeier – Castle Schwanberg, Germany

1986   G. Tinhofer, G. Schmidt – Bernried, Germany

1987   H. Göttler, H.J. Schneider – Kloster Banz/Staffelstein, Germany

1988   J. van Leeuwen – Amsterdam, The Netherlands

1989   M. Nagl – Castle Rolduc, The Netherlands

1990   R. Möhring – Berlin, Germany

1991   G. Schmidt, R. Berghammer – Fischbachau, Germany

1992   E.W. Mayr – Wiesbaden-Naurod, Germany

1993   J. van Leeuwen – Utrecht, The Netherlands

1994   G. Tinhofer, E.W. Mayr, G. Schmidt – Herrsching, Germany

1995   M. Nagl – Aachen, Germany

1996   G. Ausiello, A. Marchetti-Spaccamela – Como, Italy

1997   R. Möhring – Berlin, Germany

1998   J. Hromkovič – Smolenice, Slovak Republic

1999   P. Widmayer – Ascona, Switzerland

2000   D. Wagner – Konstanz, Germany

2001   A. Brandstädt – Boltenhagen near Rostock, Germany

2002   L. Kučera – Cesky Krumlov, Czech Republic

2003   H.L. Bodlaender – Elspeet, The Netherlands

2004   J. Hromkovič, M. Nagl – Bad Honnef, Germany

2005   D. Kratsch – Metz, France

2006   F.V. Fomin – Bergen, Norway

Program Committee

Andreas Brandstädt	University of Rostock, Germany
Jianer Chen	Texas A&M University, USA
Derek Corneil	University of Toronto, Canada
Fedor V. Fomin	University of Bergen, Norway (chair)
Frédéric Havet	CNRS, Sophia-Antipolis, France
Pinar Heggernes	University of Bergen, Norway
Juraj Hromkovic	ETH, Switzerland
Arie Koster	ZIB, Germany
Daniel Kral	Charles University, Czech Republic
Dieter Kratsch	University of Metz, France
Ludek Kucera	Charles University, Czech Republic
Alberto Marchetti-Spaccamela	Università di Roma "La Sapienza", Italy
Haiko Müller	University of Leeds, UK
Naomi Nishimura	University of Waterloo, Canada
Hartmut Noltemeier	Universität Würzburg, Germany
Dimitrios M. Thilikos	Universitat Politècnica de Catalunya, Spain
Dorothea Wagner	Universität Karlsruhe (TH), Germany
Peter Widmayer	ETH, Switzerland
Gerhard J. Woeginger	Eindhoven University of Technology, The Netherlands
Xiao Zhou	Tohoku University, Japan

Additional Reviewers

Andreas Tuscherer	Hajo Broersma	Zdenek Dvorak
Benjamin Hiller	Tiziana Calamoneri	Sandor Fekete
Tobias Achterberg	Ehsan Chiniforooshan	Jiri Fiala
Nadine Baumann	Serafino Cicerone	Dimitris Fotakis
Michael Baur	Sabine Cornelsen	Daniele Frigioni
Luca Becchetti	Bruno Courcelle	Marco Gaertler
Marc Benkert	Peter Damaschke	Omer Gimenez
Anne Berry	Daniel Delling	Robert Goerke
Jim Bim	Camil Demetrescu	Illya Hicks
Hans-Joachim Boecken-hauer	Danny Dyer	Patricia M. Hill
	Debora Donato	Vit Jelinek
Vincenzo Bonifaci	Feodor F. Dragan	Tommy Jensen

Jan Kara
Menelaos Karavelas
Bastian Katz
Lefteris M. Kirousis
Joachim Kneis
Ekkehard Koehler
Stavros G. Kolliopoulos
Jan Kratochvil
Bernhard Kron
Sven Krumke
Joachim Kupke
Sascha Kurz
Van Bang Le
Mathieu Liedloff
Martin Mares
Daniel Marx
Ross McConnell
Sascha Meinert

Daniel Meister
Jozef Miskuf
Tobias Moemke
Martin Noellenburg
Sang-il Oum
Ondrej Pangrac
Charis Papadopoulos
Christophe Paul
Daniel Paulusma
Marc Pfetsch
Artem Pyatkin
Bert Randerath
Michael Rao
Guido Schaefer
Thomas Schank
Ingo Schiermeyer
Jean-Sebastien Sereni
Maria Serna

Jordi Petit i Silvestre
Christian Sloper
Frits Spieksma
Ulrike Stege
Lorna Stewart
Karol Suchan
Jan Arne Telle
Ioan Todinca
Berthold Voecking
Tomas Vyskocil
Annegret Wagler
Egon Wanke
Alexander Wolff
Jozef Siran
David Wood
Florian Zickfeld

Table of Contents

Treewidth: Characterizations, Applications, and Computations

Hans L. Bodlaender

Institute of Information and Computing Sciences, Utrecht University, P.O. Box
80.089, 3508 TB Utrecht, The Netherlands
hansb@cs.uu.nl

Abstract. This paper gives a short survey on algorithmic aspects of
the treewidth of graphs. Some alternative characterizations and some
applications of the notion are given. The paper also discusses algorithms
to compute the treewidth of given graphs, and how these are based on
the different characterizations, with an emphasis on algorithms that have
been experimentally tested.

1 Introduction

For approximately a quarter of a century, the notion of treewidth is used in many
graph algorithmic and graph theoretic studies. In the 1980's, several researchers
invented independently notions that were strongly related, or equivalent: partial k-trees (Arnborg and Proskurowski, e.g., [3,7]), treewidth and tree decompositions (Robertson and Seymour [56]), clique trees (Lauritzen and Spiegelhalter [47]), recursive graph classes (Borie [26,27]), k-terminal recursive graph
classes (Wimer [64,65]), decomposition trees (Lautemann [48]), and context-free
graph grammars (Lengauer and Wanke [49]). See also [61]. Of these, the notions
treewidth and *tree decompositions* became the most used (followed by *partial
k-trees*).

This short survey discusses some applications of the notion of treewidth, and
some recent experimental work on computing the treewidth of a given graph.
Some of the algorithms to compute the treewidth are based on a different characterization of the notion. A few of such equivalent notions are also briefly surveyed.

Other surveys on treewidth are e.g., [13,15,16,43,54,55].

2 Definitions and Characterizations

We assume the reader to be familiar with standard notions from graph theory.
Throughout this paper, $n = |V|$ denotes the number of vertices of graph $G =
(V, E)$. A graph $G = (V, E)$ is *chordal*, if every cycle in G of length at least
four has a chord, i.e., there is an edge connecting two non-consecutive vertices
in the cycle. A *triangulation* of a graph $G = (V, E)$ is a graph $H = (V, F)$ that
contains G as subgraph ($F \subseteq E$) and is chordal. A triangulation H is a *minimal
triangulation* of G if there does not exist a triangulation H' of G with H' a

F.V. Fomin (Ed.): WG 2006, LNCS 4271, pp. 1–14, 2006.
© Springer-Verlag Berlin Heidelberg 2006

proper subgraph of H. A set of vertices S is a *separator*, if $G[V - S]$ is not connected. A separator is an *inclusion minimal separator*, if it does not contain another separator as proper subset.

Definition 1. *A* tree decomposition *of a graph* $G = (V, E)$ *is a pair* $(\{X_i \mid i \in I\}, T = (I, F))$ *with* $\{X_i \mid i \in I\}$ *a collection of subsets of* V, *called* bags, *and* $T = (I, F)$ *a tree, such that for all* $v \in V$, *there exists an* $i \in I$ *with* $v \in X_i$, *for all* $\{v, w\} \in E$, *there exists an* $i \in I$ *with* $v, w \in X_i$, *and for all* $v \in V$, *the set* $I_v = \{i \in I \mid v \in X_i\}$ *forms a connected subgraph (subtree) of* T.

The width *of tree decomposition* $(\{X_i \mid i \in I\}, T = (I, F))$ *equals* $\max_{i \in I} |X_i| - 1$. *The* treewidth *of a graph* G, $\mathbf{tw}(G)$, *is the minimum width of a tree decomposition of* G.

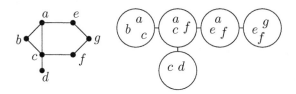

Fig. 1. A Graph and a Tree Decomposition of Width 2

An example of a tree decomposition is shown in Figure 1.

A permutation π of the vertices of a graph is called an *elimination order*. Given an elimination order π of the graph $G = (V, E)$, the *fill-in graph* of G with respect to π is constructed as follows: for $i = 1$ to n, we add an edge between each pair of higher numbered vertices of the i'th vertex in the order. An example is shown in Figure 2. When we 'eliminate" vertices with number 1, 3, 5, 6, or 7, no edges are added. The middle graph is obtained when we eliminate the vertex with number 2; the last one when we eliminate the vertex with number 4.

Elimination orderings and triangulations give alternative characterizations of treewidth. See e.g., [15] for a proof.

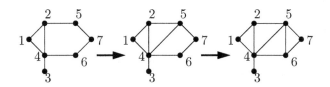

Fig. 2. Obtaining the Fill-in Graph

Theorem 1. *Let* $G = (V, E)$ *be a graph, and* $k < |V|$. *The following are equivalent.*

- *G has treewidth at most k.*
- *There is an elimination ordering π of G, such that each vertex has at most k higher numbered neighbors in the fill-in graph with respect to π.*
- *G has a triangulation with maximum clique size at most k + 1.*
- *G has a minimal triangulation with maximum clique size at most k + 1.*

Several algorithms to compute the treewidth of a graph exploit these alternative characterizations. There is an interesting connection between alternative characterizations of chordal graphs, and alternative characterizations of treewidth. Compare the following classic results with Theorem 1.

Theorem 2 (See [37,58,39]). *Let G be a graph. The following are equivalent.*

- *G is chordal.*
- *G is the intersection graph of subtrees of a tree, i.e., G has a tree decomposition such that each bag is a clique.*
- *G has a perfect elimination scheme, i.e., a permutation of the vertices such that for each vertex, its higher numbered neighbors form a clique.*

3   Applications

In this section, a number of algorithmic applications of treewidth are discussed.

3.1   Problems Restricted to Graphs of Small Treewidth

Many problems that are NP-hard (and some that are PSPACE-hard or #P-hard) on arbitrary graphs become linear or polynomial time solvable when the inputs are restricted to graphs with some constant upper bound on the treewidth. These include many of the most famous graph problems, like HAMILTONIAN CIRCUIT, INDEPENDENT SET, VERTEX COVER, etc. Most well known is the result of Courcelle [33] that each problem that can be formulated in *Monadic Second Order Logic* (MSOL) can be solved in linear time on graphs of bounded treewidth. For extensions of this result, see e.g., [5,27]. Also, several problems have tailor-made algorithms that solve them in linear or polynomial time, assuming a constant upper bound on the treewidth of the input graphs. There is a large number of papers with such a result, e.g., [7,40,45,63,66]. Such algorithms, sometimes based on a notion strongly related to treewidth (pathwidth or branchwidth) have also been used successfully in experimental settings: for the FREQUENCY ASSIGNMENT problem and for constraint satisfaction (Koster [42,45]); for the TRAVELING SALESMAN PROBLEM (Cook and Seymour [32]); for problems on planar graphs (Dorn, see [1]); for problems on graphs of small pathwidth (Pönitz and Tittmann, see e.g., [34]).

These algorithms usually use dynamic programming and have the following structure. First, a tree decomposition of small width is constructed. Then, one bag is chosen as root. In a bottom-up order, for each bag of the tree decomposition, a table is computed. Given the table of the root bag, one can find the answer to the problem quickly. Exploiting that bags are (usually) separators, for computing a table, only some local information on the vertices in the bag, and the tables of the children of the bag are needed. The time to compute a bag typically is exponential on the size of the bag (and its children), but does not depend on the size of the graph. Thus, when the treewidth is bounded by a constant, the algorithm uses linear time. As the algorithm uses time, exponential in the width of the used tree decomposition, this motivates the research for efficient algorithms to compute the treewidth of graphs. See Section 4.

3.2   Probabilistic Networks

Probabilistic networks are the underlying technology of many modern decision support systems. In a probabilistic network, we have a directed acyclic graph, and for each vertex of the graph, a table of conditional probabilities. Each vertex represents a statistical variable, and the (in)dependencies are modeled by the graph structure. A central problem in the use of such networks is *inference*: given values for some *observed* variables, we want to compute the probability distributions for the other variables. This problem is #P-hard. The most used method to solve this problem is as follows. First, the *moralized graph* is build: we add an edge between each pair of vertices that are tail of edges with the same head, and then drop all directions of edges. Then, the algorithm of Lauritzen and Spiegelhalter [47] solves the problem in linear time when the moralized graph has small treewidth. The latter appears to be the case for many probabilistic networks from real-world domains.

3.3   Electrical Networks

In a recent book on graph theory, Bollobás [25] describes the theory of computing the resistance of electrical networks. This theory traces back to a paper by Kirchhoff from 1847 [41]. We have a graph (possibly with parallel edges) with two special vertices, which we will call s and t. Each edge has a resistance: a positive number. Given a potential difference between s and t, we ask how much electrical current will flow through the network, and how much the resistance of the entire network will be.

Three laws that allow to transform networks to smaller, equivalent networks are given. The first law (*series rule*) allows us to remove vertices $\neq s, t$ that have degree 2. The second law (*parallel rule*) allows us to remove parallel edges. The third law (*star-triangle rule* or Y-Δ *rule*) allows us to remove vertices $\neq s, t$ of degree three. See Figure 3. In the first rule, $R = R_1 + R_2$; in the second, we have $\frac{1}{R} = \frac{1}{R_1} + \frac{1}{R_2}$, and in the third, $S = R_1 R_2 + R_1 R_3 + R_2 R_3$.

When we can apply the rules until we have only a single edge from s to t, then these three rules allow us to compute the resistance of the electrical network. We can use the notion of treewidth to determine for which networks there exists such a series of applications of these three rules that yield a single edge. Note that the order in which we apply rules to vertices matters. Consider the graph shown in Figure 3.3. Two different orders of selecting vertices for reduction

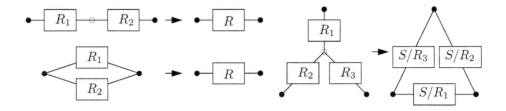

Fig. 3. Series, Parallel, and Star-Triangle Rules for Reduction of Electrical Networks

are used. Removal of parallel edges is assumed and not explicitly shown. If we first eliminate c, then we obtain a clique with five vertices, and no rule applies. However, selecting vertices in the order c, b, d, e yields a single edge $\{s,t\}$. We may assume that the graph $G' = (V, E \cup \{s,t\}$ is a biconnected graph.

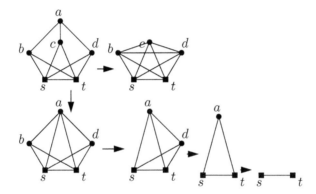

Fig. 4. Different Orders of Applying Reduction Rules of Electrical Networks

(A biconnected component of G' that does not contain both s and t is irrelevant for the computation of the resistance of G.) The next result shows that we can check in linear time (cf. Section 4) if there is an order in which an electrical network can be reduced to a single edge using the rules, and if so, find such an order.

Proposition 1. *Let* $G' = (V, E \cup \{s,t\})$ *be a biconnected graph. There exists a series of rule applications that reduces G to the single edge $\{s,t\}$, if and only if G' has treewidth at most three.*

Proof. Suppose we have an series of rule applications that reduces G to a single edge $\{s,t\}$. Now, take the elimination ordering of G, that puts vertices in the order in which they are removed, and then ends with s, t. For each vertex $v \notin \{s,t\}$, its neighbors (two or three) at the moment it is removed are higher numbered neighbors in the fill-in graph, and thus we have an elimination ordering of G (and of G') such that each vertex has at most three higher numbered neighbors. So G and G' have treewidth at most three.

Suppose G' has treewidth at most three. G' hence is a subgraph of a chordal graph H with maximum clique size four (see Section 2). Repeat the following step. As H is chordal, H has two non-adjacent simplicial vertices (see [39]. A vertex is simplicial if its neighborhood is a clique). As $\{s, t\}$ is an edge in G' and hence in H, there is a vertex $v \notin \{s, t\}$ that is simplicial. As v with its neighbors forms a clique, v has degree at most three. As G' is biconnected, v has degree two or three. So, we can apply the series or star-triangle rule to v, and then remove possibly created parallel edges. Repeat on $H[V - \{v\}]$ until we have the single edge $\{s, t\}$. □

4   Computations

As discussed in Section 3.1, there are many algorithms that first find a tree decomposition, and then use it to solve the problem at hand. As the second step usually is exponential in the width of the tree decomposition, there is a need for efficient algorithms to compute the treewidth of a given graph, and to find tree decompositions with optimal or close to optimal width. In this section, such algorithms will be discussed. Many papers have been written on this topic. Here we focus on algorithms that have been experimentally evaluated.

As for any graph parameter, we can classify the algorithms to compute treewidth into *exact algorithms*, *upper bound algorithms*, and *lower bound algorithms*. In addition, *preprocessing* is an important technique, which is in several cases of great help.

4.1   Upper Bounds

In this section, we discuss a number of algorithms that give upper bounds on the treewidth of the input graphs. Some algorithms have a guaranteed approximation ratio (and hence, can be seen to be lower bound algorithms as well). A typical example are the algorithms by Amir [2]. (See e.g., also [10].) Other heuristics do not have such a guarantee, but often give tree decompositions with close to optimal width. *Construction heuristics* take a graph, and build a tree decomposition (or, a different representation, e.g., an elimination ordering.) *Improvement heuristics* take a tree decomposition (or elimination ordering), and stepwise try to change it to one with smaller and smaller width.

Construction Heuristics. Some construction heuristics, like the algorithms of Amir [2] use a technique, known as *nested dissection*. Here, repeatedly separators are constructed in specific subgraphs of the input graph. Other construction heuristics build an elimination ordering of the graph. As discussed in Section 2, a permutation of the vertices give us an upper bound on the treewidth. Thus, we can use any algorithm or heuristic to build a permutation of the vertices as construction heuristic. Often used heuristics are the *Minimum Degree* and

Minimum Fill-In heuristics (explained below), and algorithms that build elimination orderings that are used for the recognition of chordal graphs: Maximum Cardinality Search, Lexicographic Breadth First Search [62,11,58].

The *Minimum Degree* heuristic is here explained in terms of tree decompositions. It can also be seen as a heuristic to obtain an elimination ordering. A short proof of the following folklore fact can e.g., be found in [24].

Lemma 1. *Let $W \subseteq V$ induce a clique in $G = (V, E)$. Let $(\{X_i \mid i \in I\}, T = (I, F))$ be a tree decomposition of G. There is an $i \in I$ with $W \subseteq X_i$.*

The *Minimum Degree* heuristic works as follows. If $|V| = 1$, we take a trivial tree decomposition with one bag. Otherwise, choose a vertex $v \in V$ with minimum degree. Let G' be the graph, obtained by turning the set of neighbors of v into a clique, and then removing v. Recursively, build a tree decomposition of G'. This tree decomposition must contain a bag i which contains all neighbors of v (as this set is a clique in G'). Now, add a new bag which consists of v and its neighbors and make it adjacent to i. This is a tree decomposition of G.

In the *Minimum Degree* heuristic, we have chosen a vertex of minimum degree: this makes that the bag with this vertex is as small as possible. Other, related heuristics, make different choices for v. The *Minimum Fill-In* heuristic chooses a vertex v, such that the number of edges added when turning v's neighborhood into a clique is as small as possible. In [8,31], some alternative manners to choose v are investigated.

Improvement Heuristics. The *Minimum Separating Vertex Sets* heuristic of Koster [42] starts with a trivial tree decomposition: one bag containing all vertices, and refines this stepwise, using minimum separators. Other currently proposed improvement heuristics for treewidth use a form of local search. Clautiaux et al. [31] use tabu search. As solution space, they take the set of elimination orderings. Two solutions are neighboring, if they represent different triangulations (compare Theorem 1), and are obtained by moving one vertex to a different position in the ordering. Graph theoretic arguments provide a fast test to see if the triangulation changes. Recently, Koster, Marchal and van Hoesel [44] investigated local search algorithms based on 'flipping' edges in triangulations.

The representation by triangulations can also be used for a *postprocessing* step in combination with many heuristics. Take a tree decomposition of the input graph G, obtained by some heuristic. Transform this into a triangulation H of G. If this is not a minimal triangulation, we can obtain a subgraph H' of H that is a minimal triangulation of G, e.g., with the algorithm of [12]. The corresponding tree decomposition never has a larger treewidth compared to the first one, but sometimes has a smaller treewidth.

4.2   Lower Bound Heuristics

Approximation algorithms with a guaranteed approximation ratio give both an upper bound and a lower bound. There are also a number of heuristics that provide only lower bounds. Lower bound heuristics have several uses: if a lower bound matches the upper bound provided by a heuristic, we know we have the exact treewidth; if we have a very large lower bound for the treewidth, we know an approach using tree decompositions and dynamic programming as discussed in Section 3.1 will yield slow algorithms; lower bounds can be used to stop branches in a branch and bound algorithm (see Section 4.3.)

If G has treewidth k, then it has a vertex of degree at most k (consider the first vertex of an elimination order). So, the minimum degree is a simple lower bound on the treewidth. It can be improved using the following observation: the treewidth of a subgraph of G is at most the treewidth of G. Thus, the following algorithm, which computes the *degeneracy* of a graph, yields a lower bound on the treewidth: Set $k = 0$. While G is not empty, select a vertex v of minimum degree in G. Set k to the maximum of k and the degree of v, then remove v and its incident edges from G and repeat.

The *contraction* of an edge also does not increase the treewidth: if G' is obtained from G by contracting edge $\{v, w\}$ to x, then a tree decomposition of G can be transformed to one of G' by replacing each occurrence of v or w in a bag by x. Thus, instead of deleting a vertex, we contract it to one of its neighbors. Two strategies perform well here: contracting to a neighbor of minimum degree, or contracting to a neighbor that has few common neighbors. See [23].

Instead of using the minimum degree, one can use different bounds, and possibly also combine these with vertex deletion or contraction [46]: the one-but-smallest degree; if G is not complete, the minimum over all pairs v, w of non-adjacent vertices of the maximum degree of v and w (Ramachandramurthi [52,53]); the maximum number of visited neighbors of a vertex when it is visited by the Maximum Cardinality Search algorithm (Lucena [50], see also [19]). Also, one can run an exact algorithm on a graph, obtained by contracting edges.

A clever method to improve lower bounds, based on adding edges to G was found by Clautiaux et al. [30]. Combining these LBN and LBP-methods with contraction gives further improvement to the lower bounds for many cases [22].

Seymour and Thomas gave a 'min-max' characterization of treewidth, using the notion of *brambles* [59]. This was used in [18] to obtain a new type of lower bound method for treewidth. This method appears to work well in particular for graphs that are planar or 'close to planar'.

4.3   Exact Algorithms

In theory, for each fixed k, there is a linear time algorithm that tests if a given graph G has treewidth at most k, and if so, finds a tree decomposition of width at most k [14]. However, an experimental evaluation has shown that this algorithm is much too slow in practice [57]. Fortunately, some other algorithms compute for many practical cases the treewidth exactly of (not too large) graphs.

One approach is to build an elimination ordering with a branch and bound algorithm. At each branching step of the algorithm, the next vertex in the elimination ordering is chosen. Several rules, e.g., lower bounds, are used to cut off some branches of the decision tree. See [38,9]. Note that the different representation of treewidth by elimination orderings is again of great use.

Recently, a dynamic programming of the style of the classic Held-Karp algorithm for the TRAVELING SALESMAN PROBLEM has been given [17]. Let for a set $S \subseteq V$ of vertices $TWDP(S)$ be the minimum over all elimination orderings that start with the vertices in S in some order, the minimum over all vertices in S of their number of higher numbered neighbors in the fill-in graph. Clearly, by Theorem 1, the treewidth of G equals $TWDP(V)$. One can show that for all $V \subseteq S, V \neq \emptyset, TWDP(S) = \min_{v \in S} \max(TWDP(S - \{v\}), |\{w \in V - S \mid$ there is a path from v to w using only vertices in $\{v\} \cup (V - S)\}|$. Using this in a dynamic programming algorithm with a few addition optimizations leads to an algorithm that can compute the treewidth for graphs with $30 - 60$ vertices.

Shoikhet and Geiger [60] has shown that an algorithm of Arnborg et al. [4] can be used to compute the treewidth. This algorithm builds a tree decomposition, and also uses a form of dynamic programming.

If $k \leq 3$, then there are relatively simple and extremely fast algorithms to test if the treewidth is at most k, and if so, find the corresponding tree decompositions. These are based on reduction. Arnborg and Proskurowski [6] give six rules, illustrated in Figure 5. A graph G has treewidth at most three, if and only if it is reduced to the empty graph by repeated application of these rules. This gives a linear time algorithm to test if a graph has treewidth at most three, see [51]. It is also possible to construct a tree decomposition of width at most three if existing. Also, the order in which the vertices are eliminated by the rules gives an elimination ordering where each vertex has three higher numbered neighbors in the fill-in graph; see also [21].

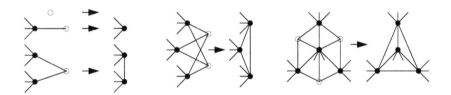

Fig. 5. The Six Reduction Rules for Treewidth Three

4.4   Preprocessing

For many real-world problems, preprocessing is an extremely important technique. For treewidth, two approaches have been used: reduction (or simplification), and splitting (the divide step of divide and conquer). With these approaches, often significant reductions in problem size can be obtained.

Reduction. In [21], the six rules of Arnborg and Proskurowski for treewidth three are used and generalized for preprocessing graphs for computing treewidth.

Besides a graph G, the algorithm maintains a variable *low*, that gives a lower bound on the treewidth of the original input graph. Each rule modifies G and possibly *low*, such that the maximum of *low* and the treewidth of G does not change. Each rule also decreases the size of G. For many graphs from applications, these rules give significant reductions in the size of G. After the preprocessing step, another (e.g., an exact algorithm from Section 4.3) algorithm is used to compute the treewidth of G. The hope is that the size reduction by the preprocessing helps to significantly reduce the time used by this exact algorithm. This technique is also known as *simplification*. Generalizations of the rules were given in [35].

Splitting. A different form of preprocessing is obtained by using *safe separators*, i.e., splitting the graph in different parts. As in a divide and conquer algorithm, the treewidth of each part can be computed separately, and the treewidth of the original graph is the maximum of the treewidth of the parts. Let for a graph G' and vertex set S, $G' + clique(S)$ be the graph, obtained by making S a clique in G', i.e., adding an edge between each pair of non-adjacent vertices in S. In [20], a separator $S \subseteq V$ is safe (for treewidth) in a graph $G = (V, E)$ if the treewidth of G equals the maximum over all connected components W of $G[V - S]$ of the treewidth of $G[W \cup S] + clique(S)$. Thus, if we have a safe separator S, we have as parts all graphs $G[W \cup S] + clique(S)$, for all connected components W of $G[V - S]$. In [20], it is shown that the following sets are safe separators, and can be found, if existing, in polynomial time.

- Separators of size at most one.
- Inclusion minimal separators of size two.
- Inclusion minimal separators S of size at least three, such that no component of $G[V - S]$ contains at least $|V| - 4$ vertices.
- Separators that are a clique.
- Inclusion minimal separators S for which there is a $v \in S$ with $S - \{v\}$ a clique.

5   Conclusions

In this survey, different characterizations and applications of treewidth, and methods to compute the treewidth of a graph were discussed. Experimental and theoretical research benefit from each other. Many of the experimentally tested algorithms are based on interesting combinatorial insights, and the experimental algorithms give rise to new and interesting theoretical questions.

I believe that for testing algorithms experimentally, it is of great importance to make a good selection of the graphs on which to test the algorithms. As random graphs have in general properties that do not need to hold for the graphs encountered in real life applications, I believe one should not rely on only testing the algorithms on randomly generated graphs.

There is a growing number of examples of the fact, that in several cases, the treewidth of graphs can be used to practically solve real-life problems. Of course, small treewidth of the graphs at hand is needed, but appears that in several problem domains, graphs of small treewidth are sufficiently often encountered.

These is much room for additional work. One intriguing question is whether the algorithm of Bouchitté and Todinca [28,29], or its form of Fomin et al. [36] can be used in a practical setting to compute the treewidth of a given graph.

Acknowledgment

I am very grateful to many colleagues for discussions, help, and cooperations. In particular, I thank Arie Koster without whom most of my work reported here would not have been possible.

References

1. J. Alber, F. Dorn, and R. Niedermeier. Experimental evaluation of a tree decomposition based algorithm for vertex cover on planar graphs. *Disc. Appl. Math.*, 145:210–219, 2005.
2. E. Amir. Efficient approximation for triangulation of minimum treewidth. In *Proceedings of the 17th Conference on Uncertainty in Artificial Intelligence*, pages 7–15, 2001.
3. S. Arnborg. Efficient algorithms for combinatorial problems on graphs with bounded decomposability – A survey. *BIT*, 25:2–23, 1985.
4. S. Arnborg, D. G. Corneil, and A. Proskurowski. Complexity of finding embeddings in a k-tree. *SIAM J. Alg. Disc. Meth.*, 8:277–284, 1987.
5. S. Arnborg, J. Lagergren, and D. Seese. Easy problems for tree-decomposable graphs. *J. Algorithms*, 12:308–340, 1991.
6. S. Arnborg and A. Proskurowski. Characterization and recognition of partial 3-trees. *SIAM J. Alg. Disc. Meth.*, 7:305–314, 1986.
7. S. Arnborg and A. Proskurowski. Linear time algorithms for NP-hard problems restricted to partial k-trees. *Disc. Appl. Math.*, 23:11–24, 1989.
8. E. H. Bachoore and H. L. Bodlaender. New upper bound heuristics for treewidth. In S. E. Nikoletseas, editor, *Proceedings of the 4th International Workshop on Experimental and Efficient Algorithms WEA 2005*, pages 217–227. Springer Verlag, Lecture Notes in Computer Science, vol. 3503, 2005.
9. E. H. Bachoore and H. L. Bodlaender. A branch and bound algorithm for exact, upper, and lower bounds on treewidth. In S.-W. Cheng and C. K. Poon, editors, *Proceedings 2nd International Conference on Algorithmic Aspects in Information and Management, AAIM 2006, Lecture Notes in Computer Science, vol. 4041*, pages 255–266, 2006.
10. A. Becker and D. Geiger. A sufficiently fast algorithm for finding close to optimal clique trees. *Artificial Intelligence*, 125:3–17, 2001.
11. A. Berry, J. R. S. Blair, and P. Heggernes. Maximum cardinality search for computing minimal triangulations. In L. Kučera, editor, *Proceedings 28th Int. Workshop on Graph Theoretic Concepts in Computer Science, WG'02*, pages 1–12. Springer Verlag, Lecture Notes in Computer Science, vol. 2573, 2002.
12. J. R. S. Blair, P. Heggernes, and J. Telle. A practical algorithm for making filled graphs minimal. *Theor. Comp. Sc.*, 250:125–141, 2001.

13. H. L. Bodlaender. A tourist guide through treewidth. *Acta Cybernetica*, 11:1–23, 1993.
14. H. L. Bodlaender. A linear time algorithm for finding tree-decompositions of small treewidth. *SIAM J. Comput.*, 25:1305–1317, 1996.
15. H. L. Bodlaender. A partial k-arboretum of graphs with bounded treewidth. *Theor. Comp. Sc.*, 209:1–45, 1998.
16. H. L. Bodlaender. Discovering treewidth. In P. Vojtáš, M. Bieliková, and B. Charron-Bost, editors, *SOFSEM 2005: Theory and Practive of Computer Science: 31st Conference on Current Trends in Theory and Practive of Computer Science*, pages 1–16. Springer-Verlag, Lecture Notes in Computer Science 3381, 2005.
17. H. L. Bodlaender, F. V. Fomin, A. M. C. A. Koster, D. Kratsch, and D. M. Thilikos. On exact algorithms for treewidth. In Y. Azar and T. Erlebach, editors, *Proceedings 14th Annual European Symposium on Algorithms ESA 2006*, pages 672–683. Springer Verlag, Lecture Notes in Computer Science, vol. 4168, 2006.
18. H. L. Bodlaender, A. Grigoriev, and A. M. C. A. Koster. Treewidth lower bounds with brambles. In G. S. Brodal and S. Leonardi, editors, *Proceedings 13th Annual European Symposium on Algorithms, ESA2005*, pages 391–402. Springer-Verlag, Lecture Notes in Computer Science, vol. 3669, 2005.
19. H. L. Bodlaender and A. M. C. A. Koster. On the Maximum Cardinality Search lower bound for treewidth. In J. Hromkovič, M. Nagl, and B. Westfechtel, editors, *Proc. 30th International Workshop on Graph-Theoretic Concepts in Computer Science WG 2004*, pages 81–92. Springer-Verlag, Lecture Notes in Computer Science 3353, 2004.
20. H. L. Bodlaender and A. M. C. A. Koster. Safe separators for treewidth. *Disc. Math.*, 306:337–350, 2006.
21. H. L. Bodlaender, A. M. C. A. Koster, and F. v. d. Eijkhof. Pre-processing rules for triangulation of probabilistic networks. *Computational Intelligence*, 21(3):286–305, 2005.
22. H. L. Bodlaender, A. M. C. A. Koster, and T. Wolle. Contraction and treewidth lower bounds. In S. Albers and T. Radzik, editors, *Proceedings 12th Annual European Symposium on Algorithms, ESA2004*, pages 628–639. Springer, Lecture Notes in Computer Science, vol. 3221, 2004.
23. H. L. Bodlaender, A. M. C. A. Koster, and T. Wolle. Contraction and treewidth lower bounds. *Journal of Graph Algorithms and Applications*, 10:5–49, 2006.
24. H. L. Bodlaender and R. H. Möhring. The pathwidth and treewidth of cographs. *SIAM J. Disc. Math.*, 6:181–188, 1993.
25. B. Bollobás. *Modern Graph Theory*. Graduate Texts in Mathematics, Springer, New York, 1998.
26. R. B. Borie. *Recursively Constructed Graph Families*. PhD thesis, School of Information and Computer Science, Georgia Institute of Technology, 1988.
27. R. B. Borie, R. G. Parker, and C. A. Tovey. Automatic generation of linear-time algorithms from predicate calculus descriptions of problems on recursively constructed graph families. *Algorithmica*, 7:555–581, 1992.
28. V. Bouchitté and I. Todinca. Treewidth and minimum fill-in: Grouping the minimal separators. *SIAM J. Comput.*, 31:212–232, 2001.
29. V. Bouchitté and I. Todinca. Listing all potential maximal cliques of a graph. *Theor. Comp. Sc.*, 276:17–32, 2002.

30. F. Clautiaux, J. Carlier, A. Moukrim, and S. Négre. New lower and upper bounds for graph treewidth. In J. D. P. Rolim, editor, *Proceedings International Workshop on Experimental and Efficient Algorithms, WEA 2003*, pages 70–80. Springer Verlag, Lecture Notes in Computer Science, vol. 2647, 2003.

31. F. Clautiaux, A. Moukrim, S. Négre, and J. Carlier. Heuristic and meta-heuristic methods for computing graph treewidth. *RAIRO Operations Research*, 38:13–26, 2004.

32. W. Cook and P. D. Seymour. Tour merging via branch-decomposition. *INFORMS J. on Computing*, 15(3):233–248, 2003.

33. B. Courcelle. The monadic second-order logic of graphs I: Recognizable sets of finite graphs. *Information and Computation*, 85:12–75, 1990.

34. K. Dohmen, A. Pönitz, and P. Tittmann. A new two-variable generalization of the chromatic polynomial. *Disc. Math. and Theor. Comp. Sc.*, 6:69–90, 2004.

35. F. v. d. Eijkhof, H. L. Bodlaender, and A. M. C. A. Koster. Safe reduction rules for weighted treewidth. To appear in Algorithmica.

36. F. V. Fomin, D. Kratsch, and I. Todinca. Exact (exponential) algorithms for treewidth and minimum fill-in. In *Proceedings of the 31st International Colloquium on Automata, Languages and Programming, ICALP 2004*, pages 568–580, 2004.

37. F. Gavril. The intersection graphs of subtrees in trees are exactly the chordal graphs. *J. Comb. Theory Series B*, 16:47–56, 1974.

38. V. Gogate and R. Dechter. A complete anytime algorithm for treewidth. In *Proceedings of the 20th Annual Conference on Uncertainty in Artificial Intelligence UAI-04*, pages 201–208, Arlington, Virginia, USA, 2004. AUAI Press.

39. M. C. Golumbic. *Algorithmic Graph Theory and Perfect Graphs*. Academic Press, New York, 1980.

40. K. Jansen and P. Scheffler. Generalized coloring for tree-like graphs. In *Proceedings 18th International Workshop on Graph-Theoretic Concepts in Computer Science WG'92*, pages 50–59, Berlin, 1993. Springer Verlag, Lecture Notes in Computer Science, vol. 657.

41. G. Kirchhoff. Über die Auflösung der Gleichugen, auf welche man bei der Untersuchung der linearen Verteilung galvanischer Ströme geführt wird. *Ann. Phys. Chem.*, 71:497–508, 1847.

42. A. M. C. A. Koster. *Frequency Assignment - Models and Algorithms*. PhD thesis, Univ. Maastricht, Maastricht, The Netherlands, 1999.

43. A. M. C. A. Koster, H. L. Bodlaender, and S. P. M. van Hoesel. Treewidth: Computational experiments. In H. Broersma, U. Faigle, J. Hurink, and S. Pickl, editors, *Electronic Notes in Discrete Mathematics*, volume 8, pages 54–57. Elsevier Science Publishers, 2001.

44. A. M. C. A. Koster, B. Marchal, and C. P. M. van Hoesel. Local search algorithms for treewidth. Work in progress, 2006.

45. A. M. C. A. Koster, S. P. M. van Hoesel, and A. W. J. Kolen. Solving partial constraint satisfaction problems with tree decomposition. *Networks*, 40(3):170–180, 2002.

46. A. M. C. A. Koster, T. Wolle, and H. L. Bodlaender. Degree-based treewidth lower bounds. In S. E. Nikoletseas, editor, *Proceedings of the 4th International Workshop on Experimental and Efficient Algorithms WEA 2005*, pages 101–112. Springer Verlag, Lecture Notes in Computer Science, vol. 3503, 2005.

47. S. J. Lauritzen and D. J. Spiegelhalter. Local computations with probabilities on graphical structures and their application to expert systems. *The Journal of the Royal Statistical Society. Series B (Methodological)*, 50:157–224, 1988.

48. C. Lautemann. Decomposition trees: structured graph representation and efficient algorithms. In *Proceedings CAAP'88*, pages 28–39. Springer Verlag, Lecture Notes in Computer Science, vol. 299, 1988.

49. T. Lengauer and E. Wanke. Efficient analysis of graph properties on context-free graph languages. In *Proceedings of the 15th International Colloquium on Automata, Languages and Programming, ICALP'88*, pages 379–393. Springer Verlag, Lecture Notes in Computer Science, vol. 317, 1988.

50. B. Lucena. A new lower bound for tree-width using maximum cardinality search. *SIAM J. Disc. Math.*, 16:345–353, 2003.

51. J. Matoušek and R. Thomas. Algorithms for finding tree-decompositions of graphs. *J. Algorithms*, 12:1–22, 1991.

52. S. Ramachandramurthi. A lower bound for treewidth and its consequences. In E. W. Mayr, G. Schmidt, and G. Tinhofer, editors, *Proceedings 20th International Workshop on Graph Theoretic Concepts in Computer Science WG'94*, pages 14–25. Springer Verlag, Lecture Notes in Computer Science, vol. 903, 1995.

53. S. Ramachandramurthi. The structure and number of obstructions to treewidth. *SIAM J. Disc. Math.*, 10:146–157, 1997.

54. B. A. Reed. *Tree width and tangles, a new measure of connectivity and some applications*, volume 241 of *LMS Lecture Note Series*, pages 87–162. Cambridge University Press, Cambridge, UK, 1997.

55. B. A. Reed. *Algorithmic aspects of tree width*, pages 85–107. CMS Books Math. / Ouvrages Math. SMC, 11. Springer, New York, 2003.

56. N. Robertson and P. D. Seymour. Graph minors. II. Algorithmic aspects of tree-width. *J. Algorithms*, 7:309–322, 1986.

57. H. Röhrig. Tree decomposition: A feasibility study. Master's thesis, Max-Planck-Institut für Informatik, Saarbrücken, Germany, 1998.

58. D. J. Rose, R. E. Tarjan, and G. S. Lueker. Algorithmic aspects of vertex elimination on graphs. *SIAM J. Comput.*, 5:266–283, 1976.

59. P. D. Seymour and R. Thomas. Graph searching and a minimax theorem for tree-width. *J. Comb. Theory Series B*, 58:239–257, 1993.

60. K. Shoikhet and D. Geiger. A practical algorithm for finding optimal triangulations. In *Proc. National Conference on Artificial Intelligence (AAAI '97)*, pages 185–190. Morgan Kaufmann, 1997.

61. M. M. Sysło. NP-complete problems on some tree-structured graphs: A review. In M. Nagl and J. Perl, editors, *Proc. WG'83 International Workshop on Graph Theoretic Concepts in Computer Science*, pages 342–353, Linz, West Germany, 1983. University Verlag Rudolf Trauner.

62. R. E. Tarjan and M. Yannakakis. Simple linear time algorithms to test chordiality of graphs, test acyclicity of graphs, and selectively reduce acyclic hypergraphs. *SIAM J. Comput.*, 13:566–579, 1984.

63. J. A. Telle and A. Proskurowski. Algorithms for vertex partitioning problems on partial k-trees. *SIAM J. Disc. Math.*, 10:529 – 550, 1997.

64. T. V. Wimer. *Linear Algorithms on k-Terminal Graphs*. PhD thesis, Dept. of Computer Science, Clemson University, 1987.

65. T. V. Wimer, S. T. Hedetniemi, and R. Laskar. A methodology for constructing linear graph algorithms. *Congressus Numerantium*, 50:43–60, 1985.

66. X. Zhou, K. Fuse, and T. Nishizeki. A linear algorithm for finding $[g, f]$-colorings of partial k-trees. *Algorithmica*, 27:227–243, 2000.

Locally Injective Graph Homomorphism: Lists Guarantee Dichotomy

Jiří Fiala and Jan Kratochvíl

Department of Applied Mathematics and
Institute for Theoretical Computer Science (ITI)[*],
Faculty of Mathematics and Physics, Charles University,
Malostranské nám. 2/25, 118 00 Prague, Czech Republic
{fiala, honza}@kam.mff.cuni.cz

Abstract. We prove that in the List version, the problem of deciding the existence of a locally injective homomorphism to a parameter graph H performs a full dichotomy. Namely we show that it is polynomially time solvable if every connected component of H has at most one cycle and NP-complete otherwise.

1 Introduction

We consider finite undirected graphs without loops or multiple edges. A recently intensively studied notion, for its algebraic motivation and connections as well as being a natural generalization of graph coloring, is the notion of *graph homomorphisms*. A homomorphism $f : G \to H$ from a graph G to a graph H is an edge-preserving vertex mapping, i.e., a mapping $f : V(G) \to V(H)$ such that $f(x)f(y) \in E(H)$ whenever $xy \in E(G)$. (For a recent monograph on graph homomorphisms the reader is referred to [14].) It follows from the definition that the neighborhood of every vertex is mapped into the neighborhood of its image, formally $f(N_G(x)) \subseteq N_H(f(x))$ for all $x \in V(G)$. Properties of these restricted mappings, local constraints, lead to the definition of *locally constrained homomorphisms*. The homomorphism f is called *locally injective* (*bijective, surjective,* resp.) if for every $x \in V(G)$, the restricted mapping $f : N_G(x) \to N_H(f(x))$ is injective (bijective, surjective, resp.). All three of these notions have been studied on their own with different motivations. Locally surjective homomorphisms correspond to so called role assignment graphs studied in sociological applications [11], locally bijective ones correspond to graph covers well known from topological graph theory [2,15] and theory of local computation [1,3]. Locally injective homomorphisms are closely related to generalized $L(2, 1)$-labelings of graphs and the Frequency Assignment Problem [8,9].

From the computational complexity point of view we are interested in the decision problem if an input graph G allows a homomorphism of certain type into a fixed target graph H. As these problems are parametrized by the graph H, we

---

[*] Supported by the Ministry of Education of the Czech Republic as project 1M0021620808.

use the notation H-Hom (when asking for the existence of any homomorphism), H-LIHom, H-LBHom and H-LSHom (when asking for locally injective, bijective or surjective homomorphisms, respectively).

The complexity of H-Hom is fully understood and dichotomy was proved by Hell and Nešetřil in [13]. The problem is polynomially solvable if H is bipartite and NP-complete otherwise. The complexity of H-LSHom was studied in [17] and completed by Fiala and Paulusma in [11]. This problem (for connected graph H) is solvable in polynomial time if H has at most 2 vertices and NP-complete otherwise. Several papers have been devoted to studying the complexity of locally injective and bijective homomorphisms, but only partial results are known [8,10,16]. Though conjectured at least for the case of locally bijective homomorphisms, the Polynomial/NP-completeness dichotomy has been proved in neither of the last two cases. However, this question is natural especially in view of the fact that locally constrained homomorphisms can be expressed as Constraint Satisfaction Problem, for which the dichotomy was conjectured in the fully general case by Feder and Vardy [6].

The CSP view also suggests considering the List versions of the problems, since lists correspond to unary relations (cf. next section). The input to the List version of a homomorphism problem is a graph G together with lists $L(u) \subseteq V(H)$ of admissible targets for every vertex $u \in V(G)$. The question is if G allows a (locally injective, bijective etc.) homomorphism $f : G \to H$ such that $f(u) \in L(u)$ for every $u \in V(G)$. We refer to these problems as List-H-Hom, List-H-LIHom etc. A deep result of Hell et al. gives a full characterization of the case of general homomorphisms [5,12]. The problem List-H-Hom is polynomially solvable for so called double circular arc graphs H and NP-complete otherwise.

Setting the lists to the entire vertex set of the target graph, one immediately sees that H-Hom $\propto$ List-H-Hom, H-LIHom $\propto$ List-H-LIHom, H-LBHom $\propto$ List-H-LBHom and H-LSHom $\propto$ List-H-LSHom. Thus in each case the borderline between polynomial and NP-complete instances (dichotomy assumed) of the List version will lie within the easy instances of the non-List one. This is well seen in the above mentioned case of general homomorphisms and also in the case of the locally surjective ones — List-H-LSHom remains polynomial for graphs H with at most two vertices. The trouble with the locally injective and locally bijective homomorphisms is that the full characterization of the non-List versions is not known. Nevertheless, lists do help! The purpose of this paper is to show that in the case of locally injective homomorphisms, lists guarantee a full dichotomy.

Theorem 1. *The* List-H-LIHom *problem is solvable in linear time if the graph H contains at most one cycle in each connected component, and it is* NP-complete *otherwise.*

The paper is structured as follows. In the next section we quickly describe the connection to the Constraint Satisfaction Problem. In Section 3 we give the argument for the polynomial part of our theorem. The technical reductions for the NP-hardness part are presented in Section 4 and the proof is summarized in Section 5. The final section contains concluding remarks.

2   Locally Constrained Homomorphisms as CSP

The CSP is parametrized by a fixed template $\mathcal{X} = (X; S_1, \ldots, S_k)$, where S_i's are relations on a finite set X, the arity of S_i being n_i, $i = 1, 2, \ldots, k$. The input of the $\mathcal{X}$-CSP is a structure $\mathcal{U} = (U; R_1, \ldots, R_k)$, where U is a (large) set and R_i is an n_i-ary relation on U, for $i = 1, 2, \ldots, k$. The question is whether there exists a mapping (in fact, a structural homomorphism) $f : U \to X$ such that for every i and every n_i-tuple $(u_1, \ldots, u_{n_i}) \in U^{n_i}$, $(u_1, \ldots, u_{n_i}) \in R_i$ implies $(f(u_1), \ldots, f(u_{n_i})) \in S_i$. Feder and Vardy [6] conjecture that for every template $\mathcal{X}$, this problem is either polynomial time solvable or NP-complete. This is known for binary structures [19], and for many special cases (cf. e.g., Bulatov et al. [4]).

Though in natural structures one tends to overlook unary relations, from the point of view of the formal definition they form a fully coherent part of the picture. And they correspond to lists. A unary relation is just a subset of the ground set. If a vertex $u \in U$ belongs to unary relations $R_{i_1}, \ldots, R_{i_t}$, then the constraints given by the unary relations of $\mathcal{X}$-CSP merely say that $f(u) \in \bigcap_{j=1}^{t} S_{i_j}$, which is equivalent to setting the list of admissible images of u to $L(u) = \bigcap_{j=1}^{t} S_{i_j}$.

The general homomorphism problem H-Hom is obviously a CSP problem — the template is H itself, and the input structure is G (edges of both graphs are considered as symmetric binary relations). We will show that also locally injective and bijective homomorphisms can be expressed as CSP. Given a graph H with h vertices, set

$$D = \{(x, y)| \ x \neq y \in V(H)\},$$

$$D_i = \{(x)| \ \deg_H(x) = i, x \in V(H)\}, \ i = 0, 1, \ldots, h - 1.$$

Here D is the symmetric binary relation containing all pairs of distinct vertices and D_i's are unary relations controlling the degrees. We derive the following templates from H:

$$\mathcal{L}_H = (V(H); S_1 = E(H), S_2 = D)$$

and

$$\mathcal{B}_H = (V(H); S_1 = E(H), S_2 = D, S_{3+i} = D_i, i = 0, \ldots, h - 1).$$

Observation 1. *For every graph* H, H-LIHom $\propto \mathcal{L}_H$-*CSP and* H-LBHom $\propto \mathcal{B}_H$-*CSP.*

Proof. For an input graph G, define $\mathcal{U} = (V(G); R_1 = E(G), R_2 = \{(x, y)|; x \neq y \ \wedge \ \exists z : xz, yz \in E(G)\})$. Then U is a feasible instance for $\mathcal{L}_H$-CSP if and only if G allows a locally injective homomorphism into H — the relations R_1 and S_1 control that a candidate vertex mapping is a graph homomorphism, and R_2 with S_2 control the local injectivity. For the locally bijective case, just add $R_{3+i} = \{u \in V(G)| \ \deg_G(u) = i\}$ for each $i = 0, 1, \ldots, h - 1$, to ensure that this mapping is locally bijective on every neighborhood.

Observe, however, that this does not mean that the Feder-Vardy conjecture would imply dichotomy for H-LIHom or H-LBHom. Our observation is useful only when the corresponding CSP problem is polynomially solvable, which is unfortunately only seldom (whenever H has at least three vertices, 3-COLORABILITY $\propto \mathcal{L}_H$-CSP $\propto \mathcal{B}_H$-CSP and both problems are NP-complete). The point is that the inputs of $\mathcal{L}_H$-CSP (or $\mathcal{B}_H$-CSP) derived from G are not arbitrary. In view of this becomes the fact that adding lists endorses dichotomy on List-H-LIHom even more interesting.

3   The Polynomial Case

Lemma 1. *For any connected graph H containing at most one cycle, the* List-H-LIHom *problem is solvable in linear time.*

Proof. Without loss of generality we may assume that the input graph G is connected, since otherwise we can treat each component separately.

If H is a tree, then any connected graph G that allows a locally injective homomorphism $G \to H$ is a subtree of H. Since H is fixed, the graph G itself must have a bounded number of vertices and the problem H-LIHom is solvable in constant time even with lists being incorporated.

Let H have exactly one cycle. We first note that the number of connected graphs G of diameter at most $2 \operatorname{diam}(H)$ that allow a locally injective homomorphism to H is bounded (each such graph has at most $1 + \Delta^{2\operatorname{diam}(H)}$ vertices, where Δ is the maximum vertex degree of H). Hence, for such instances, the List-H-LIHom problem can be decided in constant time.

For the rest of the proof suppose that $\operatorname{diam}(G) > 2 \operatorname{diam}(H)$. Denote by C_l the unique cycle of H. We distinguish two cases:

- G contains a cycle, say C_k: Then this cycle must be mapped onto C_l. This might happen in at most $2l$ ways. Some of these $2l$ ways may be further excluded by the list constraints. If we fix a mapping $C_k \to C_l$, it remains to decide whether this mapping can be extended to the remaining vertices of G. For every vertex u of C_k, we can solve this question in constant time, since this is equivalent to the tree List-H-LIHom problem. (Now the instance is the component of $G \setminus E_{C_k}$ containing u and the parameter is the component of $H \setminus E_{C_l}$ containing the image of u. Both are trees.)
- G is a tree: A necessary condition for G to allow a locally injective homomorphism to H is that G contains a path P_k of length $k = \operatorname{diam}(G) - 2 \operatorname{diam}(H)$ such that the components of $G \setminus E_{P_k}$ stemming from the inner vertices of P_k map locally injectively to the components of $H \setminus E_{C_l}$ and the two trees of diameter $\operatorname{diam}(H)$ hanging on the termini of P_k map locally injectively to H. We first find P_k by $\operatorname{diam}(H)$ iterations of peeling off vertices of degree one. Then, as in the above case, we try all $2l$ possibilities of a locally injective mapping $P_k \to C_l$ and exclude those which do not satisfy the list constraints. Finally for each such partial mapping, we check if it can be extended to the entire G by solving at most $k+1$ List-H-LIHom problems of constant size.

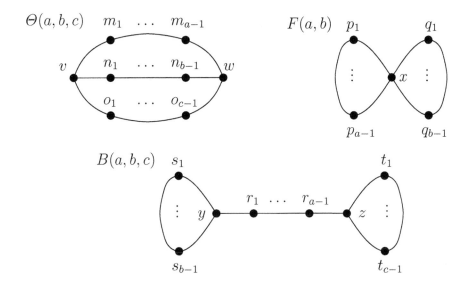

Fig. 1. The basic three types of graphs

4   Auxiliary NP-Hardness Reductions

In this section we show NP-hardness of the List-H-LIHom problem for three basic types of graphs depicted in Fig. 1 (they are informally called the Theta graph, the Flower graph and the Weight graph). The problem we use for our NP-hardness reductions is EDGE-PRECOLORING EXTENSION. It has been shown NP-complete even when restricted to cubic bipartite graphs [7]. The input of this variant consists of a cubic bipartite graph together with a partial coloring of its edges by three colors — say red, blue and white[1]. The question is whether this partial coloring can be extended to the entire edge set such that each vertex is incident with edges of all three colors.

For positive integers a, b, c, where $b, c \geq 2$, let $\Theta(a, b, c)$ be the graph on $a + b + c - 1$ vertices consisting of two vertices of degree three connected by paths of lengths a, b, and c (cf. Fig. 1 top left).

Lemma 2. *For arbitrary positive integers a, b, c, where $b, c \geq 2$, the* List-$\Theta(a, b, c)$-LIHom *problem is* NP-*hard.*

Proof. If $a = b = c \geq 2$, we can reduce the EDGE-PRECOLORING EXTENSION problem directly. Let G be the instance of the EDGE-PRECOLORING EXTENSION problem. We replace each edge $e \in E(G)$ by a path P^e of length a. We further associate colors red, blue and white with the inner vertices of the three paths of $\Theta(a, a, a)$ in such a way that vertices of the first path represent color red, of the second path color blue, and those of the third one the white color. If the edge

---

[1] The favorite tricolor for Czechs — as well as for several other nations.

e was precolored by a color α, we assign the vertices of P^e lists consisting of vertices of the path in $\Theta(a, a, a)$ representing the color α. For the other vertices we let the lists be the whole vertex set of $\Theta(a, a, a)$.

Straightforwardly, proper 3-edge colorings of G are in one-to-one correspondence with valid locally injective homomorphisms to $\Theta(a, a, a)$. (Such homomorphisms must in fact be locally bijective.) The local conditions on vertices of degree three represent the condition that the three colors on the edges of the original graph must be distinct.

For the case of $a \leq b \leq c$, $a < c$, the construction is slightly more sophisticated. The edge colors will be represented by sequences of images of the vertices along the paths representing the edges. We use vertex lists to enforce that only three feasible patterns may exist.

Denote first by v, w the two vertices of degree three in $\Theta(a, b, c)$. The inner vertices along the path of length a from v to w will be denoted by $m_1, m_2, \ldots, m_{a-1}$ (this set may be empty), along the path of length b in the same direction by $n_1, \ldots, n_{b-1}$, and along the remaining path by $o_1, \ldots, o_{c-1}$.

We define three sequences of length $a + b + c + 1$ by

$$R = (v, m_1, \ldots, m_{a-1}, w, n_{b-1}, \ldots, n_1, v, o_1, \ldots, o_{c-1}, w)$$
$$B = (v, n_1, \ldots, n_{b-1}, w, o_{c-1}, \ldots, o_1, v, m_1, \ldots, m_{a-1}, w)$$
$$W = (v, o_1, \ldots, o_{c-1}, w, m_{a-1}, \ldots, m_1, v, n_1, \ldots, n_{b-1}, w)$$

Let G be an instance of the EDGE-PRECOLORING EXTENSION problem and let V^1, V^2 be its two classes of the bipartition. We replace each edge $e \in E_G$ by a path P^e of length $a + b + c$ and call the resulting graph G'.

The vertices of each P^e are denoted by $u_1^e, \ldots, u_{a+b+c+1}^e$ in such a way that the first vertex u_1^e belongs to V^1 and the last vertex $u_{a+b+c+1}^e \in V^2$. We define the lists of the vertices in G' so that for each $e \in E(G)$ and each $i = 1, 2, \ldots, a + b + c + 1$, we set

$$L(u_i^e) = \begin{cases} \{R_i, B_i, W_i\} & \text{if } e \text{ is not precolored} \\ \{R_i\} & \text{if } e \text{ is precolored } red \\ \{B_i\} & \text{if } e \text{ is precolored } blue \\ \{W_i\} & \text{if } e \text{ is precolored } white. \end{cases}$$

We claim that G' allows a locally injective homomorphism to $\Theta(a, b, c)$ that respects all list constraints if and only if the edge precoloring of the original graph G can be extended to a proper 3-edge coloring of G.

Suppose first that $f : G' \to \Theta(a, b, c)$ is a locally injective homomorphism. We prove that only the sequences given by R, B and W may appear along any path P^e. The local injectivity constraints then imply that the derived edge-coloring is proper.

Assume first the case $a < b < c$. Let P^e be a path in G'. By list constraints we have $f(u_1^e) = v$ for any such P^e. Assume that $f(u_2^e) = m_1$ (or $f(u_2^e) = w$ if $a = 1$). Then the mapping f is uniquely determined for the next $a - 1$ vertices of P^e due to the local constraints. As the subsequence $v, o_{c-1}, \ldots, o_1, w$ starts at a different position in the sequence B it cannot be used as the further extension of f on P^e. In other words the mapping f has to follow the sequence R for the next $b-1$ vertices as well as for the final segment of $c-1$ vertices. For the other two possibilities, i.e., when $f(u_2^e) = n_1$ or o_1, resp., we involve similar arguments to conclude that the only feasible pattern of the mapping f along P^e is given by the sequences B or W, respectively.

It remains to consider the case when $b = a$ or $b = c$ and $a < c$. Without loss of generality assume $a = b$, i.e., the target graph is $\Theta(a, a, c)$. By the arguments presented so far, it might be possible that a mapping of some P^e may follow on the first a vertices the sequence R, while after u_{a+1}^e it continues along B. The crucial observation here is that if $f(u_2^e) = o_1$ then the whole path must be mapped according to the sequence W. Viewed from the other side also if $f(u_{a+b+c}^e) = o_{c-1}$, then the mapping of P^e follows the pattern R. Then the pattern W is used for $|V^1|$ edges, hence giving every vertex of the side V^2 a neighbor mapped onto n_{b-1}. Similarly the pattern R is used for $|V^2| = |V^1|$ edges, giving each vertex on the V^1 side a neighbor mapped onto m_1. For the paths P^e corresponding to the remaining matching in G, the local injectivity constraints imply that $f(u_2^e) = n_1$ and $f(u_{a+b+c}^e) = m_{a-1}$, and hence only the pattern B may be used along these paths.

For the reverse implication, assume $E(G)$ is properly colored. We define the mapping $f : G' \to \Theta(a, b, c)$ such that for every $e \in E(G)$ and all $i = 1, 2, \ldots, a + b + c + 1$, $u_i^e = R_i$ if e is red, and analogously with B_i representing blue and with W_i representing white colors. On every vertex of G' the mapping preserves the list constraints and is locally injective: on the inner vertices of any P^e since the patterns R, B, W respect local injectivity constraints, and on the vertices of degree three since the coloring is proper.

For integers $a, b \geq 3$, let $F(a, b)$ be the graph on $a + b - 1$ vertices consisting of two cycles C_a and C_b sharing exactly one vertex (cf. Fig. 1 top right).

Lemma 3. *For arbitrary integers $a, b \geq 3$, the* List-$F_{a,b}$-LIHom *problem is* NP-*hard.*

Proof. Assume x is the vertex of $F(a, b)$ of degree four and that the inner vertices of the two cycles are $p_1, \ldots, p_{a-1}$ and $q_1, \ldots, q_{b-1}$.

We extend the proof of the previous lemma for the NP-hardness of the List-$\Theta(a, a, b)$-LIHom problem. Let G' be the graph constructed from an instance G of the EDGE-PRECOLORING EXTENSION problem. Define lists L' such that for every $u \in V_{G'}$ (see Fig. 2):

$$x \in L'(u) \iff v \in L(u) \lor w \in L(u)$$
$$p_i \in L'(u) \iff m_i \in L(u) \lor n_{a-i} \in L(u)$$
$$q_i \in L'(u) \iff o_i \in L(u).$$

We claim that G' allows a locally injective homomorphism $f : G' \to \Theta(a, a, b)$ respecting the list constraints L if and only if there exists a locally injective $g : G' \to F(a, b)$ respecting the list constraints L'.

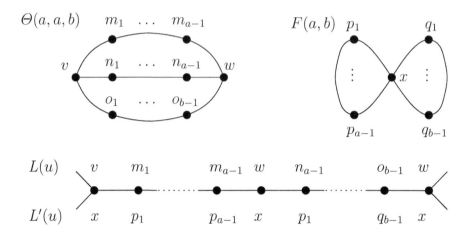

Fig. 2. The reduction for List-$F(a,b)$-LIHom. The lower part shows the new lists along a path P^e (for e precolored red).

As $\Theta(a,a,b)$ can be mapped locally injectively onto $F(a,b)$, the "only if" implication is straightforward (taking into account also the way how the new lists L' are constructed from L). For the opposite direction, note that the lists L' defined above together with local constraints assure that along each P^e the vertex x will be involved exactly four times (twice on the end). (This follows by a simple case analysis when $b \neq 2a$, and a little more subtle argument analogous to the case $a = b < c$ of the proof of Lemma 2 works for $b = 2a$. The latter is omitted because of space limitations.) We modify a feasible $g : G' \to F(a,b)$ by a series of substitutions. All odd occurrences of x along every P^e will be replaced by v and even occurrences by w. We further replace each pattern $v, p_1, \ldots, p_{a-1}, w$ by the pattern $v, m_1, \ldots, m_{a-1}, w$. Similarly, we replace patterns $v, p_{a-1}, \ldots, p_1, w$ by $v, n_1, \ldots, n_{a-1}, w$ and also all patterns $v, q_1, \ldots, q_{b-1}, w$ by $v, o_1, \ldots, o_{b-1}, w$. In this way we obtain a feasible locally injective homomorphism $f : G' \to \Theta(a,a,b)$.

For positive integers a, b, c, where $b, c \geq 3$, let $B(a,b,c)$ be the only graph on $a + b + c - 1$ vertices consisting of two disjoint cycles C_b and C_c connected by a path of length a.

Lemma 4. *For arbitrary positive integers a, b, c, where $b, c \geq 3$, the* List-$B(a,b,c)$-LIHom *problem is* NP-*hard.*

Proof. Let y, z be the two vertices of $B(a,b,c)$ of degree three. Denote the inner vertices of the path P_a by $r_1, \ldots, r_{a-1}$ (this set may be empty), and let $s_1, \ldots, s_{b-1}$ be the inner vertices of C_b and $t_1, \ldots, t_{c-1}$ the inner vertices of C_c.

We extend the proof of the previous lemma for the NP-hardness of the List-$F(b, 2a + c)$-LIHom. Let G' be the graph constructed from an instance G of the EDGE-PRECOLORING EXTENSION problem. Define lists L'' such that for every

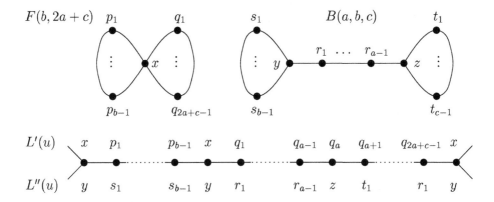

Fig. 3. The reduction for List-$B(a,b,c)$-LIHom. The lower part shows the new lists along a path P^e (for e precolored red).

$u \in V_{G'}$ (see Fig. 3):

$$y \in L''(u) \iff x \in L'(u)$$
$$s_i \in L''(u) \iff p_i \in L'(u)$$
$$r_i \in L''(u) \iff q_i \in L'(u) \vee q_{2a+c-i} \in L'(u)$$
$$z \in L''(u) \iff q_a \in L'(u) \vee q_{a+c} \in L'(u)$$
$$t_i \in L''(u) \iff q_{a+i} \in L'(u)$$

We claim that G' allows a locally injective homomorphism $g : G' \to F(b, 2a + c)$ respecting the list constraints L' if and only if there exist a locally injective homomorphism $h : G' \to B(a, b, c)$ with list constraints L''.

The core argument is that for any homomorphism g feasible for the instance (G', L'), every vertex mapped onto x has at most one neighbor mapped onto one of q_1, q_{2a+c-1}, and so the feasible homomorphisms $g : C' \to F(b, 2a + c)$ are in a one-to-one correspondence with locally injective homomorphisms $h : G' \to B(a, b, c)$ obeying the list constraints L''.

5   Proof of Theorem 1

Observation 2. *If a graph H is an induced subgraph of a graph H', then List-H-LIHom $\propto$ List-H'-LIHom.*

Proof. Given an input (G, L) of List-H-LIHom, it suffices to use it as an instance of List-H'-LIHom. Since the lists $L(u), u \in V(G)$ do not contain vertices of $H' \setminus H$, any feasible homomorphism $G \to H'$ uses only vertices of H, and since H is an induced subgraph of H', such a mapping is a homomorphism into H. In the opposite direction, any feasible (locally injective) homomorphism $G \to H$ is clearly a (locally injective) homomorphism to H'.

Proof of Theorem 1

The polynomial-time algorithm. Assume each component H_j of H contains at most one cycle. Then for each component G_i of G and every H_j we test whether G_i allows a locally injective homomorphism (we may restrict lists of the vertices in G_i to subsets of H_j). Each such a subproblem can be decided independently. The overall problem has an affirmative answer if and only if for every G_i, there exists at least one H_j allowing a locally constrained homomorphism. The overall computational complexity remains linear.

The NP-complete part. The membership of the List-H-LIHom problem in NP is obvious. Assume H contains a component with at least two cycles. Let H' be a smallest induced subgraph of H containing at least two cycles.

If H' contains exactly two cycles, these two cycles must be edge-disjoint (the symmetric difference of the two cycles would yield a new cycle otherwise). If the two cycles are further vertex-disjoint, then H' is isomorphic to some $B_{a,b,c}$. If the cycles share one vertex, then H' is isomorphic to some $F_{a,b}$.

Otherwise H' has two intersecting cycles. If the graph is not isomorphic to some $\Theta(a, b, c)$, then between some of the three paths of $\Theta(a, b, c)$ there exists an induced edge. Then either H' contains a smaller $\Theta(a', b', c')$ (contradicting the minimality of H') or H' is isomorphic to K_4.

We have proved in Section 4 that for each of the graphs $B(a, b, c)$, $F(a, b)$, and $\Theta(a, b, c)$, the List-H-LIHom problem is NP-hard. For the case of $H' = K_4$, the NP-hardness follows from the trivial reduction H-LIHom $\propto$ List-H-LIHom and the fact that K_4-LIHom is NP-complete [8]. The conclusion then follows by Observation 2.

6   Concluding Remarks

6.1 It is interesting that the case of locally surjective homomorphisms does not seem expressible as a CSP problem. Yet a full dichotomy holds both for List and non-List versions.

6.2 An irritating open problem is an analog of Observation 2 for locally bijective homomorphisms. A direct proof would hinge on an involved garbage collection.

6.3 For many instances of the H-LIHom problem for $H = B(a, b, c)$, $F(a, b)$, and $\Theta(a, b, c)$, we can prove NP-hardness also for the non-List version. However, a full characterization even for these simple graphs is not known.

6.4 Recently Daniel Marx [18] showed that the EDGE-PRECOLORING EXTENSION problem remains NP-complete for planar bipartite 3-regular graphs. Our reductions in the proofs of Lemmas 2-4 preserve planarity, and since D. Marx suggested [personal communication] a simple planarity preserving reduction for List-K_4-LIHom, we can conclude that whenever a connected graph H contains at least two cycles, the List-H-LIHom problem is NP-complete even for planar inputs. However, the complexity of K_4-LIHom and K_4-LBHom (without lists) is still an interesting open problem.

Acknowledgment

We would like to thank P. Hell for tempting us to consider the List versions of the locally constrained homomorphism problems.

References

1. ANGLUIN, D. Local and global properties in networks of processors. *Proceedings of the 12th ACM Symposium on Theory of Computing* (1980), 82–93.
2. BIGGS, N. Constructing 5-arc transitive cubic graphs. *Journal of London Mathematical Society II. 26* (1982), 193–200.
3. BODLAENDER, H. L. The classification of coverings of processor networks. *Journal of Parallel Distributed Computing 6* (1989), 166–182.
4. BULATOV, A. A., JEAVONS, P., AND KROKHIN, A. A. Classifying the complexity of constraints using finite algebras. *SIAM J. Comput. 34*, 3 (2005), 720–742.
5. FEDER, T., HELL, P., KLEIN, S., AND MOTWANI, R. List partitions. *SIAM J. Discrete Math. 16*, 3 (2003), 449–478.
6. FEDER, T., AND VARDI, M. Y. The computational structure of momotone monadic SNP and constraint satisfaction: A sudy through datalog and group theory. *SIAM Journal of Computing 1* (1998), 57–104.
7. FIALA, J. NP completeness of the edge precoloring extension problem on bipartite graphs. *Journal of Graph Theory 43*, 2 (2003), 156–160.
8. FIALA, J., AND KRATOCHVÍL, J. Complexity of partial covers of graphs. In *Algorithms and Computation, 12th ISAAC '01, Christchurch, New Zealand* (2001), no. 2223 in Lecture Notes in Computer Science, Springer Verlag, pp. 537–549.
9. FIALA, J., AND KRATOCHVÍL, J. Partial covers of graphs. *Discussiones Mathematicae Graph Theory 22* (2002), 89–99.
10. FIALA, J., KRATOCHVÍL, J., AND PÓR, A. On the computational complexity of partial covers of theta graphs. *Electronic Notes in Discrete Mathematics 19* (2005), 79–85.
11. FIALA, J., AND PAULUSMA, D. A complete complexity classification of the role assignment problem. *Theoretical Computer Science 1*, 349 (2005), 67–81.
12. HELL, P. From homomorphisms to colorings and back. in *Topics in Discrete Mathematics*, Dedicated to J. Nešetřil on the occasion of his 60th birthday, (M. Klazar, J. Kratochvíl, J. Matoušek, M. Loebl, R. Thomas and P. Valtr, eds.), Springer, 2006, pp. 407–432.
13. HELL, P., AND NEŠETŘIL, J. On the complexity of H-colouring. *Journal of Combinatorial Theory B 48* (1990), 92–110.
14. HELL, P., AND NEŠETŘIL, J. *Graphs and Homomorphisms*. Oxford University Press, 2004.
15. HLINĚNÝ, P. $K_{4,4} - e$ has no finite planar cover. *Journal of Graph Theory 21*, 1 (1998), 51–60.
16. KRATOCHVÍL, J., PROSKUROWSKI, A., AND TELLE, J. A. Covering regular graphs. *Journal of Combinatorial Theory B 71*, 1 (Sept 1997), 1–16.

17. KRISTIANSEN, P., AND TELLE, J. A. Generalized *H*-coloring of graphs. In *Algorithms and Computation, 11th ISAAC '01, Taipei, Taiwan* (2000), no. 1969 in Lecture Notes in Computer Science, Springer Verlag, pp. 456–466.

18. MARX, D. NP-completeness of list coloring and precoloring extension on the edges of planar graphs. *Journal of Graph Theory 49* 4 (2005) 313-324.

19. SCHAEFER, T. J. The complexity of the satisfability problem. In *Proceedings of the 10th Annual ACM Symposium on Theory of Computing* (1978), ACM, pp. 216–226.

Generalised Dualities
and Finite Maximal Antichains[*]

Jan Foniok[1], Jaroslav Nešetřil[1], and Claude Tardif[2,**]

[1] Department of Applied Mathematics
and
Institute of Theoretical Computer Science (ITI)[***]
Charles University
Malostranské nám. 25, 118 00 Praha 1, Czech Republic
{foniok, nesetril}@kam.mff.cuni.cz
[2] Department of Mathematics and Computer Science
Royal Military College of Canada
PO Box 17000, Station 'Forces'
Kingston, Ontario K7K 7B4 Canada
Claude.Tardif@rmc.ca

Abstract. We fully characterise the situations where the existence of a homomorphism from a digraph G to at least one of a finite set $\mathcal{H}$ of directed graphs is determined by a finite number of forbidden subgraphs. We prove that these situations, called *generalised dualities*, are characterised by the non-existence of a homomorphism to G from a finite set of forests.

Furthermore, we characterise all finite maximal antichains in the partial order of directed graphs ordered by the existence of homomorphism. We show that these antichains correspond exactly to the generalised dualities. This solves a problem posed in [1]. Finally, we show that it is NP-hard to decide whether a finite set of digraphs forms a maximal antichain.

1 Introduction and Previous Results

Several classical colouring problems (such as bounding the chromatic number of graphs with given properties) can be treated more generally and sometimes more efficiently in the context of graphs and homomorphisms between them. Recall that, given graphs $G = (V, E)$, $G' = (V', E')$, a *homomorphism* is any mapping $f : V \to V'$ which preserves edges:

$$xy \in E \Rightarrow f(x)f(y) \in E' \ .$$

This is denoted by $f : G \to G'$. For a recent introduction to the topic of graphs and their homomorphisms, we refer the reader to the book [2].

---

[*] This research was partially supported by the EU network COMBSTRU.
[**] The third author's research is supported by grants from NSERC and ARP.
[***] Institute for Theoretical Computer Science is supported as project 1M0021620808 by the Ministry of Education of the Czech Republic.

Let H be a fixed graph (sometimes called a template). For an input graph G, the H-*colouring problem* asks whether there exists a homomorphism $G \to H$. Such a homomorphism is also called an H-*colouring*; the K_k-colouring problem is simply the question whether $\chi(G) \leq k$. Of course, the complexity of the H-colouring problem depends on H. This complexity was determined for undirected graphs in [3]. However, already for directed graphs the problem is unsolved.

The H-colouring problem is also (and perhaps more often) called the *constraint satisfaction problem (CSP(H))*. This is particularly used when the problem is generalised to relational structures and their homomorphisms, as these structures can encode arbitrary constraints. This setting, originally motivated by problems from Artificial Intelligence, leads to the important problem of dichotomy, general heuristic algorithms (consistency check) and, more recently, to an interesting and fruitful algebraic setting (pioneered by Bulatov, Jeavons and Krokhin, see [4,5]).

Further work in the area of CSP complexity led to the following dichotomy conjecture.

Conjecture 1 ([6]). Let H be a finite relational structure. Then CSP(H) is either solvable in polynomial time or NP-complete.

Some particular instances of CSP were studied intensively. This includes the case when the graphs for which there exists an H-colouring are determined by well-described forbidden subgraphs (see [7,8]) and as a special case, when they are determined by a finite family of forbidden subgraphs. Of course, in these cases we get polynomial instances of CSP.

A pair (F, D) of directed graphs is called a *duality pair* if for every directed graph G, we have $F \to G$ if and only if $G \nrightarrow D$. Here, and from now on, $A \to B$ denotes the fact that there exists a homomorphism from A to B. The duality relationship is denoted by the equation

$$F\!\to\; = \;\nrightarrow\! D$$

where $F\!\to$ denotes the class of graphs admitting a homomorphism from F and $\nrightarrow\! D$ the class of graphs not admitting a homomorphism to D. The dualities in the category of directed graphs are characterised in [9,10]:

Theorem 2 ([9,10]). *Given a directed graph F, there exists a directed graph D_F such that (F, D_F) is a duality pair if and only if F is homomorphically equivalent to an orientation of a tree. For a tree F, such a digraph D_F is unique up to homomorphism equivalence.*

We say that A and B are homomorphically equivalent if both $A \to B$ and $B \to A$. The unique D such that (F, D) is a duality pair is called *the dual* of the tree F. We use the notation $D = D(F)$.

Here we generalise the notion of a duality pair: for two finite sets $\mathcal{F}, \mathcal{D}$ of graphs, we say that $(\mathcal{F}, \mathcal{D})$ is a *generalised duality* if for any graph G, there exists $F \in \mathcal{F}$ such that $F \to G$ if and only if $G \to D$ for no $D \in \mathcal{D}$; briefly

$$\bigcup_{F \in \mathcal{F}} F\!\to\; = \bigcap_{D \in \mathcal{D}} \nrightarrow\! D \;.$$

We shall assume that there exist no homomorphisms between any two distinct elements of $\mathcal{F}$ as well as between any two distinct elements of $\mathcal{D}$.

The special case $|\mathcal{D}| = 1$ is characterised by the following theorem proved in [10].

Theorem 3 ([10]). *Let $\mathcal{F} = \{F_1, F_2, \ldots, F_m\}$ be a finite nonempty set of digraphs. The pair $(\mathcal{F}, \{D\})$ is a generalised duality if and only if $D = \prod_{i=1}^m D_i$ and (F_i, D_i) is a duality pair for $i = 1, 2, \ldots, m$.*

When $|\mathcal{D}| = 1$, the generalised duality $(\mathcal{F}, \mathcal{D})$ is called a *finitary homomorphism duality* in [10]. The theorem states that the finitary dual is the product of the duals of the trees $F_1, \ldots, F_m$; this product will be denoted by $D(F_1, \ldots, F_m)$ or $D(M)$ if $M = \{F_1, \ldots, F_m\}$.

The product meant here is the usual product in the category of digraphs and homomorphisms, see e.g. [2, Section 2.1].

Notice that the following two pairs are also cases of generalised dualities: $(\emptyset, \{\mathbf{1}\})$, $(\{\mathbf{0}\}, \emptyset)$, where $\mathbf{0}$ is the digraph consisting of a single vertex and no edges and $\mathbf{1}$ is the digraph with one vertex with a loop.

The relation $\to$ induces a partial order $\mathcal{C}$ on the classes of homomorphic equivalence of graphs. This order is called the *homomorphism order*. The homomorphism order is actually a distributive lattice, with the disjoint union (the sum) of graphs being the supremum and the categorical product being the infimum. The standard order-theoretic terminology is applied here.

Particular properties of the homomorphism order, that were studied, are density (solved for undirected graphs by Welzl [11] and for directed graphs by Nešetřil and Tardif [10]) and the description of finite maximal antichains.

Earlier results characterise all maximal antichains of size 1 and 2 in the homomorphism order of directed graphs.

Theorem 4 ([12]). *The only maximal antichains of size 1 in the homomorphism order of directed graphs are directed paths of length 0, 1, and 2 and a single vertex with a loop.*

Theorem 5 ([1]). *The maximal antichains of size 2 in the homomorphism order of directed graphs are precisely the pairs $\{T, D_T\}$, where T is a tree different from P_0, P_1 and P_2, and D_T is its dual.*

2   Generalised Dualities

In this section, we characterise all generalised dualities. We restrict ourselves to the case $|\mathcal{F}| \geq 2$, as the other cases are described in the previous section. First, we present a construction of generalised dualities from a family of forests.

2.1   The Construction

Let $\mathcal{F} = \{F_1, F_2, \ldots, F_m\}$ be an arbitrary fixed nonempty finite set of core (cf. [2]) forests that are pairwise incomparable.

Consistently with the above notation, let $\mathcal{F}_c = \{C_1, \ldots, C_n\}$ be the set of all distinct connected components of the graphs in $\mathcal{F}$; each of these components is a core tree.

A subset $M \subseteq \mathcal{F}_c$ is a *quasitransversal* if it satisfies

(T1) M is an antichain, i.e. for every $C \neq C' \in M$ we have $C \parallel C'$, and
(T2) M *supports* $\mathcal{F}$, i.e. for every $F \in \mathcal{F}$ there exists $C \in M$ such that $C \to F$.

For two quasitransversals M, M' we define $M \preceq M'$ if and only if for every $C' \in M'$ there exists $C \in M$ such that $C \to C'$. Note that this order is different from the homomorphism order of forests corresponding to the quasitransversals. On the other hand, we have:

Lemma 6. *Let M, M' be two quasitransversals. Then $D(M) \to D(M')$ if and only if $M \preceq M'$*

Lemma 7. *The relation $\preceq$ is a partial order on the set of all quasitransversals.*

A quasitransversal M is a *transversal* if

(T3) M is a maximal quasitransversal in $\preceq$.

Set $\mathcal{D} = \mathcal{D}(\mathcal{F}) = \{D(M) : M \text{ is a transversal}\}$.
We have:

Theorem 8. *The pair $(\mathcal{F}, \mathcal{D})$ is a generalised duality.*

Before outlining the proof, we give three of examples.

Example. First, let $\mathcal{F} = \{T_1, T_2, \ldots, T_n\}$ be a set of pairwise incomparable trees and $D_1, D_2, \ldots, D_n$ their respective duals. By (T2), every transversal contains all these trees. Therefore there exists only one transversal $M = \{T_1, T_2, \ldots, T_n\}$ and $\mathcal{D} = \{D_1 \times D_2 \times \ldots \times D_n\}$. This situation shows how the finitary duality is a special case of the generalised duality.

Now, let T_1, T_2, T_3 and T_4 be pairwise incomparable trees with duals D_1, D_2, D_3, D_4. Let $\mathcal{F} = \{T_1 + T_2, T_1 + T_3, T_4\}$. Then we have two transversals $\{T_1, T_4\}$ and $\{T_2, T_3, T_4\}$; and $\mathcal{D} = \{D_1 \times D_4, D_2 \times D_3 \times D_4\}$.

Finally, let $T_1 \to T_3$ and $\mathcal{F} = \{T_1 + T_2, T_3 + T_4\}$. The transversals are $\{T_1\}$, $\{T_2, T_3\}$ and $\{T_2, T_4\}$. Hence $\mathcal{D} = \{D_1, D_2 \times D_3, D_2 \times D_4\}$.

Proof (of Theorem 8). Let X be a digraph such that $X \to D$ for some $D \in \mathcal{D}$. We want to prove that $F_i \nrightarrow X$ for $i = 1, \ldots, m$. For contradiction, assume that $F_i \to X$ for some i. Let M be the transversal for which $D(M) = D$. By (T2), there exists $C \in M$ such that $C \to F_i \to X$, therefore $X \nrightarrow D(C)$. That is a contradiction with the assumption that $X \to D \to D(C)$.

Now, let X be a digraph such that $F_i \nrightarrow X$ for $i = 1, \ldots, m$. We want to prove that there exists $D \in \mathcal{D}$ such that $X \to D$. Let C_{j_i} be a component of F_i such that $C_{j_i} \nrightarrow X$ for $i = 1, \ldots, m$. Let $M' = \min_{\to}\{C_{j_i} : i = 1, \ldots, m\}$; by $\min_{\to} S$ we mean the set of all elements of S that are minimal with respect to the homomorphism order $\to$. Because M' is a quasitransversal, there exists a transversal M such that $M' \preceq M$. We have that $C \nrightarrow X$ for each $C \in M$, and therefore $X \to D(M) \in \mathcal{D}$. $\qquad\square$

2.2   The Characterisation

We will now prove that all generalised dualities are of the above form.

Theorem 9. *If $(\mathcal{F}, \mathcal{D})$ is a generalised duality, then all elements of $\mathcal{F}$ are forests and $\mathcal{D} = \mathcal{D}(\mathcal{F})$; in particular, $\mathcal{D}$ is uniquely determined by $\mathcal{F}$.*

Proof. The proof consists of five steps. Suppose that $\mathcal{F} = \{A_1, A_2, \ldots, A_m\}$ and $\mathcal{D} = \{D_1, D_2, \ldots, D_p\}$. Consistently with the above notation, let $\mathcal{F}_c = \{C_1, C_2, \ldots, C_n\}$ be the set of all distinct connected components of the structures in $\mathcal{F}$. Quasitransversals and transversals are defined in the same way as above; we note that neither for the definition nor for proving Lemma 7 do we need the fact that the elements of $\mathcal{F}_c$ are trees.

For a quasitransversal M, let $\overline{M} = \{C' \in \mathcal{F}_c : C \in M \Rightarrow C \nrightarrow C'\}$.

Fact 1. *If $M \subseteq \mathcal{F}_c$ is a transversal, then there exists a unique digraph $D \in \mathcal{D}$ that satisfies*

1. *$C \nrightarrow D$ for every $C \in M$,*
2. *$C' \rightarrow D$ for every $C' \in \overline{M}$.*

Proof. If $\overline{M} = \emptyset$, let $D \in \mathcal{D}$ be arbitrary. Otherwise set $S = \sum_{C' \in \overline{M}} C'$. Because $(\mathcal{F}, \mathcal{D})$ is a generalised duality, either there exists $F \in \mathcal{F}$ such that $F \rightarrow S$ or there exists $D \in \mathcal{D}$ such that $S \rightarrow D$. If $F \rightarrow S$, by (T2) there exists $C \in M$ satisfying $C \rightarrow F \rightarrow S$, and since C is connected, $C \rightarrow C'$ for some $C' \in \overline{M}$, which is a contradiction with the definition of $\overline{M}$. Therefore there exists $D \in \mathcal{D}$ that satisfies $S \rightarrow D$. It can be checked that such D satisfies both (1) and (2) and that such a graph must be unique. □

For a transversal M, the unique $D \in \mathcal{D}$ satisfying the conditions (1) and (2) above is denoted by $d(M)$.

Fact 2. *$\mathcal{D} = \{d(M) : M \text{ is a transversal}\}$.*

Proof. Let $D \in \mathcal{D}$. We want to show that $D = d(M)$ for a transversal M. Let $M' = \min_{\rightarrow}\{C' \in \mathcal{F}_c : C' \nrightarrow D\}$ be the set of all $\mathcal{C}$-minimal components that are not homomorphic to D. The set M' is a quasitransversal: if some $F \in \mathcal{F}$ is not supported by M', then all its components are homomorphic to D, and also $F \rightarrow D$, a contradiction.

Let M be a transversal such that $M' \preceq M$. To prove that $D = d(M)$, it suffices (by the uniqueness part of Fact 1) to check conditions (1) and (2). □

Fact 3. *For two distinct transversals M_1, M_2, we have*

   (a) $\overline{M_1} \cap M_2 \neq \emptyset$,
   (b) $d(M_1) \nrightarrow d(M_2)$.

Proof. (a) By (T3), $M_1 \npreceq M_2$, so there exists $C_2 \in M_2$ such that $C_1 \nrightarrow C_2$ for any $C_1 \in M_1$. Obviously $C_2 \in \overline{M_1} \setminus \overline{M_2} \subseteq \overline{M_1}$. Since we chose $C_2 \in M_2$, we have $C_2 \in \overline{M_1} \cap M_2$.

(b) Let $C_2 \in \overline{M_1} \cap M_2$, as above. Then $C_2 \rightarrow d(M_1)$ and $C_2 \nrightarrow d(M_2)$. Consequently $d(M_1) \nrightarrow d(M_2)$. □

Fact 4. *If M is a transversal, then the pair $(M, \{d(M)\})$ is a finitary homomorphism duality, and consequently $d(M) = D(M)$.*

Proof. We can prove that for a digraph G, the following statements are equivalent:

(1) $G \in \bigcap_{C \in M}(C \nrightarrow)$
(2) $C \nrightarrow G$ for any $C \in M$
(3) $C \nrightarrow G + \sum_{\check{C} \in \overline{M}} \check{C}$ for any $C \in M$
(4) $G + \sum_{\check{C} \in \overline{M}} \check{C} \to d(M)$
(5) $G \to d(M)$
(6) $G \in (\to d(M))$

The equivalence $(1) \Leftrightarrow (6)$ is precisely the definition of finitary duality. $\square$

Fact 5. *Each component $C \in \mathcal{F}_c$ is a tree.*

For the proof, we use the following

Theorem 10 ([10]). *Let A and C be relational structures such that $A < C$, and C is a connected structure that is not a tree. Then there exists a structure X such that $A < X < C$.*

Proof (of Fact 5). Using Theorem 10, it can be proved that if some component $C \in \mathcal{F}_c$ is not a tree, then there exists a digraph X such that $X < C$ and X is homomorphic to exactly the same elements of $\mathcal{F}_c$ as C, and moreover for any $C' \in \mathcal{F}_c$, $C' \neq C$, we have $C' \to C$ if and only if $C' \to X$. Then if G is created by replacing C with X in some $F \in \mathcal{F}$, the graph G violates the definition of generalised duality (no $F \in \mathcal{F}$ is homomorphic to G and G is homomorphic to no $D \in \mathcal{D}$). $\square$

We finish the proof of Theorem 9. All elements of $\mathcal{F}$ are forests by virtue of Fact 5. The set $\mathcal{D}$ is uniquely determined as a consequence of Fact 2 and due to Fact 4 and Theorem 3 it is determined by the transversal construction. $\square$

3   Finite Maximal Antichains

First, we discuss when a generalised duality forms a maximal antichain; precisely, for what families $\mathcal{F}$ of incomparable forests is $\mathcal{Q} = \mathcal{F} \cup \mathcal{D}(\mathcal{F})$ a maximal antichain in the homomorphism order of digraphs.

Obviously, if a generalised duality forms an antichain, then it is maximal. It is also evident that $F \nrightarrow D$ for any $F \in \mathcal{F}$, $D \in \mathcal{D}$. So, a generalised duality does not form an antichain if and only if there exist $D \in \mathcal{D}$ and $F \in \mathcal{F}$ such that $D \to F$.

Let $P_1 = (\{1, 2\}, \{(1, 2)\})$ be the digraph consisting of a single edge (the path of length 1). If $P_1 \in \mathcal{F}_c$, then obviously $\mathcal{F} = \{P_1\}$ and $\mathcal{D} = \{\mathbf{0}\}$. So for the rest, we can assume that $P_1 \notin \mathcal{F}_c$.

Let $P_2 = (\{1, 2, 3\}, \{(1, 2), (2, 3)\})$ be the directed path of length 2.

Lemma 11. *Let $\mathcal{F}$ be a set of pairwise incomparable core forests. Then $\mathcal{F} \cup \mathcal{D}(\mathcal{F})$ is not an antichain if and only if $\mathcal{F} = \{P_1\}$ or $\mathcal{F} = \{P_2\}$.*

Proof. We have just observed that if $\mathcal{F} \cup \mathcal{D}(\mathcal{F})$ is not an antichain, there exist $D \in \mathcal{D}$ and $F \in \mathcal{F}$ such that $D \to F$. Fix such F and D.

Let A be a digraph. If there exists a tree T such that $A \to T$, we say that A is *balanced*. It is easy to see that A is balanced if and only if it is homomorphic to a forest.

Since F is a forest, we have that D is balanced. Moreover, by Theorem 9, $D = D(M)$ for a transversal $M \subseteq \mathcal{F}_c$.

Let Z_s be the antidirected orientation of a cycle of length $2s+1$ such that Z_s contains a unique directed path of length 2. Since Z_s is not balanced, $Z_s \nrightarrow D = D(M)$; therefore (by the definition of finitary duality) for every positive integer s there exists $C \in M$ such that $C \to Z_s$. Hence there exist some s and $C \to Z_s$ such that $|V(C)| < 2s+1$. Since any proper subgraph of Z_s is homomorphic to P_2, we get that $C \to P_2$. This finishes the proof as the other implication is evident. □

We have now observed that only two generalised dualities that are not antichains exist: $(\{P_1\}, \{\mathbf{0}\})$ and $(\{P_2\}, \{P_1\})$. Let us now consider the question when a maximal antichain is not a generalised duality.

Notice that a finite maximal antichain $\mathcal{Q}$ is formed from a generalised duality if and only if there exist disjoint sets $\mathcal{F}, \mathcal{D}$ such that $\mathcal{Q} = \mathcal{F} \cup \mathcal{D}$ and for an arbitrary digraph X there exists $F \in \mathcal{F}$ such that $F \to X$ or there exists $D \in \mathcal{D}$ such that $X \to D$.

Lemma 12. *Let $\mathcal{Q}$ be a finite maximal antichain in the homomorphism order of digraphs. Then the following are equivalent:*

1. *$\mathcal{Q}$ is not formed from a generalised duality, i.e. whenever $\mathcal{Q} = \mathcal{F} \cup \mathcal{D}$, the pair $(\mathcal{F}, \mathcal{D})$ is not a generalised duality,*
2. *$\mathcal{Q}$ is one of the sets $\{P_1\}$ or $\{P_2\}$.*

Proof. Both sets $\{P_1\}$ and $\{P_2\}$ are obviously finite maximal antichains not formed from a generalised duality.

The other implication is proved by splitting $\mathcal{Q}$ into $\mathcal{F}$ and $\mathcal{D}$ in a suitable way, which allows us to show that all elements of $\mathcal{F}$ are balanced. But since $(\mathcal{F}, \mathcal{D})$ is not a generalised duality, we can use an argument similar to the above proof to show that $\mathcal{F}$ is one of the sets $\{P_1\}$ or $\{P_2\}$. □

Thus we come to the correspondence between generalised dualities and maximal antichains. This solves a problem posed in [1], where maximal antichains of size 2 were characterised.

Theorem 13. *The correspondence*

$$(\mathcal{F}, \mathcal{D}) \;\mapsto\; \mathcal{Q} = \mathcal{F} \cup \{D \in \mathcal{D} : D \nrightarrow F \text{ for any } F \in \mathcal{F}\}$$

is a one-to-one correspondence between generalised dualities and finite maximal antichains in the homomorphism order of directed graphs.

Proof. Follows immediately from Theorem 4 and Lemmas 11 and 12. □

4 Extensions

4.1 MAC Decidability

We are interested in the following decision problem, called the *MAC decision problem*: given a finite nonempty set $\mathcal{Q}$ of digraphs, decide whether $\mathcal{Q}$ is a maximal antichain. The results of the previous section allow us to state the following result, whose proof will appear in the full version of the paper.

Theorem 14. *The MAC decision problem is decidable. Moreover, it is NP-hard.*

Another consequence of Theorem 13 is the following.

Theorem 15. *Let $\mathcal{Q}$ be a finite maximal antichain in $\mathcal{C}$. An element of $\mathcal{Q}$ that is comparable with an input structure A can be found in polynomial time.*

4.2 Duality Decidability

Using a recent result of [13], we can deduce that it is decidable whether for a set $\mathcal{H}$ of digraphs there exists a set $\mathcal{F}$ of digraphs such that $(\mathcal{F}, \mathcal{H})$ is a generalised duality.

It is easy to see that $\mathcal{H}$ is the right-hand side of a generalised duality if and only if each structure in $\mathcal{H}$ is a finitary dual and they are pairwise incomparable. The former is decidable (and even in NP) due to [13], the latter is obviously in NP. It also follows from [13] that in general, the problem is NP-complete.

4.3 GCSP Dichotomy

As an analogy to CSP, we define *GCSP*, the *generalised constraint satisfaction problem*, as the following: given a finite set $\mathcal{H}$ of digraphs, decide for an input digraph G whether there exists $H \in \mathcal{H}$ such that $G \to H$.

Note that if $(\mathcal{F}, \mathcal{D})$ is a generalised duality, then GCSP($\mathcal{D}$) is polynomially solvable.

As in Conjecture 1, one could ask whether there is a dichotomy for GCSP. However, this problem is not very captivating, as the positive answer to the dichotomy conjecture for CSP would imply a positive answer here as well:

Theorem 16. *Let $\mathcal{H}$ be a finite nonempty set of pairwise incomparable digraphs.*

1. If CSP(H) is tractable for all $H \in \mathcal{H}$, then GCSP(H) is tractable.
2. If CSP(H) is NP-complete for some $H \in \mathcal{H}$, then GCSP(H) is NP-complete.

Thus from the complexity (and dichotomy) point of view, generalised CSP is equivalent to CSP. But their first-order definability is another matter: it is both interesting and more involved.

4.4 First-Order Definable GCSP

We remark that GCSP($\mathcal{H}$) is first-order definable if and only if there exists a set $\mathcal{F}$ such that $(\mathcal{F}, \mathcal{H})$ is a generalised duality. This result is an extension of a similar theorem for CSP contained in [14], and its proof follows the same way.

5   Summary and Concluding Remarks

In Sect. 2, we characterised all the generalised dualities $(\mathcal{F}, \mathcal{D})$ in the category of directed graphs: the set $\mathcal{D}$ such that $(\mathcal{F}, \mathcal{D})$ is a generalised duality exists if and only if $\mathcal{F}$ is a finite family of forests; if this is the case, $\mathcal{D}$ is determined uniquely (up to homomorphic equivalence).

In Sect. 3 we described all finite maximal antichains in the homomorphism order of directed graphs. They all appear to be formed from generalised dualities by taking all maximal elements of $\mathcal{F} \cup \mathcal{D}$.

We mention here that the result on generalised dualities extends to the fully general setting of relational structures with relations of arbitrary arity. Similarly, Theorem 13 can be generalised for relational structures with one relation of arbitrary arity. The maximal antichains which are not of the form $\mathcal{F} \cup \mathcal{D}$ for a generalised duality $(\mathcal{F}, \mathcal{D})$ are $\{P_1\}$ and $\mathcal{S}$, where $\mathcal{S}$ is the set of all core trees with two edges. Both these results will appear in the full version of this paper.

Let us note that the characterisation of finite maximal antichains is hard and interesting for infinite graphs. It has been proved in [15] that for every countable infinite graph G, G not equivalent to K_1, K_2, K_ω, there exists a graph H incomparable with G. There are also infinitely many maximal antichains, however, as pointed out in [15], all maximal antichains seem to contain a finite graph.

We believe that the interplay of order theoretic notions (such as maximal antichain) and descriptive complexity notions (such as generalised duality and first order definability) leads to further insight into the structure of CSP. For example, the duality theorems present a rich supply of non-trivial CSP problems, for which polynomial algorithms exist.

References

1. Nešetřil, J., Tardif, C.: On maximal finite antichains in the homomorphism order of directed graphs. Discuss. Math. Graph Theory **23** (2003) 325–332
2. Hell, P., Nešetřil, J.: Graphs and Homomorphisms. Volume 28 of Oxford Lecture Series in Mathematics and Its Applications. Oxford University Press (2004)
3. Hell, P., Nešetřil, J.: On the complexity of H-coloring. J. Combin. Theory Ser. B **48** (1992) 92–119
4. Bulatov, A., Krokhin, A., Jeavons, P.G.: Classifying the complexity of constraints using finite algebras. SIAM J. Comput. **34** (2005) 720–742
5. Cohen, D., Jeavons, P.: The complexity of constraint languages. In Rossi, F., van Beek, P., Walsh, T., eds.: Handbook of Constraint Programming. Elsevier (2006)
6. Feder, T., Vardi, M.Y.: The computational structure of monotone monadic SNP and constraint satisfaction: A study through Datalog and group theory. SIAM J. Comput. **28** (1998) 57–104
7. Hell, P., Nešetřil, J., Zhu, X.: Duality and polynomial testing of tree homomorphisms. Trans. Amer. Math. Soc. **348** (1996) 1281–1297
8. Nešetřil, J., Pultr, A.: On classes of relations and graphs determined by subobjects and factorobjects. Discrete Math. **22** (1978) 287–300
9. Komárek, P.: Good characterisations in the class of oriented graphs. PhD thesis, Czechoslovak Academy of Sciences, Prague (1987) In Czech.

10. Nešetřil, J., Tardif, C.: Duality theorems for finite structures (characterising gaps and good characterisations). J. Combin. Theory Ser. B **80** (2000) 80–97
11. Welzl, E.: Color families are dense. Theoret. Comput. Sci. **17** (1982) 29–41
12. Nešetřil, J., Zhu, X.: Path homomorphisms. Math. Proc. Cambridge Philos. Soc. **120** (1996) 207–220
13. Larose, B., Loten, C., Tardif, C.: A characterisation of first-order constraint satisfaction problems. Submitted (2006)
14. Atserias, A.: On digraph coloring problems and treewidth duality. In: Proceedings of the 20th IEEE Symposium on Logic in Computer Science (LICS'05), IEEE Computer Society (2005) 106–115
15. Nešetřil, J., Shelah, S.: On the order of countable graphs. European J. Combin. **24** (2003) 649–663

Chordal Deletion Is Fixed-Parameter Tractable

Dániel Marx

Institut für Informatik,
Humboldt-Universität zu Berlin,
Unter den Linden 6, 10099
Berlin, Germany
dmarx@informatik.hu-berlin.de

Abstract. It is known to be NP–hard to decide whether a graph can be made chordal by the deletion of k vertices. Here we present a uniformly polynomial-time algorithm for the problem: the running time is $f(k) \cdot n^\alpha$ for some constant α not depending on k and some f depending only on k. For large values of n, such an algorithm is much better than trying all the $O(n^k)$ possibilities. Therefore, the chordal deletion problem parameterized by the number k of vertices to be deleted is fixed-parameter tractable. This answers an open question of Cai [2].

1 Introduction

A graph is chordal if it does not contain an induced cycle of length greater than 3. It can be decided in linear time whether a graph is chordal [9]. However, it is NP-complete to decide whether a graph can be made chordal by the deletion of k vertices [6], by the deletion of k edges [7], or by the addition of k edges [10] (if k is part of the input).

In this paper we investigate these problems from the parameterized complexity point of view. Parameterized complexity deals with problems where the input has a distinguished part k (usually an integer) called the *parameter*. A parameterized problem is called *fixed-parameter tractable* if there is an algorithm with running time $f(k) \cdot n^\alpha$, where $f(k)$ is an arbitrary function and α is a positive constant independent of k. It turns out that several NP-hard problems, such as MINIMUM VERTEX COVER and LONGEST PATH, are fixed-parameter tractable. The function $f(k)$ is usually exponential, thus if the parameter k can be arbitrary, then the algorithms are not polynomial (as expected). However, for small fixed values of k, fixed-parameter tractable problems have low-degree polynomial algorithms, which are sometimes even practically feasible. For more background, the reader is referred to the monograph of Downey and Fellows [3].

If k is a fixed constant, then the three chordal deletion/completion problems can be solved in polynomial time by complete enumeration. For example, in the vertex deletion problem we can try all the $O(n^k)$ possible sets of size k and check whether their removal makes the graph chordal. Moreover, in [1,5] it is shown that it can be decided in $O(4^k/(k+1)^{3/2} \cdot (n+m))$ or $O(k^2 nm + k^6 2^{4k})$ time whether a graph with n vertices and m edges can be made chordal by adding

F.V. Fomin (Ed.): WG 2006, LNCS 4271, pp. 37–48, 2006.
© Springer-Verlag Berlin Heidelberg 2006

k edges. Therefore, chordal edge completion (which is also called the *minimum fill-in problem*) is fixed-parameter tractable. The main result of the paper is that chordal vertex deletion is also fixed-parameter tractable. This answers an open question of Cai [2].

Theorem 1. *Chordal vertex deletion is fixed-parameter tractable with parameter k, the number of vertices to be deleted.*

The *iterative compression* method introduced in [8] allows us to concentrate on the following easier problem: given a set X of $k + 1$ vertices such that $G \setminus X$ is chordal, find k vertices whose deletion makes G chordal. To solve this "solution compression" problem, we first determine the size of the maximum clique in the chordal graph $G \setminus X$. If the clique size $G \setminus X$ is small, then $G \setminus X$ (and hence the slightly larger G) has small treewidth. Using standard techniques, the problem can be solved in linear time for graphs with bounded treewidth. On the other hand, we show that if there is a large clique in $G \setminus X$, then the clique contains "irrelevant" vertices that can be removed from the graph without changing the solvability of the problem.

The paper is organized as follows. Section 2 reviews some basic facts on chordal graphs. In Section 3 we show how the iterative compression method of [8] can be applied to our problem. Section 4 discusses how we can reduce the size of the cliques to make our graph a bounded treewidth graph.

2   Chordal Graphs

A graph is *chordal* if it does not contain a cycle of length greater than 3 as an induced subgraph. This is equivalent to saying that every cycle of length greater than 3 contains at least one chord, i.e., an edge connecting two vertices not adjacent in the cycle. A chordless cycle of length greater than 3 will be called a *hole*. Chordal graphs can be also characterized as the intersection graphs of subtrees of a tree (see e.g., [4]):

Theorem 2. *The following two statements are equivalent:*

1. *$G(V, E)$ is chordal.*
2. *There exists a tree $T(U, F)$ and a subtree $T_v \subseteq T$ for each $v \in V$ such that $u, v \in V$ are neighbors in $G(V, E)$ if and only if $T_u \cap T_v \neq \emptyset$ (i.e., T_u and T_v have a common node).*

For clarity, we will use the word "vertex" when we refer to the graph $G(V, E)$, and "node" when referring to $T(U, F)$. The tree T together with the subtrees T_v is called the *tree decomposition* of G. A tree decomposition of G can be found in polynomial time (see [4,9]). We say that a vertex v contains node x if T_v contains node x. Similarly, node x can be reached from vertex v is shorthand for saying that there is a vertex v' that contains x and there is a path between v and v'. For an arbitrary node x of T, the vertices whose tree contain x induce a clique. Moreover, for every clique K, there is a node of T such that every $v \in K$ contains this node (cf. [4]). The following easy observation will be used repeatedly:

Proposition 3. *Let x, y, z be vertices in $G(V, E)$ such that $xy, xz \in E$ but $yz \notin E$. If there is a walk T in $G \setminus x$ from y to z such that in T only y and z are neighbors of x, then $T \cup x$ contains a hole of length at least 4.*   □

The problem studied in this paper is formally defined as follows:

CHORDAL VERTEX DELETION

    Input: A graph $G(V, E)$ and an integer k

Parameter: k

        Task: Is there a set $X \subseteq V$ of size at most k such that $G \setminus X$ is chordal?

If the deletion of $X \subseteq V$ makes the graph chordal, then we say that X is a *hole cover*. It turns out that the deletion problem is very different from the edge completion problem. The algorithms in [1,5] for chordal edge completion use the standard method of bounded search trees (with some non-trivial optimizations). The techniques rely on the fact that the graph cannot contain a large hole, otherwise the graph could not be made chordal by adding k edges. In the deletion problem we cannot make this assumption: it is possible that the graph can be made chordal by deleting few vertices, even if there are large holes.

3   Iterative Compression

Reed, Smith and Vetta [8] have shown that the BIPARTITE VERTEX DELETION problem (make the graph bipartite by deleting k vertices) is fixed-parameter tractable. They introduced the method of *iterative compression* that can be used in the case of the chordal deletion problem as well. The idea is that it is sufficient to show that the following easier problem is fixed-parameter tractable:

HOLE COVER COMPRESSION

    Input: A graph G, an integer k, and a size $k + 1$ hole cover W.

Parameter: k

    Task: Is there a size k hole cover X of G?

This problem is easier: the extra input W gives us useful information on G. In particular, we know that $G \setminus W$ is chordal, our algorithm builds on this fact.

Assume that we have an algorithm with running time $f(k)n^\alpha$ for HOLE COVER COMPRESSION, then CHORDAL VERTEX DELETION can be solved as follows. Let v_1, v_2, ..., v_n be an ordering of the vertices, and let G_i be the graph induced by v_1, ..., v_i. We try to find a size k hole cover for each G_i. Graph G_k trivially has a size k hole cover. Now assume that G_i has a size k hole cover W. Clearly, $W \cup v_{i+1}$ is a size $k+1$ hole cover of G_{i+1}. Therefore, the compression algorithm can be used to find a size k hole cover for G_{i+1}. If there is such a hole cover, then we can proceed to G_{i+2}. Otherwise the answer is no, we can conclude that supergraph G of G_{i+1} cannot have a size k hole cover either. The algorithm calls the compression method at most n times, thus the total running time is $f(k)n^{\alpha+1}$, which shows that the problem is fixed-parameter tractable.

The compression algorithm will be described later in this section. Our algorithm is somewhat weaker than the one defined in the above scheme. First, it looks for only a W-avoiding hole cover, i.e., a hole cover that is disjoint from W. Furthermore, the compression algorithm either finds a W-avoiding hole cover W', or returns a set N, whose size can be bounded by a function of k, such that every W-avoiding size k hole cover contains at least one vertex of N. (Such a set N will be called a *necessary* set.) If the compression algorithm finds that $N = \emptyset$ is necessary, then this means that there is no W-avoiding hole cover.

If the compression algorithm returns a necessary set N, then we can conclude that every size k hole cover contains at least one vertex of $N \cup W$. Therefore, we branch into $|N \cup W|$ directions: for each vertex v of $N \cup W$, we check whether there is a size $k - 1$ hole cover of $G \setminus v$. Thus the problem can be reduced to at most b_k subproblems with smaller parameter values, where b_k depends only on k. The overall algorithm is the following:

CHORDAL VERTEX DELETION(G, k)

1. Set $i := k$ and let W be the vertices of G_k.
2. *Invariant condition:* W is a size k hole cover of G_i. If $i = n$, then return "W is a size k hole cover of G."
3. Set $W := W \cup v_{i+1}$, now W is a size $k + 1$ hole cover of G_{i+1}.
4. Call HOLE COVER COMPRESSION(G_{i+1}, k, W).
   - If the answer is a size k hole cover W' of G_{i+1}, then let $W := W'$, $i := i + 1$, and go to Step 2.
   - If the answer is a set N, then let $T := N \cup W$.
5. For each vertex $v \in T$, call CHORDAL VERTEX DELETION$(G \setminus v, k - 1)$.
   - If the answer is yes for some $v \in T$, and W is a size $k - 1$ hole cover of $G \setminus v$, then answer "$W \cup v$ is a size k hole cover of G."
   - If the answer is no for every $v \in T$, then answer "No."

CHORDAL VERTEX DELETION calls the compression algorithm at most n times, and may make at most b_k recursive calls to CHORDAL VERTEX DELETION with parameter $k - 1$ (where b_k is the maximum size of T). Therefore, if the compression algorithm runs in $f(k)n^{\alpha}$ time (which will be shown in the next section), and CHORDAL VERTEX DELETION runs in $g(k - 1)n^{\alpha+1}$ time for parameter $k - 1$, then for parameter k the algorithm runs in $g(k)n^{\alpha+1}$ time, for some appropriate constant $g(k)$. Thus by induction, we have a $g(k)n^{\alpha+1}$ time algorithm for every k, proving Theorem 1.

Now let us turn our attention to the HOLE COVER COMPRESSION algorithm itself. Assume that a size $k + 1$ hole cover W of G is given. Let $V_0 = V \setminus W$ and denote by G_0 the chordal graph $G \setminus W$. If the size of the maximum clique in V_0 is c, then the treewidth of the chordal graph G_0 is $c - 1$, and the treewidth of G is at most $c - 1 + k + 1$. With standard techniques (applying Courcelle's Theorem or dynamic programming) we can show that CHORDAL VERTEX DELETION is linear-time solvable for graphs with bounded treewidth (details omitted).

Lemma 4. *For every k and w,* CHORDAL VERTEX DELETION *can be solved in linear time for graphs with treewidth at most w.* □

In Section 4, we present a method of reducing the clique size of G_0 to a constant depending only on k. A vertex $v \in V$ is *irrelevant* if every W-avoiding size k hole cover of $G \setminus v$ is also a hole cover of G. If we identify an irrelevant vertex v, then the problem can be reduced to finding a size k hole cover in $G \setminus v$. We show that if there is a clique K in G_0 whose size is greater than some constant c_k, then the problem can be reduced to a simpler form. More precisely, for a large clique K the clique reduction algorithm does one of the following:

- *Identifies an irrelevant vertex $v \in K$.* In this case v can be deleted. If the maximum clique size is still larger than c_k, then the algorithm can be applied again. Otherwise we can use the algorithm of Lemma 4.
- *Identifies a set N of constant size such that every W-avoiding size k hole cover contains at least one vertex of N.* As mentioned above, it is possible that the compression algorithm returns a necessary set.

In the following, it is assumed that in the graph G every hole of size 4 or 5 is completely contained in W. If there is a hole H of size at most 5 that has vertices outside W, then every W-avoiding hole cover has to contain at least one vertex of $H \setminus W$, thus the compression algorithm can return $H \setminus W$ as a necessary set of constant size. Testing whether such a hole H exists can be easily done in polynomial time (e.g., by complete enumeration).

In summary, the compression algorithm makes the following steps:

HOLE COVER COMPRESSION(G, k, W)

1. By complete enumeration, determine if there is a hole H of length at most 5 that is not completely contained in W.
   - If there is such a hole H, then return "$H \setminus W$ is a necessary set."
2. If the clique size of $G \setminus W$ is at most c_k, then use the algorithm of Lemma 4.
3. If $G \setminus W$ has a clique K of size more than c_k, then call the clique reduction algorithm for K.
   - If the result is an irrelevant vertex v, then delete v from G, and go to Step 2.
   - If the result is a necessary set N, then return "N is a necessary set."

The clique reduction method is described in the following section.

4   Clique Reduction

Henceforth it is assumed that $W = \{w_1, w_2, \ldots, w_{k+1}\}$ is a hole cover of G. As in the previous section, let $V_0 = V \setminus W$ and denote by G_0 the chordal graph $G \setminus W$. In this section we show that if there is a large clique K in G_0, then K contains a irrelevant vertex.

In the rest of the paper, we prove several lemmas that state certain properties of the instance. However, these properties do not always hold, but in this case the compression algorithm can identify and return a necessary set. We use the expression "We can make sure" to mean that there is a polynomial-time algorithm that finds a necessary set if the statement does not hold.

4.1   Labeling

If a vertex $v \in V_0$ is the neighbor of some vertex $\ell \in W$, then we say that v *has label* ℓ. A vertex can have more than one label, the labels of a given vertex form a subset of W. The following property will be used repeatedly:

Proposition 5. *If P is a path connecting u and v, vertices u and v are not neighbors, they have label ℓ, and the internal vertices of P do not have label ℓ, then every W-avoiding hole cover has to contain at least one vertex of P.*

Proof. If X is a W-avoiding hole cover disjoint from P, then $\ell u P v \ell$ contains a hole in $G \setminus X$ (Prop. 3), a contradiction.   □

In Lemma 7 we give a bound on the number of independent labeled vertices. To prove Lemma 7, we need the following technical result:

Lemma 6. *Let B be a connected set of vertices in G_0 without label t, and let A be a set of vertices with label t in the neighborhood of B. We can make sure that for every vertex $z \in B$, if z and its neighbors are deleted from G_0, then at most $(k+1)^2$ components of the remaining graph can contain vertices from A.*

Proof. Consider the tree decomposition of G_0, let T_v be the subtree of T that corresponds to v. Assume that after deleting z and its neighbors, components $C_1, \ldots, C_{(k+1)^2+1}$ contain vertices of A, let $a_i \in A$ be in C_i. The vertices in C_i are not neighbors of z, hence their subtrees do not intersect T_z. Let p_i be the node closest to T_z that is contained in some vertex of C_i.

Let Z contain those neighbors of z that do not have label t. We show that if a vertex $v \in Z$ contains p_i for at least $k+2$ different values of i, then v is a necessary vertex. Assume without loss of generality that v contains $p_1, \ldots, p_{k+2}$. If X is a W-avoiding size k hole cover, then without loss of generality it can be assumed that X does not contain p_1, p_2 or any of the vertices in C_1 and C_2. Since p_1 can be reached from a_1 using only the vertices in C_1, and v contains p_1, thus there is a path P_1 that connects a_1 and v. The internal vertices of P_1 may have label t, let a'_1 be the t-labeled vertex of P_1 closest to v, and let P'_1 be the subpath of P_1 connecting a'_1 and v. We can define P'_2 and a'_2 in a similar way. Now $a'_1 \in C_1$ and $a'_2 \in C_2$ are not neighbors, they both have label t, and the internal vertices of the path $a'_1 P'_1 v P'_2 a'_2$ do not have label t. Hole cover X does not intersect this path, which is a contradiction by Prop. 5.

For each component C_i, there is at least one vertex of Z that contains node p_i. This follows from the fact that vertex a_i is in the neighborhood of B, thus there is a path between a_i and z whose internal vertices do not have label t. For each component C_i, select a $z_i \in Z$ that contains p_i. We have seen in the previous paragraph that the same vertex of Z can be z_i for at most $k+1$ different values of i. Therefore, if we have more than $(k+1)^2$ components, then without loss of generality it can be assumed that $z_1, \ldots, z_{k+2}$ are distinct vertices. We claim that z is a necessary vertex. Assume that X is a W-avoiding size k hole cover that does not contain z. Without loss of generality, it can be assumed that X does not contain z_1, z_2, or any of the vertices in C_1 and C_2. As in the previous paragraph, there is a path P_1' that connects a t-labeled vertex $a_1' \in C_1$ and vertex z_1 such that only a_1' has label t in this path. The path P_2' is similarly defined in C_2. Now X contains none of the vertices in the path $a_1' P_1' z_1 z z_2 P_2' a_2'$, thus X is not a hole cover by Prop. 5. $\square$

Lemma 7. *Let B be a connected subset of V_0 such that no vertex in B has label t. We can make sure that there can be at most k^4 independent vertices with label t in the neighborhood of B.*

Proof. Let $I = \{v_1, v_2, \ldots, v_{k^4+1}\}$ be an independent set of vertices with label t in the neighborhood of B. We show that there is no W-avoiding size k hole cover in G. Assume that X is such a hole cover. It has to contain at least one vertex of B, otherwise $v', v'' \in I \setminus X$ can be connected by a path P whose internal vertices are in B, which is not possible by Prop. 5. For each vertex v_i we can select an $x_i \in X \cap B$ such that v_i and x_i are connected by a path P_i whose internal vertices are in $B \setminus X$. Since $|X| = k$, there has to be more than k^3 vertices v_i such that the corresponding vertices x_i are the same. Assume without loss of generality that $x_1 = x_2 = \cdots = x_{k^3+1} = x$. We claim that if x and its neighbors are deleted, then vertices $v_1, \ldots, v_{k^3+1}$ are separated. By Lemma 6, is not possible (for $k \geq 3$).

Assume that v_1 and v_2 are connected by a path P that does not go through the neighborhood of x. Let y_1 (resp., y_2) be the neighbor of x on P_1 (resp., P_2). If y_1 and y_2 are the same or neighbors, then the t-labeled vertices v_1 and v_2 are connected by the walk $v_1 P_1 y_1 y_2 P_2 v_2$ in $G \setminus X$, which is not possible by Prop. 5. If y_1 and y_2 are not neighbors, then $y_1 P_1 v_1 P v_2 P_2 y_2$ is a walk connecting y_1 and y_2 such that its internal vertices are not neighbors of x. Therefore, by Prop. 3, there is a hole in G_0, contradicting the fact that W is a hole cover. $\square$

4.2   Dangerous Vertices

Let us fix a clique K of G_0. A vertex $v \in V_0 \setminus K$ is called a *t-dangerous vertex* (for K) if v has label t and there is a path P from v to a vertex $u \in K$ such that only v has label t on the path. Vertex v is a *t*-dangerous vertex* if v has label t and there is a path P from v to a vertex $u \in K$ such that v and u are not neighbors, u also has label t, and the internal vertices of the path do not have label t. Vertex u is the *t-witness* (*t*-witness*) of v, the path P is a *t-witness* (*t*-witness*) *path* of v. A vertex v can be t-dangerous for more than one $t \in W$, or it can be t- and t^*-dangerous at the same time.

The name dangerous comes from the observation that if there is a hole in G that goes through the clique K, then the hole has to go through a dangerous vertex as well. For example, if a hole starts in $t \in W$, goes to the t-labeled neighbor $v \in V_0$ of t, goes to a t-labeled vertex $u \in K$ via a path $P \subseteq V_0$, and returns to t, then v is a t^*-dangerous vertex, u is its witness, and P is the witness path. Thus when we delete vertices to make the graph chordal, our aim is to destroy as many witness paths as possible and to make many vertices non-dangerous. It will turn out that if a clique is large, then it contains many vertices whose deletion does not affect the dangerous vertices, thus there is no use of deleting them.

First we bound by k^4 (resp., k^6) the number of independent t-dangerous (resp., t^*-dangerous) vertices. Since G_0 is chordal (hence perfect), it follows that these vertices can be covered by k^4 (resp., k^6) cliques.

Lemma 8. *We can make sure that there are at most k^4 independent t-dangerous vertices.*

Proof. Consider the subgraph G'_0 of G_0 induced by those vertices that do not have label t. The clique K contains vertices only from one connected component of G'_0, let B be this component. Clearly, every t-dangerous vertex is a neighbor of B in G_0. Therefore, by Lemma 7, the size of an independent set of t-dangerous vertices can be at most k^4. □

Lemma 9. *We can make sure that there are at most k^6 independent t^*-dangerous vertices.*

Proof. Consider the subgraph G'_0 of G_0 induced by the vertices without label t. Let $C_1, \ldots, C_c$ be the connected components of G'_0. The internal vertices of a witness path for a t^*-dangerous vertex are completely contained in one of these components. Let A be a set of independent t^*-dangerous vertices, and let $A_i \subseteq A$ contain a t^*-dangerous vertex v if and only if v has a witness path with internal vertices only in C_i.

If $|A_i| > k^4$, then we are ready by using Lemma 7 for connected subgraph C_i. For each t^*-dangerous vertex $v_j \in A$, fix a witness u_j and a witness path P_j. Select an arbitrary path P_j and throw away all the other paths that use the same component as P_j. Repeat this until every path is either selected or thrown away. In each step we select one path and throw away less than k^4 paths, thus $|A| > k^6$ implies that we can select more than k^2 paths. Thus assume without loss of generality that the paths $P_1, \ldots, P_{k^2+1}$ do not intersect each other outside the clique K. If a vertex $u \in K$ is contained in more than $k + 1$ of these paths, then u is a necessary vertex. To see this, notice that for each $i = 1, \ldots, k + 1$ a W-avoiding size k hole cover has to contain at least one vertex of each P_i (Prop. 5). Since $P_1, \ldots, P_{k+1}$ intersect each other only in u, this is only possible if X contains u. Therefore, it can be assumed that every $u \in K$ is contained in at most k of the paths $P_1, \ldots, P_{k^2+1}$. Now a simple counting argument shows that there are $k + 1$ pairwise disjoint paths. This means that there is no W-avoiding size k hole cover, as it would have to intersect all these $k + 1$ disjoint paths. □

4.3 Marking the Clique

In the next two lemmas, we show that for a clique Q of dangerous vertices, there is only a constant (i.e., depending only on k) number of vertices in K whose deletion can make a vertex of Q non-dangerous. For every other vertex $u \in K$, if v is t-dangerous, then $v \in Q$ remains t-dangerous in $G \setminus u$. Even more is true: if X is a size k set and $v \in Q$ is t-dangerous in $G \setminus X$, then v remains t-dangerous in $G \setminus (X \cup u)$ as well.

Lemma 10. *Let Q be a clique of t-dangerous vertices. For every k, there is a constant d_k, such that we can mark d_k vertices in K such that if X is a set of size k and $v \in Q$ has an unmarked t-witness u in $G_0 \setminus X$, then v has a marked t-witness u' in $G_0 \setminus (X \cup u)$.*

Proof. Consider the tree decomposition of the chordal graph G_0, let T_z be the subtree corresponding to a vertex z. Since Q and K are cliques, thus there are two nodes x and y such that every vertex of Q contains x, and every vertex of K contains y. Consider the unique path connecting x and y in the tree, and identify the nodes of the path with the integers $1, 2, \ldots, n$, where $x = 1$ and $y = n$. Let $u_1, u_2, \ldots$ be an ordering of the vertices not having label t in K such that if a_i denotes the smallest node of T_{u_i} on this path, then the sequence a_i is non-decreasing.

We mark the vertices $u_1, \ldots, u_{k+1}$ (or up to u_j, if there are only j such vertices). Assume that the witness u of v is the vertex u_i in this ordering. If $i \leq k + 1$, then u_i is marked, and we are ready. Otherwise there is a marked vertex $u_{i'}$ for some $i' \leq k + 1 < i$ that is not contained in X. By the way the vertices are ordered, the witness path that goes from v to u_i has to go through the neighborhood of $u_{i'}$. Thus there is a path from v to $u_{i'}$, showing that $u_{i'}$ is a t-witness of v in $G \setminus (X \cup u)$. $\square$

The next lemma proves a similar statement for t^*-dangerous vertices. However, now the marking procedure is more complicated. The reason for this complication is that a t^*-witness for v has to satisfy two requirements: the witness has to be reachable from v (thus it has to be close to the clique Q), but it should not be a neighbor of v (thus it should not be too close to Q). The proof will appear in the full version of the paper.

Lemma 11. *Let Q be a clique of t^*-dangerous vertices. For every k, there is a constant d_k^* such that we can mark d_k^* vertices in K such that if X is a set of size k and $v \in Q$ has an unmarked t^*-witness in $G_0 \setminus X$, then v has a marked t^*-witness as well in $G_0 \setminus X$.* $\square$

In the next three lemmas, we extend Lemma 10 and Lemma 11 to apply not only for a clique Q of dangerous vertices, but for all dangerous vertices. Moreover, we extend it by requiring witnesses that satisfy certain other properties as well.

Let $F \subseteq W$ be a set of labels. An F-*free* vertex is a vertex that does not have any of the labels in F. Assume that a t-dangerous vertex v has a witness u with a corresponding witness path P. If the vertices in $P \setminus v$ are F-free, then we say that u is an F-*free witness*. Moreover, let $\ell \in F$, and assume that the vertices in $P \setminus (u \cup v)$ do not have labels from F, vertex u has label ℓ but it does not have any other label from F. In this case u is said to be an ℓ-*labeled F-free witness*.

Notice that an ℓ-labeled F-free witness is *not* an F-free witness, since it has label ℓ, which is not allowed for F-free witnesses.

By Lemma 8 and 9, there are no large independent sets of dangerous vertices. Since G_0 is chordal, it follows that the number of cliques required to cover the dangerous vertices is a constant dependening only on k. The proofs of the following lemmas are based on this observation, and on the fact that number of different sets F that we have to consider depends also only on k. Details omitted.

Lemma 12. *For every k, there is a constant $c_k^{(1)}$ such that we can mark $c_k^{(1)}$ vertices in K such that for every size k vertex set X, set of labels F, and label $t \in F$, if in $G_0 \setminus X$ there is a t-dangerous vertex v with an unmarked F-free witness u, then v has an F-free marked witness u' in $G_0 \setminus (X \cup u)$.*

Lemma 13. *For every k, there is a constant $c_k^{(2)}$ such that we can mark $c_k^{(2)}$ vertices in K such that for every size k vertex set X, set of labels F, and labels $\ell, t \in F$, if in $G_0 \setminus X$ there is a t-dangerous vertex v with an unmarked ℓ-labeled F-free witness u, then v has such a marked witness in $G_0 \setminus (X \cup u)$ as well.*

Lemma 14. *For every k, there is a constant $c_k^{(3)}$ such that we can mark $c_k^{(3)}$ vertices in K such that for every size k vertex set X and label $t \in W$, if in $G \setminus X$ a t^*-dangerous vertex v has an unmarked witness u, then v has a marked witness in $G_0 \setminus (X \cup u)$ as well.*

4.4   Fragments of a Hole

Let H be a hole in G. Since $G \setminus W$ is chordal, H has to contain at least one vertex of W. Hence $H \setminus W$ is a set of paths $P_1, P_2, \ldots, P_s$, the set $F = H \cap W$ together with this collection of paths will be called the *fragments* of the hole H. The internal vertices of every path P_i are F-free. Moreover, each end point has exactly one label from F. The only exception is that if a path consists of only a single vertex, then it contains exactly two labels from F. A label in F can appear only on at most two vertices in the fragments: if a vertex of W is in the hole, then at most two of its neighbors can belong to the hole. However, the neighbors of a vertex in W can also be in W, thus it is possible that a label in F appears on only one or on none of the paths.

The following two easy lemmas show that if we have the fragments of a hole, and a path is replaced with some new path satisfying certain requirements, then the new collection of paths also induces a hole. Proofs are omitted.

Lemma 15. *Let F, P_1, $\ldots$, P_s be the fragments of the hole H. Let P_1' be a path that has the same end points as P_1, and whose internal vertices are F-free. There is a hole in the graph induced by the vertices of F, P_1', P_2, $\ldots$, P_s.*

Lemma 16. *Let F, P_1, $\ldots$, P_s be the fragments of the hole H. Assume that the length of P_1 is at least 1. Let x and y be the end points of P_1, and let ℓ_x and ℓ_y be their labels in F, respectively. Let P_1' be a path with the following properties:*

- *the end points of P_1' are x and y' where y' is a vertex that has label ℓ_y, but does not have any other label from F,*

- *the internal vertices of P'_1 are F-free,*
- *if $\ell_x = \ell_y$, then x and y' are not neighbors.*

There is a hole in the graph induced by the vertices of F, P'_1, P_2, ..., P_s.

To show that a vertex $u \in K$ is irrelevant, we have to show that every hole cover of $G \setminus (X \cup u)$ is a hole cover in $G \setminus X$. That is, if X is a size k set and there is a hole H in $G \setminus X$ going through u, then there is a hole H' in $G \setminus (X \cup u)$. The idea is to look at the fragments of H: if path P_1 is going through u, then we find a path P'_1 avoiding u, and use Lemma 15 or 16 to obtain the hole H'.

Lemma 17. *For every k, there is a constant c_k such that we can make sure that every clique of size greater than c_k contains an irrelevant vertex.*

Proof. Given a clique K, mark the vertices according to Lemma 12, 13, and 14. Moreover, for each $F \subseteq W$ and $\ell \in F$, we consider those vertices that have label ℓ, but do not have any other label from F, and we mark $k + 1$ of these vertices (if there are less than $k+1$ such vertices for a given F and ℓ, then all of them are marked). We argue that any unmarked vertex is irrelevant. Since the number of marked vertices depends only on k, the lemma follows.

Let u be an unmarked vertex. To show that u is irrelevant, assume that $X \subseteq V_0$ has size at most k, and H is a hole in $G \setminus X$ containing u. We have to show that $G \setminus X$ contains a hole avoiding u.

Let F, P_1, ..., P_s be the fragments of H. Since the paths of the fragments are independent (i.e., the vertices on two different paths are not neighbors), without loss of generality it can be assumed that u is in P_1 and only P_1 intersects the clique K. Let x and y be the two end vertices of P_1, let their labels be ℓ_x and ℓ_y, respectively. Path P_1 can contain at most one other vertex of K besides u. We consider several cases depending on which combination of $x = y$, $u = x$, $u = y$, $\ell_x = \ell_y$, $|K \cap P_1| = 1$ holds:

Case 1: P_1 consists of only a single vertex ($x = y = u$). Details omitted.
In the remaining cases we assume that $x \neq y$ and w.l.o.g $u \neq x$.

Case 2: P_1 consists of two vertices x, $y = u$, and P_1 is completely contained in K. Since u is not marked, there are $k + 1$ marked vertices in K that have label ℓ_y but do not have any other label from F. At least one of these vertices are not in X, let u' be such a vertex. If we replace $P_1 = \{x, u\}$ with the path $P'_1 = \{x, u'\}$, then by Lemma 16 there is a hole not containing u.

In the remaining cases we assume without loss of generality that end point x is not in K. The following 4 cases handle the situation when y is in K.

Case 3: $|K \cap P_1| = 1$, $u = y$, $\ell_x \neq \ell_y$. Vertex x is an ℓ_x-dangerous vertex for K, and u is an ℓ_y-labeled F-free witness for x in $G_0 \setminus X$. By the way the vertices are marked (see Lemma 13) there is another ℓ_y-labeled F-free witness u' in $G_0 \setminus (X \cup u)$. Let P'_1 be the witness path corresponding to u'. Now F, P'_1, P_2, ..., P_s satisfy Lemma 16, thus there is a hole not containing u.

Case 4: $|K \cap P_1| = 2$, $u = y$, $x \notin K$, $\ell_x \neq \ell_y$. Since u is not marked, thus there are $k + 1$ marked vertices in K that have label ℓ_y but do not have any other label from F. At least one of these vertices are not in X, let u' be such a vertex.

Let P_1' the same as P_1, but replace the last vertex u with u'. Now Lemma 16 is satisfied, hence there is a hole not containing u.

Case 5: $|K \cap P_1| = 2$, $y \neq u$. Vertex x is ℓ_x-dangerous in $G_0 \setminus X$ and u is an F-free witness for x. By Lemma 12, there is a marked F-free witness u' in $G_0 \setminus (X \cup u)$; let P' be the corresponding witness path. The internal vertices of $P_1' := xP'u'y$ are F-free, thus by Lemma 15 there is a hole in $G_0 \setminus (X \cup u)$.

Case 6: $u = y$, $\ell_x = \ell_y$. In this case $s = 1$: the hole consists of only $\ell_x \in W$ and the path P_1. Vertex x is an ℓ_x^*-dangerous vertex in $G_0 \setminus X$ for K, and u is a witness for x. By the way the vertices are marked (see Lemma 14) in $G_0 \setminus (X \cup u)$ there is another witness $u' \in K$ of x. Let P_1' be the witness path corresponding to u'. It is clear that F, P_1' satisfies Lemma 16.

Case 7: $x, y \notin K$. Vertex x (resp., y) is an ℓ_x-dangerous (resp., ℓ_y-dangerous) vertex in $G_0 \setminus X$ for K, and u is an F-free witness for both x and y. By the way the vertices are marked (see Lemma 12) in $G \setminus X$ there is another F-free witness u_x' (resp., u_y') for x (resp., y). Let P_x (resp., P_y) be the witness path of u_x' (resp., u_y'). Let P_1' be the path $xP_xu_x'u_y'P_yy$ going from x to y. The internal vertices of P_1' are F-free, thus by Lemma 15, $G \setminus X$ has a hole avoiding u.   □

References

1. L. Cai. Fixed-parameter tractability of graph modification problems for hereditary properties. *Inform. Process. Lett.*, 58(4):171–176, 1996.
2. L. Cai. Parameterized complexity of vertex colouring. *Discrete Appl. Math.*, 127:415–429, 2003.
3. R. G. Downey and M. R. Fellows. *Parameterized complexity.* Springer, 1999.
4. M. C. Golumbic. *Algorithmic graph theory and perfect graphs.* Academic Press, New York, 1980.
5. H. Kaplan, R. Shamir, and R. E. Tarjan. Tractability of parameterized completion problems on chordal, strongly chordal, and proper interval graphs. *SIAM J. Comput.*, 28(5):1906–1922, 1999.
6. J. M. Lewis and M. Yannakakis. The node-deletion problem for hereditary properties is NP-complete. *J. Comput. System Sci.*, 20(2):219–230, 1980.
7. A. Natanzon, R. Shamir, and R. Sharan. Complexity classification of some edge modification problems. *Discrete Appl. Math.*, 113(1):109–128, 2001.
8. B. Reed, K. Smith, and A. Vetta. Finding odd cycle transversals. *Operations Research Letters*, 32(4):299–301, 2004.
9. D. J. Rose, R. E. Tarjan, and G. S. Lueker. Algorithmic aspects of vertex elimination on graphs. *SIAM J. Comput.*, 5(2):266–283, 1976.
10. M. Yannakakis. Computing the minimum fill-in is NP-complete. *SIAM J. Algebraic Discrete Methods*, 2(1):77–79, 1981.

A Fixed-Parameter Algorithm for the Minimum Weight Triangulation Problem Based on Small Graph Separators

Christian Knauer[1] and Andreas Spillner[2]

[1] Institute of Computer Science, Freie Universität Berlin
christian.knauer@inf.fu-berlin.de
[2] Institute of Computer Science, Friedrich-Schiller-Universität Jena
spillner@minet.uni-jena.de

Abstract. We present a fixed-parameter algorithm which computes for a set P of n points in the plane in $O(2^{c\sqrt{k}\log k} \cdot k\sqrt{k}n^3)$ time a minimum weight triangulation. The parameter k is the number of points in P that lie in the interior of the convex hull of P and $c = (2+\sqrt{2})/(\sqrt{3}-\sqrt{2}) < 11$.

1 Introduction

In this paper we outline how small graph separators might help to reduce the running time of fixed-parameter algorithms for certain hard geometric problems. For an introduction to the concept of fixed-parameter algorithms we refer the reader to the book of Downey and Fellows [6]. We will focus on the Minimum Weight Triangulation problem (MWT). The input for MWT consists of a set P of n points in the plane and we want to compute a triangulation of the point set P such that the total edge length is minimum. As a recent breakthrough this problem was shown to be NP-hard by Mulzer and Rote [19]. Suggestions how to cope with MWT include heuristics [4,13] and approximation algorithms [14,20,21].

Hoffmann and Okamoto [11] were the first to show that MWT is fixed-parameter tractable with respect to the number k of *inner points* in P which are the points in P lying in the interior of the convex hull of P. They give an algorithm running in $O(6^k n^5 \log n)$ time. In [23] it is shown that the approach of Hoffmann and Okamoto can be improved to yield an algorithm running in $O(2^k kn^3 + n^3)$ time. Independently, Grantson et al. presented algorithms for MWT running in $O(4^k kn^4)$ [8] and $O(k!kn^3)$ [9] time. Other geometric problems that have been successfully parameterized with the number of inner points include the Traveling Salesman Problem (TSP) [3] and the problem of finding certain types of optimal convex partitions [7,10,24].

In [11] it was posed as an open problem to find a fixed-parameter algorithm for MWT running in $2^{o(k)}poly(n)$ time. We will give such an algorithm based on small separators for planar graphs. We decided to concentrate on the problem MWT because for this problem our ideas can be explained most easily. However, we expect the basic idea to work also for TSP by applying techniques similar to

F.V. Fomin (Ed.): WG 2006, LNCS 4271, pp. 49–57, 2006.
© Springer-Verlag Berlin Heidelberg 2006

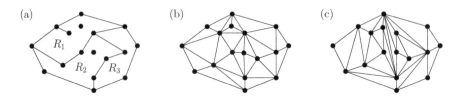

Fig. 1. A point set, a partition and two triangulations of this point set

those presented in [12]. Graph separators have been used before in the design of fixed-parameter algorithms. One approach, as followed by Alber et al. [2], is to reduce the input to a linear size problem kernel and then apply the separator result to this problem kernel. Unfortunately, for MWT no reduction to such a problem kernel is known.

The paper is structured as follows. In Section 2 we outline the basic strategy underlying the known fixed-parameter algorithms for MWT. In Section 3 we show that the existence of small graph separators implies that the aforementioned basic strategy yields a much better running time than the methods used for analysis so far suggest. We conclude in Section 4.

2 The Basic Strategy

In this section we outline how the known fixed-parameter algorithms attack MWT. To ease argumentation we assume that no three points in P are collinear but the algorithm can be adapted to work without this assumption.

Let $CH(P)$ denote the convex hull of P. A *partition* of P is a set E of straight line segments with endpoints in P, called *edges*, such that edges do not cross each other. Two edges *cross* if they share a point which is not an endpoint of both edges. We will only consider partitions E of P that contain all the edges forming the boundary of $CH(P)$. The edges in such an E partition the interior of $CH(P)$ into a finite set $\mathcal{R}(E)$ of regions. A *region* is a connected component of the set of those points of the interior of $CH(P)$ that do not belong to any edge in E. In Figure 1(a) we give an example of a set of points, drawn as solid disks, and the edges of a partition E of this point set. The edges in E partition the interior of the convex hull into the set $\mathcal{R}(E) = \{R_1, R_2, R_3\}$ of regions.

A partition T of P is called a *triangulation* of P if all the regions in $\mathcal{R}(T)$ are empty triangles. A region of a partition is *empty* if it does not contain a point of P in its interior. There can be more than one triangulation of P. In Figures 1(b) and (c) we show two different triangulations of the same point set. A triangulation T of P is called a *minimum weight triangulation* of P if the sum of the lengths of the edges in T is minimum.

To find a minimum weight triangulation of P we can simply do an exhaustive search among all triangulations of P and keep track of the one with minimum total length encountered so far. To this end, we start with an arbitrary edge e of the convex hull of P. For every triangulation T of P there is a unique triangle

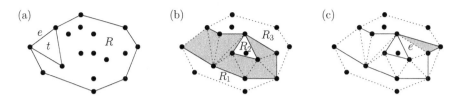

Fig. 2. The exhaustive search for a minimum weight triangulation

in $\mathcal{R}(T)$ that has edge e in its boundary. Thus, we do a branching on the set of empty triangles with vertices in P and edge e in its boundary. An example of one such triangle t is given in Figure 2(a). The edges of the convex hull together with the edges of t form a partition E of P. In our example there are exactly two regions in $\mathcal{R}(E)$ one of which is the empty triangle t. The other region $R \neq t$ in $\mathcal{R}(E)$ is a topological disk with some points of P in its interior.

At this point we have to extend our notion of a partition such that we can also speak of the partition of a region $R \in \mathcal{R}(E)$ of a partition E of P: We call a set of edges $E'_{|R}$ a *partition of* R if there is a partition E' of P such that $E \subseteq E'$ and $E'_{|R}$ contains exactly those edges in E' that are contained in the closure of region R. Now we can also speak of triangulations of a region R. Coming back to our example in Figure 2(a), in the particular branch of the recursion tree of the algorithm we search through all triangulations of P that contain the edges of triangle t, i.e., all triangulations of region R.

In general, we have a partition E of P and for one of the regions in $\mathcal{R}(E)$ a particular triangulation is fixed. We will call this region *fixed* and all other regions in $\mathcal{R}(E)$ we will call *open*. The goal is to compute minimum weight triangulations for all open regions in $\mathcal{R}(E)$. Throughout the algorithm we will maintain the following invariant: Every open region R is a topological disk with possibly some points of P in its interior and the part of the boundary of R that is contained in the interior of the convex hull $CH(P)$ of P is connected. In Figure 2(b) we show an example of the situation after several branchings have been performed. The fixed region is indicated by shading. It remains to find minimum weight triangulations for the open regions R_1, R_2 and R_3, each. The parts of the boundaries of these regions that are contained in the interior of $CH(P)$ are drawn with solid lines.

Let's focus on an open region R. In order to maintain our invariant we select an edge e that is part of the boundary of R but not an edge of $CH(P)$. Then we branch on the set of empty triangles with vertices in P that are contained in region R and have edge e in its boundary. For an example consider the region R_3 in Figure 2(b). In Figure 2(c) we show a possible choice for edge e and one of the empty triangles on which we do the branching. The triangle is indicated by shading. As a result we obtain two new smaller open regions, both contained in R_3, for which we have to compute a minimum weight triangulation, each.

So, the open regions that we encounter during our search for a minimum weight triangulation can be viewed as the subproblems we have to deal with. In order to do this efficiently we build on the following property.

Property 1. An open region is essentially defined by the part of its boundary that is contained in the interior of $CH(P)$. Since this part is connected it forms a possibly closed polygonal path with pairwise non-crossing edges. All vertices of such a polygonal path are inner points, except at most two that are outer points.

Since we repeatedly run into the same subproblems it is worthwhile to try to employ the dynamic programming technique, i.e., we store minimum weight triangulations for subproblems, once computed, in a table where we can look them up in constant time, each. However, this will only pay off, if we are able to give a suitable bound on the total number of subproblems encountered by our algorithm. This can be done as follows. Due to Property 1 it suffices to bound the number of polygonal paths that define the open regions encountered as subproblems. Suppose the polygonal path has exactly l vertices that are inner points. There are $\binom{k}{l}$ possible ways to select these l inner points. For the at most two vertices on the polygonal path that are outer points there are at most n^2 possible choices. By a fundamental result of Ajtai et al. [1] the number of crossing-free polygonal paths on top of the selected vertices is in $O(d^l)$ for some constant $d > 1$. The current record holders are Sharir and Welzl [22] with $d < 87$. Thus, we can bound the number of subproblems asymptotically by $\sum_{l=0}^{k} \binom{k}{l} n^2 d^l = (d+1)^k \cdot n^2$.

A subproblem can easily be processed as outlined above. After selecting edge e in constant time we consider at most n triangles. For each of these triangles we must do some checking, for example whether the triangle is empty. This can easily be done in $O(k)$ time for each triangle. Hence, we can process a subproblem in $O(kn)$ time. Since there are at most $O(88^k \cdot n^2)$ subproblems at all we obtain a running time in $O(88^k \cdot kn^3)$.

Note that Hoffmann and Okamoto [11] obtained a much better result based on the observation that we can restrict to subproblems that are defined by polygonal paths that are monotone with respect to the x-axis. The number of these so-called x-monotone paths is in $O(2^k n^2)$. Employing some additional arguments of geometric flavor it is possible to obtain an algorithm running in $O(2^k kn^3)$ time [23]. However, it seems hard to substantially improve upon this result relying on the concept of x-monotone paths only. So we tried a different approach: find a good upper bound on the number of vertices of the polygonal paths that define subproblems. And this is the point where small graph separators come into play. Unfortunately, it seems that our approach does not mix well with the concept of x-monotone paths.

3   Employing Graph Separators

In this section we show that in the basic strategy outlined in Section 2 we can restrict to subproblems which are defined by a crossing-free polygonal path with only $O(\sqrt{k})$ vertices. In Section 3.1 we briefly review the separator results we want to employ in Section 3.2.

3.1   Simple Cycle Separators

Given a planar graph $G = (V, E)$ a vertex separation is a partition of V into three sets A, B and S such that no edge in E has one endpoint in A and the other endpoint in B. The set S is called the *separator*. A trivial vertex separation would be $S = V$ and $A = B = \emptyset$. But we are interested in vertex separations where S is small and A contains roughly the same number of vertices as B. This can be achieved by a famous result of Lipton and Tarjan [16]: every planar graph with n vertices admits a vertex separation such that $|S| \leq 2\sqrt{2n}$ and $max\{|A|, |B|\} \leq \frac{2}{3}n$.

To use vertex separators in the design of algorithms is not a new idea. Lipton and Tarjan already have employed there result in that way [17]. We will follow Lingas [15] and use the related result of Miller [18], formulated as the following theorem for future reference.

Theorem 1. *Every triangulated planar graph admits a vertex separation such that $|S| \leq \beta\sqrt{n}$, $max\{|A|, |B|\} \leq \alpha n$ and the vertices of S lie on a simple cycle.*

The constants in Miller's result are $\alpha = 2/3$ and $\beta = 2\sqrt{2}$. We can even achieve $\beta = (2 + \sqrt{2})/\sqrt{3} < 1.9712$ as shown by Djidjev and Venkatesan [5].

3.2   Employing Simple Cycle Separators

A standard way of employing graph separators is to devise a divide and conquer algorithm where the input is split into smaller subproblems by a suitable separator. In Section 1 we have already pointed out that we could follow this approach in the design of fixed-parameter algorithms if the problem under consideration admitted a reduction to a linear size problem kernel.

In this paper we will only use the existential result of Theorem 1 in the analysis of the running time of the algorithm implicit in the description of the simple strategy in Section 2. Note that we will not compute graph separators at any point of this algorithm. We will refer to this algorithm as the *basic algorithm*, for short. The core result of our analysis is summarized in the following lemma.

Lemma 1. *In the algorithm for MWT outlined in Section 2 it is sufficient to consider only subproblems defined by crossing-free polygonal paths with at most $c\sqrt{k}$ vertices that are inner points for some constant $c > 0$.*

Proof: As before, let P denote the given set of points. We will prove a slightly more general result. In order to give a precise formulation we introduce some more notation. For a partition E of P and a region $R \in \mathcal{R}(E)$ we denote by R_{in} and R_{bd} the set of points in P lying in the interior and on the boundary of R, respectively. Furthermore we extend the notion of fixed and open regions introduced in Section 2 to situations where we want to find a minimum weight triangulation of a region $R \in \mathcal{R}(E)$.

Claim. For every partition E of P we can compute for every region $R \in \mathcal{R}(E)$ a minimum weight triangulation with the basic algorithm in such a way that at any time the number of points in R_{in} on the boundary of the fixed region is at most $c\sqrt{|R_{in}|}$.

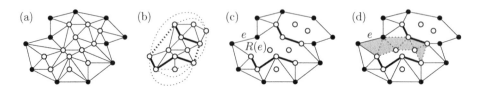

Fig. 3. Using a simple cycle separator to obtain the skeleton

We will do induction on the number l of points in R_{in}. It is clear that for a region R such that $R_{in} = \emptyset$ the fixed region cannot contain points of R_{in} on its boundary. Hence, the claim is trivially true for $l = 0$.

Now suppose we have a region R that contains $l > 0$ points of P in its interior. Let T denote a fixed minimum weight triangulation of R. An example is given in Figure 3(a). The points in R_{bd} and R_{in} are drawn as solid disks and empty circles, respectively. To keep our argumentation clear we introduce the graph $G(T)$ whose vertices correspond to the points of P contained in $R_{in} \cup R_{bd}$ and whose edges correspond to the edges in the triangulation T. Let $G_{in}(T)$ denote the subgraph of $G(T)$ that is induced by those vertices of $G(T)$ that correspond to points in R_{in}. The graph $G_{in}(T)$ is planar but not necessarily triangulated. To apply Theorem 1 we consider an arbitrary triangulation $G_{in}^{\Delta}(T)$ (in the graph theoretic sense) of $G_{in}(T)$. In Figure 3(b) edges of $G_{in}(T)$ are drawn with solid lines. The dotted edges are added to obtain a triangulated planar graph.

Now, in $G_{in}^{\Delta}(T)$ we can find a simple cycle separator C such that $|C| \leq \beta\sqrt{l}$. In Figure 3(b) the edges of the separating cycle are drawn heavier. Let $G_{sk}(T)$ denote the subgraph of $G(T)$ that is induced by the vertices in C and those vertices of $G(T)$ that correspond to points in R_{bd}. We call $G_{sk}(T)$ the *skeleton* of $G(T)$, for short. The edges in the skeleton correspond to a partition E_{sk} of R. Due to our construction every region in $\mathcal{R}(E_{sk})$ has at most αl points of P in its interior. Furthermore every region in $\mathcal{R}(E_{sk})$ has at most $\beta\sqrt{l}$ points of R_{in} on its boundary. In our example we obtain the partition E_{sk} shown in Figure 3(c).

Now imagine we run the basic algorithm on region R. Although we do not know what the partition E_{sk} looks like exactly, its existence helps us to guide the algorithm. As outlined in Section 2 we start with an arbitrary edge e on the boundary of R. There is a unique region $R(e) \in \mathcal{R}(E_{sk})$ that has edge e in its boundary. If we knew region $R(e)$ we could first compute a minimum weight triangulation for $R(e)$. Since $R(e)$ contains at most $\alpha l < l$ points of P in its interior, according to the induction hypothesis, this could be done with the basic algorithm in such a way that the fixed region has always at most $c\sqrt{\alpha l}$ points of $R_{in}(e)$ on its boundary. In Figure 3(d) a fixed region possibly encountered by the algorithm is indicated by shading.

This suggests the following strategy to compute a minimum weight triangulation for region R. We start at edge e and try to keep the fixed region a subregion of $R(e)$. Of course, we cannot tell exactly when we leave region $R(e)$. However, as soon as there are more than $c\sqrt{\alpha l} + \beta\sqrt{l}$ points of R_{in} on the boundary of the fixed region we know that we are definitely outside $R(e)$. This can be seen as follows: As long as we stay within region $R(e)$ the boundary of the fixed region can only contain points of R_{in} which are in $R_{in}(e) \cup R_{bd}(e)$. By the induction hypothesis we can manage to have at most $c\sqrt{\alpha l}$ points from $R_{in}(e)$ in the boundary of the fixed region. The number of points in $R_{in} \cap R_{bd}(e)$ is bounded by the size of the separator C which is at most $\beta\sqrt{l}$.

Now the indicated argumentation can be iterated: Imagine that at some point we have region $R(e)$ as the fixed region. We continue with a region in $\mathcal{R}(E_{sk})$ adjacent to $R(e)$ and so on. The number of points of R_{in} in the boundary of the fixed region will always be at most the size of the separator C plus the bound obtained from the induction hypothesis, which is $\beta\sqrt{l} + c\sqrt{\alpha l}$. Thus, as soon as we encounter a fixed region with more than this number of points from R_{in} on its boundary we know that we do not have to further pursue this branch of the recursion tree and track back.

It remains to devise a suitable value for constant c. We need $\beta\sqrt{l}+c\sqrt{\alpha l} \leq c\sqrt{l}$. This can be achieved with $c = \beta/(1-\sqrt{\alpha})$. This finishes our inductive proof of the claim above. Since inner points on the defining polygonal path of a subproblem are inner points on the boundary of the fixed region this gives us Lemma 1. ∎

The result of Lemma 1 leads to an improved bound on the running time of the basic algorithm as stated in the following theorem.

Theorem 2. *We can compute a minimum weight triangulation of a set P of n points in the plane in $O(2^{c\sqrt{k}\log k} \cdot k\sqrt{k}n^3)$ time. The parameter k is the number of inner points in P and $c = (2+\sqrt{2})/(\sqrt{3}-\sqrt{2}) < 10.7421$.*

Proof: By Lemma 1 it suffices to consider subproblems that are defined by crossing-free polygonal paths with at most $c\sqrt{k}$ inner points on it. Such a polygonal path can be described by the sequence of its vertices. When there are l inner points on the polygonal path then there are at most k^l possible sequences. For the at most two outer points on the polygonal path there are at most n^2 possible choices. Hence, the number of defining polygonal paths and thus also the number of subproblems encountered by the algorithm can asymptotically be bounded by $\sum_{l=0}^{\lfloor c\sqrt{k}\rfloor} k^l n^2 = n^2 (k^{\lfloor c\sqrt{k}\rfloor+1} - 1)/(k-1) \in O(k^{\lfloor c\sqrt{k}\rfloor}n^2) \subseteq O(2^{c\sqrt{k}\log k}n^2)$. In Section 2 we have outlined that using dynamic programming a subproblem can be processed in $O(kn)$ time. However, it is no longer sufficient to consider only one arbitrary edge. In order to stay inside the fixed region we must consider every edge on the defining polygonal path. Since such a path has $O(\sqrt{k})$ edges a subproblem can be processed in $O(nk\sqrt{k})$ time. This gives us an upper bound of $O(2^{c\sqrt{k}\log k} \cdot k\sqrt{k}n^3)$ on the running time of the algorithm where $c = \beta/(1-\sqrt{\alpha}) = (2+\sqrt{2})/[\sqrt{3}(1-\sqrt{2/3})] = (2+\sqrt{2})/(\sqrt{3}-\sqrt{2})$. ∎

Remark 1. It seems that employing the combinatorial bound $O(d^l)$ on the number of crossing-free polygonal paths on top of l vertices [1,22] does not help to get rid of the $\log k$-factor in the exponent in the bound on the running time stated in Theorem 2.

4   Concluding Remarks

The main contribution of this paper is that we have outlined a way to beat the 2^k term in the bound on the running time of previously known algorithms for MWT. We expect the general idea to work also for other hard geometric problems such as TSP. However, our results as stated here are mostly of theoretical interest since $2^k < 2^{c\sqrt{k}\log k}$ for all $k \leq 20000$.

References

1. M. Ajtai, V. Chvátal, M. Newborn, and E. Szemerédi. Crossing-free subgraphs. *Annals of Discrete Mathematics*, 12:9–12, 1982.
2. J. Alber, H. Fernau, and R. Niedermeier. Graph separators: a parameterized view. *Journal of Computer and System Sciences*, 67:808–832, 2003.
3. V. G. Deineko, M. Hoffmann, Y. Okamoto, and G. J. Woeginger. The traveling salesman problem with few inner points. *Operations Research Letters*, 34(1):106–110, 2006.
4. M. T. Dickerson, S. A. McElfresh, and M. H. Montague. New algorithms and empirical findings on minimum weight triangulation heuristics. In *Proc. Symposium on Computational Geometry*, pages 238–247. ACM Press, 1995.
5. H. N. Djidjev and S. M. Venkatesan. Reduced constants for simple cycle graph separation. *Acta Informatica*, 34:231–243, 1997.
6. R. G. Downey and M. R. Fellows. *Parameterized Complexity*. Springer, 1999.
7. M. Grantson. Fixed-parameter algorithms and other results for optimal partitions. Licentiate Thesis, Lund University, Sweden, 2004.
8. M. Grantson, C. Borgelt, and C. Levcopoulos. A fixed parameter algorithm for minimum weight triangulation: Analysis and experiments. Technical Report LU-CS-TR:2005-234, Lund University, Sweden, 2005.
9. M. Grantson, C. Borgelt, and C. Levcopoulos. Minimum weight triangulation by cutting out triangles. In *Proc. International Symposium on Algorithms and Computation*, volume 3827 of *LNCS*, pages 984–994. Springer, 2005.
10. M. Grantson and C. Levcopoulos. A fixed-parameter algorithm for the minimum number convex partition problem. In *Proc. Japanese Conference on Discrete and Computational Geometry*, volume 3742 of *LNCS*, pages 83–94. Springer, 2004.
11. M. Hoffmann and Y. Okamoto. The minimum weight triangulation problem with few inner points. *Computational Geometry*, 34(3):149–158, 2006.
12. R. Z. Hwang, R. C. Chang, and R. C. T. Lee. The searching over separators strategy to solve some NP-hard problems in subexponential time. *Algorithmica*, 9:398–423, 1993.
13. Y. Kyoda, K. Imai, F. Takeuchi, and A. Tajima. A branch-and-cut approach for minimum weight triangulation. In *Proc. International Symposium on Algorithms and Computation*, volume 1350 of *LNCS*, pages 384–393. Springer, 1997.
14. C. Levcopoulos and D. Krznaric. Quasi-greedy triangulations approximating the minimum weight triangulation. In *Proc. Symposium on Discrete Algorithms*, pages 392–401. ACM Press, 1996.
15. A. Lingas. Subexponential-time algorithms for minimum weight triangulation and related problems. In *Proc. Canadian Conference on Computational Geometry*, 1998.

16. R. J. Lipton and R. E. Tarjan. A separator theorem for planar graphs. *SIAM Journal on Applied Mathematics*, 36:177–189, 1979.
17. R. J. Lipton and R. E. Tarjan. Applications of a planar separator theorem. *SIAM Journal on Computing*, 9:615–627, 1980.
18. G. L. Miller. Finding small simple cycle separators for 2-connected planar graphs. *Journal of Computer and System Sciences*, 32:265–279, 1986.
19. W. Mulzer and G. Rote. Minimum weight triangulation is NP-hard. In *Proc. Symposium on Computational Geometry*, pages 1–10. ACM Press, 2006.
20. D. A. Plaisted and J. Hong. A heuristic triangulation algorithm. *Journal of Algorithms*, 8:405–437, 1987.
21. J. Remy and A. Steger. A quasi-polynomial time approximation scheme for minimum weight triangulation. In *Proc. Symposium on Theory of Computing*, pages 316–325. ACM Press, 2006.
22. M. Sharir and E. Welzl. On the number of crossing-free matchings, (cycles, and partitions). In *Proc. Symposium on Discrete Algorithms*, pages 860–869. ACM Press, 2006.
23. A. Spillner. A faster algorithm for the minimum weight triangulation problem with few inner points. In *Proc. Algorithms and Complexity in Durham Workshop*, pages 135–146. KCL Publications, 2005.
24. A. Spillner. Optimal convex partitions of point sets with few inner points. In *Proc. Canadian Conference on Computational Geometry*, pages 34–37, 2005.

Divide-and-Color[*]

Joachim Kneis, Daniel Mölle, Stefan Richter, and Peter Rossmanith

Department of Computer Science, RWTH Aachen University, Fed. Rep. of Germany
{kneis, moelle, richter, rossmani}@cs.rwth-aachen.de

Abstract. We introduce *divide-and-color*, a new technique for the solution of hard graph problems. It is a combination of the well-known *divide-and-conquer* paradigm and *color-coding* [2]. Our approach first randomly colors all edges or nodes of a graph black and white, and then solves the problem recursively on the two induced parts.

We demonstrate this technique by giving new randomized algorithms for the solution of two important problems. These yield runtime bounds of $O^*(4^k)$ for finding a simple path of length k and $O^*(4^{(h-1)k})$ for finding k edge-disjoint (resp. vertex-disjoint) copies of a graph H with h edges (resp. h nodes) in a given graph. Derandomization gives deterministic algorithms for these problems with running times $O^*(2^{4k})$ and $O^*(2^{4hk})$, respectively.

All these results significantly improve over the currently known best bounds. In particular, our generic algorithms beat specialized ones that have been designed to find k triangles or paths of length two.

1 Introduction

As of today, it is commonly assumed that there are no polynomial-time algorithms for NP-complete problems. On the other hand, these problems occur over and over again in real-life applications, and thus simply need to be dealt with. This contradictory situation has led to the development of several relaxed notions of "solving," such as approximation and randomization. Another approach is called parameterized complexity.

In classical NP-completeness theory, run-time bounds are derived with regard to the worst case taken over all instances of the same size. Informally, the paradigm of parameterized complexity [5] tries to look closer in order to find properties that better measure the hardness of a problem. An instance of length n may still be easy to solve if it fulfills a certain property. If this property can be quantified by a number k—such as the treewidth of a graph, the number of literals that may be set to true in a propositional formula, or the maximum length of a simple path—this number can then be seen as a parameter describing the hardness of an instance.

More precisely, if there exists an algorithm that solves the problem in $O(f(k) \cdot poly(n))$ time for some function f, we say that the problem is fixed-parameter tractable. The corresponding complexity class is called FPT. Since appropriately

---

[*] Supported by the DFG under grant RO 927/6-1 (TAPI).

chosen parameters can be very small in instances occuring in practice, this paradigm often yields rather applicable algorithms. On the other hand, the values of $f(k)$ can be rather large for certain problems. For example, the best parameterized algorithm for DOMINATING SET OF QUEENS known today has a running time of $O^*(225^k)$ [3].

There are a lot of techniques for designing FPT algorithms in the literature, among them kernelization, bounded search trees, dynamic programming, treewidth methods and color-coding. The best runtime bounds for the parameterized versions of some graph problems have been obtained using the latter method, which has been introduced by Alon et al. [2]. The underlying idea of this technique is to color the nodes or edges of a graph randomly in order to ease the detection of certain subgraphs. For instance, a path of length k may be hard to find[1], but after coloring the nodes with k colors, a *colorful* path of length $k-1$—i.e., a path that consists of k nodes having k different colors—can be detected quite quickly using dynamic programming.

The runtime bounds obtained via color-coding are usually of the form $O^*(c^k)$ for some constant c. Note that this is only exponential in k and thus much better than, e.g., bounds resulting from Courcelle's famous theorem [4]. The constant c, however, can be very large, as detailed in Section 2. Derandomizing the resulting algorithms [2,6] leads to bounds of the same form with even larger constants.

In this paper, we present a new and more efficient color-coding technique that can also be derandomized with less dramatic effects on the runtime bound. The crucial idea lies in using only two colors combined with a divide-and-conquer approach. Recursively solving the problem on two subgraphs induced by the coloring turns out to be easier than employing a dynamic programming algorithm as used in classical color-coding.

For a graph G, let $V[G]$ denote the set nodes in G. Similarly, let $E[G]$ denote the set of edges. We apply the new technique to several graph problems, obtaining new runtime bounds of $O^*(2^{2k})$ for LONGEST PATH and $O^*(2^{2(h-1)k})$ for H-GRAPH PACKING, where $h = |V[H]|$, as well as H-GRAPH EDGE-PACKING where $h = |E[H]|$, in the randomized case. The respective bounds for the derandomized case are $O^*(2^{4k})$ and $O^*(2^{4hk})$. Let us first formally define the problems that are to play a rôle in what follows.

Definition 1. *The problem* LONGEST PATH *is defined as follows:*

Input: *A graph G and a number k*
Parameter: k
Question: Is there a simple path of length k in G?

The classical color-coding algorithm by Alon et al. [2] solves LONGEST PATH in time $O^*((2e)^k) = O^*(2^{2.45k})$ (randomized) or $2^{O(k)}$ with large hidden

---

[1] One way of solving this problem is to look at a depth-first search tree of the graph. If there is a back-edge spanning a path of length k, we are done. Otherwise, there is a tree decomposition of width k, and the problem can be solved using Courcelle's Theorem [4]. Unfortunately, this method yields extremely large values of $f(k)$ even for small k.

constants (deterministic). We improve these running times to $O^*(2^{2k})$ and $O^*(2^{4k})$, respectively.

Definition 2. *The problem* H-GRAPH PACKING *is defined as follows:*

> Input: A graph G and a number k
> Parameter: k
> Question: Are there k vertex-disjoint instances of H in G?

Definition 3. *The problem* H-GRAPH EDGE-PACKING *is defined as follows:*

> Input: A graph G and a number k
> Parameter: k
> Question: Are there k edge-disjoint instances of H in G?

As of today, the best runtime bounds published for H-GRAPH PACKING and H-GRAPH EDGE-PACKING are $O^*(2^{2.45hk})$ for the randomized and $O^*(2^{10hk})$ for the deterministic case [6]. We take these to $O^*(2^{2(h-1)k})$ and $O^*(2^{4hk})$, respectively.

As to our knowledge, specialized algorithms have been designed for at least two variants of H-GRAPH PACKING and H-GRAPH EDGE-PACKING, the first being $K_{2,1}$-PACKING [10] and the second EDGE-DISJOINT TRIANGLE PACKING [9]. The corresponding runtime bounds are $O^*(2^{5.301k})$ and $O^*(2^{\frac{9}{2}k\log k+\frac{9}{2}k})$.[2] Our randomized algorithms improve both of these to $O^*(2^{4k})$. Moreover, the derandomized version yields a bound of $O^*(2^{12k})$ for EDGE-DISJOINT TRIANGLE PACKING.

2   Color-Coding

Unfortunately, the papers on color-coding that have been published so far omit a lot of details, especially regarding hidden factors in the respective runtime bounds. This holds for the results on both the problem of finding a path of length k and the problem of finding k occurrences of a fixed subgraph. In the seminal paper on color-coding [2], for example, Alon *et al.* considered k a mere constant. In subsequent papers [6,9], k was already interpreted as a parameter, but the runtime bounds were stated without giving explicit constants for the basis of the exponential function. This section is dedicated to a more precise analysis of these bounds.

Let us first examine the algorithm for H-GRAPH EDGE-PACKING by Fellows *et al.* [6]: Given a graph G and a number k, does G contain k edge-disjoint occurrences of some fixed subgraph H? Let h denote the number of edges in H. For the analysis, fix a set M of k edge-disjoint occurrences of H in an instance.

---

[2] The latter algorithm is based on kernelization. There also is a $2^{O(k)}$-algorithm based on color-coding. However, the constants hidden in the O-notation are so large that the kernelization-based algorithm is considered more practical.

If coloring all the edges in G randomly using hk colors leaves all the edges in M colored differently, the following algorithm will find these k occurrences of H (or, possibly, some other k occurrences):

For each $S \subseteq \{1, \ldots, hk\}$ where $|S|$ is a multiple of h, use dynamic programming to find out whether G contains $|S|/h$ occurrences of H whose edges have exactly the colors in S. If this is eventually the case for $S = \{1, \ldots, hk\}$, the graph G indeed contains k edge-disjoint occurrences of H. This computation takes $O^*(2^{hk})$ steps. The probability of coloring all the hk edges in M with different colors is $(hk)!/(hk)^{hk} = \Omega(\sqrt{k}e^{-hk})$. In order to make the failure probability exponentially small, it thus suffices to repeat the entire procedure $O^*(e^{hk})$ times. This gives a total running time of $O^*((2e)^{hk}) = O^*(5.44^{hk})$.

A deterministic algorithm can be obtained via derandomization, as described by Alon *et al.* [2]. It is possible to construct a family $\mathcal{F}$ of hash functions $\{1, \ldots, n\} \to \{1, \ldots, k\}$ such that for every $S \subseteq \{1, \ldots, n\}$ with $|S| = k$ there is one $f \in \mathcal{F}$ that maps S onto $\{1, \ldots, k\}$, i.e., is bijective on S. There are such families of size $2^{O(k)} \log n$ that can be efficiently constructed and evaluated [2]. Instead of coloring the edges randomly, we can consecutively use all members of a family of hash functions $f : E \to \{1, \ldots, hk\}$. This leads to a running time of $2^{O(hk)}n^{O(h)}$.

The constants in the above bound are rather large, but Fellows *et al.* have proposed two ideas on how to improve the running time [6]. The first one is to use more colors in order to increase the probability that all the edges in a solution are assigned different colors. This, however, leads to a more costly dynamic programming phase. The second idea regards the family of hash functions employed: Alon *et al.* did not aim to minimize the size of $\mathcal{F}$ with respect to k. Instead, efficient evaluation was a criterion, but is not needed in this context. For example, a size of $O^*(2^{4hk})$ would be possible at the expense of using $6hk$ colors, leading to an overall running time of only $O^*(2^{10hk})$ [6,7,11].

Let us now briefly discuss the original algorithm [2] for LONGEST PATH: We color the nodes randomly using k colors. For a fixed path of length k, there are k^k possible colorings, $k!$ of which make the path *colorful* by assigning k different colors to the k nodes in the path. Hence, the path is colorful with a probability of e^{-k}. A colorful path of length k can be detected in $O^*(2^k)$ steps using dynamic programming. On the other hand, it takes e^k iterations to obtain an exponentially small probability that the path is never colorful. Consequently, this randomized algorithm for LONGEST PATH has a runtime bound of $O^*((2e)^k) = O^*(5.44^k)$.

3   Algorithms Based on Divide-and-Color

The basic idea of our new approach is to use only two colors—say, black and white—and to solve a reduced instance of the problem on each of the two induced subgraphs recursively. These two solutions must then be combined into a solution for the original instance. Intuitively, this method has three major advantages:

Firstly, when coloring in stages, a wrong coloring may be amended in the current stage, whereas in classical color-coding, one would have to start from scratch. Secondly, in opposition to classical color-coding, there is no more need for solving the complicated problem addressed by the dynamic programming algorithm. This is because the recursive approach eventually reduces the problem to trivial instances. Thirdly, derandomization becomes a lot easier in the new approach, because we can use almost k-wise independent random variables [1] instead of universal families of hash functions.

3.1   Finding a Path of Length k

Instead of solving the decision problem LONGEST PATH directly, we use an embedding into the following closely related search problem:

Definition 4. *Given a graph* $G = (V, E)$, *we write* $u \xrightarrow{k} v$ *for two nodes* $u, v \in V$ *whenever there is a simple path of length* $k - 1$ *from* u *to* v.

The problem EXTENDED LONGEST PATH *is defined as follows:*

Input: *A graph* $G = (V, E)$ *and a number* k
Parameter: k
Output: *The set* $\{ (u, v) \in V^2 \mid u \xrightarrow{k} v \}$.

We say an algorithm solves EXTENDED LONGEST PATH *with error probability* p *if for every graph* $G = (V, E)$ *and every* $u, v \in V$ *with* $u \xrightarrow{k} v$ *its result contains* (u, v) *with probability at least* p *on input* G, k.

Lemma 1. *Algorithm L from Table 1 solves* EXTENDED LONGEST PATH *with error probability at most* $1/4$ *in time* $O^*(4^k)$.

Proof. It is easy to see that if Algorithm L returns a pair $\{u, v\}$, then indeed $u \xrightarrow{k} v$. On the other hand, if G contains such a path, L does not return the corresponding pair $\{u, v\}$ with some probability p_k.

Assume there is a simple k-node path from u to v in G. The probability that the first $\lceil k/2 \rceil$ nodes are black and that the other $\lfloor k/2 \rfloor$ nodes are white is 2^{-k}. In that case, $L(G_1)$ and $L(G_2)$ do not contain pairs that allow the algorithm to insert $\{u, v\}$ into R with probability at most $2p_{\lceil k/2 \rceil}$. After $3 \cdot 2^k$ iterations, the probability that $\{u, v\} \notin R$ is at least

$$\left(1 - 2^{-k} + 2^{-k+1} p_{\lfloor k/2 \rfloor} \right)^{3 \cdot 2^k},$$

because with probability $1 - 2^{-k}$ the coloring is bad and with probability at most $2^{-k+1} p_{\lceil k/1 \rceil}$ the coloring is good, but the recursive detection of the black or white subpath fails. We show that $p_k \leq 1/4$ for every k: Obviously $p_1 = 0$, and for $k \geq 2$ we get the inequality $p_k \leq (1 - 2^{-k} + 2^{-k+1}/4)^{3 \cdot 2^k} = (1 - 2^{-(k+1)})^{\frac{3}{2} \cdot 2^{k+1}} \leq e^{-3/2} < 1/4$ by induction on k.

Let T_k denote the number of recursive calls issued by Algorithm L. Observe that the recursive calls issued in the innermost for-loop have the same parameters for all u, v, w, x. It thus suffices to perform these two calls only once for each

Table 1. Algorithm L

Input: A graph $G = (V, E)$ and a number k Parameter: k

Output: The set $\{ (u, v) \in V^2 \mid u \xrightarrow{k} v \}$.

if $k = 1$ **then return** $\{ (v, v) \mid v \in V \}$ **fi**; **for**
$3 \cdot 2^k$ times **do**
  Choose some $V' \in 2^V$ with uniform probability;
  $G_1 := G[V']$; $G_2 := G[V - V']$; $R := \emptyset$;
  for all $u, v, w, x \in V$ **do**
    if $(u, v) \in L(G_1, \lceil k/2 \rceil) \wedge \{v, w\} \in E \wedge (w, x) \in L(G_2, \lfloor k/2 \rfloor)$
    then $R := R \cup \{(u, x)\}$ **fi**
  od
od; **return** R

---

coloring and to remember the results when processing the inner for-loop. We obtain the recurrence

$$T_k \leq 3 \cdot 2^k \cdot (T_{\lceil k/2 \rceil} + T_{\lfloor k/2 \rfloor}) \leq 3 \cdot 2^{k+1} T_{\lceil k/2 \rceil}.$$

Using the simple fact that $k + \lceil k/2 \rceil + \lceil \lceil k/2 \rceil / 2 \rceil + \cdots + 1 \leq 2k + \log k$ where the sum ends when a term becomes 1, we get $T_k = O(3^{\log k} k^2 4^k) = O(k^{\log 3} k^2 4^k) = O^*(4^k)$. All other operations performed during a call to the algorithm only take polynomial time.

Algorithm L thus finds a paths of length k with probability at least $3/4$ if one exists. Iterating it linearly often yields an exponentially small error probability while keeping the running time at $O^*(4^k)$. $\qquad\square$

Theorem 1. LONGEST PATH *can be solved with exponentially small error probability in time* $O^*(4^k)$.

Proof. Applying Algorithm L a linear number of times solves the more general problem EXTENDED LONGEST PATH with exponentially small error probability. $\qquad\square$

3.2   Graph Packing and Graph Edge-Packing

Lemma 2. *Algorithm P from Table 2 solves H-*GRAPH EDGE-PACKING *with failure probability* $1/4$ *in time* $O^*(4^{(h-1)k})$, *where* $h = |E[H]|$.

Proof. Assume that H has $h > 1$ edges. In line 6 of the algorithm, E is partitioned into E' and $E - E'$. Let us call the edges in E' *black* and the edges in $E - E'$ *white*. Fix a solution, i.e., a set of k edge-disjoint copies of H in G. For each copy of H, the probability that all its edges have the same color is 2^{-h+1}. The probability that all copies are monochromatic is $2^{-(h-1)k}$ and the probability that exactly $\lceil k/2 \rceil$ copies are black and $\lfloor k/2 \rfloor$ copies are white is at least $2^{-(h-1)k}/k$.

Table 2. Algorithm P

Input: A graph $G = (V, E)$ and a number k
Parameter: k
Question: Does G contain k edge-disjoint copies of H?

if $k \leq 1$ **then**
    if there are k or more edge-disjoint copies of H **then return** *true*
    else return *false* **fi**
fi;
for $3 \cdot 2^{(h-1)k}$ times **do**
    Choose some $E' \in 2^E$ with uniform probability;
    $G_1 := (V, E'); G_2 := (V, E - E')$;
    if $P(G_1, \lceil k/2 \rceil) \wedge P(G_2, \lfloor k/2 \rfloor)$ **then return** *true* **fi**
od;
return *false*

Let p_k be the probability that Algorithm P does not find the fixed solution. Then

$$p_k \leq \left(1 - \frac{2^{-(h-1)k}}{k} + \frac{2^{-(h-1)k}}{k}\left(p_{\lceil k/2 \rceil} + p_{\lfloor k/2 \rfloor}\right)\right)^{3 \cdot 2^{-(h-1)k}}.$$

We show by induction on k that $p_k \leq 1/4$. For $k \leq 1$ we have $p_k = 0$ and if $k > 1$ then

$$p_k \leq \left(1 - \frac{2^{-(h-1)k}}{k} + \frac{2^{-(h-1)k}}{2k}\right)^{3 \cdot 2^{(h-1)k}} \leq \left(1 - 2^{-(h-1)k-1}\right)^{3 \cdot 2^{(h-1)k}} < e^{-3/2}.$$

The runtime analysis is very similar to that of Algorithm L.   □

Theorem 2

1. *The problem H-GRAPH EDGE-PACKING can be solved with exponentially small error probability in $O^*(2^{2(h-1)k})$ steps where $h = |E[H]|$.*
2. *The problem H-GRAPH PACKING can be solved with exponentially small error probability in $O^*(2^{2(h-1)k})$ steps where $h = |V[H]|$.*

Proof. For H-GRAPH EDGE-PACKING simply repeat the 1/4-error-probability algorithm linearly often in order to amplify the success probability. H-GRAPH PACKING can be solved very similarly. Essentially, we have to color nodes instead of edges.   □

Corollary 1. *We can solve* EDGE-DISJOINT TRIANGLE PACKING *in* $O^*(2^{4k})$ *steps and* $K_{1,s}$-PACKING *in* $O^*(2^{2sk})$ *with exponentially small failure probability.*

4   Derandomization

The common object of the divide-and-color algorithms introduced in the last section lies in finding some subset S of nodes or edges with some property and given size. The first step towards this goal always consists in coloring all nodes or edges black and white. Progress will be made whenever this global coloring implies one particular coloring of S. For example, when looking for a path of length k, we want the first half of this path black, and the second one white. In the randomized approaches detailed above, we color the nodes or edges randomly and repeat this very often, so that the right coloring of S is hit with high probability.

In order to make failure impossible, we have to cycle through a family of node or edge colorings deterministically. Doing so will succeed when we make sure that *every* possible coloring of S is hit at least once. Since we do not know S, however, we need to hit every coloring for any set of size $|S|$ at least once. The notion of (almost) k-wise independence will ease this task for us.

Definition 5. [1] *A set $X \subseteq \{0,1\}^n$ is k-wise independent if when $x_1 \dots x_n$ is chosen uniformly from X, then for any k positions $i_1 < \cdots < i_k$ and any k-bit string y, we have $\Pr[x_{i_1} \dots x_{i_k} = y] = 2^{-k}$. It is (ϵ, k)-independent if $|\Pr[x_{i_1} \dots x_{i_k} = y] - 2^{-k}| \leq \epsilon$.*

If we had an $X \subseteq \{0,1\}^n$ that is $(2^{-|S|-1}, |S|)$-independent, we could color a set of n nodes or edges according to each element in X with the guarentee that S is colored in every possible way at least once. Fortunately, Alon *et al.* have found a construction that does the job.

Proposition 1. [1] *Let $n = 2^t - 1$ and let k be an odd integer. Then it is possible to construct n bits that are (ϵ, k)-independent using $2(\lceil \log \frac{1}{\epsilon} + \log(1 + (k-1)t/2) \rceil)$ bits.*

Moreover, their construction works in quasi-linear time. Notice that the number of bits in the seed of an $(2^{-|S|-1}, |S|)$-independent set X can be bounded by $2|S| + 2\log|S| + 2\log\log n + 2$. The cardinality of X is thus no more than $O(4^{|S|} \cdot |S|^2 \cdot \log^2 n)$, and X can be constructed in $O^*(4^{|S|})$ time.

Theorem 3. *Algorithm P' solves H-GRAPH EDGE-PACKING in time $O^*(16^{hk})$, where $h = |E[H]|$.*

Proof. There are $O(4^{hk}(hk)^2 \log^2 |E|) = O^*(4^{hk})$ possibilities to choose s. A runtime analysis similar to that for Algorithm L shows that the total number of recursive calls is $O^*(16^{hk})$.

The correctness follows from the fact that for a solution consisting of k copies of H with a total of hk edges, E' eventually contains exactly the edges of $\lceil k/2 \rceil$ many of those copies. □

Analogously, we can prove the following theorem.

Theorem 4. *There is a deterministic algorithm solving LONGEST PATH in time $O^*(16^k)$.*

Table 3. Algorithm P′

Input: A graph $G = (V, E)$, $E = \{e_1, \ldots, e_m\}$ and a number k
Parameter: k
Question: Does G contain k edge-disjoint copies of H?

if $k \leq 1$ **then**
    if there are k or more edge-disjoint copies of H **then return** *true*
    else return *false* **fi**
fi;
for all $s \in \{0,1\}^{\lceil 2hk + 2\log(hk) + 2\log\log|E|\rceil + 2}$ **do**
    Expand s into $x \in \{0,1\}^{|E|}$ as in Proposition 1;
    $E' := \{ e_i \in E \mid x_i = 1 \}$;
    $G_1 := (V, E')$; $G_2 := (V, E - E')$;
    if $P(G_1, \lceil k/2 \rceil) \wedge P(G_2, \lfloor k/2 \rfloor)$ **then return** *true* **fi**
od;
return *false*

---

5   Conclusion

We have introduced a new technique and applied it to design faster algorithms
for several problems. However, we did not exhaust the method's full potential.
In particular, we think that the cost incurred by derandomization could be decreased. While almost k-wise independent random variables seem to be the right
tool for derandomizing the LONGEST PATH algorithm, a weaker property suffices for packing problems. In the case of TRIANGLE PACKING, for instance, we
used almost $3k$-independent random variables to make sure we encounter every
possible coloring of the $3k$ edges of k triangles. It would be enough, however,
if for every possible way of grouping $3k$ edges into k triangles, exactly half of
these triangles were colored black and the others white at least once. Designing
appropriate sample spaces thus seems to be a promising goal.

Remark. Some of the results presented in this paper, such as a randomized
$O(4^k poly(n))$ algorithm for LONGEST PATH, have also been obtained by Liu et
al. using a different approach. The respective paper was accepted for IWPEC
2006 and will appear later this year [8].

Acknowledgement. We thank Martin Dietzfelbinger for pointing us in the direction of almost k-wise independent random variables and Ryan Williams for
helpful comments regarding the runtime analysis.

References

1. N. Alon, O. Goldreich, J. Håstad, and R. Peralta. Simple constructions of almost k-
wise independent random variables. *Journal of Random structures and Algorithms*,
3(3):289–304, 1992.

2. N. Alon, R. Yuster, and U. Zwick. Color-coding. *J. ACM*, 42(4):844–856, 1995.
3. M. Cesati. Compendium of parameterized problems, 2005. Available online at `http://bravo.ce.uniroma2.it/home/cesati/research/compendium.ps`.
4. B. Courcelle, J. A. Makowsky, and U. Rotics. Linear time solvable optimization problems on graphs of bounded clique width. In *Proc. of 24th WG*, number 1517 in LNCS, pages 1–16. Springer, 1998.
5. R. G. Downey and M. R. Fellows. *Parameterized Complexity*. Springer-Verlag, 1999.
6. M. R. Fellows et al. Faster fixed-parameter tractable algorithms for matching and packing problems. In *Proc. of 12th ESA*, number 3221 in LNCS, pages 311–322. Springer, 2004.
7. M. Fredman, J. Komlos, and E. Szemeredi. Storing a sparse table with $O(1)$ worst case access time. In *Proc. of 23d FOCS*, pages 165–169, 1982.
8. Y. Liu, S. Lu, J. Chen, and S.-H. Sze. Greedy localization, color-coding, and dynamic programming: Improved algorithms for matching and packing problems. In *Proc. of 2nd IWPEC*, number 4169 in LNCS. Springer, 2006. To appear.
9. L. Mathieson, E. Prieto, and P. Shaw. Packing edge disjoint triangles: A parameterized view. In *Proc. of 1st IWPEC*, number 3162 in LNCS, pages 127–137. Springer, 2004.
10. E. Prieto and C. Sloper. Looking at the stars. In *Proc. of 1st IWPEC*, number 3162 in LNCS. Springer, 2004.
11. C. F. Slot and P. van Emde Boas. On tape versus core; an application of space efficient hash functions to the invariance of space. *Elektronische Informationsverarbeitung und Kybernetik*, 21(4/5):246–253, 1985.

Listing Chordal Graphs and Interval Graphs

Masashi Kiyomi[1], Shuji Kijima[2,*], and Takeaki Uno[1]

[1] National Institute of Informatics, 2-1-2 Hitotsubashi, Chiyoda-ku, Tokyo 101-8430, Japan
masashi@grad.nii.ac.jp, uno@nii.jp
[2] Department of Mathematical Informatics, Graduate School of Information Science and
Technology, The University of Tokyo, Bunkyo-ku, Tokyo 113-8656, Japan
kijima@misojiro.t.u-tokyo.ac.jp

Abstract. We propose three algorithms for enumeration problems; given a graph
G, to find every chordal supergraph (in K_n) of G, to find every interval supergraph (in K_n) of G, and to find every interval subgraph of G in K_n. The algorithms are based on the reverse search method. A graph is chordal if and only
if it has no induced chordless cycle of length more than three. A graph is an interval graph if and only if it has an interval representation. To the best of our
knowledge, ours are the first results about the enumeration problems to list every
interval subgraph of the input graph and to list every chordal/interval supergraph
of the input graph in polynomial time. The time complexities of the first algorithm
is $O((n + m)^2)$ for each output graph, and those for the rest two algorithms are
$O(n^3)$ for each output graph, where m is the number of edges of input graph
G. We also show that a straight-forward depth-first search type algorithm is not
appropriate for these problems.

1 Introduction

Listing all the objects that satisfy a specified property, with no duplications, is called
"*enumeration*". For example, the enumeration of substrings contained by a string "abcab" is "a", "b", "c", "ab", "bc", "ca", "abc", "bca", "cab", "abca", "bcab" and "abcab".
Enumeration has many applications in such areas as data mining, optimization, and statistics.

Since enumeration problems often require us to output very many objects, we need
to reduce the time used to output each object in order to keep the total time reasonably
short. We also need to keep the total memory space reasonably small, as well as in the
case of solving other computational problems. However, straight-forward algorithms for
an enumeration often take much time and space, requiring time exponential in output
size and space exponential in input size. Developing an output-sensitive enumeration
algorithm using a small amount of memory is an important task. When we enumerate
exponentially many objects (in input size) in output-sensitive computational time with
a small memory, we have to avoid duplicate outputs without storing previously output
objects in memory, since the size of the storage must be exponentially large. Further,
simple search strategies may fail with some problems. For example, branch-and-bound
type algorithms are not effective if the subproblems related to the bounding operations

---

* JSPS Research Fellow.

F.V. Fomin (Ed.): WG 2006, LNCS 4271, pp. 68–77, 2006.
© Springer-Verlag Berlin Heidelberg 2006

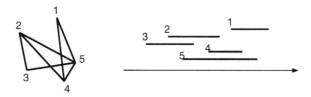

Fig. 1. An interval graph and its interval representation

are difficult. However, efficient algorithms for enumerations have been provided for some problems, such as enumerations of vertices of a polytope, all cells in a hyperplane arrangement, spanning trees of a graph [1], maximal cliques of a graph [12] and perfect elimination orderings of a chordal graph [5].

We have addressed the supergraph enumeration problems and the subgraph enumeration problem related to chordal graphs and interval graphs: enumerating all chordal/interval graphs (in K_n) that contain the given graph, and enumerating all interval graphs that are included in the given graph. Related to these problems, in [10], the authors previously proposed an algorithm for chordal subgraph enumeration using techniques different from those of the techniques used in this paper.

A graph is chordal if and only if it has no induced chordless cycle of length more than three. A graph G is an interval graph if and only if there is a one-to-one correspondence between its vertices and a set of intervals on the real line, such that two vertices are adjacent if and only if the corresponding intervals have an intersection. The set of intervals is called an interval representation of G (Fig. 1). Chordal graphs have many applications, for example, to matrix computation [3] or to relational databases[2]. Chordal graphs are also used for modeling some systems in computer science and in the social sciences [16]. Enumerating chordal graphs may be applied to the enumeration of these models. Interval graphs also have many applications in such diverse areas as archeology, biology, and scheduling [8].

In this paper, we define the adjacencies between chordal graphs and between interval graphs. Two adjacent chordal/interval graphs differ only in one of their edges. We can thus traverse the graphs by removing or adding one edge. We propose efficient polynomial delay algorithms traversing along the adjacencies based on reverse search. The complexities of the algorithms are $O(n^3)$ time and $O(n^2)$ space for the supergraph enumerations, and $O((n+m)^2)$ time and $O(n+m)$ space for the subgraph enumeration.

The organization of the paper is as follows. Section 2 provides the preliminaries. Then, we show some properties of chordal graphs and interval graphs to develop the search trees in Section 3. Section 4 describes the algorithms we propose, and Section 5 concludes the paper.

2   Preliminaries

2.1   Terms and Notations

Graph $G = (V, E)$ is a subgraph of $G' = (V', E')$ iff $V \subseteq V'$ and $E \subseteq E'$, and graph G is a supergraph of G' iff G' is a subgraph of G.

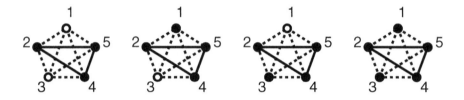

Fig. 2. All the graphs in K_5 whose edge set is $\{(2,4),(2,5),(4,5)\}$. The filled vertices are in the vertex set, and non-filled vertices are not in the vertex set.

Given a graph G that is a subgraph of the n-vertex complete graph K_n, we call finding (without duplications) all subgraphs of G that have a given specified property as *subgraph enumeration*. The property is, for example, that the graph is chordal, or that the graph is an interval graph. Similarly, given graph G included in K_n, we call finding (without duplications) all subgraphs of K_n that are supergraphs of G and have specified properties as *supergraph enumeration*. To be more specific, when the property is that the graph is chordal (an interval graph), we call the problems as *chordal (interval) subgraph/supergraph enumeration*. Given an edge set E, it is easy to enumerate every graph whose edge set is E, that may contain some isolated vertices. (See 2)). Therefore, we enumerate graphs by enumeration of edge sets. For simplicity, given an edge set E, we denote by $G\langle E\rangle$ the graph whose vertex set is vertices incident to edges in E and whose edge set is E.

We denote by $G + e$ the graph obtained by adding edge e to the edge set of graph G, and we denote by $G - e$ the graph obtained by removing edge e from that edge set.

2.2 Difficulty of Super/Subgraph Enumeration

Enumerating chordal/interval graphs involves some difficulties. Chordal graphs in all graphs with n vertices are exponentially few in ratio to non-chordal graphs [17]. Since interval graphs are chordal graphs [11], this fact is also true for interval graphs. Thus, if we enumerate them by generating all graphs included in K_n and determine whether they satisfy the definitions of these graph classes, the algorithm takes an exponentially long time in n to output a graph.

We can consider branch-and-bound type algorithms. Let us consider a chordal subgraph enumeration algorithm: given a graph $G = (V, E)$, we enumerate all chordal subgraph of G. At every level of the search, we choose an edge e in E, and divide the problem into two subproblems: an enumeration of chordal subgraphs containing edge e, and an enumeration of those not containing edge e. We stop dividing the problem when we know that there is no chordal subgraph in a subproblem. In this way, we can enumerate all chordal subgraphs of G by outputting all leaves of the search tree. The problem is that deciding whether or not there is a chordal subgraph in a subproblem is difficult. Let A be the current graph in the algorithm and B be the input graph. The problem to decide whether or not there is a graph belonging to specified graph class that contains A and included in B is known as "graph sandwich problem". The graph sandwich problem for chordal graphs is proved to be NP-complete [9]. The algorithm thus possibly takes an exponentially long time at the bounding phase, or

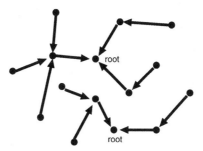

Fig. 3. Spanning forest on the objects to be enumerated. Paths from all leaves aim at the roots.

the algorithm visits exponentially many subgraphs of G that are not chordal, unless $P = NP$. Hence, the algorithm may take an exponentially long time to output a chordal subgraph. Golumbic et al. also showed in [9] that a graph sandwich problem for interval graphs, deciding whether or not there is an interval graph included in one given graph and containing another given graph, is NP-complete. The interval subgraph enumeration thus has a similar difficulty. We use the reverse search method to overcome this difficulty.

2.3 Reverse Search

For efficient enumerations, a good search method is necessary. The reverse search method suits this requirement. The reverse search is a sophisticated depth-first search type scheme for enumerations and was originally developed by Avis and Fukuda in [1]. Because of its simplicity and efficiency, this search method has been used in many algorithms for problems in many fields [1,13,14,12].

Let $\mathcal{F}$ be the set of objects that we want to enumerate. For example, $\mathcal{F}$ is the set of chordal/interval graphs included in/containing a given graph in our problems. We define a parent-child relation by determining a parent for each object, except for some specified objects called *root objects*, which do not have parents. The definition of the parents has to satisfy that no object is a proper ancestor of itself, that is, by starting from an object x and moving in continuing succession to its parent and parent's parent, we never come to the start object x again. The graph representation of the relation induces a set of disjoint rooted trees spanning all objects, in which paths from all leaves aim to the roots. We illustrate an example of the graph representation in Fig. 3. Each object to be enumerated is drawn by a point, and an object and its parent are connected by a directed arrow.

Tracing each edge in the reverse direction enables us to perform a depth-first search to visit all objects. We can thus find all objects without duplications by using two subalgorithms: one algorithm to find all root objects, and another to find all children of each object. We often have to store the nodes that we have visited in memory to avoid visiting them twice when performing depth-first search algorithms. In contrast with such cases, the reverse search algorithms automatically avoid duplications by limiting the base structure for searching to tree (forest) structure.

2.4   Chordal Graphs and Interval Graphs

Chordal graphs have the important property of having at least one *simplicial vertex*, and if the chordal graph has more than one vertex, there are at least two simplicial vertices in it [7], where a vertex is simplicial iff its neighbors induce a clique. Eliminating a simplicial vertex from a chordal graph is well known to result in another chordal graph, where an elimination of a vertex means removing the vertex and edges incident to it. Thus, we can iteratively eliminate simplicial vertices from a chordal graph until the graph has no vertices. The vertex ordering by which we eliminate the simplicial vertices is called a *perfect elimination ordering*. It is known that a graph has a perfect elimination ordering if and only if the graph is chordal. Given a graph G, we can decide whether or not graph G is chordal in time linear in the input size [15].

Interval graphs also have interesting properties. We can find an interval representation of a given interval graph, and thus decide whether or not the graph is an interval graph, in time linear in the input size [4,6]. In general, an interval graph has many non-isomorphic interval representations. The class of interval graphs is a proper subclass of that of chordal graphs [11]. Although this fact means that we can solve optimization problems on interval graphs with algorithms for chordal graphs, this is not the case for enumeration problems.

3   Preparation for Our Algorithms

All algorithms presented in this paper are based on the reverse search method. We explain how to define the parent-child relations for the reverse searches, in this section.

In the chordal supergraph enumeration, we define the parent H of a chordal graph H' such that the difference between H and H' is always one, i.e., we can obtain, in reverse searching, a child by removing an edge. Given chordal graph H, with such a parent-child relation, checking all the candidates of children to determine whether or not they are children of H requires only polynomial time, since there must be only $O(n^2)$ of candidates of children of H. Therefore, defining parents in such a way, we can enumerate each chordal supergraph of the given graph in polynomial time. Defining such parent-child relations also enables us to find the root easily. We can add at least an edge to every chordal graph not equal to K_n keeping the chordality, as we show in the next section. The root of the chordal supergraph enumeration is thus simply K_n. In a similar way, we can define parents for the interval supergraph enumeration and the interval subgraph enumeration. To show that we can define parents in such a way and that the root is K_n (the empty graph in the case of the subgraph enumeration), and to make our algorithm efficient, we must first prove some lemmas about edge deletions and edge additions.

Lemma 1. *Given an n-vertex chordal graph $H = (V, E) \neq K_n$ and any vertex $v \in V$ of degree smaller than $n - 1$, there exists a vertex $v' \in V$ not adjacent to v such that* *graph $H + (v, v')$ is chordal. Moreover, we can find such vertex v' in O(n+m) time, where m is $|E|$.*

Proof. We can find a perfect elimination ordering of H where the vertex v is located at the tail [15]. We denote such a perfect elimination ordering by $P = (p_1, p_2, \ldots, p_n)$, where $p_n = v$. Let p_k be the last vertex in P that is not adjacent to v. We show that p_k is the desired vertex v' by proving that P is a perfect elimination ordering of $H' = H + (v, p_k)$ and H' is thus chordal.

We denote by H_j the subgraph of H induced by $\{p_j, p_{j+1}, \ldots, p_n\}$, and we denote by H'_j the subgraph of H' induced by $\{p_j, p_{j+1}, \ldots, p_n\}$. We denote by $N_i(v)$ (resp. $N'_i(v)$) the set of neighbors of vertex v in H_i (resp. H'_i).

Vertex p_i ($i = 1, 2, \ldots, k-1, k+1, k+2, \ldots, n$) is a simplicial vertex of H'_i, for $N_i(p_i)$ and $N'_i(p_i)$ are identical. Since p_k is a simplicial vertex of H_k, $N_k(p_k)$ induces a clique in H_k and also in H'_k. Since all the vertices of $N_k(p_k)$ are adjacent to vertex p_n, $N'_k(p_k) = N_k(p_k) \cup \{p_n\}$ also induces a clique in H'_{p_k}. Hence, vertex p_k is a simplicial vertex of H'_{p_k}. P is thus a perfect elimination ordering of H'.

We can obtain a perfect elimination ordering P such that v is located at the tail in O($n + m$) time [15]. To find p_k from the ordering P, we need O(n) time. Thus, we can find the desired v' in O($n + m$) time. □

Lemma 2. *Given an n-vertex interval graph $I = (V, E) \neq K_n$ and any vertex $v \in V$ of degree smaller than $n - 1$, there exists a vertex $v' \in V$ not adjacent to v such that graph $I + (v, v')$ is an interval graph. Moreover, we can find such vertex v' in O(n+m) time, where m is $|E|$.*

Proof. We denote an interval representation of I by $(I_1, I_2, \ldots, I_n)$, where each interval I_i ($i = 1, 2, \ldots, n$) corresponds to vertex $v_i \in V$. We can assume without loss of generality that all of the end points of the intervals are distinct. Let vertex v_j be the vertex v.

Since the degree of vertex v is smaller than $n - 1$, the corresponding interval $I_j = [l_j, r_j]$ does not intersect some intervals. We can assume without loss of generality that there are some intervals at the right-side of I_j that do not intersect I_j, since the symmetric intervals also form an interval representation of interval graph I. Let $I_k = [l_k, r_k]$ be the interval such that the difference between r_j and the left end is the smallest among such intervals. We change the interval I_j to $[l_j, l_k + \epsilon]$, where ϵ is a sufficiently small number (Figure 4). Then, the resulting interval representation is that of an interval graph $I + (v, v_k)$. Hence, the vertex v_k is the desired vertex v'.

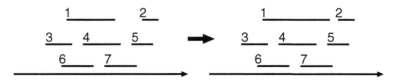

Fig. 4. We can add a new edge incident to vertex 1

We can obtain an interval representation $(I_1, I_2, \ldots, I_n)$ of I in $O(n+m)$ time [4,6], and to find I_k from the representation, we need $O(n)$ time. We can thus find the desired v' in $O(n+m)$ time.   □

Lemma 3. *Given an n-vertex connected interval graph $I = (V, E)$ ($|E| \geq 1$), there exists an edge $e \in E$ such that $G\langle E \setminus \{e\}\rangle$ is a connected interval graph. Moreover, we can find the edge e in $O(n+m)$ time, where m is $|E|$.*

Proof. We denote an interval representation of I by $(I_1 = [l_1, r_1], I_2 = [l_2, r_2], \ldots, I_n = [l_n, r_n])$, where each interval $I_i (i = 1, 2, \ldots, n)$ corresponds to vertex $v_i \in V$. We can assume without loss of generality that all of the end points of the intervals are distinct. Let l_j be the smallest among l_i $(i = 1, 2, \ldots, n)$.

Since I is connected, $I_j = [l_j, r_j]$ intersects some intervals. Let $I_k = [l_k, r_k]$ be the interval that intersects I_j and the value l_k be the largest among such intervals. Setting r_j to $l_k - \epsilon$ produces a new interval graph that we obtain by removing an edge between vertices v_j and v_k. If the interval graph is connected, (v_j, v_k) is the desired edge. Otherwise, the graph is divided into two connected interval graphs, where the left-hand one contains I_j. Then, doing the process recursively with the right-hand intervals, we can find the desired edge eventually. At the end of the recursion, it is possible that the graph is divided into an isolated vertex and the rest of the graph. However, in the sense of an edge set, the graph is connected.

It takes $O(n+m)$ time to obtain an interval representation of I. We can find every k at every iteration by sweeping intervals from l_1 until we find the desired vertex. Each interval I_i $(i = 1, 2, \ldots, n)$ is swept at most once. Therefore, we can find all k's at every iteration in $O(n)$ time. Thus, the time complexity to find e is $O(n+m)$.   □

4   Enumeration Algorithms

4.1   Supergraph Enumeration of Chordal/Interval Graphs

First, we present our algorithm for the chordal supergraph enumeration. We define the root chordal graph and the parent of other chordal graphs. Let A be an algorithm obtained from the proof of Lemma 1; A inputs a chordal graph H, and outputs an edge $e^*(H) = (v, v')$ such that v is the youngest vertex with degree smaller than $n - 1$ and $H + (e^*(H))$ is chordal. We can transform H to be canonical in some way, since H has vertex indices. Thus, we can assume that A outputs the identical edge for the identical inputs. Hence, we can define $e^*(H)$ uniquely for any $H \neq K_n$. We define the parent of a chordal graph $H \neq K_n$ as $H + e^*(H)$. The root chordal graph is the complete graph K_n.

Lemma 4. *For every chordal graph $H \neq K_n$, the parent always exists and is unique.*

Lemma 5. *The parent-child relation induces a rooted tree, whose root is the complete graph K_n.*

We describe an outline of our algorithm below. We just have to call the procedure with the argument $H = K_n$.

procedure enum_super_chordal(H, G)
H : **chordal graph**, G : **graph**;
begin
  output H;
  for every edge $e \in E(H) \setminus E(G)$ **do**
    if $H - e$ is chordal **and**
      the parent of $H - e$ is H **then**
        enum_super_chordal($H - e, G$);
  end for
end.

The algorithm performs a depth-first back tracking to enumerate all the chordal supergraphs included in K_n. From the lemmas above, we can easily confirm that this algorithm finds without duplications all the chordal supergraphs of G. We discuss the complexity in the theorem below.

Theorem 1. *There is an algorithm for the chordal supergraph enumeration, that finds every chordal supergraph of a given graph in $O(n^3)$ time and $O(n^2)$ space.*

Proof. Given a chordal graph H, let v^* be the youngest vertex that has a degree smaller than $n - 1$.

For edge e incident to a vertex younger than v^* such that $H - e$ is chordal, the parent of $H - e$ is always H, due to the definition of the parent. If more than one vertices are younger than v^*, these vertices induces a clique, since their degrees are $n - 1$. We consider any two of such vertices, v_1 and v_2. From the symmetric property, for any vertex v and edge $e_2 = (v, v_2)$, $H - e_2$ is chordal if and only if edge $e_1 = (v, v_1)$ satisfies that $H - e_1$ is chordal. Thus, we have to check chordality on only $O(n)$ graphs, to determine, for every edge e incident to vertices younger than v^*, whether or not $H - e$ is chordal. We can do these checks in $O(n^3)$ time, since we can check chordality of a graph in $O(n^2)$ time.

For edge $e = (v^*, v^+)$, where v^+ is elder than v^*, such that $H - e$ is chordal, we can check whether or not the parents of $H - e$ is H in $O(n^2)$ time, since we can obtain the parent of $H - e$ in $O(n^2)$ time by Lemma 1. Hence, for all edges $e = (v^*, v^+)$, we can check whether or not $H - e$ is a child of H, i.e., $H - e$ is chordal and the parent of $H - e$ is H, in $O(n^3)$ time.

For edge e whose end points are elder than v^*, the parent of $H - e$ is never H, since we can add to $H - e$ an edge whose one end point is v^* keeping chordality by Lemma 1.

Hence, the total time complexity to check for every candidate of a child of H whether or not it is really a child of H is $O(n^3)$. □

The algorithm and the analysis technique also can be applied to the case of the interval supergraph enumeration similarly, since Lemma 2 plays the same role of Lemma 1. We can thus enumerate interval graphs which contain given graph G, where the time complexity for finding every interval supergraph is $O(n^3)$.

Theorem 2. *There is an algorithm for the interval supergraph enumeration, that finds every interval supergraph of a given graph in $O(n^3)$ time and $O(n^2)$ space.*

Notice that these algorithms and the analysis can be applied to connected chordal/interval supergraph enumeration, since the parent of a connected chordal/interval graph is connected.

4.2   Subgraph Enumeration of Interval Graphs

Similarly to the algorithm for the supergraph enumeration, we can develop an algorithm for interval subgraph enumeration.

First, we explain the case of the connected interval graph enumeration. We define the root interval graph and a parent-child relation. The parent interval graph of connected interval graph I that is not the empty graph is a connected interval graph obtained by removing edge $e^*(I)$ from I, where $e^*(I)$ is an edge obtained by the algorithm derived from Lemma 3. We define the root as the empty graph.

Now, we present our algorithm below. We just have to call the procedure with the argument $I = (\phi, \phi)$, i.e., the empty graph.

> **procedure** enum_sub_interval(I, G)
> I : **interval graph**, G : **graph**;
> **begin**
>   **output** I;
>   **for** every edge $e \in E(G) \setminus E(I)$ **do**
>   **if** $I + e$ is connected interval **and**
>   the parent of $I + e$ is I **then**
>   enum_sub_interval($I + e, G$);
>   **end for**
> **end**.

For graph I, we have to check $O(n + m)$ candidates of the children. We can find the parent of the given interval graph in $O(n + m)$ time, by Lemma 3. Thus, computational time of this algorithm to find each connected interval graph included in graph G is $O((n+m)^2)$. In contrast to the case of the supergraph enumerations, we can not remove an edge incident to an arbitrary vertex keeping the intervality. Thus, the time complexity of this algorithm for finding each connected interval subgraph is larger than that in the case of supergraph enumeration.

The (not necessarily connected) interval subgraph enumeration algorithm is straightforward from that of the connected version. With some deterministic algorithm, we can easily obtain an ordering of the connected components of an interval graph I. Let $I^1, I^2, \ldots, I^s$ be such an ordering of the connected components of I. We define a parent of I as (the parent of I^1) $\cup I^2 \cup, \ldots, \cup I^s$.

Theorem 3. *There is an algorithm for the (connected) interval subgraph enumeration, that finds every (connected) interval subgraph of a given graph in $O((n + m)^2)$ time.*

5   Conclusion

We focused on adding and removing edges on chordal and interval graphs. We showed that we can define the rooted tree structures by them on all the chordal/interval supergraphs of a given graph, and on all the interval subgraph of a given graph. Using the structures, we introduced algorithms for interval subgraph enumeration, and chordal/interval supergraph enumerations. The time complexity to find every chordal or interval graph containing a given graph are both $O(n^3)$ for each. The time complexity to find every interval graph included in a given graph is $O((n + m)^2)$ for each.

References

1. Avis, D., and Fukuda, K.: Reverse search for enumeration, Discrete Applied Mathematics **65** (1996) 21–46
2. Beeri, C., Fagin, R., Maier, D., and Yanakakis, M.: On the desirability of acyclic database schemes, Journal of the ACM **30** (1983) 479–513
3. Blair, J. R. S., and Peyton, B.: An introduction to chordal graphs and clique trees, Graph Theory and Sparse Matrix Computation, **IMA56** (1993) 1–29
4. Booth, K. S., and Lueker, G. S.: Testing for the consecutive ones property, interval graphs, and graph planarity using PQ tree algorithms, Journal of Computing and System Sciences **13** (1976) 335–379
5. Chandran, L. S., Ibarra, L., Ruskey, F., and Sawada, J.: Fast generation of all perfect elimination orderings of a chordal graph, Theoretical Computer Science, **307** (2003) 303–317
6. Corneil, D. G., Olariu, S., and Stewart L.: The ultimate interval graph recognition algorithm?, Proceedings of 9th Annual ACM-SIAM Symposium on Discrete Algorithms, ACM, 1998, 175–180
7. Dirac, G. A.: On rigid circuit graphs, Abhandl. Math. Seminar Univ. Hamburg **25** (1961) 71–76
8. Golumbic, M. C.: Algorithmic graph theory and perfect graphs, Academic Press, New York, 1980
9. Golumbic, M. C., Kaplan, H., and Shamir R.: Graph sandwich problems, Journal of Algorithms **19** (1995) 449–473
10. Kiyomi, M., and Uno, T.: Generating chordal graphs included in given graphs, IEICE Transactions on Information and Systems **E89-D** (2006) 763–770
11. Leckerkerker, C. G., and Boland, J. C.: Representation of a finite graph by a set of intervals on the real line, Fund Math 51 (1962), 45–64
12. Makino, and K. Uno, T.: New algorithms for enumerating all maximal cliques, Lecture Notes in Computer Science **3111** (2004) 260–272.
13. Nakano, S.: Enumerating floorplans with n rooms, Lecture Notes in Computer Science **2223** (2001) 107–115.
14. Nakano, S.: Efficient generation of triconnected plane triangulations, Computational Geometry Theory and Applications **27(2)** (2004) 109–122.
15. Rose, D. J., Tarjan, R. E., and Lueker, G. S.: Algorithmic aspects of vertex elimination on graphs, SIAM Journal on Computing **5** (1976) 266–283
16. Whittaker, J.: Graphical models in applied multivariate statistics, Wiley, New York, 1990
17. Wormald, N. C.: Counting labelled chordal graphs, Graphs and Combinatorics **1** (1985) 193–200

A Branch-and-Reduce Algorithm for Finding a Minimum Independent Dominating Set in Graphs

Serge Gaspers[1] and Mathieu Liedloff[2]

[1] Department of Informatics, University of Bergen, N-5020 Bergen, Norway
serge.gaspers@ii.uib.no
[2] Laboratoire d'Informatique Théorique et Appliquée, Université Paul
Verlaine - Metz, 57045 Metz Cedex 01, France
liedloff@univ-metz.fr

Abstract. A dominating set $\mathcal{D}$ of a graph $G = (V, E)$ is a subset of vertices such that every vertex in $V \setminus \mathcal{D}$ has at least one neighbour in $\mathcal{D}$. Moreover if $\mathcal{D}$ is an independent set, i.e. no vertices in $\mathcal{D}$ are pairwise adjacent, then $\mathcal{D}$ is said to be an independent dominating set. Finding a minimum independent dominating set in a graph is an NP-hard problem. We give an algorithm computing a minimum independent dominating set of a graph on n vertices in time $O(1.3575^n)$. Furthermore, we show that $\Omega(1.3247^n)$ is a lower bound on the worst-case running time of this algorithm.

1 Introduction

During the last years the interest in the design of exact exponential time algorithms has been growing significantly. Nice surveys have been written on this subject. In one due to Woeginger [18], the author emphasizes the major techniques used to design exact exponential time algorithms. We also refer the reader to the recent survey of Fomin et al. [7] discussing some new techniques in the design of exponential time algorithms. In particular they discuss Measure & Conquer and lower bounds.

The Minimum Independent Dominating Set problem (MIDS) is also known as Minimum Maximal Independent Set, since every independent dominating set is a maximal independent set. This problem asks for a set of minimum cardinality that is both independent and dominating. Whereas Maximum Independent Set and Minimum Dominating Set have been studied very deeply in the field of exact algorithms, the best known exact algorithm for MIDS trivially enumerates all maximal independent sets.

Known Results. A set $\mathcal{I} \subseteq V$ of a graph $G = (V, E)$ is independent if no two vertices in $\mathcal{I}$ are adjacent. The problem of finding a Maximum Independent Set (MIS) of a graph was among the first problems shown to be NP-hard.

It is known that a MIS of a graph on n vertices can be computed in $O(1.4423^n)$ time by combining a result due to Moon and Moser, who showed in 1965 [13]

F.V. Fomin (Ed.): WG 2006, LNCS 4271, pp. 78–89, 2006.
© Springer-Verlag Berlin Heidelberg 2006

that the number of maximal independent sets of a graph is upper bounded by $3^{n/3}$, and a result due to Johnson, Yannakakis and Papadimitriou, providing in [11] a polynomial delay algorithm to generate all maximal independent sets. Moreover many exact algorithms for this problem have been published, starting in 1977 by an $O(1.2600^n)$ algorithm by Tarjan and Trojanowski [17]. The best known algorithms for MIS until now are an $O(1.2108^n)$ algorithm by Robson [15] in 1986, a very long algorithm of running time $O(1.1889^n)$ by Robson [16] in 2001 and a very simple algorithm with running time $O(1.2210^n)$ by Fomin et al. [5] in 2006.

A set $\mathcal{D} \subseteq V$ of a graph $G = (V, E)$ is dominating if every vertex in $V \setminus \mathcal{D}$ has at least one neighbour in $\mathcal{D}$. The problem of finding a Minimum Dominating Set (MDS) of a graph is well known to be NP-hard.

Until recently, the only known exact exponential time algorithm to solve MDS asked for trivially enumerating the 2^n subsets of vertices. The year 2004 saw a particular interest in providing some faster algorithms for solving this problem. Indeed, three papers with exact algorithms for MDS were published. In [8] Fomin et al. present an $O(1.9379^n)$ time algorithm, in [14] Randerath and Schiermeyer establish an $O(1.8899^n)$ time algorithm and Grandoni [9] obtains an $O(1.8026^n)$ time algorithm.

By now, the fastest published algorithm is due to Fomin et al. [6]. They use the Measure & Conquer approach to obtain an algorithm with running time $O(1.5263^n)$ and using polynomial space. By applying a memorization technique they show that this running time can be reduced to $O(1.5137^n)$ when allowing exponential space usage.

A natural and well studied combination of these two problems asks for a subset of vertices of minimum cardinality that is both dominating and independent. This problem is called Minimum Independent Dominating Set (MIDS).

It has been established that a minimum independent dominating set $(MIDS)$ can be found in polynomial time for several graph classes like interval graphs [2], chordal graphs [4], cocomparability graphs [12] and AT-free graphs [1], whereas the problem remains NP-complete for bipartite graphs [3] and comparability graphs [3]. Concerning inapproximability results, Halldórsson established in [10] that there is no constant $\epsilon > 0$ such that MIDS can be approximated within a factor of $n^{1-\epsilon}$ in polynomial time, assuming $P \neq NP$.

To the best of our knowledge, the only paper giving an exact exponential time algorithm for MIDS has been written by Randerath and Schiermeyer [14]. They use the result due to Moon and Moser [13] as explained previously and an algorithm enumerating all the maximal independent sets to obtain an $O(1.4423^n)$ time algorithm for MIDS.

Our Results. In this paper we present an $O(1.3575^n)$ time algorithm for solving MIDS using the Measure & Conquer approach to analyze its running time. As the bottleneck of the algorithm in [14] are the vertices of degree two, we use some nice tricks to handle them more efficiently such as marking some vertices and a sophisticated reduction rule described in section 3.1. Combined with some elaborated branching rules, this enables us to lower bound shrewdly the progress made by the algorithm at each branching step, and thus to obtain an algorithm which improves the best known result from $O(1.4423^n)$ to $O(1.3575^n)$. Furthermore, we obtain a very close lower bound of $\Omega(1.3247^n)$ on the running time of our algorithm, which is very rare for non trivial exponential time algorithms.

2   Preliminaries

Let $G = (V, E)$ be an undirected and simple graph. For a vertex $v \in V$ we denote by $N(v)$ the neighbourhood of v and by $N[v] = N(v) \cup \{v\}$ the closed neighbourhood of v. The degree $d(v)$ of v is the cardinality of $N(v)$. For a given subset of vertices $S \subseteq V$, $G[S]$ denotes the subgraph of G induced by S, $N(S)$ denotes the set of neighbours in $V \setminus S$ of vertices in S and $N[S] = N(S) \cup S$. We also define $N_S(v)$ as $N(v) \cap S$ and $d_S(v)$ (called the S-degree of v) as the cardinality of $N_S(v)$. In the same way, given two subsets of vertices $S \subseteq V$ and $X \subseteq V$, we define $N_S(X) = N(X) \cap S$.

A *clique* is a set $S \subseteq V$ of pairwise adjacent vertices. A graph $G = (V, E)$ is called *bipartite* if V admits a partition into two independent sets. A bipartite graph $G = (V, E)$ is a *complete bipartite graph* if every vertex of one independent set is adjacent to every vertex of the other independent set. A *connected component* of a graph is a maximal subset of vertices inducing a connected subgraph.

In a branch-and-reduce algorithm the current problem is divided into smaller ones such that an optimal solution, if one exists, occurs in at least one subproblem. If the algorithm considers only one subproblem in a given case, we refer to a reduction rule, otherwise to a branching rule.

Consider a vertex $u \in V$ of degree two with two non adjacent neighbours v_1 and v_2. In such a case, a branch-and-reduce algorithm will typically branch into three subcases when considering u: either u or v_1 or v_2 are in the solution set. In the third branch, one can consider that v_1 is not in the solution set as this is already considered by the second branch. In order to memorize that v_1 is not in the solution set but still needs to be dominated, we mark v_1.

Definition 1. *A marked graph $G = (F, M, E)$ is a triple where $F \cup M$ denotes the set of vertices of G and E denotes the set of edges of G. The vertices in F are called* free *vertices and the ones in M* marked *vertices.*

Definition 2. *Given a marked graph $G = (F, M, E)$, an independent dominating set $\mathcal{D}$ of G is a subset of free vertices, i.e. $\mathcal{D} \subseteq F$, such that $\mathcal{D}$ is an independent dominating set of the graph $G' = (F \cup M, E)$.*

Remark 1. It is possible that such an independent dominating set does not exist in a marked graph, namely if a marked vertex has no free neighbours.

To close this section we introduce the notion of an *induced marked subgraph*.

Definition 3. *Given a marked graph $G = (F, M, E)$ and two subsets $S, T \subseteq (F \cup M)$, an induced marked subgraph $G[S, T]$ is the marked graph $G' = (S \cap (F \cup M), T \cap (F \cup M), E')$ where $E' \subseteq E$ are the edges of G with both end points in $S \cup T$.*

Note that notions like neighbourhood and degree in a marked graph $G = (F, M, E)$ are the same as in the corresponding simple graph $G = (F \cup M, E)$.

3 Computing a *MIDS* on Marked Graphs

In this section we present an algorithm solving MIDS on marked graphs.

¿From the previous definitions it follows that a subset $\mathcal{D} \subseteq V$ is a $MIDS$ of a graph $G' = (V, E)$ if and only if $\mathcal{D}$ is a $MIDS$ of the marked graph $G = (V, \emptyset, E)$. Hence the algorithm of this section is able to solve the problem on simple graphs as well.

Given a marked graph $G = (F, M, E)$, consider the graph $G[F]$ induced by its free vertices. In the following subsection we introduce a reduction rule which deletes a connected component of $G[F]$ which is a clique.

3.1 Eliminating Cliques in $G[F]$

Consider the function **RedClique**. (See next page.)

Lemma 1. *Let $G = (F, M, E)$ be a marked graph and C a connected component of $G[F]$ which is a clique. The function* **RedClique** *computes in polynomial time a marked graph $G' = (F', M', E')$ such that:*

(i)   the size of a $MIDS$ of G is equal to the size of a $MIDS$ of G' plus one,
(ii) $F' = F \setminus C$,
(iii) no edge of $E' - E$ has both end points in F', i.e. the function adds no edge between two free vertices.

Proof. First, note that whenever there is a clique component C in $G[F]$, every independent dominating set contains exactly one vertex of C. Indeed, at least one vertex of C has to be taken in the independent dominating set to dominate C and at most one vertex in C can be taken because the solution has to be an independent set.

If $|C| = 1$, the unique vertex in C must be part of the $MIDS$. So, the function just deletes C and its neighbourhood. By now we assume that $|C| \geq 2$.

If there is a vertex $v \in C$ with no marked neighbour, then we will not choose this vertex in the $MIDS$. As a matter of fact, every vertex in C dominates C. So, a vertex in C which also dominates some marked vertices is always a better choice than a vertex that does not. Consequently, the function just deletes v and calls itself recursively on the clique component $C - \{v\}$.

Assume now that $|C| \geq 2$ and that every vertex in C has at least one neighbour in M. Then, the function will create one new marked vertex $h_{i,j}$ for every two vertices $h_i, h_j \in N(C)$ that do not share a same neighbour in C. It replaces $N[C]$ by these new marked vertices. A vertex $h_{i,j}$ will be adjacent to a vertex $v \in F \setminus C$ iff h_i or h_j was adjacent to v. So, when all vertices $h_{i,j}$ will be dominated by vertices in $F \setminus C$ in G', at least all the vertices in $N(C)$ except the neighbours

---

Function RedClique($G = (F, M, E), C \subseteq F$)
Input: A marked graph $G = (F, M, E)$ and a clique $C \subseteq F$ such that C is a
      connected component of $G[F]$.
Output: A marked graph $G' = (F', M', E')$ s.t. G' has the properties
        defined in **Lemma 1.**

  if $|C| = 1$ **then**
    | $G' \leftarrow G[F \setminus C, M \setminus N(C)]$;
  else
    if $\exists v \in C$ s.t. $N_M(v) = \emptyset$ **then**
      | $G' \leftarrow$ **RedClique**$(G[F - \{v\}, M], C - \{v\})$
    else
      let $N(C) = \{h_1, h_2, \ldots, h_k\}$
      $H \leftarrow \emptyset$
      for $i \leftarrow 1$ *to* $k - 1$ **do**
        for $j \leftarrow i + 1$ *to* k **do**
          if $N_C(h_i) \cap N_C(h_j) = \emptyset$ **then**
            | **add to** H **a new marked vertex** $h_{i,j}$

      $G' = (F', M', E') \leftarrow G[F \setminus C, M \setminus N(C)]$
      $M' \leftarrow M' \cup H$
      foreach $h_{i,j} \in H$ **do**
        foreach $v \in N_F(N[C])$ s.t. $\{v, h_i\} \in E$ *or* $\{v, h_j\} \in E$ **do**
          | $E' \leftarrow E' \cup \{v, h_{i,j}\}$

  return G'

---

of a unique vertex $u \in C$ are dominated in G. It is then clear which vertex of C will be in the $MIDS$. And whenever a vertex $h_{i,j}$ is not dominated in G', no vertex of C can dominate all undominated vertices in $N(C)$ in G.

Remark that, once all these new marked vertices are dominated, it is possible to determine in polynomial time which vertex of the clique C must be added to the solution in order to obtain a $MIDS$ for the initial marked graph.

As $N[C]$ is deleted from the original graph, we have $F' = F - C$. The function does not create new edges between two free vertices because the only new edges created during the computation join free and new marked vertices. It is not hard to see that **RedClique** has polynomial running time. $\qquad\square$

3.2 The Algorithm

In this subsection, we give the algorithm **ids** computing the size of a $MIDS$ of a marked graph. The branching rules are quite complicated but it is fairly simple to check that the algorithm computes the size of a $MIDS$ (if one exists). It is not difficult to transform **ids** into an algorithm that actually outputs a $MIDS$. In the next section we prove the correctness and give a detailed analysis of **ids**.

Algorithm ids(G)
Input: A marked graph $G = (F, M, E)$.
Output: The size of a $MIDS$ of G.

if $F = M = \emptyset$ then
    return 0   (0)

if $\exists u \in M$ s.t. $d_F(u) = 0$ then
    return ∞ (1)

else if $\exists u \in M$ s.t. $d_F(u) = 1$ then
    let v be the unique free neighbour of u
    return $1 + \text{ids}(G[F \setminus N[v], M \setminus N(v)])$ (2)

else if $\exists C \subseteq F$ s.t. C is a clique $\wedge N_F(C) = \emptyset$ then
    return $1 + \text{ids}(\textbf{RedClique}(G, C))$ (3)

else if $\exists B \subseteq F$ s.t. B induces a complete bipartite graph $\wedge N_F(B) = \emptyset$ then
    let B be partitioned into two independent sets X and Y
    return $\min\{ |X| + \text{ids}(G[F \setminus N[X], M \setminus N(X)]);$ (4)
                   $|Y| + \text{ids}(G[F \setminus N[Y], M \setminus N(Y)])\}$

else if $\exists C \subseteq F$ s.t. C is a clique $\wedge |C| \geq 3 \wedge \exists! v \in C$ s.t. $d_F(v) \geq |C|$ then
    return $\min\{ 1 + \text{ids}(G[F \setminus N[v], M \setminus N(v)]);$ (5)
               $\text{ids}(G[F \setminus \{v\}, M \cup \{v\}, E])\}$

else
    choose $u \in F$ of minimum F-degree with a neighbour in F of
    maximum F-degree
    if $d_F(u) = 1$ then
       return $1 + \min\{ \text{ids}(G[F \setminus N[u], M \setminus N(u)]);$ (6)
                   $\text{ids}(G[F \setminus N[N_F(u)], M \setminus N(N_F(u))])\}$

    else if $d_F(u) = 2$ then
       let $N_F(u) = \{v_1, v_2\}$
       return $1 + \min\{ \text{ids}(G[F \setminus N[u], M \setminus N(u)]);$ (7)
                   $\text{ids}(G[F \setminus N[v_1], M \setminus N(v_1)]);$
                   $\text{ids}(G[F \setminus (N[v_2] \cup \{v_1\}), (M \cup \{v_1\}) \setminus N(v_2)])\}$

    else
       choose $v \in F$ of maximum F-degree
       return $\min\{ 1 + \text{ids}(G[F \setminus N[v], M \setminus N(v)]);$ (8)
                   $\text{ids}(G[F \setminus \{v\}, M \cup \{v\}])\}$

---

4   Correctness and Analysis of the Algorithm

Intuitively, marked vertices do not make the instance of the problem more difficult: they cannot be taken in the $MIDS$ and the only thing they are good for is to put restrictions on their free neighbours. Moreover, free vertices having only marked neighbours can be handled without branching. So, it is an advantage when the F-degree of a vertex decreases. We will therefore assign different weights to the free vertices according to their F-degree.

Let n_i denote the number of free vertices having F-degree i. For the running time analysis we consider the following measure of the size of G:

$$k = k(G) = \sum_{i \geq 0} w_i n_i \leq n$$

where the weights $w_i \in [0, 1]$. In order to simplify the running time analysis, we make the following assumptions:

- $w_0 = 0$,
- $w_i = 1$ for $i \geq 3$,
- $w_1 \leq w_2$,
- $\Delta w_1 \geq \Delta w_2 \geq \Delta w_3$ where $\Delta w_i = w_i - w_{i-1}, i \in \{1, 2, 3\}$.

Theorem 1. *Algorithm* **ids** *solves the minimum independent dominating set problem in time* $O(1.3575^n)$.

Proof. Let $P[k]$ denote the number of subproblems recursively solved to compute a solution for an instance of size k. As the time spent in each call of **ids**, excluding the time spent by the corresponding recursive calls, is polynomial, it is sufficient to show that for a valid choice of the weights, $P[k] = O(1.3575^n)$.

We will analyse the nine cases of algorithm **ids** one by one. Cases (0) to (3) are reduction rules and the other cases correspond to branching rules.

Case (0). If the set of vertices is empty, the algorithm returns 0 since it has computed an independent dominating set of the marked graph.

Case (1). If there is a marked vertex u having no free neighbour, u has no possibility to be dominated and thus the algorithm returns ∞, meaning that there is no solution for this subproblem.

Case (2). If there is a marked vertex u with only one free neighbour v, the only possibility for u to be dominated is to add v to the $MIDS$. So, $N[v]$ is deleted and the measure k decreases at least by $2w_1$.

Case (3). If there is a clique C of free vertices which are not adjacent to any other free vertices, we use the function of Lemma 1 to remove C. Since the number of free vertices decreases by $|C|$ and no new edges are added between any two free vertices, the F-degrees of the remaining free vertices do not increase. Thus the measure k decreases. (Note that the number of marked vertices and their F-degree can increase by this reduction, but these parameters do not occur in our measure.)

Case (4). If there is a subset B of free vertices such that $G[B]$ induces a complete bipartite graph and no vertex of B is adjacent to a free vertex outside B, then the algorithm branches into two subcases. Let X and Y be the two maximal independent sets of $G[B]$. Then a $MIDS$ contains either X or Y. In both cases we delete B and the marked neighbours of either X or Y. The smallest possible subset B satisfying the conditions of this case is a P_3, i.e. a path of three vertices, as any smaller complete bipartite component in F is handled by case (3). Since we only count the number of free vertices, we obtain the following recurrence:

$$P[k] \leq 2P[k - 2w_1 - w_2]. \tag{1}$$

It is clear that any complete bipartite component with more than three vertices would lead to a better recurrence.

Case (5). If there is a subset C of at least three free vertices which form a clique and only one vertex $v \in C$ has free neighbours outside C, the algorithm either includes v in the solution set or it excludes this vertex by marking it. In the first case, all the neighbours of v are deleted (including C). In the second case, v is marked and the $C - \{v\}$ clique component appears in $G[F]$. Then $C - \{v\}$ will be deleted by the reduction rule of case (3). In both cases, C is deleted and in the first case, the neighbours of v outside C are also deleted (at least one free vertex of F-degree at least one). So we have:

$$P[k] \leq P[k - w_1 - 2w_2 - w_3] + P[k - 2w_2 - w_3]. \tag{2}$$

Case (6). If there is a free vertex u such that $d_F(u) = 1$, a $MIDS$ either includes u or its free neighbour v in the solution. Vertex v cannot have F-degree one because this would have been handled by case (3). For the analysis, we consider two cases:

1. $d_F(v) = 2$. Let x denote the other free neighbour of v. Note that $d_F(x) \neq 1$ as this would have been handled by case (4). We consider again two subcases:
   (a) $d_F(x) = 2$. When u is chosen in the independent dominating set, u and v are deleted and the degree of x decreases to one. When v is chosen in the independent dominating set, u, v and x are deleted from the marked graph. So, we obtain the following recurrence for this case:

$$P[k] \leq P[k - 2w_2] + P[k - w_1 - 2w_2]. \tag{3}$$

   (b) $d_F(x) \geq 3$. Vertices u and v are deleted in the first branch, and u, v and x are deleted in the second branch. The recurrence for this subcase is:

$$P[k] \leq P[k - w_1 - w_2] + P[k - w_1 - w_2 - w_3]. \tag{4}$$

2. $d_F(v) \geq 3$. At least one free neighbour of v has F-degree at least 2. Otherwise case (4) would have been applied. Therefore the recurrence for this subcase is:

$$P[k] \leq P[k - w_1 - w_3] + P[k - 2w_1 - w_2 - w_3]. \tag{5}$$

Case (7). If there is a free vertex u such that $d_F(u) = 2$ and none of the above cases apply, the algorithm branches into three subcases. Let v_1 and v_2 be the two free neighbours of u. Either u belongs to the $MIDS$, or v_1 is taken in the $MIDS$, or v_1 is being marked and v_2 is taken in the $MIDS$. We distinguish two cases:

1. $d_F(v_1) = d_F(v_2) = 2$. In this case, due to the choice of the vertex u by the algorithm, all free vertices of this connected component T in $G[F]$ have F-degree 2. T cannot be a C_4, i.e. a cycle of 4 vertices, as this is a complete bipartite graph and would have been handled by case (4).

(a) Suppose that T is a C_5. Let the vertices of T be ordered (u, v_1, x_1, x_2, v_2). When u is taken in the $MIDS$, u, v_1, v_2 are deleted and in the next recursive call, case (3) is applied for the clique $\{x_1, x_2\}$ and thus, x_1 and x_2 will also be deleted. When v_1 is taken in the $MIDS$, three vertices are again deleted and case (3) will be applied for $\{v_2, x_2\}$. When v_2 is taken in the $MIDS$, $N[v_2]$ is deleted and v_1 becomes marked. In the next recursive call, x_1 will be taken in the $MIDS$ by case (2). In every recursive call, T is entirely deleted:

$$P[k] \leq 3P[k - 5w_2]. \tag{6}$$

(b) Suppose that T is a C_6. Let the vertices of T be ordered $(u, v_1, x_1, y, x_2, v_2)$. When u is taken in the $MIDS$ u, v_1, v_2 are deleted and in the next recursive call, case (4) will be applied for $\{x_1, y, x_2\}$ and thus, the algorithm will branch into two subcases, both deleting x_1, y and x_2. When v_1 is taken in the $MIDS$, three vertices are again deleted and case (4) will be applied for $\{v_2, x_2, y\}$. When v_2 is taken in the $MIDS$, $N[v_2]$ is deleted and v_1 becomes marked. In the next recursive call, x_1 will be taken in the $MIDS$ by case (2). Finally in each of the 5 recursive calls, T is entirely deleted, thus:

$$P[k] \leq 5P[k - 6w_2]. \tag{7}$$

(c) Suppose that T is a C_7. Let the vertices of T be labeled $(u, v_1, x_1, y_1, y_2, x_2, v_2)$ in clockwise order. When u is chosen in the $MIDS$ u, v_1, v_2 are deleted and the F-degrees of x_1, x_2 decrease by one. We obtain a similar situation when branching on v_1: three vertices are deleted and the F-degrees of two vertices decrease to one. When the algorithm chooses v_2 in the $MIDS$, v_1 is marked and x_1 must be added to the $MIDS$ by case (2) and y_2 will then be added by case (3). Consequently, the algorithm deletes the C_7 entirely and we obtain the recurrence:

$$P[k] \leq 2P[k + 2w_1 - 5w_2] + P[k - 7w_2]. \tag{8}$$

(d) Suppose now that T is a C_l, $l \geq 8$. Using the same arguments as in the previous cases, it is not hard to check that we obtain the following recurrence:

$$P[k] \leq 2P[k + 2w_1 - 5w_2] + P[k + 2w_1 - 8w_2]. \tag{9}$$

2. Without loss of generality, suppose now that $d_F(v_1) \geq 3$. We analyze two subcases:
   (a) $d_F(v_2) = 2$. In this subcase, v_1 and v_2 are not adjacent, otherwise case (5) could have been applied. Let x_3 denote the other neighbour of v_2. Recall that due to the choice of u by the algorithm $\forall y \in F$, $d_F(y) \geq d_F(u)$. If $d_F(x_3) = 2$, as previously we branch on u, v_1 and v_2, and we get the following recurrence:

$$P[k] \leq P[k + w_1 - 3w_2 - w_3] + P[k + w_1 - 4w_2 - w_3] + P[k - 3w_2 - w_3]. \tag{10}$$

And if $d_F(x_3) \geq 3$, let q denote the number of vertices in $N_F(v_1)$ with F-degree at least 3. In the worst case $q < 3$ and branching on u, v_1 and v_2, we obtain the following recurrence for $q \in \{0, 1, 2\}$:

$$P[k] \leq P[k + (2 - q)w_1 - (4 - q)w_2 - w_3] +$$
$$P[k + w_1 - (4 - q)w_2 - (1 + q)w_3] + P[k - 2w_2 - 2w_3]. \quad (11)$$

(b) $d_F(v_2) \geq 3$. If v_1 and v_2 are not adjacent, branching on u, v_1 and v_2 leads to the following recurrence:

$$P[k] \leq P[k - w_2 - 2w_3] + P[k - 3w_2 - w_3] + P[k - 3w_2 - 2w_3]. \quad (12)$$

However if v_1 and v_2 are adjacent, let $x_1 \in N_F(v_1) \setminus \{u, v_2\}$. We consider two possible cases:

    i. if $d_F(x_1) = 2$, we obtain:

$$P[k] \leq P[k + w_1 - 2w_2 - 2w_3] + 2P[k - 2w_2 - 2w_3]. \quad (13)$$

    ii. if $d_F(x_1) \geq 3$. Let $x_2 \in N_F(v_2) \setminus \{u, v_1\}$. If $d_F(x_2) = 2$, then:

$$P[k] \leq P[k + w_1 - 2w_2 - 2w_3] + P[k + w_1 - 2w_2 - 3w_3] +$$
$$P[k - 2w_2 - 2w_3]. \quad (14)$$

However if $d_F(x_2) \geq 3$ we get the following recurrence:

$$P[k] \leq P[k - w_2 - 2w_3] + 2P[k - w_2 - 3w_3]. \quad (15)$$

Case (8). In this case the algorithm either takes v in the $MIDS$ or marks it, i.e. v does not belong to the $MIDS$. We consider two cases:

1. $d_F(v) = 3$. In this case, regarding the previous rules handled by the algorithm, every free vertex has degree three. $N_F[v]$ cannot be a clique, otherwise case (3) would have been applied. So, at least two vertices in $N_F(v)$ have a neighbour outside $N_F[v]$ (remark that this could be the same vertex). This implies that if the algorithm takes v in the $MIDS$, the F-degree of at least two free vertices decreases to two in the worst case (if $|N_F(N_F[v])| = 1$ then the decrease of the measure would be higher since $\Delta w_2 + \Delta w_3 \geq 2\Delta w_3$ because of the conditions on the weights). If the algorithm marks v, then three free vertices get F-degree two. The recurrence for this case is:

$$P[k] \leq P[k + 2w_2 - 6w_3] + P[k + 3w_2 - 4w_3]. \quad (16)$$

2. $d_F(v) \geq 4$. When v is taken in the $MIDS$, at least five free vertices are deleted. When v is marked, the measure decreases by w_3. Thus we have this recurrence:

$$P[k] \leq P[k - 5w_3] + P[k - w_3]. \quad (17)$$

Finally the values of weights are computed by a random local search for minimizing the bound on the running time. Using the values $w_1 = 0.8372$ and $w_2 = 0.9644$ for the weights, one can easily verify that $P[k] = O(1.3575^n)$. $\quad \square$

The tight recurrences of the latter proof (i.e. the worst case recurrences) (15) and (16) correspond to cases where there are many vertices of F-degree 3 in the local structure the algorithm considers.

5   A Lower Bound on the Running Time of the Algorithm

In order to analyze the progress of the algorithm during the computation of a *MIDS*, we used a non standard measure. In this way we have been able to determine an upper bound on the size of the subproblems recursively solved by the algorithm, and consequently we obtained an upper bound on the worst case running time. However the use of another measure could provide a "better upper bound" without changing the algorithm but only improving the analysis.

In this section, we establish a lower bound on the worst case running time of our algorithm. This lower bound gives a really good estimation on the precision of the analysis. For example, in [6] Fomin et al. obtain a $O(1.5137^n)$ time algorithm for solving the dominating set problem and they exhibit a construction of a family of graphs giving a lower bound of $\Omega(1.2599^n)$ for its running time. They say that the upper bound of many exponential time algorithms is likely to be overestimated only due to the choice of the measure for the analysis of the running time, and they note the gap between their upper and lower bound for their algorithm. However, for our algorithm we have the following result:

Theorem 2. *Algorithm* **ids** *solves the minimum independent dominating set problem in* $\Omega(1.3247^n)$.

Proof. Due to space restriction we only give a sketch of the proof. A detailed proof will be given in the full version of this paper.

Consider the graph $G_n = (V_n, E_n)$ (see Fig. 1) defined by $V_n = \{u_i, v_i : 1 \leq i \leq n\}$ and $E_n = \{u_1, v_1\} \cup \{\{u_i, v_i\}, \{u_i, u_{i-1}\}, \{v_i, v_{i-1}\}, \{u_i, v_{i-1}\} : 2 \leq i \leq n\}$. We denote by $G'_n = (V, \emptyset, E)$ the marked graph corresponding to the graph $G_n = (V, E)$.

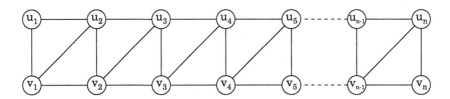

Fig. 1. Graph G_n

It is possible to show that given the graph G'_n as input, as long as the remaining graph has more than four vertices, algorithm **ids** applies case (7) in each recursive call. Moreover, without loss of generality, we can suppose that whenever **ids** would apply case (7), it chooses the vertex with smallest index.

Consider now the graph G'_n and the search tree which results of branchings using case (7) until k vertices, $1 \leq k \leq 2n$, have been removed from the given input graph G'_n (G'_n has $2n$ vertices). Denote by $L[k]$ the number of leaves in this search tree. It is not hard to see that this would lead to the recurrence

$L(k) = L(k-3) + L(k-4) + L(k-5)$ and therefore $L(k) \geq 1.3247^k$. Consequently 1.3247^n is a lower bound of the maximum number of leaves that a search tree for **ids** could give with an input graph on n vertices. □

References

1. Broersma, H., T. Kloks, D. Kratsch, and H. Müller, Independent sets in Asteroidal Triple-free graphs, *SIAM J. Discrete Math.*, **12**, (1999), pp. 276–287.
2. Chang, M.-S., Efficient algorithms for the domination problems on interval and circular-arc graphs, *SIAM J. Comput.*, **27**, (1998), pp. 1671–1694.
3. Corneil, D.-G. and Y. Perl, Clustering and domination in perfect graphs, *Discrete. Appl. Math.*, **9**, (1984), pp. 27–39.
4. Farber, M., Independent domination in chordal graphs, *Operation Research Letters*, **1**, (1982), pp. 134–138.
5. Fomin, F. V., F. Grandoni, and D. Kratsch, Measure and Conquer: A Simple $O(2^{0.288n})$ Independent Set Algorithm, *Proceedings of SODA 2006*, (2006), pp. 18–25.
6. Fomin, F. V., F. Grandoni, and D. Kratsch, Measure and conquer: Domination - A case study, *Proceedings of ICALP 2005*, *LNCS* **3380**, (2005), pp. 192–203.
7. Fomin, F. V., F. Grandoni, and D. Kratsch, Some new techniques in design and analysis of exact (exponential) algorithms, *Bulletin of the EATCS*, **87**, (2005), pp. 47–77.
8. Fomin, F. V., D. Kratsch, and G. J. Woeginger, Exact (exponential) algorithms for the dominating set problem, *Proceedings of WG 2004*, *LNCS* **3353**, (2004), pp. 245–256.
9. Grandoni, F., A note on the complexity of minimum dominating set, *J. Discrete Algorithms*, **4**, (2006), pp. 209–214.
10. Halldórsson, M. M., Approximating the Minimum Maximal Independence Number, *Inf. Process. Lett.*, **46**, (1993), pp. 169–172.
11. Johnson, D. S., M. Yannakakis, and C. H. Papadimitriou, On generating all maximal independent sets, *Inf. Process. Lett.*, **27**, (1988), pp. 119–123.
12. Kratsch, D., and L. Stewart, Domination on Cocomparability Graphs, *SIAM J. Discrete Math.*, **6**, (1993), pp. 400–417.
13. Moon, J. W., and L. Moser, On cliques in graphs, *Israel J. Math.*, **3**, (1965), pp. 23–28.
14. Randerath, B., and I. Schiermeyer, Exact algorithms for Minimum Dominating Set, Technical Report zaik-469, Zentrum fur Angewandte Informatik, Köln, Germany, April 2004.
15. Robson, J. M., Algorithms for maximum independent sets, *J. Algorithms*, **7**, (1986), pp. 425–440.
16. Robson, J. M., Finding a maximum independent set in time $O(2^{n/4})$, Technical Report 1251-01, LaBRI, Université Bordeaux I, 2001.
17. Tarjan, R. E., and A. E. Trojanowski, Finding a maximum independent set, *SIAM J. Comput.*, **6**, (1977), pp. 537–546.
18. Woeginger, G. J., Exact algorithms for NP-hard problems: A survey, *Combinatorial Optimization - Eureka, You Shrink!*, *LNCS* **2570**, (2003), pp. 185–207.

Improved Edge-Coloring with Three Colors

Łukasz Kowalik[1,2]

[1] Institute of Informatics, Warsaw University, Warsaw, Poland
[2] Max-Planck-Institute für Informatik, Saarbrücken, Germany
kowalik@mimuw.edu.pl

Abstract. We show an $O(1.344^n) = O(2^{0.427n})$ algorithm for edge-coloring an n-vertex graph using three colors. Our algorithm uses polynomial space. This improves over the previous, $O(2^{n/2})$ algorithm of Beigel and Eppstein [1]. We extend a very natural approach of generating inclusion-maximal matchings of the graph. The time complexity of our algorithm is estimated using the "measure and conquer" technique.

1 Introduction

In the problem of edge-coloring the input is an undirected graph and the task is to assign colors to the edges so that edges with a common endpoint have different colors. This is one of the most natural graph coloring problems and arises in a variety of scheduling applications.

We consider the problem of verifying whether a given graph G is edge-colorable with k colors and finding such a coloring. Let $\Delta(G)$ denote the maximum degree in graph G. Trivially at least $\Delta(G)$ colors are needed, so if $k < \Delta(G)$ the answer is "no". On the other hand, Vizing [9] proved that when $k \geq \Delta + 1$ the answer is "yes". Unfortunately, when $k = \Delta(G)$ the problem is NP-hard even for $k \geq 3$. In this paper we focus on the simplest NP-hard case: $k = 3$.

One way to solve our problem is to apply a vertex-coloring algorithm to the line graph L of G. The currently fastest algorithm for 3-vertex-coloring a given graph L is due to Beigel and Eppstein [1] and it works in $O(1.3289^{|V(L)|})$ time. For $k = 3$, since $\Delta(G) = 3$, the line graph has at most $\frac{3}{2}n$ vertices (throughout the paper n denotes the number of vertices in the input graph G), hence it yields an $O(1.532^n)$ algorithm. However, for 3-edge-coloring Beigel and Eppstein get an $O(2^{n/2}) = O(1.415^n)$-time algorithm by applying nontrivial preprocessing and their algorithm for $(3, 2)$-CSP problem.

As it was pointed out by Fomin [4], the paper of Fomin and Høie [7] implies an $O(6^{n/6}) = O(1.34801^n)$-time algorithm based on dynamic programming and path decomposition. However, such an algorithm uses exponential space.

In this paper we present a 3-edge-coloring algorithm with time complexity $O(1.344^n) = O(2^{0.427n})$. The space complexity of our algorithm is polynomial (even linear). We apply the "measure and conquer" technique. Its basic idea was introduced by Beigel and Eppstein [1], and further developed by Fomin, Grandoni and Kratsch, who recognized its power and wide applicability – they used it in analysis of very simple algorithms for the minimum dominating set [5]

F.V. Fomin (Ed.): WG 2006, LNCS 4271, pp. 90–101, 2006.
© Springer-Verlag Berlin Heidelberg 2006

and maximum independent set problems [6], obtaining the best known upper bounds on their complexity. In many papers the algorithm consists on identifying one of a large number of possible local configurations in the instance, reducing the configuration in several ways obtaining several smaller instances, and solving the problem in each of them recursively (this is called *branching*). Then the time complexity analysis is rather short and trivial. This situation is reversed in the papers of Fomin, Grandoni and Kratsch [5,6] and also in an earlier paper of Eppstein [2] on TSP in cubic graphs. In these works, the algorithm performs just a few types of different reductions, while the tedious case analysis is moved to the time complexity proof. Our algorithm follows the same approach.

2   The Algorithm and Its Correctness

For the sake of simplicity, we will describe an algorithm for *deciding* whether a given graph is 3-edge-colorable. It is straightforward to extend our description to an algorithm which *finds* a coloring if one exists and which has the same time and space complexity as the decision version. Throughout the paper we will consider only *subcubic* graphs, i.e. graphs with vertices of degree at most three, since any other graph clearly is not 3-edge-colorable.

2.1   Outline

The outline of our algorithm is as follows. Let G be the input graph. Let us call a subcubic graph *semi-cubic* when it has no 1-vertices and each pair of 2-vertices is at distance at least 3. Our algorithm finds a set $\mathcal{A}$ of semi-cubic graphs such that G is 3-edge-colorable if and only if $\mathcal{A}$ contains a 3-edge-colorable graph. Additionally, the graphs in $\mathcal{A}$ have no cycles of length smaller than 5. Generating these graphs is done using branching and $\mathcal{A}$ may have exponential size. Then we make use of the following simple fact. A matching M in graph H is called *fitting* when each connected component of $H - M$ is a path or even length cycle.

Lemma 1. *A graph is 3-edge-colorable iff it contains a fitting matching.*   □

For each of the generated semi-cubic graphs $H \in \mathcal{A}$ the algorithm verifies whether it contains a fitting matching. Again using branching, the algorithm checks a possibly exponential number of (not necessarily fitting) matchings. Then for each of them it can be verified in polynomial time whether it can be completed to a fitting matching.

   The intuition behind the above algorithm is as follows. This is an improvement of the natural algorithm which generates all maximal matchings of the input graph, and for each of them verifies (in polynomial time) whether it is fitting. (This is a counterpart of Lawler's 3-vertex-coloring algorithm [8]). Unfortunately, the number of maximal matchings may be large. However, for cubic graphs one can observe that every fitting matching is a perfect matching, and the number of perfect matchings is much lower than the number of maximal matchings. A matching in subcubic graph will be called *semi-perfect* when every 3-vertex is matched. Clearly, every fitting matching in semi-cubic graph is semi-perfect. Again, in semi-cubic graphs there are much fewer semi-perfect matchings than maximal matchings. The only problem is that the input graph is subcubic but not necessarily semi-cubic. However, it turns out that if a graph G contains a pair of 2-vertices at distance 2 (hence it is not semi-cubic), then there are two graphs such that G is 3-edge-colorable if and only if at least one of the two graphs is edge-colorable, and, what is very important, these two graphs are much smaller than G (i.e. work factor is large – see Section 3). Similarly, graphs with no 4-cycles have fewer semi-perfect matchings and one can get rid of these cycles using another large work factor reduction. That is why the set $\mathcal{A}$ is generated. To reduce the time complexity even further we notice that generating all semi-perfect matchings of semi-cubic graphs in $\mathcal{A}$ is not needed. Instead, we generate all matchings with some nice structure so that verifying whether they extend to a fitting semi-perfect matching takes only polynomial time.

2.2   Generating Semi-cubic Graphs

In this section we describe the part of our algorithm which generates a set $\mathcal{A}$ of semi-cubic graphs with neither 3- nor 4-cycles and such that the input graph is 3-edge-colorable iff $\mathcal{A}$ contains a 3-edge-colorable graph. This part is implemented as a recursive procedure EDGECOLOR – see Pseudocode 2.1. After finding each graph $H \in \mathcal{A}$, the algorithm calls function FITTINGM, described in the next subsection, which verifies whether H contains a fitting matching. Note that the input graph in procedure EDGECOLOR is allowed to have double and triple edges. This simplifies the correctness proof (one does not need to care about keeping the graph simple during reductions) but also makes our result more general.

Lemma 2. *Consider an execution of algorithm* EDGECOLOR *on an input graph* G. *Let* $\mathcal{A}$ *be the set of all graphs* H *such that* FITTINGM$(H,H,\emptyset)$ *was executed. Then* G *is 3-edge-colorable iff at least one graph in* $\mathcal{A}$ *is 3-edge-colorable.*

Proof. It suffices to prove that whenever EDGECOLOR(G) performs "**return** EDGECOLOR(G_1)" / "**return** EDGECOLOR(G_1) **or** EDGECOLOR(G_2)" then G_1 / both G_1 and G_2 are subcubic multigraphs with no self-loops, and G is 3-edge-colorable if and only if G_1 is edge colorable / at least one of graphs G_1, G_2 is edge colorable.

For an example consider the case in line 22. (The easy proofs of the other cases are left to the reader.) Let $G_1 = G - V(C) + \{x'y', u'z'\}$ and $G_2 = G - V(C) + \{x'u', y'z'\}$. G_1 and G_2 may have double edges $x'y'$, $u'z'$, $x'u'$ or $y'z'$. However, in this proof when we refer to these edges we mean the edges added to G after removing $V(C)$. Clearly both G_1 and G_2 are subcubic multigraphs. Also, both G_1 and G_2 have no self-loops since the condition in line 13 was false.

Assume there is a 3-edge-coloring of G. We will show that one of G_1, G_2 is 3-edge-colorable. Then cycle C is either colored with two or three colors. In the first case all the four edges xx', yy', zz' and uu' have the same color, say a. Hence one gets a 3-edge-coloring of G_1 by copying the colors of edges in $E(G) \cap E(G_1)$

Pseudocode 2.1. Procedure EDGECOLOR(G)

Input: subcubic multigraph G with no self-loops.
Output: TRUE if G is 3-edge-colorable, FALSE otherwise.

 1: **if** exists $v \in V(G)$ such that $\deg(v) \in \{0, 1\}$ **then**
 2:     **return** EDGECOLOR($G - v$)
 3: **else if** exists $uv \in E(G)$ such that $\deg(u) = \deg(v) = 2$ **then**
 4:     **return** EDGECOLOR($G - uv$)
 5: **else if** G contains a triple edge uv **then**
 6:     **return** EDGECOLOR($G - \{u, v\}$)
 7: **else if** G contains a double edge uv **then**
 8:     **if** $\deg(u) = 2$ or $\deg(v) = 2$ **then**
 9:         **return** EDGECOLOR($G - \{u, v\}$)
10:     **else**
11:         Let u_1 (resp. v_1) be the neighbor of u (resp. v) distinct from v (resp. u)
12:         **if** $u_1 = v_1$ **then return** FALSE **else return** EDGECOLOR($G - \{u, v\} + u_1 v_1$)
13: **else if** exists a 3-cycle C **then**
14:     Let G' be the graph obtained from G by contracting $V(C)$ into one vertex.
15:     **return** EDGECOLOR(G')
16: **else if** exists a path $xuzvy$ (possibly $x = y$) such that $\deg(u) = \deg(v) = 2$ **then**
17:     $z' \leftarrow$ the neighbor of z distinct from u and v ▷ (Note that $\deg(z) = 3$)
18:     **return** EDGECOLOR($G - \{z, v\} + uy$) **or** EDGECOLOR($G - \{u, z, v\} + xz'$)
19: **else if** exists a 4-cycle $C = xyzu$ with $\deg(x) = \deg(y) = \deg(z) = 3$, $\deg(u) = 2$
      then
20:     Let x' (resp. y', z') be the neighbor of x (resp. y, z) outside the cycle
21:     **return** EDGECOLOR($G - V(C) + x'y'$) **or** EDGECOLOR($G - V(C) + z'y'$)
22: **else if** exists a 4-cycle $C = xyzu$ with $\deg(x) = \deg(y) = \deg(z) = \deg(u) = 3$
      then
23:     Let x' (resp. y', z', u') be the neighbor of x (resp. y, z, u) outside the cycle
24:     **return** EDGECOLOR($G - V(C) + \{x'y', u'z'\}$) **or** EDGECOLOR($G - V(C) + \{x'u', y'z'\}$)
25: **else**   ▷ G is semi-cubic and has no 3-, 4-cycles
26:     **return** FITTINGM($G, G, \emptyset$)

---

from G and coloring both edges $x'y'$ and $u'z'$ with color a. In the second case one color appears in $E(C)$ twice, and each of the two other once. By symmetry, we can assume w.l.o.g that edges xy, yz, zu, ux have colors b, a, b, c. Then both yy' and zz' have color c and both uu' and xx' have color a. Hence one gets a 3-edge-coloring of G_2 by copying the colors of edges in $E(G) \cap E(G_2)$ from G, coloring edge $y'z'$ with color c and $u'x'$ with color a.

Now assume G_1 is 3-edge-colorable (by symmetry there is no need for checking G_2 separately). Now we show how to color G. The common edges of G_1 and G inherit their colors from G_1. If edges $x'y'$ and $u'z'$ have the same color, say a, then in G edges xx', yy', zz' and uu' are colored with a, xy and zu with b, and yz and ux with c. Finally assume that $x'y'$ has color a and $u'z'$ has color b. Then xx', yy' and zu are colored with a, while zz', uu' and xy are colored with b, and both xu and yz are colored with c.   □

2.3 Finding a Fitting Matching

In this section we describe a recursive procedure FITTINGM (G_0, G, M). The parameters G and G_0 are simple semi-cubic graphs, and $G \subseteq G_0$. The parameter M is a matching in G_0 such that $V(M) \cap V(G) = \emptyset$ and every 3-vertex in $V(G_0) - V(G)$ is matched. The procedure verifies whether G_0 contains a fitting matching $M' = M \cup N$ such that $N \subseteq E(G)$.

We will use the following auxiliary definitions. Any vertex in G which has degree 3 in G_0 will be called *forced*. A *switch* is a 4-path P in G such that P forms a connected component in G and the two inner vertices of P are forced, while the endvertices are not forced. An edge e in G will be called *allowed* if both of its endvertices are forced and e does not belong to a switch. The *weight* of an edge is the sum of degrees of its endvertices.

Pseudocode 2.2. Procedure FITTINGM(G_0,G,M)

1: **if** every connected component of G is a switch **then**
2:     **return** SETSWITCHES(G_0,G,M)   ▷ see the next section
3: **else if** exists a forced vertex $v \in V(G)$ such that $\deg_G(v) = 0$ **then**
4:     **return** FALSE
5: **else if** exists a non-forced vertex $v \in V(G)$ such that $\deg_G(v) = 0$ **then**
6:     **return** FITTINGM(G_0, $G - \{v\}$, M)
7: **else if** exists a forced vertex $v \in V(G)$ such that $\deg_G(v) = 1$ **then**
8:     $u \leftarrow$ the neighbor of v in G
9:     **return** FITTINGM(G_0, $G - \{u, v\}$, $M \cup \{uv\}$)
10: **else**
11:     $uv \leftarrow$ an allowed edge with the highest weight in G (it exists, see proof of Th. 1)
12:     **return** FITTINGM(G_0, $G - \{u, v\}$, $M \cup \{uv\}$) **or** FITTINGM(G_0, $G - uv$, M)

2.4 Setting Switches

Lemma 3. *Let G_0 be a simple semi-cubic graph, and let G be a subgraph of G_0 in which each connected component is a switch. Let M be a matching in G_0 such that $V(M) \cap V(G) = \emptyset$ and every 3-vertex in $V(G_0) - V(G)$ is matched. Then procedure SETSWITCHES (G_0, G, M) verifies whether G_0 contains a fitting matching M' such that $M' = M \cup N$ for some $N \subseteq E(G)$.*

Let us assume that G_0, G and M satisfy the assumptions of the above lemma. Let $s = xuvy$ be a switch and let M' be a semi-perfect matching in G_0 such that $M' = M \cup N$ for some $N \subseteq E(G)$. Observe that either $uv \in M'$ (then we will say that the switch is *closed* in M') or $xu, vy \in M$ (the switch is *open* in M'). Let $M'' = M' \oplus E(s)$, where $\oplus$ denotes the xor operation. Clearly, then M'' is a semi-perfect matching (recall that $V(M) \cap V(G) = \emptyset$). Also, if s was open in M', it is closed in M'' and vice versa. Let C be a cycle in $G_0 - M'$ and let $s = xuvy$ and $s' = x'u'v'y'$ be a pair of closed switches such that $V(s) \subseteq V(C)$ and $V(s') \subseteq V(C)$ and let P be a path in C between x and y. When P contains exactly one of the vertices x', y', we will say that the switches s, s' are *crossing*.

We will say that a set of switches S *improves* matching M' when for matching $M'' = M' \oplus \bigcup_{s \in S} E(s)$ graph $G_0 - M''$ has fewer odd cycles than $G_0 - M'$ or the total length of odd cycles in $G_0 - M''$ is larger than in $G_0 - M'$.

Pseudocode 2.3. Procedure SETSWITCHES(G_0,G,M)

1: $M' \leftarrow M$
2: **for each** switch $s = xuvy$ in G **do**
3:     $M' \leftarrow M' \cup \{xu, vy\}$ ▷ Set all the switches as open
4: **while** $G_0 - M'$ contains an odd cycle C **do**
5:     **if** there is a switch s or pair of crossing switches s_1, s_2 which improve M' **then**
6:         $M' \leftarrow M' \oplus E(s)$, (resp. $M' \leftarrow M' \oplus (E(s_1) \cup E(s_2))$
7:     **else**
8:         **return** FALSE
9: **return** TRUE

Proof (of Lemma 3). Assume that the procedure returned TRUE. Note that after the loop in lines 2–3 is performed M' is a semi-perfect matching. Hence $G_0 - M'$ is a collection of paths and cycles. Since the condition in line 4 was eventually not satisfied all the cycles in $G_0 - M'$ are even, hence M' is fitting. The condition that M' is an extension of M using edges of G is also trivially satisfied.

Now assume that FALSE is returned. Then there is an odd cycle C in $G_0 - M'$ and no switch/pair of crossing switches which improve a current matching M'.

Claim 1: Any switch $s = xuvy$ such that $V(s) \cap V(C) \neq \emptyset$ is closed and $V(s) \subseteq V(C)$.

First assume that s is open. Then $x, y \notin V(C)$ since they are of degree 2 in G_0 and they are matched by M'. Hence u or v is in $V(C)$. As $uv \notin M'$ it implies that both u and v are in $V(C)$. The connected component of $G_0 - M'$ containing x is a path, similarly for y. If it is just one path with x and y as endpoints, then either $M'' = M' \oplus E(s)$ has 1 odd cycle fewer than M' (when the path is of even length) or M'' has larger total length of odd cycles (when it is odd). This is a contradiction. Similarly, when these two paths are distinct, $M'' = M' \oplus E(s)$ has one odd cycle fewer than M' (C transforms into a path), a contradiction again.

Hence s is closed. Then $x, u \in V(C)$ or $v, y \in V(C)$. Assume w.l.o.g. $x, u \in V(C)$. Now assume that $v \notin V(C)$. Then the connected component of $G_0 - M'$ containing u (i.e. cycle C) is distinct from the connected component K_v of $G_0 - M'$ containing v. Hence, in $G_0 - [M' \oplus E(s)]$ cycle C is replaced by a path. If K_v is a cycle, then operation $M' \oplus E(s)$ splits it into a path. Also, if K_v is a path, it splits into two paths. Hence the number of odd cycles in $G_0 - [M' \oplus E(s)]$ is smaller than it was in $G_0 - M'$, a contradiction. Hence we are left with the case $v \in V(C)$. Then also $y \in V(C)$. This establishes the claim.

Let T be the set of all the switches that touch C. Now assume that there exists a fitting matching F in G_0 such that $F = M \cup N$ and $N \subseteq E(G)$, contradicting our lemma. Hence there is a set of switches S such that $M' \oplus \bigcup_{s \in S} E(s) = F$. Observe that $S \cap T \neq \emptyset$, for otherwise $G_0 - F$ contains the odd cycle C. Since among the switches in T there is no crossing pair, we can enumerate switches in $S \cap T$ from s_1 to $s_{|S \cap T|}$ so that after removing the forced vertices of any switch in $S \cap T$, the cycle C splits into two parts such that one part contains all the switches with smaller numbers and the other with larger numbers (informally, we enumerate the switches from left to right). Since each switch s in T has all vertices in C, after performing the operation $M' \oplus E(s)$ cycle C either splits into two paths (which is excluded, because then s improves M') or into a path and an odd cycle. Let us denote this resulting shorter cycle by $C(s)$. Observe that cycle $C(s)$ either contains all the switches preceding s (then $C(s)$ will be called preceding), or all the switches succeeding s (then $C(s)$ will be called succeeding). If every switch s in $T \cap S$ has cycle $C(s)$ succeeding then the cycle $C(s_{|S \cap T|})$ is odd, and it exists in $G_0 - F$, hence F is not fitting, a contradiction. Let $s_i \in T \cap S$ be a switch with preceding cycle $C(s_i)$ such that for any $j < i$, cycle $C(s_j)$ is succeeding. If $i = 1$ then the cycle $C(s_1)$ is odd, and it exists in $G_0 - F$, hence F is not fitting, a contradiction. Hence $i > 1$. Let us denote the forced vertices of s_i by u_i, v_i, and the forced vertices of s_{i-1} by u_{i-1}, v_{i-1} in such a way that they appear around cycle C in the order $u_i, u_{i-1}, v_{i-1}, v_i$. Let a, b, c and d denote the length of the path in C joining u_i with u_{i-1}, u_{i-1} with v_{i-1}, v_{i-1} with v_i, and v_i with u_i, respectively (there are always two such paths, but we mean the path which does not contain the other two forced vertices of switches s_{i-1}, s_i). Then $a + b + c + 1 \equiv 1 \pmod 2$ and $a + c + d + 1 \equiv 1 \pmod 2$, since the cycles $C(u_i)$ and $C(u_{i-1})$ have odd length. It implies that $b + d \equiv 0 \pmod 2$. As C has odd length, $a + b + c + d \equiv 1 \pmod 2$. Hence $a + c \equiv 1 \pmod 2$ and so $a + c + 2$ is odd. However, $a + c + 2$ is the length of a cycle that appears after performing operation $M' \oplus (E(s_{i-1}) \cup E(s_i))$. This cycle exists also in $G_0 - F$, which is a contradiction. It ends the proof. □

Theorem 1. *Algorithm* EDGECOLOR *correctly verifies whether a given subcubic graph is 3-edge-colorable.*

Proof. By Lemma 2 and Lemma 1 it suffices to show that FITTINGM$(G,G,\emptyset)$ correctly verifies whether the semi-cubic graph G has a fitting matching. Using Lemma 3 for the base case it easy to show by induction on $|V(G)|$ that FITTINGM(G_0,G,M), with parameters satisfying assumptions of Lemma 3, verifies whether G_0 contains a fitting matching $M' = M \cup N$ for some $N \subseteq E(G)$. The only unclear issue is why in line 11 graph G must contain an allowed edge. To see this consider any connected component K of G which is not a switch. (It exists since the condition in line 1 was false.) Let v be any vertex in K. If v is not forced then since the condition in line 5 was false v has at least one neighbor, say w, which must be forced since G_0 is semi-cubic. As the conditions in lines 3 and 7 were false $\deg_G(w) \geq 2$. Since G_0 is semi-cubic at least one of the neighbors of w, say u, is forced. Hence edge uw is allowed. It ends the proof. □

3   Time Complexity Analysis

Solving Recurrences. Our algorithm uses the "branch-and-reduce" approach, which means that it is recursively applied to the problem instance and uses two types of *rules. Reduction rules*, (see e.g. lines 1-15 of EDGECOLOR procedure) simplify the instance. *Branching rules* (see e.g. lines 18, 21 of EDGECOLOR) also simplify the instance, but in several ways, generating several smaller instances in such a way that the initial instance (graph) is 3-edge-colorable iff one of the simplified ones is. Then the problem is solved recursively for each of the smaller instances. Hence execution of procedure EDGECOLOR is a traversing of a recursion tree, with nodes corresponding to single calls of procedures EDGECOLOR, FITTINGM and SETSWITCHES. Reducing rules correspond to nodes with only one child, hence their number is only polynomially larger than the number of nodes corresponding to branching rules. It follows that in order to bound the time complexity up to a polynomial factor it suffices to focus on branching rules. Then for each branching rule we act like it was the only rule in the algorithm. Consider a branching rule generating smaller instances I_1, I_2 (there are never more ones in our algorithm) such that the size of I_1, I_2 is smaller than the initial instance by r_1 and r_2, respectively. This leads to a recurrence of the form $T(s) = T(s - r_1) + T(s - r_2)$, whose solution is the unique positive zero of the function $f(x) = 1 - x^{-r_1} - x^{-r_2}$. This zero will be called a *work factor* and denoted as $\lambda(r_1, r_2)$. After finding the zeroes of all the functions corresponding to branching rules we choose the largest one, say λ. Then the time complexity of the algorithm is $O(\lambda^n p(n))$ for some polynomial p. The intuition behind it is that in the worst case the "weakest branching rule" may apply in all nodes of the search tree. In this paper we use only numerically obtained (not sharp) upper bounds of the work factors. Hence we can omit the polynomial factor in time complexity, since $\lambda^n p(n) = O((\lambda + \epsilon)^n)$ for any $\epsilon > 0$.

Analysis Via Measure and Conquer. We apply the "measure and conquer" approach for estimating the number of nodes of search tree. This approach consists in using a carefully selected measure of the size of an instance of our problem. For example, a natural measure of the instance size in our case is the number of vertices of the graph. However, it is clear that in procedure FITTINGM, a forced vertex of degree 1 should not be counted into the instance size, because it disappears from the graph in polynomial time. We can go further: intuitively, forced 2-vertex x should contribute less to the instance size than a 3-vertex, because there are only two choices: either one or the other edge incident with x belongs to a fitting matching (or, the 2-vertex is closer to becoming a 1-vertex, which does not contribute to the size). This suggests using weights of vertices, and defining the size of the instance as the sum of vertices weights. Clearly, the time complexity (solution of a relevant recurrence) depends heavily on the size measure used. The weights are chosen in a way which minimizes the time complexity, i.e., the largest work factor. This was done by quasi-convex programming algorithm due to Eppstein [3] (a C++ implementation can be downloaded from the author's website http://www.mimuw.edu.pl/~kowalik/papers).

Theorem 2. *Algorithm* EDGECOLOR *works in* $O(1.344^n)$ *time.*

Proof. Let G be the graph passed to procedure EDGECOLOR or FITTINGM. Vertex in G may be in one of the following states. Either it is *unmarked*, or it is *marked as forced* or it is *marked as unforced*. We assume that in the input graph all the vertices are unmarked, while in the moment of calling procedure FITTINGM, all the vertices of degree 3 are marked as forced and all the other are marked as unforced. Now we define a non-standard measure $s(G)$ of the size of G as the sum of weights of vertices, which are assigned as follows. Forced 2-vertices have weight α and unforced 1-vertices have weight β. Values of these parameters will be adjusted later; at this moment let us merely put a bound $0 \leq \alpha, \beta \leq 1$. Isolated vertices and forced 1-vertices have weight 0. All the remaining vertices — i.e. unmarked vertices, forced 3-vertices and unforced 2-vertices — have weight 1. Note that the size of the instance passed to procedure EDGECOLOR is simply the number of non-isolated vertices. We observe that when there is no branching, i.e. a procedure calls another procedure just once, the size of the instance does not increase.

Now for each possible branching rule in our algorithm we are going to determine its work factor. Some of the work factors will depend on the values of parameters α and β. After identifying all work factors we will set these parameters in such a way that the largest work factor is minimized.

First we focus on branchings in procedure EDGECOLOR. There are three of them, in lines 18, 21 and 24. They have the following work factors: $\lambda(2, 3) \leq 1.325$ for the first one and $\lambda(4, 4) \leq 1.190$ for the two latter ones.

Now we consider the branching rule in procedure FITTINGM (in line 12). This will require more involved analysis. Let uv be the edge picked by the algorithm (recall it is the heaviest allowed edge in G). Assume w.l.o.g. that $\deg(u) \leq \deg(v)$.

Case 1. $\deg(u) = \deg(v) = 3$. Consider the graph with vertices u and v removed. As graph G_0 is semi-cubic at most one neighbor of u is unforced (analogously for v). An unforced neighbor decreases its weight either from 1 to β (when it has degree 2) or from β to 0 (when it is a 1-vertex). A forced 3-neighbor decreases its weight from 1 to α. Any forced 2-neighbor becomes forced 1-vertex and hence it will be matched with its unique neighbor before the next branching happens. This unique neighbor has weight α, β, or 1. Then both of them are removed from the graph so the size decreases by at least $\alpha + \min\{\alpha, \beta\}$. The last statement is not true when two neighbors of u and v are forced 2-neighbors with a common neighbor, but then the relevant call of FITTINGM returns FALSE in polynomial time (without performing any branching). The other possibilities of counting some weight reduction twice are excluded because of the lack of 3- and 4-cycles. It follows that the size of the instance is reduced by at least $2 + 2\min\{1 - \alpha, \alpha + \min\{\alpha, \beta\}\} + 2\min\{\beta, 1 - \beta, 1 - \alpha, \alpha + \min\{\alpha, \beta\}\}$. On the other hand, when only the edge uv is removed, the size reduces by $2(1 - \alpha)$. Hence we get the following upper bound on the work factor: $\lambda(2 + 2\min\{1 - \alpha, \alpha + \min\{\alpha, \beta\}\} + 2\min\{\beta, 1 - \beta, 1 - \alpha, \alpha + \min\{\alpha, \beta\}\}, 2(1 - \alpha))$.

Case 2. $\deg(u) < 3$. Since uv was an allowed edge, u is forced. As reduction rules were not applied, $\deg(u) = 2$. Consider the maximal path P of forced 2-vertices containing u. Observe that in each of the two recursive calls all the vertices of P are matched and removed from graph G without branching (in $O(|V(P)|)$ time).

Case 2.1 P is a cycle. If $|V(P)|$ is odd then within $O(|V(P)|)$ steps the function returns FALSE, because there appears a forced 0-vertex. If $|V(P)|$ is even $|V(P)| \geq 6$, since the graph contains no 4-cycles. Then in both of the recursive calls the size is reduced by at least 6α, which gives the work factor $\lambda(6\alpha, 6\alpha)$.

Case 2.2 P is a simple path, $P = v_1 \cdots v_k$. Let x and y be the neighbors of v_1 and v_k outside P, respectively. Assume w.l.o.g. that $\deg(x) \geq \deg(y)$.

Case 2.2.1. $|V(P)| \geq 3$.

Case 2.2.1.1. $\deg(x) = 1$ and $\deg(y) = 1$. In one of the two recursive calls x is matched with v_1 and v_2 with v_3. These vertices disappear, reducing the size by $\beta + 3\alpha$. The neighbor of v_3 distinct from v_2 is either y (and then it is removed as an unforced 0-vertex) or v_4 (and then it is matched with its another neighbor). This gives us further reduction in size by at least $\min\{\beta, \alpha + \min\{\alpha, \beta\}\}$. In the other recursive call v_1 is matched with v_2, and v_3 is matched with its another neighbor (either v_4 or y). A reasoning similar as before shows that the reduction in size is also $\beta + 3\alpha + \min\{\beta, \alpha + \min\{\alpha, \beta\}\}$. To sum up, this case has work factor $\lambda(\beta + 3\alpha + \min\{\beta, \alpha + \min\{\alpha, \beta\}\}, \beta + 3\alpha + \min\{\beta, \alpha + \min\{\alpha, \beta\}\})$.

Case 2.2.1.2. $\deg(x) \geq 2$. Either x is of degree 2 and then it is unforced by the definition of P or x is of degree 3 and then it must be forced. In both cases x has weight 1. First consider the recursive call where x is matched with v_1 and v_2 with v_3. Because cycles have length at least 5 neighbors of x are distinct from v_2, v_3. Consider such a neighbor $\tilde{x}$. Then $\tilde{x}$ reduces its weight either by β (if $\tilde{x}$ is unforced 1-vertex) or by $1 - \beta$ (if $\tilde{x}$ is unforced 2-vertex) or by $1 - \alpha$ (if $\tilde{x}$ is forced 3-vertex) or by α (if $\tilde{x}$ is forced 2-vertex). Hence the instance size decreases by at least 1 (for x) plus 3α (for v_1, v_2, v_3) plus $\min\{\beta, 1 - \beta, \alpha, 1 - \alpha\}$ (for $\tilde{x}$).

Now consider the other recursive call, with v_1 matched with v_2 and v_3 matched with its neighbor distinct from v_2, say $\tilde{v}_4$. The weight of x decreases either by $1 - \beta$ (when x is of degree 2) or by $1 - \alpha$ (when x is of degree 3). As before, vertices v_1, v_2, v_3 are matched which causes reduction in size of 3α. Vertex $\tilde{v}_4$ can be unforced 1- or 2-vertex or forced 3-vertex, hence it has weight at least $\min\{1, \beta\} = \beta$. To sum up, in this recursive call within $O(|V(P)|)$ steps the instance reduces its size by $\geq \min\{1 - \beta, 1 - \alpha\} + 3\alpha + \beta$ without branching.

Let us write down the work factor for this case: $\lambda(1 + 3\alpha + \min\{\beta, \alpha, 1 - \beta, 1 - \alpha\}, \min\{1 - \beta, 1 - \alpha\} + 3\alpha + \beta)$.

Case 2.2.2. $|V(P)| = 2$. Since P is not a part of a switch and there are no forced 1-vertices, at least one vertex of x, y is of degree ≥ 2.

Case 2.2.2.1. $\deg(x) = 2$ and $\deg(y) = 1$. Observe that x is unforced by the definition of P and y is unforced because forced 1-vertices are excluded. It follows that weights of x and y are 1 and β, respectively. It also implies that $\{u, v\} = \{v_1, v_2\}$. Hence allowed edges have both ends of degree 2, for otherwise the algorithm would not choose edge uv. Let z be the neighbor of x outside P. Vertex z is forced, since G_0 is semi-cubic. Hence $\deg(z) \geq 2$. The neighbors of z distinct from x are also forced, again because G_0 is semi-cubic. It follows that edges joining z and its neighbors distinct from x are allowed. Consequently z has degree 2 and its only neighbor distinct from x, say $\tilde{z}$ is also of degree 2.

In one of the recursive calls before any branching is performed, v_2 is matched with y, v_1 with x and z with $\tilde{z}$, which gives reduction in instance size of $\alpha + \beta + \alpha + 1 + 2\alpha$. In the other recursive call v_1 is matched with v_2, x reduces its weight from 1 to β and y is removed as an unforced 0-vertex, which gives reduction of $2\alpha + (1 - \beta) + \beta$. Hence we get the following work factor: $\lambda(4\alpha + \beta + 1, 2\alpha + 1)$.

Case 2.2.2.2. $\deg(x) = 3$ and $\deg(y) = 1$. Then x has weight 1 and y has weight β. Note that $\{u, v\} = \{x, v_1\}$. Hence neither of the neighbors of x is a 3-vertex, for otherwise the algorithm could choose an allowed edge with larger weight. In one of the recursive calls, v_2 is matched with y and v_1 with x. Moreover, both neighbors of x distinct from v_1 reduce their weight. The reduction is from α to 0 if such a neighbor was a forced 2-vertex, from 1 to β if it was an unforced 2-vertex and from β to 0 if it was an unforced 1-vertex. Hence the instance size reduction is $\alpha + \beta + \alpha + 1 + 2\min\{\alpha, 1 - \beta, \beta\}$. In the other recursive call, v_1 is matched with v_2, y is removed and x decreases its weight from 1 to α, which gives a size reduction of $2\alpha + \beta + (1 - \alpha)$. We get the following work factor: $\lambda(2\alpha + \beta + 1 + 2\min\{\alpha, 1 - \beta, \beta\}, \alpha + \beta + 1)$.

Case 2.2.2.3. $\deg(x) \geq 2$ and $\deg(y) \geq 2$. Then neither x nor y is a forced 2-vertex, for otherwise P would be longer. Hence both x and y have weight 1.

First we consider the recursive call where v_1 is matched with x and v_2 with y. Note that either one of x, y is of degree 3 and then it has two neighbors outside P which reduce their weights or both x and y have degree 2, each of them has a neighbor outside P which reduces its weight and these neighbors are distinct, because G_0 is semi-cubic. Hence in both cases there are 2 vertices which reduce their weights. It follows that the total reduction of size in this recursive call is $2 + 2\alpha + 2\min\{\beta, 1 - \beta, \alpha, 1 - \alpha\}$. In the other recursive call v_1 is matched with v_2 and the weight of both x and y is reduced from 1 to either α or β. It implies that the instance reduces its size by $2\alpha + 2\min\{1 - \beta, 1 - \alpha\}$. We get the following work factor: $\lambda(2 + 2\alpha + 2\min\{\beta, 1 - \beta, \alpha, 1 - \alpha\}, 2\alpha + 2\min\{1 - \beta, 1 - \alpha\})$.

Case 2.2.3. $|V(P)| = 1$. Since G_0 is semi-cubic and P is maximal, at least one of vertices x and y has degree 3. Hence $\deg(x) = 3$. Let z_1, z_2 be the neighbors of x outside P. Note that w.l.o.g. the edge uv picked by the algorithm is equal to xv_1. Since it is the heaviest edge, $\deg(z_1), \deg(z_2) \leq 2$. As G_0 is semi-cubic, at least one of z_1, z_2 (say, z_2) is forced. Hence $\deg(z_2) = 2$. Let w be the neighbor of z_2 distinct from x.

In one of the recursive calls, v_1 is matched with x, z_2 with w and both y and z_1 reduce their weight. Let $\text{red}_1(y)$ denote the reduction of weight of y in this recursive call. Vertex z_1 reduces its weight either from 1 to β (when it is unforced 2-vertex) or from α to 0 (when it is forced 2-vertex) or from β to 0 (when it is unforced 1-vertex). Hence it reduces its weight by at least $\min\{1 - \beta, \beta, \alpha\}$. Let $\text{weight}(w)$ denote the weight of w. Then the size of the instance reduces by at least $\alpha + 1 + \alpha + \text{weight}(w) + \text{red}_1(y) + \min\{1 - \beta, \beta, \alpha\}$.

In the other recursive call, v_1 is matched with y and x reduces its weight from 1 to α. Let $\text{weight}(y)$ denote the weight of y (it is either 1 or β). Then the size of the instance reduces by $\alpha + \text{weight}(y) + (1 - \alpha) = \text{weight}(y) + 1$.

Case 2.2.3.1. $\deg(y) = 1$ and $\deg(w) = 1$. Then $\text{weight}(y) = \beta$, $\text{red}_1(y) = \beta$ and $\text{weight}(w) = \beta$. We get the work factor: $\lambda(2\alpha + 2\beta + 1 + \min\{1 - \beta, \beta, \alpha\}, \beta + 1)$.

Case 2.2.3.2. $\deg(y) = 1$ and $\deg(w) \geq 2$. Then $\text{weight}(y) = \beta$, $\text{red}_1(y) = \beta$ and $\text{weight}(w) \in \{1, \alpha\}$. If $\text{weight}(w) = \alpha$, i.e. w is a forced 2-vertex, then in the first recursive call considered by us, the neighbor of w distinct from z_2 reduces its weight (by at least $\min\{1 - \beta, 1 - \alpha, \beta, \alpha\}$). Note that since there are no cycles shorter than 5 and y is of degree 1, this neighbor of w is neither of vertices z_1, v_1, y. Hence in the first recursive call either w reduces its weight from 1 to 0 or w reduces its weight from α to 0 and its neighbor reduces its weight by $\min\{1 - \beta, 1 - \alpha, \beta, \alpha\}$. In any case we get reduction of at least $\min\{1, \alpha + \min\{1 - \beta, 1 - \alpha, \beta, \alpha\}\} = \alpha + \min\{1 - \beta, 1 - \alpha, \beta, \alpha\}$. This gives the following work factor: $\lambda(3\alpha + \beta + 1 + \min\{1 - \beta, \beta, \alpha\} + \min\{1 - \beta, 1 - \alpha, \beta, \alpha\}, \beta + 1)$.

Case 2.2.3.3. $\deg(y) \geq 2$. Maximality of P implies that y is not a forced 2-vertex. Then $\text{weight}(y) = 1$, $\text{red}_1(y) \in \{1 - \alpha, 1 - \beta\}$ and $\text{weight}(w) \in \{\alpha, \beta, 1\}$. This gives a work factor: $\lambda(2\alpha + 1 + \min\{\alpha, \beta, 1\} + \min\{1 - \alpha, 1 - \beta\} + \min\{1 - \beta, \beta, \alpha\}, 2)$.

We numerically obtained the following values of parameters: $\alpha = 0.39082$ and $\beta = 0.58623$. For these values one can easily check (by finding zeroes of the 11 polynomials corresponding to cases 1 – 2.2.3.3) that the highest work factors correspond to Case 1, Case 2.1 and Case 2.2.1.1. and are bounded by 1.344. This implies that the algorithm works in time $O(1.344^{s(G)})$, where G is the input graph. This settles the theorem, since $s(G) = n$. $\square$

Acknowledgments. The author wishes to thank anonymous referees for careful reading and helpful suggestions.

References

1. R. Beigel and D. Eppstein. 3-coloring in time $O(1.3289^n)$. *J. Algorithms*, 54(2):168–204, February 2005.
2. D. Eppstein. The traveling salesman problem for cubic graphs. In *Proc. 8th Int. Workshop on Algorithms and Data Str. (WADS'03)*, volume 2748 of *LNCS*, pages 307–318, 2003.
3. D. Eppstein. Quasiconvex analysis of backtracking algorithms. In *Proc. 15th Annual ACM-SIAM Symposium on Discrete Algorithms (SODA'04)*, pages 781–790, 2004.
4. F. Fomin. Personal communication, 2006.
5. F. Fomin, F. Grandoni, and D. Kratsch. Measure and conquer: Domination – a case study. In *Proc. 32nd International Colloquium on Automata, Languages and Programming (ICALP'05)*, pages 191–203, 2005.
6. F. Fomin, F. Grandoni, and D. Kratsch. Measure and conquer: A simple $O(2^{0.288n})$ independent set algorithm. In *Proc. 17th Annual ACM-SIAM Symposium on Discrete Algorithms (SODA'06)*, pages 18–25, 2006.
7. F. Fomin and K. Høie. Pathwidth of cubic graphs and exact algorithms. *Information Processing Letters*, 97(5):191–196, 2006.
8. E. L. Lawler. A note on the complexity of the chromatic number problem. *Information Processing Letters*, (5):66–67, 1976.
9. V. G. Vizing. On the estimate of the chromatic class of a p-graph. *Diskret. Analiz*, 3:25–30, 1964.

Vertex Coloring of Comparability+ke and −ke Graphs

Yasuhiko Takenaga and Kenichi Higashide*

The University of Electro-Communications, Chofu, Tokyo, Japan
`takenaga@cs.uec.ac.jp`

Abstract. $\mathcal{F} + k$e and $\mathcal{F} - k$e graphs are classes of graphs close to graphs in a graph class $\mathcal{F}$. They are the classes of graphs obtained by adding or deleting at most k edges from a graph in $\mathcal{F}$. In this paper, we consider vertex coloring of comparability+ke and comparability−ke graphs. We show that for comparability+ke graphs, vertex coloring is solved in polynomial time for $k = 1$ and NP-complete for $k \geq 2$. We also show that vertex coloring of comparability−1e graphs is solved in polynomial time.

1 Introduction

Many graph problems are NP-complete for general graphs. It is natural to consider that if the graph problem is tractable for a graph class $\mathcal{F}$, it is also tractable for a class of graphs which are close to graphs in $\mathcal{F}$. $\mathcal{F} + k$e and $\mathcal{F} - k$e graphs are classes of graphs close to $\mathcal{F}$. They are the classes of graphs obtained by adding or deleting at most k edges from graphs in F. Recently, complexity of several problems on such graph classes has been interested in from parametric point of view [1,5,6]. In general, the problem becomes difficult as k increases. A problem with parameter k is said to be fixed parameter tractable if it can be solved in $f(k)|x|^c$ time, where f is an arbitrary function and $|x|$ is the size of input [2].

Vertex coloring problem is a very important graph problem, which is NP-complete for general graphs. Vertex coloring for parameterized graph classes are considered in [1,6]. It is shown in [1] that, if vertex coloring of $\mathcal{F}$ graphs is solved in polynomial time, vertex coloring of $\mathcal{F} + k$e graphs is fixed parameter tractable if $\mathcal{F}$ is closed under identification of nonadjacent vertices, and that of $\mathcal{F} - k$e graphs is fixed parameter tractable if $\mathcal{F}$ is closed under edge contraction. In [6], it is shown that vertex coloring of such parameterized graphs sometimes have close relations to precoloring extension problems. It can be another motivation to study vertex coloring for parameterized graph classes.

In this paper, we consider vertex coloring of comparability+ke and −ke graphs. As a comparability graph is a perfect graph, its chromatic number equals its clique number, and the vertex coloring of comparability graphs can be solved in polynomial time [4]. In addition, comparability graphs are closed neither under identification of nonadjacent vertices nor under edge contraction.

---

* Presently with PFU Ltd.

F.V. Fomin (Ed.): WG 2006, LNCS 4271, pp. 102–112, 2006.

In this paper, we first show that for comparability+ke graphs, vertex coloring is solved in polynomial time for $k = 1$ and NP-complete for $k \geq 2$. Next, we show that vertex coloring of comparability-1e graphs is solved in polynomial time.

2 Preliminaries

Let $G = (V, E)$ be an undirected graph. Then vertex coloring problem is defined as follows.

VERTEX COLORING
Input: A graph $G = (V, E)$ and a positive integer $t \leq |V|$.
Question: Is G t-colorable? That is, is there a function $f\ :\ V \to \{1, 2, \dots, t\}$ that satisfies $f(u) \neq f(v)$ for all $(u, v) \in E$.

The chromatic number of G, denoted as $\chi(G)$, is the smallest t for which G is t-colorable. The clique number of a graph G, denoted as $\omega(G)$, is the degree of the maximum complete subgraph of G.

The *modulator* of an $\mathcal{F} + ke$ graph $G = (V, E)$ is a set of edges $E_k (\subset E)$ s.t. $(V, E - E_k) \in \mathcal{F}$ and $|E_k| \leq k$. The modulator of an $\mathcal{F} - ke$ graph G is a set of edges E_k s.t. $(V, E \cup E_k) \in \mathcal{F}$ and $|E_k| \leq k$. In this paper, we assume that the modulator is given. For a fixed k, the modulator of $\mathcal{F} + ke$ or $\mathcal{F} - ke$ graphs can be found in polynomial time provided that it can be checked in polynomial time whether a graph is in class $\mathcal{F}$ or not.

A comparability graph is an undirected graph s.t. a transitive graph can be obtained by giving appropriate orientation to each edge. A directed graph is called transitive if $(u, v) \in E$ and $(v, w) \in E$, then $(u, w) \in E$ holds. An example of a comparability graph and its transitive orientation is shown in Fig.1(a) and (b). It can be recognized whether a graph is a comparability graph or not and its transitive orientation can be found in $O(\gamma|E|)$ time, where γ is the maximum degree of a vertex [3]. If a comparability graph is given, its transitive orientation is obtained in linear time [7].

A transitive graph can be represented by a Hasse diagram. If $(u, v) \in E$ and $(v, w) \in E$, (u, w) is omitted in a Hasse diagram. Fig.1(c) is the Hasse diagram representation of Fig.1(b). When there exists a path from u to v in a Hasse diagram, we call that u is an ancestor of v and v is a descendant of u, denoted as

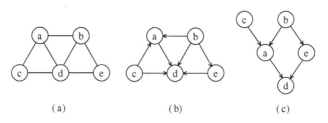

(a)   (b)   (c)

Fig. 1. An example of a comparability graph. (a) A comparability graph. (b) A transitive orientation. (c) A Hasse diagram.

$u \prec v$. In this paper, for simplicity, we draw a Hasse diagram as an undirected graph. All the edges are assumed to be downward edges.

For a transitive graph G, a function $f : V \to \{1, 2, \ldots, \omega(G)\}$ is called a leveling function if $f(u) < f(v)$ is satisfied for all $(u, v) \in E$. We define $levmin$ and $levmax$ as follows.

$$levmin(v) = \begin{cases} 1 & \text{if } v \text{ is a source} \\ \max_{(u,v) \in E} levmin(u) + 1 & \text{otherwise} \end{cases}$$

$$levmax(v) = \begin{cases} \omega(G) & \text{if } v \text{ is a sink} \\ \min_{(v,u) \in E} levmax(u) - 1 & \text{otherwise} \end{cases}$$

Then $levmin$ and $levmax$ are leveling functions. Obviously, in the corresponding Hasse diagram of G, $levmin(v)$ and $levmax(v)$ have the same value as those in G. We say that $levmin(v)$ is the level of vertex v, and that v is in level $levmin(v)$. As a maximal path in the Hasse diagram corresponds to a maximal clique in a transitive graph, $\omega(G) = \max_{v \in V} levmin(v)$. If we color vertex v by $f(v)$ for some leveling function f, it is an optimal vertex coloring of G. In this paper, we call such coloring a levelwise coloring of G by f.

3 Coloring Comparability+ke Graphs

3.1 Coloring Comparability+1e Graphs

Theorem 1. *Vertex coloring problem of comparability+1e graphs can be solved in polynomial time.*

In this section, we give a polynomial time algorithm for coloring comparability +1e graphs. In the following of this paper, to describe a comparability graph, we use the Hasse diagram of the corresponding transitive graph. However, we should note that there may exist several transitive graphs which can be obtained from a comparability graph.

Let $G = (V, E)$ be a comparability+1e graph and $E_1 = \{(a, b)\}$ be the modulator of G. Let $G_c = (V, E - E_1)$.

It is easy to see that either $\chi(G) = \chi(G_c)$ or $\chi(G) = \chi(G_c) + 1$ holds. A coloring of G_c which colors a and b with different colors is also a coloring of G. A coloring of G_c which colors a and b with the same color is not a coloring of G. Therefore, $\chi(G) = \chi(G_c)$ iff G_c has a coloring which colors a and b with different colors.

First, consider to compute a transitive orientation of G_c. If there exists a leveling function for the transitive graph s.t. a and b are in different levels, G can be colored with $\chi(G_c)$ colors by the levelwise coloring of G_c.

If $levmin(a) \neq levmin(b)$, G can be colored with $\chi(G_c)$ colors by the levelwise coloring of G_c by $levmin$.

If $levmax(a) = levmax(b) > levmin(a) = levmin(b)$, lev defined as follows is another leveling function.

$$lev(v) = \begin{cases} levmin(v) \text{ if } levmin(v) \leq levmin(a) \text{ and } v \neq b \\ levmax(v) \text{ otherwise} \end{cases}$$

It also holds that, $lev(a) \neq lev(b)$. Then, G can be colored with $\chi(G_c)$ colors by the levelwise coloring of G_c by lev.

In the following, we consider the case of $levmin(a) = levmin(b) = levmax(a) = levmax(b)(= l)$. Define the leveling function lev' as follows.

$$lev'(v) = \begin{cases} levmin(v) & \text{if } levmin(v) < l \\ l & \text{if } levmin(v) = levmax(v) = l \\ levmax(v) & \text{otherwise} \end{cases}$$

Lemma 1. *Let the vertices a, b in a transitive graph G satisfy $levmin(a) = levmin(b) = levmax(a) = levmax(b)$. There exists a coloring of G_c s.t. a and b have different colors iff a and b are not connected in either the subgraph induced by the vertices v satisfying $lev'(v) = l$ or $l - 1$, or the subgraph induced by the vertices v satisfying $lev'(v) = l$ or $l + 1$.*

Proof. ($\rightarrow$)   When a and b are undirectedly connected in the subgraph induced by the vertices v satisfying $lev'(v) = l$ or $l - 1$, from the definition of lev', all the vertices in an undirected path from a to b are included in a $\chi(G_c)$-clique. Let $U(v)$ be the set of colors used in v and its ancestors, and $L(v)$ be the set of colors used in v and its descendants. Let vertices x, y, z satisfy $lev'(x) = lev'(y) = l$, $lev'(z) = l - 1$ and $(z, x), (z, y) \in E$. Then $L(x) \cup U(z) = L(y) \cup U(z) = \{1, \ldots, \omega(G_c)\}$ holds. Thus it follows that $L(x) = L(y)$. By repeating it, we can see that $L(a) = L(b)$.

Similarly, when a and b are undirectedly connected in the subgraph induced by the vertices v satisfying $lev'(v) = l$ or $l + 1$, $U(a) = U(b)$ holds. To satisfy the two conditions, a and b must have the same color.

($\leftarrow$)   Assume w.l.o.g. that a and b are not undirectedly connected in the subgraph induced by the vertices v satisfying $lev'(v) = l$ or $l - 1$. In this case, G_c can be colored according to f defined as follows. 1) If $lev'(v) \leq l - 2$ or $lev'(v) \geq l + 1$, let $f(v) = lev'(v)$. 2) If $l - 1 \leq lev'(v) \leq l$ and v is undirectedly connected with a in the induced subgraph, let $f(v) = lev'(v)$. 3) Otherwise, if $lev'(v) = l$ let $f(v) = l - 1$ and if $lev'(v) = l - 1$ let $f(v) = l$. f is a coloring of G_c because there exists no path that includes both vertices satisfying the condition in 2) and vertices satisfying the condition in 3).

An algorithm may be easily obtained from the above proof. Clearly, all the operations in the algorithm can be executed in polynomial time.

3.2   Coloring Comparability+2e Graphs

Let $G = (V, E)$ be a comparability+2e graph and E_2 be the modulator of G. Let $G_c = (V, E - E_2)$. Then, $\chi(G_c) \leq \chi(G) \leq \chi(G_c) + 2$ holds. We show that it is difficult to know $\chi(G)$ equals which of the three possibilities.

Theorem 2. *$\chi(G_c)$-colorability problem of comparability+2e graphs is NP-complete.*

Theorem 3. $(\chi(G_c)+1)$-*colorability problem of comparability+2e graphs is NP-complete.*

The next corollary is immediate from the above theorems.

Corollary 1. *Vertex coloring problem of comparability+ke graphs is NP-complete for $k \geq 2$.*

Proof. of Theorem 2   We give a reduction from 3SAT. Let the given 3CNF formula be $F(x_1, \ldots, x_m) = \prod_{i=1}^{n} C_i$, where $C_i = c_{i1} + c_{i2} + c_{i3}$. We construct a comparability+2e graph from F. In the following of this section, we describe a comparability graph by the corresponding Hasse diagram. The description on the edges not shown in the Hasse diagram is omitted.

The comparability graph is constructed as follows.

1. Add vertices a_1, a_2, a_3, a_4 and edges $(a_1, a_2), (a_2, a_3), (a_3, a_4)$.
2. Add vertices b_1, b_2, b_3, b_4 and edges $(b_1, b_2), (b_2, b_3), (b_3, b_4)$.
3. For each clause C_i, add vertices c_i' and c_i''. We call these vertices and a_4, b_4 the lower vertices.
4. For literal c_{ij} in F, add a vertex c_{ij}. We call these vertices the middle vertices. Add edges $(c_{i1}, a_4), (c_{i1}, c_i'), (c_{i2}, c_i'), (c_{i2}, c_i''), (c_{i3}, c_i''), (c_{i3}, b_4)$.
5. For each variable x_i, add a vertex x_i. We call these vertices and a_1, b_1 the upper vertices. If vertex c_{jk} corresponds to the positive literal of x_i, add edges (a_1, c_{jk}) and (x_i, c_{jk}). If vertex c_{jk} corresponds to the negative literal of x_i, add edges (x_i, c_{jk}) and (b_1, c_{jk}).

The clique number of the comparability graph is 4. Fig.2 is an example of the comparability graph obtained by the reduction. The modulators are (a_1, b_1) and (a_4, b_4).

Now we prove the correctness of the reduction. First, we show that the comparability+2e graph is 4-colorable if F is satisfiable. We color vertices a_1, a_2, a_3, a_4

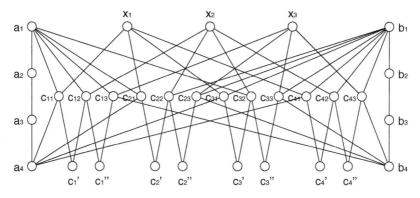

Fig. 2. The comparability graph obtained by reduction from $(x_1 + x_2 + x_3)(x_1 + \overline{x_2} + \overline{x_3})(\overline{x_1} + x_2 + \overline{x_3})(\overline{x_1} + \overline{x_2} + \overline{x_3})$

with $1, 2, 3, 4$, respectively, and color vertices b_1, b_2, b_3, b_4 with $2, 1, 4, 3$, respectively. It means that the endpoints of the modulators are colored with different colors.

Vertices c_i', c_i'' are colored according to the following rule. For a satisfying assignment of F, clause C_i has at least one satisfied literal. Choose one of the satisfied literals arbitrarily. When c_{i1} is chosen, color vertices c_i' and c_i'' with 3. When c_{i2} is chosen, color vertex c_i' with 4 and vertex c_i'' with 3. When c_{i3} is chosen, color vertices c_i' and c_i'' with 4. Color vertex x_i with 1 if $x_i = 1$ in the satisfying assignment, and with 2 if $x_i = 0$.

We show that each vertex c_{ij} can be colored with either $1, 2, 3$ or 4. A vertex c_{ij} has four adjacent vertices; two upper vertices and two lower vertices. If literal c_{ij} is satisfied, adjacent upper vertices has the same color. Therefore, c_{ij} can be colored with either 1 or 2. If literal c_{ij} is not satisfied, adjacent lower vertices has the same color. Therefore, c_{ij} can be colored with either 3 or 4. Therefore, the comparability$+2e$ graph is 4-colorable.

At last, we show that F is satisfiable if the comparability$+2e$ graph is 4-colorable. Consider a 4-coloring of the comparability$+2e$ graph. As a_1 and b_1 must have different colors, at least one of the pairs $< a_1, x_i >$ and $< x_i, b_1 >$ do not have the same color. If $< a_1, x_i >$ has the same color, let $x_i = 1$. If $< x_i, b_1 >$ has the same color, let $x_i = 0$. If both pairs have different colors, choose the value of x_i arbitrarily.

We will verify that the above assignment to the variables satisfy F. Consider lower vertices a_4, c_i', c_i'' and b_4. As a_4 and b_4 must have different colors, at least one of the pairs $< a_4, c_i' >, < c_i', c_i'' >$ and $< c_i'', b_4 >$ do not have the same color. If $< a_4, c_i' >$ does not have the same color, two upper vertices adjacent to c_{i1} must have the same color. It means that literal c_{i1} is satisfied. Similarly, c_{i2} is satisfied if $< c_i', c_i'' >$ does not have the same color, and c_{i3} is satisfied if $< c_i'', b_4 >$ does not have the same color. Therefore, at least one literal is satisfied in C_i. As it holds for any clause in F, F is satisfied by the above assignment.

Proof. of Theorem 3   Similarly to the proof of Theorem 2, we give a reduction from 3SAT. For a 3CNF formula $F = \prod_{i=1}^{n} C_i$, we construct a comparability$+2e$ graph. The comparability graph is constructed as follows.

1. Add vertices a_1, a_2, b_1 and b_2.
2. For each variable x_i, add vertices x_i, $\overline{x_i}$, x_i' and edges (a_1, x_i), (x_i', x_i), $(x_i', \overline{x_i})$, $(b_1, \overline{x_i})$.
3. For each clause C_i, add vertices c_{i1}, c_{i2}, c_{i3}, c_1', c_1'' and edges (c_{i1}, a_2), (c_{i1}, c_i'), (c_{i2}, c_i'), (c_{i2}, c_i''), (c_{i3}, c_i''), (c_{i3}, b_2). Vertices c_{i1}, c_{i2} and c_{i3} correspond to the literals in C_i.
4. When c_{ij} corresponds to a positive literal x_k, add an edge $(\overline{x_k}, c_{ij})$. When c_{ij} corresponds to a negative literal $\overline{x_k}$, add an edge (x_k, c_{ij}). That is, there exists an edge between the literals which have the different values.

The clique number of the comparability graph is 4. Fig.3 is an example of the comparability graph obtained by the reduction. The modulators are (a_1, b_1) and (a_2, b_2).

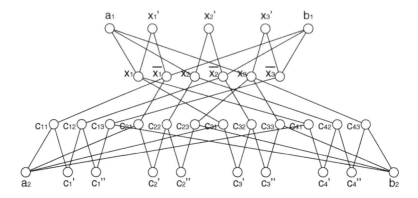

Fig. 3. The comparability graph obtained by reduction from $(x_1 + x_2 + x_3)(x_1 + \overline{x_2} + \overline{x_3})(\overline{x_1} + x_2 + \overline{x_3})(\overline{x_1} + \overline{x_2} + \overline{x_3})$

Now we prove the correctness of the reduction. First, we show that the comparability+2e graph is 5-colorable when F is satisfiable. We color vertices a_1, a_2, b_1, b_2 with $1, 4, 2, 3$, respectively. Consider a satisfying assignment for F. If $x_i = 1$ in the assignment, color vertices $x_i, x_i', \overline{x_i}$ with $5, 2, 1$, respectively. If $x_i = 0$ in the assignment, color vertices $x_i, x_i', \overline{x_i}$ with $2, 1, 5$, respectively. (Fig.4)

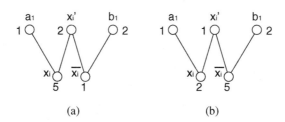

Fig. 4. (a) Coloring when $x_i = 1$. (b) Coloring when $x_i = 0$.

As clause C_i has at least one satisfied literal, choose one of the satisfied literals arbitrarily. When c_{i1} is chosen, color vertices $c_{i1}, c_i', c_i'', c_{i2}, c_{i3}$ with $5, 3, 3, 4, 4$, respectively. When c_{i2} is chosen, color vertices $c_{i2}, c_i', c_i'', c_{i1}, c_{i3}$ with $5, 4, 3, 3, 4$, respectively. When c_{i3} is chosen, color vertices $c_{i3}, c_i', c_i'', c_{i1}, c_{i2}$ with $5, 4, 4, 3, 3$, respectively. (Fig.5)

Now we verify that a 5-coloring is obtained by the above rules. Colors $1, 2$ are used only for vertices $a_1, b_1, x_i, \overline{x_i}, x_i'$. Fig.4 shows that neither 1 nor 2 appear twice in a path. Colors $3, 4$ are used only in vertices $a_2, b_2, c_{ij}, c_i', c_i''$. Fig.5 shows that neither 3 nor 4 appear twice in a path. Color 5 is used only in vertices $x_i, \overline{x_i}, c_{jk}$. Only the vertices whose corresponding literal is satisfied has color 5. On the other hand, among vertices $x_i, \overline{x_i}$ and c_{jk}, an edge exists only when two literals cannot be satisfied at the same time. Therefore, 5 does not appear twice in a path.

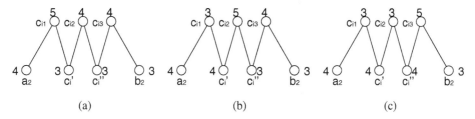

Fig. 5. (a) Coloring when c_{i1} is satisfied. (b) Coloring when c_{i2} is satisfied. (c) Coloring when c_{i3} is satisfied.

At last, we show that F is satisfiable if the comparability+2e graph is 5-colorable. Consider a 5-coloring of the comparability+2e graph. As a_1 and b_1 are adjacent, they must have different colors. Then, for each i, at least one of the pairs of vertices $< a_1, x_i' >$ and $< x_i', b_1 >$ must have different colors. If $< a_1, x_i' >$ does not have the same color, let $x_i = 1$. If $< x_i', b_1 >$ does not have the same color, let $x_i = 0$. Note that, three different colors are used in a triple $< a_1, x_i', x_i >$ if $x_i = 1$, and three different colors are used in a triple $< b_1, x_i', \overline{x_i} >$ if $x_i = 0$. We call the triples upper triples.

Similarly, as a_2 and b_2 must have different colors, for each i, at least one of the pairs $< a_2, c_i' >$, $< c_i', c_i'' >$ and $< c_i'', b_2 >$ must have different colors. It follows that three different colors are used in at least one of the triples $< a_2, c_{i1}, c_i' >$, $< c_i', c_{i2}, c_i'' >$ and $< c_i'', c_{i3}, b_2 >$. We call the triples lower triples.

Assume that three different colors are used both in an upper triple including $x_i (\overline{x_i}$ resp.) and a lower triple including c_{ij}. Then, the edge (x_i, c_{ij}) (the edge $(\overline{x_i}, c_{ij})$ resp.) does not exist. If the edge exists, 6 colors are necessary to color the comparability+2e graph. Remember that the edge (x_i, c_{ij}) or $(\overline{x_i}, c_{ij})$ exists iff the literals cannot be satisfied at the same time. Hence, if three different colors are used in a lower triple including c_{ij}, literal c_{ij} is satisfied. As each clause in F has at least one satisfied literal, F is satisfied by the above assignment.

4   Coloring Comparability$-$1e Graphs

On the vertex coloring of comparability$-ke$ graphs, we show the following result for $k = 1$.

Theorem 4. *Vertex coloring problem of comparability$-$1e graphs can be solved in polynomial time.*

In this section, we give the proof of Theorem 4. Let $G = (V, E)$ be a comparability$-$1e graph and $E_1 = \{(a, b)\}$ be the modulator of G. Let $G_c = (V, E \cup E_1)$. If $\omega(G) = \omega(G_c)$, then $\chi(G) = \omega(G_c)$ holds. If $\omega(G) = \omega(G_c) - 1$, then $\omega(G_c) - 1 \le \chi(G) \le \omega(G_c)$ holds. In the latter case, there may exist an $(\omega(G_c) - 1)$-coloring of G. In the $(\omega(G_c) - 1)$-coloring, a and b have the same color.

We first show that it is not difficult to check if $\omega(G) = \omega(G_c)$ or not.

Lemma 2. *The equality $\omega(G) = \omega(G_c) - 1$ holds iff there exists no vertex v ($v \neq a, b$) such that $levmin(v)$ is equal to $levmin(a)$ or $levmin(b)$ and $levmin(v) = levmax(v)$.*

Proof. First, observe that, for a vertex v in G_c, the size of the maximum clique including v is $levmin(v) + (\omega(G_c) - levmax(v))$. Thus, if $levmin(v) = levmax(v)$, v is included in an $\omega(G_c)$-clique of G.

$(\rightarrow)$ $\omega(G) = \omega(G_c) - 1$ holds iff all the maximum cliques of G' includes the modulator (a, b).

Assume w.l.o.g. that there exists a vertex v ($v \neq a$) such that $levmin(v) = levmax(v) = levmin(a)$. As $levmin(v) = levmax(v)$, v is included in a maximum clique of G_c. However, as $levmin(v) = levmin(a)$, v and a are not in the same clique. Therefore, G_c has a maximum clique which does not include (a, b). It means that $\omega(G) = \omega(G_c)$.

$(\leftarrow)$ Assume that there exists no vertex v ($v \neq a, b$) such that $levmin(v)$ is equal to $levmin(a)$ or $levmin(b)$ and $levmin(v) = levmax(v)$. Then, all the maximum cliques of G_c must contain a and b. Thus, the modulator (a, b) is included in all the maximum cliques of G_c. That is, $\omega(G) = \omega(G_c) - 1$ holds.

The condition of this lemma is checked easily using $levmin$ and $levmax$ of each vertex. In the following, we consider only the graphs satisfying $\omega(G) = \omega(G_c) - 1$.

Even though $\omega(G) = \omega(G_c) - 1$ holds, it is not always possible to color G with $\omega(G_c) - 1$ colors. We consider how to compute if G is $(\omega(G_c) - 1)$-colorable or not. To consider the coloring of G, we first obtain a transitive orientation of G_c. Let G_t be the obtained transitive graph. G_t is represented as a Hasse diagram $H = (V, E_H)$. In the following, we assume w.l.o.g. that $a \prec b$ in H. In a Hasse diagram, all the vertices in a path must have different colors. However, in this case, as we consider the coloring of G, not G_c, we admit that a and b have the same color in H.

We consider to modify the graph without changing its chromatic number. Let w, x, y, z be the vertices satisfying the following conditions: $(w, x), (y, x)$, $(w, z) \in E_H$, w and x are in the same $(\omega(G_c) - 1)$-clique of G_c, and neither $a \preceq w \prec z \preceq b$ nor $a \preceq y \prec x \preceq b$ holds (see Fig.6). Let $H' = (V, E_H \cup \{(y, z)\})$ and let G' be the comparability–1e graph represented by H' when the modulator is added.

Lemma 3. $\chi(G') = \chi(G)$.

Proof. As G' is obtained by adding edges to G, a coloring of G' is also a coloring of G. That is, $\chi(G') \geq \chi(G)$ holds. We show that if G is $(\omega(G_c) - 1)$-colorable, then G' is also $(\omega(G_c) - 1)$-colorable.

Consider a $(\omega(G_c) - 1)$-coloring of G. As $(w, x) \in E_H$ and w and x are in the same $(\omega(G_c) - 1)$-clique of G_c, $U(w) \cap L(x) = \emptyset$ and $U(w) \cup L(x) = \{1, \ldots, \omega(G_c) - 1\}$ hold. Similarly, as neither $a \preceq w \prec z \preceq b$ nor $a \preceq y \prec x \preceq b$ holds, $U(y) \cap L(x) = \emptyset$ and $U(w) \cap L(z) = \emptyset$ also hold. Therefore, we can see that $U(y) \subseteq U(w)$ and $L(z) \subseteq L(x)$ hold. It follows that $U(y) \cap L(z) = \emptyset$. It means that even when (y, z) is added to H, no path of the Hasse diagram contain the vertices with the same color (except two endpoints of the modulator).

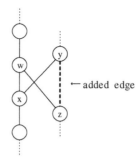

Fig. 6. Add an edge to generate H'

Add edges satisfying the above condition as far as possible. Let the resulting Hasse diagram be H_+ and the corresponding comparability$-1e$ graph be G_+. Lemma 3 shows that $\chi(G_+) = \chi(G)$. Let $V_a = \{v \mid a \prec v \text{ in } H_+\}$ and $V_b = \{v \mid v \prec b \text{ in } H_+\}$.

Lemma 4. *There exists an $(\omega(G_c) - 1)$-coloring of G_+ iff there exists no $(\omega(G_c) - 1)$-clique that does not include an endpoint of the modulator and whose all vertices are in $V_a \cup V_b$.*

Proof. The vertices in V_a (V_b resp.) cannot be colored with the same color as a and b because the vertices in V_a (V_b resp.) and a (b resp.) are on the same path in H_+. If there exists an $(\omega(G_c) - 1)$-clique of G_+ that does not include an endpoint of the modulator and whose all vertices are in $V_a \cup V_b$, there exists no $(\omega(G_c) - 1)$-coloring of G_+ because only $\omega(G_c) - 2$ colors can be used to color the clique.

Otherwise, we can color all the vertices using colors $\{1, 2, \ldots, \omega(G_c) - 1\}$ in the following manner. Color a and b with 1. In each $(\omega(G_c) - 1)$-clique, color the vertex of the smallest level not in $V_a \cup V_b$ with 1. From the assumption, each $(\omega(G_c) - 1)$-clique includes at least one vertex colored with 1. In addition, we can show that no clique includes more than one vertex colored with 1 as follows.

Assume that a vertex d ($d \neq a, b$) is colored with 1 by following the above rule. Then, either $levmin(d) = 1$ or there exists a vertex e satisfying $(e, d) \in H_+$, $e \in V_b$ and d, e are in the same $(\omega(G_c) - 1)$-clique. Also there exists a vertex f satisfying $(e, f) \in H_+$ and $f \in V_b \cup \{b\}$. From the construction of H_+, for each vertex g s.t. $(g, d) \in H_+$, an edge (g, f) must exist in H_+. It means that any ancestor of d is included in V_b. Therefore, no two vertices in a $(\omega(G_c) - 1)$-clique can be colored with 1.

Color the other vertices with the following rule: if v is a source, color v with 2, and otherwise color v with the minimum number which is not used in the ancestors of v in H_+. The coloring rule approves that no two vertices in a clique has the same color.

In addition, it is not difficult to show by induction on $levmin(v)$ that i) $levmin(v) - 1$ colors are used for v and its ancestors if vertex b is included in them, and ii) $levmin(v)$ colors are used for v and its ancestors otherwise. It means that $\omega(G_c) - 1$ colors are sufficient to color G.

An algorithm may be easily obtained from the above proofs. Now we consider the complexity of the algorithm. To obtain H_+, an edge to be added can be found in polynomial time by checking all the possibilities of w and x. Also, the number of added edges is less than $|V|^2$. Thus, H_+ can be obtained in polynomial time. It is easy to see that all the other operations can be executed in polynomial time.

5   Conclusion

In this paper, we have considered the complexity of coloring comparability+ke and $-k$e graphs. For comparability+ke graphs, we have shown that it is tractable for $k = 1$ and NP-complete for $k \geq 2$. For comparability$-k$e graphs, we have shown that it is tractable for $k = 1$. One of our future works is to clarify the complexity of coloring comparability$-k$e graphs for $k \geq 2$.

Acknowledgements

This research was partially supported by the Scientific Grant-in-Aid from Ministry of Education, Science, Sports and Culture of Japan.

References

1. L. Cai, Parameterized Complexity of Vertex Colouring, Discrete Applied Mathematics, 127, 3, 415-429, 2003.
2. R.G. Downer, M.R. Fellows, Parameterized Complexity, Springer-Verlag New York, 1997.
3. M. C. Golumbic, Algorithmic Graph Theory and Perfect Graphs, Annals of Discrete Mathematics 57, Elsevier, 2nd Edition, 2004.
4. M. Grotschel, L. Lovasz, and A. Schrijver, Polynomial Algorithms for Perfect Graphs, Annals of Discrete Mathematics, 21, 325-356, 1984.
5. J. Guo, F. Hüffner and R. Niedermeier, A Structural View on Parameterizing Problems: Distance from Triviality, IWPEC 2004, LNCS 3162, 162-173, 2004.
6. D. Marx, Parameterized Coloring Problems on Chordal Graphs, Theoretical Computer Science, 351, 3, 407-424, 2006.
7. R.M. McConnell and J.P. Spinrad, Modular Decomposition and Transitive Orientation, Discrete Math., 201, 189-241, 1999.

Convex Drawings of Graphs with Non-convex Boundary*

Seok-Hee Hong[1] and Hiroshi Nagamochi[2]

[1] School of Information Technologies, University of Sydney and NICTA
shhong@it.usyd.edu.au
[2] Department of Applied Mathematics and Physics,
Kyoto University
nag@amp.i.kyoto-u.ac.jp

Abstract. In this paper, we study a new problem of finding a convex drawing of graphs with a *non-convex* boundary. It is proved that every triconnected plane graph whose boundary is fixed with a star-shaped polygon admits a drawing in which every inner facial cycle is drawn as a convex polygon. Such a drawing, called an *inner-convex drawing*, can be obtained in linear time.

1 Introduction

Graph drawing has attracted much attention over the last ten years due to its wide range of applications, such as VLSI design, software engineering and bioinformatics. Two- or three-dimensional drawings of graphs with a variety of aesthetics and edge representations have been extensively studied (see [1]). One of the most popular drawing conventions is the *straight-line drawing*, where all the edges of a graph are drawn as straight-line segments. Every planar graph is known to have a planar straight-line drawing [3]. A straight-line drawing is called a *convex drawing* if every facial cycle is drawn as a convex polygon. Note that not all planar graphs admit a convex drawing. Tutte [8] gave a necessary and sufficient condition for a triconnected plane graph to admit a convex drawing. Thomassen [7] gave a necessary and sufficient condition for a biconnected plane graph to admit a convex drawing. Based on this result, Chiba et al. [2] presented a linear time algorithm for finding a convex drawing (if any) for a biconnected plane graph with a specified convex boundary. Tutte [8] also showed that every triconnected plane graph with a given boundary drawn as a convex polygon admits a convex drawing using the polygonal boundary. That is, when the vertices on the boundary are placed on a convex polygon, inner vertices can be placed on suitable positions so that each inner facial cycle forms a convex polygon.

However, not much attention has been paid to the problem of finding a convex drawing with a *non-convex* boundary. In this paper, a straight-line drawing is called an *inner-convex drawing* if every inner facial cycle is drawn as a convex

---

* This research was partially supported by the Scientific Grant-in-Aid from Ministry of Education, Culture, Sports, Science and Technology of Japan.

F.V. Fomin (Ed.): WG 2006, LNCS 4271, pp. 113–124, 2006.

polygon, and is simply called a *convex drawing* if no confusion arises. One can easily observe that not every triconnected plane graph has a convex drawing if its boundary is drawn as a non-convex polygon. For example, Fig. 1 shows three examples of plane graphs which have no convex drawing; the inner facial cycle f_1 in Fig. 1(b) (resp., one of the inner facial cycles f_1 and f_2 in Figs. 1(a) and (c)) cannot be drawn as a convex polygon.

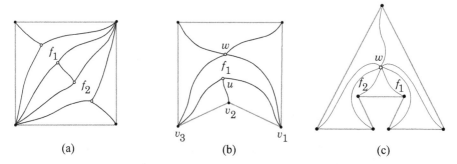

Fig. 1. (a) A biconnected plane graph with a convex boundary; (b) An internally triconnected plane graph with a star-shaped boundary; (c) A triconnected plane graph with a non star-shaped boundary

No characterization is known for any class of plane graphs that have inner-convex drawings with non-convex boundaries. In this paper, we prove that every triconnected plane graph has an inner-convex drawing if its boundary is fixed with a star-shaped polygon P, i.e., a polygon P whose kernel (the set of all points from which all points in P are visible) is not empty. This is an extension of the classical result by Tutte [8] since any convex polygon is a star-shaped polygon. Our proof gives a linear time algorithm for computing an inner-convex drawing of a triconnected plane graph with a star-shaped boundary.

This paper is organized as follows: Section 2 reviews basic terminology and proves an important property of triconnected plane graphs. Section 3 proves that a triconnected plane graph has an archfree tree, a spanning tree with a special property. Section 4 reviews the necessary and sufficient condition for a plane graph with a boundary drawn as a convex polygon to admit a convex drawing. Section 5 discusses how to draw a triconnected plane graph with a star-shaped boundary as an inner-convex drawing. Section 6 concludes.

2   Preliminaries

Throughout the paper, a graph stands for a simple undirected graph. Let $G = (V, E)$ be a graph. The set of edges incident to a vertex $v \in V$ is denoted by $E(v)$. The degree of a vertex v in G is denoted by $d_G(v)$ (i.e., $d_G(v) = |E(v)|$). For a subset $X \subseteq E$ (resp., $X \subseteq V$), $G - X$ denotes the graph obtained from G by removing the edges in X (resp., the vertices in X together with the edges in $\cup_{v \in X} E(v))$. A vertex (resp., a pair of vertices) in a connected graph is called a *cut vertex* (resp., a *cut pair*) if its removal from G results in a disconnected graph. A connected graph is called *biconnected* (resp., *triconnected*) if it is simple and has no cut vertex (resp., no cut pair). We say that a cut pair $\{u, v\}$ *separates* two vertices s and t if s and t belong to different components in $G - \{u, v\}$.

A graph $G = (V, E)$ is called *planar* if its vertices and edges are drawn as points and curves in the plane so that no two curves intersect except for their endpoints, where no two vertices are drawn at the same point. In such a drawing, the plane is divided into several connected regions, each of which is called a *face*. A face is characterized by the cycle of G that surrounds the region. Such a cycle is called a *facial cycle*. A set F of facial cycles in a drawing is called an *embedding* of a planar graph G. A *plane* graph $G = (V, E, F)$ is a planar graph $G = (V, E)$ with a fixed embedding F of G, where we always denote the outer facial cycle in F by f_o. A vertex (resp., an edge) in f_o is called an *outer vertex* (resp., an *outer edge*), while a vertex (resp., an edge) not in f_o is called an *inner vertex* (resp., an *inner edge*). A path Q between two vertices s and t in G is called *inner* if every vertex in $V(Q) - \{s, t\}$ is an inner vertex. The region enclosed by a facial cycle $f \in F$ may be denoted by f for simplicity. The set of vertices, set of edges and set of facial cycles of a plane graph G may be denoted by $V(G)$, $E(G)$ and $F(G)$, respectively.

A biconnected plane graph G is called *internally triconnected* if, for any cut pair $\{u, v\}$, u and v are outer vertices and each component in $G - \{u, v\}$ contains an outer vertex. Note that every inner vertex in an internally triconnected plane graph must be of degree at least 3. For a cut pair $\{u, v\}$ of an internally triconnected plane graph $G = (V, E, F)$, if u and v are not adjacent and there is an inner facial cycle $f \in F$ such that $\{u, v\} \in V(f)$, we say that f *separates* two vertices s and t if the cut pair $\{u, v\}$ separates them. For example, the graph in Fig. 1(a) is biconnected but not internally triconnected, the graph in Fig. 1(b) is internally triconnected but not triconnected, and the graph in Fig. 1(c) is triconnected. We then observe the following:

Lemma 1. *Let G be an internally triconnected plane graph. Then G has an inner path connecting two outer vertices s and t if and only if no facial cycle separates s and t.* □

We next show a key property of a triconnected plane graph.

Lemma 2. *Every triconnected plane graph $G = (V, E, F)$ has a spanning tree T such that each vertex $v \in V(f_o)$ is a leaf of T. Such a tree can be found in linear time.*

Proof. Since G has no cut pair, there is an inner path between any two vertices in $V(f_o)$ by Lemma 1. Let $V(f_o) = \{v_1, v_2, \ldots, v_p\}$, where vertices $v_1, v_2, \ldots, v_p$ appear in this order when we traverse f_o in the clockwise order. For each $v \in V(f_o)$, let $e_v \in E(v)$ be the edge that appears after edge $(v = v_i, v_{i+1})$ when we visit the edges in $E(v)$ around $v = v_i$ in the clockwise order. To prove the lemma, it suffices to show that $G^* = G - \cup_{v \in V(f_o)}(E(v) - \{e_v\})$ remains connected, since any spanning tree T of the graph satisfies the condition of the lemma, and it is immediate to see that such a tree can be computed in linear time.

To show the connectedness of G^*, we define an inner path Q_v from $v = v_i$ to v_{i-1} as follows. Let $E(v) = \{(v, u_1 = v_{i+1}), (v, u_2), (v, u_3), \ldots, (v, u_{h-1}), (v, u_h = v_{i-1})\}$, where $(v, u_1), \ldots, (v, u_h)$ appear in this order when we visit the edges in $E(v)$ in the clockwise order around v, and $f_j \in F$, $j = 1, 2, \ldots, h$ be the facial cycle that contains edges (v, u_j) and (v, u_{j+1}), where $V(f_j) \cap V(f_o) = \{v_i\}$ and $f_j \neq f_{j'}$ for $j \neq j'$ by the triconnectivity of G. Then there is an inner path Q_v from $v = v_i$ to v_{i-1} which consists of subpaths $f_j - v$, $j = 2, 3, \ldots, h - 1$. That is, path Q_v and edge $(v = v_i, u_h = v_{i-1})$ surround the union of faces f_j, $j = 2, 3, \ldots, h - 1$. Note that $G - (E(v) - \{e_v\})$ contains Q_v. Hence if G has an inner path Q from $v \in V(f_o)$ to a vertex $w \in V$ that does not use edge e_v, then Q must use an edge $(v, u_j) \in E(v) - \{e_v\}$ and the subpath from v to u_j along Q_v and the subpath from u_j to w along Q give rise to an inner path from v to w without using any edge in $E(v) - \{e_v\}$. By noting that $E(v) \cap E(Q_{v'}) = \emptyset$ for any two $v, v' \in V(f_o)$, this implies that any two vertices in $V(f_o)$ are connected by an inner path in $G^* = G - \cup_{v \in V(f_o)}(E(v) - \{e_v\})$ and that G^* contains a tree T' that connects all vertices in $V(f_o)$ (note that each vertex $v \in V(f_o)$ is a leaf in T' since the degree of v is 1 in $G - \cup_{v \in V(f_o)}(E(v) - \{e_v\})$).

To complete the proof for the connectedness of G^*, we show that an arbitrary vertex $u \in V - V(T')$ is connected to a vertex in $V(T')$ in G^*. There are adjacent vertices $w_1, w_2 \in V(f_o)$ such that u is located in the region enclosed by edge (w_1, w_2) and the path T'_{w_1, w_2} between w_1 and w_2 along T'. No vertex in this region is incident to $E(v)$ with $v \notin \{w_1, w_2\}$. Hence, if u is not connected to any vertex in T' in G^*, then $\{w_1, w_2\}$ would be a cut pair which separates u and a vertex not in the region, contradicting the triconnectivity of G. Therefore, $G^* = G - \cup_{v \in V(f_o)}(E(v) - \{e_v\})$ remains connected, as required.   □

3   Archfree Paths and Archfree Trees

We say that a facial cycle f *arches* a path Q in a plane graph if there are two distinct vertices $a, b \in V(Q) \cap V(f)$ such that the subpath $Q_{a,b}$ of Q between a and b is not a subpath of f. A path Q is called *archfree* if no inner facial cycle f arches Q. Note that any subpath of a facial cycle in a triconnected plane graph is an archfree path.

Let Q be an inner path that is contained in an inner path Q' between two outer vertices s' and t' in a plane graph $G = (V, E, F)$, and let s and t be the end vertices of Q, where Q and Q' are viewed as directed paths from s' to t', as shown in Fig. 2. The outer facial cycle f_o consists of subpath f'_o from s' to t' and subpath f''_o from t' to s' when we walk along f_o in the clockwise order.

We say that an inner facial cycle $f \in F$ is *on the left side* if f is surrounded by f'_o and Q', and that f *arches* Q *on the left side* if f is on the left side of Q. The case of the right side is defined symmetrically.

For example, facial cycles f, f_1 and f_2 in Fig. 2 arch path Q on the left side, where Q is displayed as thick lines. Now we modify Q into a path $L(Q)$ from

s to t such that no inner facial cycle arches $L(Q)$ on the left side. Let F_Q be the set of all inner facial cycles $f \in F$ that arch Q on the left side, but are not contained in the region enclosed by Q and any other $f' \in F$. For example, facial cycle f_1 in Fig. 2 is enclosed by Q and f, and thereby $f_1 \notin F_Q$. The *left-aligned path* $L(Q)$ of Q is defined as an inner path from s to t obtained by replacing subpaths of Q with subpaths of cycles in F_Q as follows. For each $f \in F_Q$, let a_f and b_f be the first and last vertices in $V(f) \cap V(Q)$ when we walk along path Q from s to t, and f_Q be the subpath from a_f to b_f obtained by traversing f in the anticlockwise order. Let $L(Q)$ be the path obtained by replacing the subpath from a_f to b_f along Q with f_Q for all $f \in F_Q$ (see Fig. 2 for an example of $L(Q)$). The following is then observed:

Lemma 3. *Given an inner path Q, the left-aligned path $L(Q)$ of Q can be constructed in $O(|E_Q| + |L(Q)|)$ time, where E_Q is the set of all edges incident to a vertex in Q.* □

The *right-aligned path* $R(Q)$ of Q is defined symmetrically to the left-aligned path.

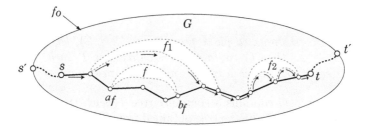

Fig. 2. Construction of the left-aligned path $L(Q)$ from an inner path Q between s and t, where thick lines show Q and the path following the arrows shows $L(Q)$

Lemma 4. *Let $G = (V, E, F)$ be an internally triconnected plane graph, and Q be an inner path from a vertex s to a vertex t. Then the left-aligned path $L(Q)$ is an inner path from s to t, and no inner facial cycle arches $L(Q)$ on the left side. Moreover, if no inner facial cycle arches Q on the right side, then $L(Q)$ is an archfree path.*

Proof. Since G is internally triconnected, we can extend Q to an inner path between two outer vertices s' and t'. Let F_Q be the set of all inner facial cycles $f \in F$ that arch Q on the left side, but are not enclosed by Q and any other $f' \in F$. It is clear that $L(Q)$ is an inner path, since a subpath of Q is replaced with a subpath of $f \in F_Q$ which is not adjacent to any outer vertex. Assume that an inner facial cycle $f^* \in F - F_Q - \{f_o\}$ arches $L(Q)$ on the left side. Let a_{f^*} and b_{f^*} be the first and last vertices in $V(f^*)$ along path $L(Q)$. Note that neither a_{f^*} or b_{f^*} is contained in subpath f_Q for any $f \in F_Q$ since face f contains no edge. Then both a_{f^*} and b_{f^*} are in $V(Q)$. This, however, implies that f^* arches Q on the left side in G, contradicting the choice of F_Q.

We consider the case where Q has no inner facial cycle that arches Q on the right side. Assume that an inner facial cycle $\hat{f} \in F - F_Q - \{f_o\}$ arches $L(Q)$ on the right side. Let $a_{\hat{f}}$ and $b_{\hat{f}}$ be the first and last vertices in $V(\hat{f})$ along path $L(Q)$. Note that both $a_{\hat{f}}$ and $b_{\hat{f}}$ belong to Q or a subpath f_Q with $f \in F_Q$, since otherwise Q would intersect the interior of face $\hat{f}$. Since no inner facial cycle arches Q on the right side by the assumption on Q, both $a_{\hat{f}}$ and $b_{\hat{f}}$ must belong to subpath f_Q for some $f \in F_Q$. This implies that $a_{\hat{f}}$ and $b_{\hat{f}}$ is a cut pair. Since G is internally triconnected, it must hold $a_{\hat{f}}, b_{\hat{f}} \in V(f_o)$. However, it is clear that $G - \{a_{\hat{f}}, b_{\hat{f}}\}$ has at least three components, contradicting the internal triconnectivity of G.   □

Corollary 1. *For any inner path Q from s to t in an internally triconnected plane graph G, the right-aligned path $R(L(Q))$ of the left-aligned path $L(Q)$ is an archfree path.*   □

A path in a tree T is called a *base path* if it is a maximal induced path in T, i.e., end vertices v and internal vertices u (if any) in the path satisfy $d_T(v) \neq 2$ and $d_T(u) = 2$, respectively. A tree T in a plane graph G is called *archfree* if every base path is an archfree path in G.

Lemma 5. *For a triconnected plane graph $G = (V, E, F)$, let S be a subset of $V(f_o)$ with $|S| \geq 2$. Then G contains an archfree tree T such that the set of leaves of T is equal to S. Such a tree T can be obtained in linear time.*

Proof. By Lemma 2, G contains a spanning tree T_1 such that each vertex $v \in V(f_o)$ is a leaf of T_1. Let T_2 be the tree obtained from T_1 by removing all vertices that are not in the path between any two vertices in S. Note that T_2 is a tree such that the set of leaves of T_2 is S and each base path of T_2 is an inner path of G. Choose a vertex $s \in S$ as the root of T_2 with each base path as a directed path from the root to leaves. Let T_3 be the tree obtained from T_2 by replacing each base path Q of T_2 with its left-aligned path $L(Q)$. Then T_4 be the tree obtained from T_3 by replacing each base path Q of T_3 with its right-aligned path $R(Q)$. From Corollary 1, T_4 is an archfree tree whose leaf set is S.

By Lemma 2, spanning tree T_1 can be computed in linear time. It is clear that T_2 can be constructed from T_1 in linear time. By Lemma 3, T_3 can be computed in $O(\sum\{|E_Q| + |L(Q)| \mid \text{base paths } Q \text{ in } T_2\}) = O(\sum_{f \in F} |E(f)|) = O(|V|)$ time. Analogously T_4 can be obtained in $O(|V|)$ time.   □

4   Convex Drawing with a Convex Boundary

For three points a_1, a_2, and a_3 in the plane, the line segment whose end points are a_i and a_j is denoted by (a_i, a_j), and the angle (a_1, a_2, a_3) formed by line segments (a_1, a_2) and (a_2, a_3) is defined by the central angle of a circle with center a_2 when we traverse the circumference from a_1 to a_3 in the clockwise order (note that $(a_1, a_2, a_3) + (a_3, a_2, a_1) = \pi$).

A polygon P is given by a sequence $a_1, a_2, \ldots, a_p$ $(p \geq 3)$ of points, called *apices*, and edges (a_i, a_{i+1}), $i = 1, 2, \ldots, p$ (where $a_{p+1} = a_1$) such that no two line segments (a_i, a_{i+1}) and (a_j, a_{j+1}), $i \neq j$ intersect each other except at apices. Let $A(P)$ denote such a sequence of apices of a polygon P, where $A(P)$ may be used to denote the set of the apices in $A(P)$. The *inner angle* $\theta(a_i)$ of an apex a_i is the angle (a_{i+1}, a_i, a_{i-1}) formed by line segments (a_i, a_{i+1}) and (a_{i-1}, a_i), and an apex a_i is called *convex* (resp., *concave* and *flat*) if $\theta(a_i) < \pi$ (resp., $\theta(a_i) > \pi$ and $\theta(a_i) = \pi$). A polygon P has no apex a with $\theta(a) = 0$ since no two adjacent edges on the boundary intersect each other. Thus, the interior of a polygon has a positive area. A polygon P is called *convex* if it has no concave apex. A *k-gon* is a polygon with exactly k apices, some of which may be flat or concave. A *side* of a polygon is a maximal line segment in its boundary, i.e., a sequence of edges $(a_i, a_{i+1}), (a_{i+1}, a_{i+2}), \ldots, (a_{i+h-1}, a_{i+h})$ such that a_i and a_{i+h} are non-flat apices and the other apices between them are flat. A k-gon has at most k sides.

A *straight-line drawing* of a graph $G = (V, E)$ in the plane is an embedding of G in the two dimensional space $\Re^2$ such that each vertex $v \in V$ is drawn as a point $\psi(v) \in \Re^2$ and each edge $(u, v) \in E$ is drawn as a straight-line segment $(\psi(u), \psi(v))$, where $\Re$ is the set of reals. Hence, a straight-line drawing of a graph $G = (V, E)$ is defined by a function $\psi : V \to \Re^2$. A straight-line drawing ψ of a plane graph $G = (V, E, F)$ is called an *inner-convex drawing* (or simply a convex drawing) if every inner facial cycle is drawn as a convex polygon. A convex drawing ψ of a plane graph $G = (V, E, F)$ is called a *strictly convex drawing* if it has no flat apex $\psi(v)$ for any vertex $v \in V$ with $d_G(v) \geq 3$. We say that a drawing ψ of a graph G is *extended* from a drawing ψ' of a subgraph G' of G if $\psi(v) = \psi'(v)$ for all $v \in V(G')$.

Let $G = (V, E, F)$ be a plane graph with an outer facial cycle f_o, and P be a $|V(f_o)|$-gon. A drawing ϕ of f_o on P is a bijection $\phi : V(f_o) \to A(P)$ such that the vertices in $V(f_o)$ appear along f_o in the same order as the corresponding apices in sequence $A(P)$.

Lemma 6. [2,7] *Let $G = (V, E, F)$ be a biconnected plane graph. Then a drawing ϕ of f_o on a convex polygon P can be extended to a convex drawing of G if and only if the following conditions (i)-(iii) hold:*

(i) For each inner vertex v with $d_G(v) \geq 3$, there exist three paths disjoint except v, each connecting v and an outer vertex;

(ii) Every cycle of G which has no outer edge has at least three vertices v with $d_G(v) \geq 3$; and

(iii) Let $Q_1, Q_2, \ldots, Q_k$ be the subpaths of f_o, each corresponding to a side of P. The graph $G - V(f_o)$ has no component H such that all the outer vertices adjacent to vertices in H are contained in a single path Q_i, and there is no inner edge (u, v) whose end vertices are contained in a single path Q_i. □

Since every inner vertex of degree 2 must be drawn as a point sub-dividing a line segment in any convex drawing, we can assume without loss of generality that a given biconnected plane graph has no inner vertex of degree 2. Then Lemma 6 can be restated as follows.

Lemma 7. *Let $G = (V, E, F)$ be a biconnected plane graph which has no inner vertex with degree 2. Then a drawing ϕ of f_o on a convex polygon P can be extended to a convex drawing of G if and only if the following conditions* (a) *and* (b) *hold:*

(a) *G is internally triconnected.*

(b) *Let $Q_1, Q_2, \ldots, Q_k$ be the subpaths of f_o, each corresponding to a side of P. Each Q_i is an archfree path in G.*   □

5   Convex Drawing with a Star-Shaped Boundary

A kernel $K(P)$ of a polygon P is the set of all points from which all points in P are visible. The boundary of a kernel, if any, is a convex polygon. A polygon P is called *star-shaped* if $K(P) \neq \emptyset$. Throughout the paper, we assume that for a given star-shaped polygon, its kernel has a positive area.

Let ϕ be a drawing of the outer facial cycle f_o of a plane graph G on a star-shaped polygon P, and let $S = \{v_1, v_2, \ldots, v_p\}$ be a subset of $V(f_o)$, where the vertices $v_1, v_2, \ldots, v_p$ in S appear in this order when we traverse f_o in the clockwise order (where $v_{p+1} = v_1$). A subset S is *valid* if S contains all vertices $v \in V(f_o)$ such that $\phi(v)$ is a concave apex of P and for any point $a \in K(P)$, the angle $(\phi(v_i), a, \phi(v_{i+1}))$ formed by line segments $(\phi(v_i), a)$ and $(a, \phi(v_{i+1}))$ is less than π. Obviously $S = V(f_o)$ is valid if P is a star-shaped polygon.

For a tree T of G whose leaf set is $S = \{v_1, v_2, \ldots, v_p\}$, we denote by $H = T + f_o$ the plane subgraph of G obtained by joining T and f_o, i.e., $V(H) = V(T) \cup V(f_o)$, $E(H) = E(T) \cup E(f_o)$ and $F(H) = \{f_1, f_2, \ldots, f_p\}$, where f_i is the cycle consisting of the path between v_i and v_{i+1} along T and the subpath from v_i to v_{i+1} in f_o. We now describe an important lemma.

Lemma 8. *Let $G = (V, E, F)$ be a triconnected plane graph, and ϕ be a drawing of the outer facial cycle f_o on a star-shaped polygon P. For drawing ϕ, let S be a valid subset of $V(f_o)$, and T be an archfree tree whose leaf set is S. Then ϕ can be extended to a strictly convex drawing ψ of $H = T + f_o$. Such a drawing ψ can be obtained in linear time.*

Proof. Let K be a circle contained in $K(P)$, where the center of K is denoted by c. It suffices to show that the lemma holds only for the case where T contains no vertex v with $d_T(v) = 2$, since two edges in T that are adjacent at such a vertex v can be replaced with a single edge and the eliminated vertex v can be re-inserted in a line segment in a straight-line drawing. See Fig. 3(a), which illustrates $H = T + f_o$ with no vertex v with $d_T(v) = 2$ and K centered at a point c. Now T has at least three leaves (since S is valid) but no vertex v with $d_T(v) = 2$. We call a non-leaf vertex in T a *fringe vertex* if it has no more than one neighbor u with $d_T(u) \geq 2$. If T has exactly one fringe vertex, i.e., T is a star centered at a vertex v^*, then ψ with $\psi(u) = \phi(u)$, $u \in S$ and $\psi(v^*) = c$ is a strictly convex drawing of H since each facial cycle f_i is drawn as a triangle whose sides do not intersect with any side of P by the validity of S.

We now consider the case where T has at least two fringe vertices. Each fringe vertex u has at least two adjacent leaves and all adjacent leaves appear consecutively along f_o, when we visit the leaves of T along P in the clockwise order. We denote its first leaf (resp., last leaf) by a_u (resp., b_u). For example, vertex v in Fig. 3(a) has $a_v = v_1$ and $b_v = v_h$.

A fringe vertex u is called *wide* in a drawing ϕ if the angle $(\phi(a_u), c, \phi(b_u))$ formed by line segments $(c, \phi(a_u))$ and $(c, \phi(b_u))$ is no less than π (for example, vertex u in Fig. 3(a) is a wide fringe vertex). Note that T has at most one wide fringe vertex.

We prove that a drawing ϕ of f_o on P can be extended to a strictly convex drawing ψ of $H = T + f_o$ such that ψ satisfies the following two conditions:

$$\text{all non-leaf vertices of } T \text{ are drawn strictly inside } K, \tag{1}$$

$$\begin{array}{l}\text{the angle } (\psi(a_u), u, \psi(b_u)) < \pi \text{ for all fringe vertices } u \\ \text{except for a wide fringe vertex.}\end{array} \tag{2}$$

We prove the lemma by induction on the number of vertices in a tree T. In the base case, T has exactly two fringe vertices u_1 and u_2, where the wide fringe vertex (if any) is denoted by u_1. Draw vertex u_1 as the center c of K, and vertex u_2 as a point at the intersection of K and triangle u_1, a_{u_2}, b_{u_2}. By the choice of u_1, the resulting drawing ψ of $H = T + f_o$ is strictly convex satisfying (1) and (2).

We now assume that the lemma holds for any tree T with at most k vertices. Let T be a tree with $k + 1$ vertices. We prove that the lemma holds for T. Let $v^* \in V(T)$ be a fringe vertex which is not wide, and $v_1, v_2, \ldots, v_h$ be the leaves adjacent to v^*, which appear in this order when we traverse f_o in the clockwise order, as shown in Fig. 3(a). We distinguish two cases, $h \geq 3$ and $h = 2$.

Case-1. $h \geq 3$. Let T' be the tree obtained from T by removing leaves $v_2, v_3, \ldots, v_{h-1}$, $S' = S - \{v_2, v_3, \ldots, v_{h-1}\}$, f_o' be the cycle obtained from f_o by replacing its subpath from v_1 to v_h with an edge (v_1, v_h). Let P' be the polygon obtained from P by replacing the edges between v_1 and v_h with a single edge, and ϕ' be the resulting drawing of f_o' on P' (see Fig. 3(b)). Note that v^* remains a non-wide fringe vertex in T' and hence, S' is valid in P'. By the inductive hypothesis, ϕ' can be extended to a strictly convex drawing ψ' of $H' = T' + f_o'$ that satisfies (1) and (2) (see Fig. 3(b)). We show that the original drawing ϕ of f_o on P can be extended to a strictly convex drawing ψ of $H = T + f_o$ satisfying (1) and (2). Such a drawing ψ is obtained from ϕ and ψ' by setting $\psi(u) = \psi'(u)$, $u \in V(H')$ and $\psi(v_i) = \phi(v_i)$, $i = 2, 3, \ldots, h - 1$. In the resulting drawing ψ, each deleted edge (v^*, v_i), $i = 2, 3, \ldots, h - 1$ is drawn as a line segment $(\psi'(v^*), \phi(v_i))$ (see Fig. 3(c)). We see that each of new facial cycle $\{v^*, v_{i-1}, v_i\}$, $i = 2, 3, \ldots, h-1$ is drawn as a triangle, which does not intersect with P since $\psi'(v^*)$ is in K and v remains as a non-wide fringe vertex in T by (2). Therefore, ϕ is a strictly convex drawing of H which satisfies (1) and (2).

Case-2. $h = 2$. Let w be the unique neighbour of v^* with $d_T(w) \geq 3$, and $e_a = (w, u_a)$ (resp., $e_b = (w, u_b)$) be the edge incident to w that appears immediately

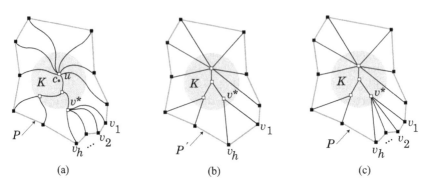

Fig. 3. (a) Reducing the number of leaves adjacent to v^* in tree T; (b) Re-inserting the deleted leaves in a convex drawing of $H' = T' + f'_o$ with boundary P'; (c) A convex drawing of $H = T + f_o$ with boundary P

before edge (w, v^*) (resp., immediately after edge (w, v^*)) when we walk around w in the clockwise order. Since $d_T(w) \geq 3$, we have $e_a \neq e_b$ (see Fig. 4(a)).

Let T' be the tree obtained from T by contracting edge (v^*, w) into a vertex w. Obviously the set of leaves in T' is S, which remains valid in P. By the inductive hypothesis, a drawing ϕ of f_o on P can be extended to a strictly convex drawing ψ' of $H' = T' + f_o$ that satisfies (1) and (2) (see Fig. 4(b)). We show that the original drawing ϕ of f_o on P can be extended to a strictly convex drawing ψ of $H = T + f_o$ satisfying (1) and (2). Such a drawing ψ is obtained from ϕ and ψ' by setting $\psi(u) = \psi'(u)$, $u \in V(H') - \{v^*\}$ and choosing $\psi(v^*)$ of v as follows. Let f_a (resp., f_b) be the facial cycle containing edges e_a and (w, v_1) (resp., e_b and (w, v_h)), and L_a (resp., L_b) be the half line which starts at $\phi'(w)$ in the direction from $\phi'(u_a)$ to $\phi'(w)$ (resp., from $\phi'(u_b)$ to $\phi'(w)$). See Fig. 4(b). To keep the faces f_a and f_b strictly convex after re-inserting edge (v^*, w), the position $\psi(v^*)$ of v^* must be in the region R enclosed by the boundary of K and the two lines L_a and L_b. Since $e_a \neq e_b$, we can choose a position $\psi(v^*)$ of v strictly inside K so that the resulting drawing ψ is strictly convex that satisfies (1) and (2). This completes the proof for the existence of a desired convex drawing ψ of $H = T + f_o$.

It is clear that the above inductive proof gives an algorithm for computing a desired drawing ϕ of H. We now show that it runs in linear time. The kernel $K(P)$ can be computed in linear time [6], and a circle K in $K(P)$ can be chosen in linear time. The operation in Case-1 can be executed in $O(1)$ time by placing v on a point in R sufficiently closed to $\phi'(w)$. The operation in Case-2 can be executed in $O(1)$ time per edge to be deleted. Following the above construction, a desired convex drawing ϕ of $H = T + f_o$ can be constructed in linear time. This completes the proof. □

We are now ready to prove Theorem 1.

Theorem 1. *Every drawing ϕ of the outer facial cycle f_o of a triconnected plane graph $G = (V, E, F)$ on a star-shaped polygon can be extended to a convex drawing ψ_G of G. Such a drawing ψ_G can be computed in linear time.*

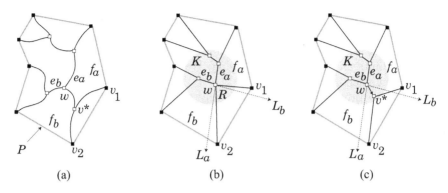

Fig. 4. (a) Contracting edge (v^*, w) in tree T; (b) Re-inserting a contracted edge (v^*, w) in a convex drawing ψ' of $H' = T' + f_o$; (c) A convex drawing of $H = T + f_o$ with boundary P

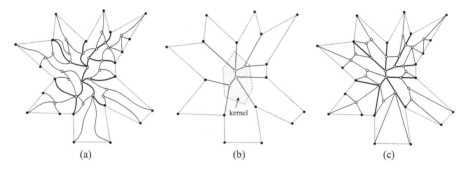

Fig. 5. (a) A triconnected plane graph $G = (V, E, F)$ with a star-shaped boundary P, where an archfree tree T is denoted by thick black lines; (b) A convex drawing of $H = T + f_o$ with boundary P; (c) A convex drawing of G with boundary P

Proof. Choose a valid subset $S = \{v_1, v_2, \ldots, v_p\}$ of $V(f_o)$ (for example $S = V(f_o)$), and compute an archfree tree T of G whose leaf set is S, which can be computed in linear time by Lemma 5 (see Fig. 5(a)). By Lemma 8, we can compute a strictly convex drawing ψ of $H = T + f_o$ as an extension of drawing ϕ of f_o on P in linear time (see Fig. 5(b)). Now each facial cycle f_i of H is drawn as a strictly convex polygon P_i. Let G_i be the subgraph of G that consists of vertices and edges in the face f_i (including those on the boundary of f_i). Since each side of P_i is an archfree path and ψ is a strictly convex drawing, G_i with the boundary P_i satisfies condition (b) of Lemma 7. It is clear that G_i with P_i also satisfies condition (a) of Lemma 7, since otherwise G with P would violate (a), contradicting the triconnectivity of G. Hence, each plane graph G_i with boundary P_i has a convex drawing ψ_i, and such a drawing ψ_i can be computed in $O(|V(G_i)| + |E(G_i)|)$ time by using the algorithm in [2]. Therefore, after computing convex drawings ψ_i for all G_i and placing them in the corresponding faces in ψ, we obtain a convex drawing ψ_G of G which is an extension of drawing ϕ of f_o on P (see Fig. 5(c)). The above algorithm runs in linear time. ☐

6   Conclusion

In this paper, it is proved that every triconnected plane graph with a star-shaped boundary admits an inner-convex drawing. We present a linear time algorithm for computing such a convex drawing.

Recently, we proved that every internally triconnected hierarchical-st plane graph with a convex boundary satisfying condition (b) in Lemma 7 admits a convex drawing [4].

It is left open to find a characterization for a plane graph with a star-shaped boundary to admit an inner-convex drawing.

References

1. G. Di Battista, P. Eades, R. Tamassia and I. G. Tollis, *Graph Drawing: Algorithms for the Visualization of Graphs*, Prentice-Hall, 1998.
2. N. Chiba, T. Yamanouchi and T. Nishizeki, Linear algorithms for convex drawings of planar graphs, Progress in Graph Theory, Academic Press, pp. 153-173, 1984.
3. I. Fáry, On straight line representations of planar graphs, Acta Sci. Math. Szeged, 11, pp. 229-233, 1948.
4. S.-H. Hong and H. Nagamochi, Convex drawings of hierarchical plane graphs, Proc. of AWOCA 2006, to appear.
5. T. Nishizeki and N. Chiba, *Planar Graphs: Theory and Algorithms*, North-Holland Mathematics Studies 140/32, 1988.
6. F. P. Preparata and M. I. Shamos, *Computational Geometry: An Introduction*, Springer, 1993.
7. C. Thomassen, Plane representations of graphs, in Progress in Graph Theory, J. A. Bondy and U. S. R. Murty (Eds.), Academic Press, pp. 43-69, 1984.
8. W. T. Tutte, Convex representations of graphs, Proc. of London Math. Soc., 10, no. 3, pp. 304-320, 1960.

How to Sell a Graph:
Guidelines for Graph Retailers

Alexander Grigoriev[1], Joyce van Loon[1,*], René Sitters[2], and Marc Uetz[1]

[1] Maastricht University, Quantitative Economics,
P.O. Box 616, NL–6200 MD Maastricht, The Netherlands
{a.grigoriev, j.vanloon, m.uetz}@ke.unimaas.nl
[2] Max-Planck Institute für Informatik, Algorithms and Complexity,
Stuhlsatzenhausweg 85, 66123 Saarbrücken, Germany
sitters@mpi-inf.mpg.de

Abstract. We consider a profit maximization problem where we are asked to price a set of m items that are to be assigned to a set of n customers. The items can be represented as the edges of an undirected (multi)graph G, where an edge multiplicity larger than one corresponds to multiple copies of the same item. Each customer is interested in purchasing a bundle of edges of G, and we assume that each bundle forms a simple path in G. Each customer has a known budget for her respective bundle, and is interested only in that particular bundle. The goal is to determine item prices and a feasible assignment of items to customers in order to maximize the total profit. When the underlying graph G is a path, we derive a fully polynomial time approximation scheme, complementing a recent NP-hardness result. If the underlying graph is a tree, and edge multiplicities are one, we show that the problem is polynomially solvable, contrasting its APX-hardness for the case of unlimited availability of items. However, if the underlying graph is a grid, and edge multiplicities are one, we show that it is even NP-complete to approximate the maximum profit to within a factor $n^{1-\varepsilon}$.

Keywords: Pricing problems, tollbooth problem, highway problem, computational complexity, dynamic programming, fully polynomial time approximation scheme.

1 Introduction

We consider a profit maximization problem that is defined on a multi graph. Given is a simple undirected graph $G = (V, I)$ with $|I| = m$ edges, and given are (integral) edge multiplicities c_i, $i \in I$. Each edge can be thought of as an *item* for sale, and the edge multiplicity determines how many copies of the corresponding item are available. We will mainly discuss problems where all c_i's are finite, but notice that most of the related work addresses problems with unlimited availability of items. There is a set of n *customers* $J = \{1, \ldots, n\}$ each of which

---

* Supported by METEOR, the Maastricht Research School of Economics of Technology and Organizations.

F.V. Fomin (Ed.): WG 2006, LNCS 4271, pp. 125–136, 2006.

is interested in purchasing exactly one *bundle* $I_j \subseteq I$ of items. This is also known as *single-minded* customers [13]. We assume that each bundle I_j forms a simple *path* in the graph G. Each customer $j \in J$ has an integral *budget* (or valuation) b_j, which is the largest amount that a customer is willing to pay for her bundle I_j.

A feasible solution to the problem consists of an allocation of items to customers and a vector of prices $p = (p_1, \ldots, p_m)$, one for each item, such that

1. each item $i \in I$ is sold to no more than c_i customers,
2. each customer $j \in J$ is either assigned the entire bundle I_j, or nothing,
3. a bundle I_j assigned to a customer j must be affordable, $\sum_{i \in I_j} p_i \leq b_j$.

Within a given solution, let us call a customer a *winner* if she gets assigned her bundle I_j. Let us denote by W the set of winners. We call a set of winners W *feasible* with respect to the availability of items whenever $\sum_{j \in W} |I_j \cap \{i\}| \leq c_i$, for all $i \in I$. Note that for any feasible set of winners W, there exists a price vector p such that all customers in W can indeed afford their respective bundles. The optimization problem consists of finding a feasible set of winners and a vector of prices such that none of the constraints $(1) - (3)$ is violated, and such that the total profit $\Pi(W, p) = \sum_{j \in W} \sum_{i \in I_j} p_i$ is maximized.

One usually distinguishes between solutions that are *envy-free* and those that are not. In the setting with single minded customers considered here, envy-freeness requires that if a customer is not a winner, then the total price of that bundle must exceed her budget. However, we mostly address problems without requiring envy-freeness.

Notice that, in contrast to the classical auction literature, we assume that we *know* the budget of every customer. At first sight this assumption may seem infeasible, yet there are good reasons for studying this type of problems. An understanding of how items should be priced under known budgets, for example, may be useful also for the more difficult problem with unknown budgets. This connection was recently made concrete by Goldberg and Hartline [9]; they reduced a mechanism design problem with unknown budgets to the underlying profit maximization problem with known budgets. Furthermore, data on customer valuations is nowadays collected at large scale, for example via specifically designed web sites; see e.g. [8]. Hence, the assumption of known budgets is reasonable in many settings. Finally, the underlying combinatorial pricing problems have their own appeal, and results with respect to their computational tractability have been obtained only recently [1,3,5,6,8,11,12].

Recall that we address problems where the bundle I_j of any customer $j \in J$ corresponds to a path in an underlying simple graph G. We consider three different types of underlying graphs, namely paths, trees, and grids. The problem on a path (with unlimited availability of items) was recently discussed by Guruswami et al. [11]. They call it the 'highway problem', motivated by the question to find optimal tolls for the usage of a single highway. In their setting, the availability of items is unlimited. To motivate the problem on a path with *limited* availability of items, consider the rental of a set of identical objects over discrete time periods; e.g. houses in holiday parks on the basis of weeks. Customers are interested in renting an object in consecutive periods. Since all objects are identical, in any time period they need to have the same price. But prices may vary from period to period. We can thus interpret a time period as an edge in a path, with edge multiplicity equal to the number of objects available in that period. Notice that envy-freeness is not necessarily an issue, since the manager can freely decide to whom to rent the objects.

1.1   Related Work

The profit maximization problem in which the customer's bundles are paths of an arbitrary graph G, and the availability of items is unlimited, is the 'tollbooth problem' addressed by Guruswami et al. [11]. They show that the problem is APX-hard even if the graph is a star, all budgets are equal to one, and the bundles contain at most 2 items. Another APX-hardness proof for the same problem was given by Briest and Krysta [5]; in contrast to the result of [11] it is also valid if several problem parameters are constantly bounded.

The tollbooth problem with the restriction that the underlying graph is a path is the 'highway problem' introduced by Guruswami et al. [11]. NP-hardness of this problem was recently shown by Bodlaender and Penninkx [4] and Briest and Krysta [5]. Guruswami et al. [11] furthermore propose a polynomial time dynamic programming algorithm when the budgets are bounded by a constant, and a pseudo-polynomial time dynamic programming algorithm when the bundle sizes are bounded by a constant.

When bundles are not paths in a graph, but arbitrary subsets of the given set of items, Demaine et al. [6] show that the problem (again with unlimited availability of items) is hard to approximate to within a (semi-)logarithmic factor. For the same problem, there exists an approximation scheme with almost linear running time, given that the number of distinct items is constant, by Hartline and Koltun [12]. Moreover, Balcan and Blum [3] derive a $\mathcal{O}(k)$-approximation algorithm, given that each customer is interested in bundles of at most k items.

Finally, independently in [3] and [5], two FPTAS's are presented for the problem where the customers' bundles are *nested*. That is, for any two bundles I_a and I_b it holds that $I_a \subseteq I_b$, $I_b \subseteq I_a$ or $I_a \cap I_b = \emptyset$. Notice that such an instance can be interpreted as a problem where the bundles are subpaths of a single path, by ordering the items appropriately. (The converse is not true, however.)

1.2   Our Results

In Section 2, we address the problem where the underlying graph G is a path, and there is an upper bound C on the availability of any item. We propose a dynamic programming algorithm that computes an optimal solution in time $\mathcal{O}(n^{2C} B^{2C} m)$. Here, $B = \max_j b_j$ is an upper bound on the budgets. The same problem with unlimited supply of items allows a dynamic programming algorithm with a computation time of $\mathcal{O}(B^{B+2} n^{B+3})$, see [11]. Based on our dynamic programming algorithm, we moreover derive an FPTAS for that problem, given that the maximum availability of any item C is constant. In contrast to previous results in that direction [3,5,12], this FPTAS does neither require a constant number of items, nor nested bundles or bounded budgets.

In Section 3 we address the problem where the availability of any item is exactly one. For the case that graph G is a path, the problem reduces to finding a maximum weight independent set in an interval graph; thus it is polynomially solvable [14]. When we generalize from a path to a tree, we can show that the problem remains polynomially solvable. When the underlying graph G is a grid, however, we show that it is NP-complete to approximate the maximum profit to within $n^{1-\varepsilon}$, for any $\varepsilon > 0$. (Recall that n is the number of customers.)

2   Selling a Multi Path

In this section, we restrict the underlying graph $G = (V, I)$ to be a path. We first discuss some preliminaries. Thereafter, we present a dynamic programming approach and a fully polynomial time approximation scheme for the case where the edge multiplicities c_i are bounded by some constant C.

2.1   Preliminaries

It is not hard to see that the profit maximization problem on a multi path is polynomially solvable if either the set of winners W is given, or if the vector of prices p is given.

Lemma 1. *The profit maximization problem on a multi path is polynomially solvable if the vector of prices $p = (p_1, \ldots, p_m)$ is given.*

Proof. If the vector of prices $p = (p_1, \ldots, p_m)$ is given, we only need to find a feasible set of winners that maximizes the total revenue. Whenever the items are available in unlimited supply, this is trivial and the set of winners is just $W := \{j \mid \sum_{i \in I_j} p_i \leq b_j\}$. For the case of limited supply, let W' be the set of customers for which the bundle is affordable, given the price vector p. For any item i, we can not sell more than c_i copies. Let a_{ij} be equal to 1 if item $i \in I_j$ for customer j, and 0 otherwise. We find a profit-maximizing feasible subset of winners by solving the following linear program, where $x_j = 1$ iff customer j is a winner.

$$\max \sum_{j \in W'} \left(\sum_{i \in I_j} p_i \right) x_j$$
$$\text{s.t.} \sum_{j \in W'} a_{ij} x_j \leq c_i \quad \forall i \in I$$
$$0 \leq x_j \leq 1 \quad \forall j \in W'$$

The constraint matrix of this linear program has the consecutive ones property, that is, all entries that are 1 appear consecutively in any column. This because the bundles I_j of any customer j consist only of consecutive items. A consecutive ones matrix is totally unimodular [15]. Hence, the corresponding polyhedron only has integral vertices, and the LP yields an integral optimal solution.   □

If on the other hand a feasible set of winners $W \subseteq J$ is given, we find an optimal price vector $p = (p_1, \ldots, p_m)$ by solving the following linear program.

$$\max \sum_{j \in W} \sum_{i \in I_j} p_i$$
$$\text{s.t.} \quad \sum_{i \in I_j} p_i \leq b_j \ \forall j \in W$$
$$p_i \geq 0 \qquad \forall i \in I$$

Since this constraint matrix has the consecutive ones property, too, we obtain the following.

Lemma 2 (See also [11, Lemma 5.1]). *The profit maximization problem on a multi path is polynomially solvable if a feasible set of winners $W \subseteq J$ is given. Moreover, since the budgets b_j are integral, there exists an optimal, integral price vector.*

2.2 Complexity

Bodlaender and Penninkx [4] as well as Briest and Krysta [5] recently showed the following.

Theorem 1 ([4,5]). *The profit maximization problem on a multi path is (weakly) NP-hard, even if the edge multiplicity is at most three.*

For their reduction it suffices, but it is also necessary, that the availability of any item is 3. If we restrict the availability of any item to be at most 2, the complexity remains open.

2.3 Dynamic Programming Algorithm

Recall that for each item $i \in I$ there are c_i copies available, and for convenience of notation let

$$C \geq \max_{i \in I} c_i$$

be an upper bound on the availability of any item. We show that we can solve this problem in time $\mathcal{O}(n^{2C} B^{2C} m)$ by finding a longest path in an acyclic digraph.

We create an m-layered digraph with an additional source s and sink t (layers 0 and $m+1$, respectively). There are arcs only from layer i to $i+1$, for $i = 0, \ldots, m$. Hence, in any $s - t$ path, there are exactly $m + 2$ nodes. In every node in layer i (corresponding to item i), we store all winners j that purchase item i. Moreover, we store the respective total amounts all these winners spend on all items in their respective bundles up to and including item i. Any node x (more precisely, the path $s - x$) represents a feasible partial solution. Arcs from node x of layer i to node y of layer $i+1$ are only introduced if the path $s - y$ represents a feasible extension of the partial solution represented by the path $s - x$. The weight on an arc that connects a node of layer i to a node of layer $i + 1$ is equal to the profit earned on item $i + 1$, that is, the total amount that the corresponding winners pay for item $i + 1$. Therefore, the weight of the longest $s - t$ path in the digraph will equal the maximum total profit, and the set of winners can be reconstructed from the longest $s - t$ path, too. The algorithm below shows a more formal description.

Algorithm 1. Dynamic Programming Algorithm

Input: Profit maximization problem on a multi path with maximal availability
　　　　of any item $\leq C$
Output: Assignment of items to customers and item prices p_i
begin (construction of digraph D)
    nodes: For each item $i \in I$, we introduce a layer of nodes: Denote by J^i the
    set of customers with $i \in I_j$. By $K^i = (j_1, j_2, \ldots, j_k)$ we denote any (sorted)
    subset of J^i of cardinality k, where $k \leq \min\{c_i, |J^i|\} \leq C$. Define
    $H_j := \{0, 1, \ldots, b_j\}$ as the possible total amount customer $j \in K^i$ can spend
    for items $\{1, \ldots, i\} \cap I_j$. Let $h^i \in H_{j_1} \times H_{j_2} \times \cdots \times H_{j_k}$ be a vector denoting
    how much each customer j spends for items $\{1, \ldots, i\} \cap I_j$, for each $j \in K^i$. If
    $K^i = \emptyset$, we let $h^i = 0$. Let all such pairs (K^i, h^i) be the nodes in layer i of D,
    for $i = 1, \ldots, m$. Let s and t denote source and sink. To unify notation,
    assume $s = (\emptyset, 0)$ and $t = (\emptyset, 0)$;
    arcs: Insert an arc from node (K^i, h^i) to node (K^{i+1}, h^{i+1}) if:
    (1) For all $j \in K^i$ with $i + 1 \in I_j$, $j \in K^{i+1}$, and for all $j \in J^i \setminus K^i$ with
    $i + 1 \in I_j$, $j \notin K^{i+1}$.
    (2) There exists a unique integral value $d \geq 0$ such that $d = h_j^{i+1} - h_j^i$ for all
    $j \in K^i \cap K^{i+1}$, and $d = h_j^{i+1}$ for all $j \in K^{i+1} \setminus K^i$.
    We furthermore connect source node s to all nodes (K^1, h^1), and we connect
    all nodes (K^m, h^m) to sink node t.
    arc lengths: For an arc a^i that connects (K^{i-1}, h^{i-1}) and (K^i, h^i): If
    $K^i = \emptyset$, we let the length of arc a^i be $\ell(a^i) = 0$, and if $K^i \neq \emptyset$, we let the
    length of arc a^i be $\ell(a^i) = d|K^i|$, where d is the (unique) value from
    condition (2) above.
end
solution: Compute the longest $s - t$ path $\mathcal{P}$ in digraph D. Whenever for
customer j we have that $j \in K^i$ with $(K^i, h^i) \in \mathcal{P}$, customer j gets assigned
item i. The price p_i for item i equals $\ell(a^i)/|K^i|$, where a^i is the arc from path $\mathcal{P}$
that connects nodes (K^{i-1}, h^{i-1}) and (K^i, h^i);

Theorem 2. *The dynamic programming algorithm outputs an optimal solution for an instance of the profit maximization problem on a multi path in* $\mathcal{O}(n^{2C} B^{2C} m)$ *time.*

Proof. The algorithm assigns items to customers and computes prices for all items according to an $s - t$ path $\mathcal{P}$. This solution is feasible if each customer j either gets the complete bundle I_j or nothing, no item $i \in I$ is assigned to more than c_i customers, and if for every customer j who gets her bundle I_j, we have that $\sum_{i \in I_j} p_i \leq b_j$.

Let an $s - t$ path $\mathcal{P}$ be fixed, let a^i be the arcs on $\mathcal{P}$, $\ell(a^i)$ the length of these arcs, and abusing notation let (W^i, h^i) be the nodes on $\mathcal{P}$. Set W^i is thus the set of customers that get assigned item i.

By definition of the nodes of the digraph, no customer j will be assigned an item not from her bundle I_j. For any customer j, consider an item $i \in I_j$ such that $j \in W^i$. That is, customer j gets assigned item i by the algorithm. By condition (1) of the digraph construction, all other items of bundle I_j must be assigned to customer j as well. Next, for any node (W^i, h^i), we have by definition that $|W^i| \leq c_i$, hence no item is oversold. Finally, let us consider the budget constraint of customer j. We know that $I_j = \{k, \ldots, k'\}$ for some $k \leq k'$. We have that

$$\sum_{i \in I_j} p_i = \sum_{i=k}^{k'} p_i = \sum_{i=k}^{k'} \frac{\ell(a^i)}{|W^i|} = h_j^k + \sum_{i=k+1}^{k'} (h_j^i - h_j^{i-1}) = h_j^{k'} \leq b_j.$$

The third equality holds due to condition (2) of the digraph construction, and the last inequality holds because $h_j^{k'} \in H_j = \{0, 1, \ldots, b_j\}$.

Now we know that any $s - t$ path $\mathcal{P}$ in D defines a feasible solution, and $W := \bigcup_{i \in I} W^i$ denotes the set of winners. The length of a path is

$$\sum_{a^i \in \mathcal{P}} \ell(a^i) = \sum_{i \in I} p_i |W^i| = \sum_{j \in W} \sum_{i \in I_j} p_i.$$

In other words, the path length defines the profit of the corresponding solution, thus the longest path yields an optimal solution.

To arrive at the computation time of $\mathcal{O}(n^{2C} B^{2C} m)$, we only need to estimate the size of the digraph D. For every item $i \in I$, there are at most $\mathcal{O}(n^C)$ different sets K^i and at most $\mathcal{O}(B^C)$ different vectors h^i. Thus, per item $i \in I$, we have at most $\mathcal{O}(n^C B^C)$ nodes (K^i, h^i). For any $i \in I$, every node (K^i, h^i) is connected to at most $\mathcal{O}(n^C B^C)$ nodes (K^{i+1}, h^{i+1}). So, per item, there are at most $\mathcal{O}(n^{2C} B^{2C})$ arcs, which means that there are at most $\mathcal{O}(n^{2C} B^{2C} m)$ arcs in D. The computation time to find the longest path in D is linear in the number of arcs, since D is acyclic [2]. $\qquad \square$

Notice that the solution constructed by the dynamic programming algorithm need not be envy-free.

2.4   FPTAS

We next show how to turn the dynamic programming algorithm into a fully polynomial time approximation scheme (FPTAS); that is, an algorithm that computes a solution with profit at least $(1 - \varepsilon)$ times the optimum profit, in time polynomial in the input and $1/\varepsilon$. To that end, we just apply the dynamic programming algorithm on a rounded instance in which $K := \frac{\varepsilon B}{2n^2}$ for any $\varepsilon > 0$ and the customers' budgets are $b_j' := \lfloor b_j / K \rfloor$.

Lemma 3. *For every solution (W, p) of the original instance, there exists a solution (W, p'') of the rounded instance with $\Pi(W, p) > \frac{1}{K} \Pi(W, p'') - mn$.*

Proof. Let (W, p) be a feasible solution of the original instance with profit $\Pi(W, p)$. Let $p_i'' = \lfloor p_i / K \rfloor$, $i = 1, \ldots, m$. Note that $p_i/K - 1 < p_i'' \leq p_i/K$. For the original instance we have for every winner $j \in W$, $\sum_{i \in I_j} p_i \leq b_j$, and it follows that

$$\sum_{i \in I_j} p_i'' = \sum_{i \in I_j} \left\lfloor \frac{p_i}{K} \right\rfloor \leq \left\lfloor \frac{\sum_{i \in I_j} p_i}{K} \right\rfloor \leq \left\lfloor \frac{b_j}{K} \right\rfloor = b_j'.$$

Hence, the same set of customers W can be made winners in the rounded instance. Then the capacity constraint is satisfied as well, and the solution (W, p'') is feasible. Finally, we have

$$\Pi(W, p'') = \sum_{j \in W} \sum_{i \in I_j} p_i'' > \sum_{j \in W} \sum_{i \in I_j} \left(\frac{p_i}{K} - 1\right) \geq \frac{1}{K}\Pi(W, p) - mn. \qquad \square$$

Lemma 4. *For every solution (W', p') of the rounded instance, there exists a solution $(W', \tilde{p})$ of the original instance with $\Pi(W', \tilde{p}) = K\Pi(W', p')$.*

Proof. Let (W', p') be a solution in the rounded instance with revenue $\Pi(W', p')$. Let $\tilde{p}_i = p_i' K$ be prices in the original instance, $i = 1, \ldots, m$. (This is integer because p_i' and K are integer.) Then the budget constraint for every customer $j \in W'$ is satisfied, because

$$\sum_{i \in I_j} \tilde{p}_i = K \sum_{i \in I_j} p_i' \leq Kb_j' = K \left\lfloor \frac{b_j}{K} \right\rfloor \leq b_j.$$

Hence, we can make the same set W' of customers winners, and solution $(W', \tilde{p})$ is feasible for the original instance. The revenue can be written as

$$\Pi(W', \tilde{p}) = \sum_{j \in W'} \sum_{i \in I_j} \tilde{p}_i = \sum_{j \in W'} \sum_{i \in I_j} (p_i' K) = K \sum_{j \in W'} \sum_{i \in I_j} p_i' = K\Pi(W', p'). \qquad \square$$

We can now combine Lemmas 3 and 4 to obtain an FPTAS.

Theorem 3. *There exists an FPTAS for the profit maximization problem on a multi path.*

Proof. Let (W, p) and (W', p') be the optimal solutions in the original and rounded instances, respectively. Consider solution $(W', \tilde{p})$ for the original instance, where $\tilde{p}_i = Kp_i'$, $i = 1, \ldots, m$, and solution (W, p'') for the rounded instance, where $p_i'' = \lfloor p_i/K \rfloor$. An application of the previous two lemmas now yields

$$\Pi(W', \tilde{p}) = K\Pi(W', p') \geq K\Pi(W, p'') > K\left(\frac{1}{K}\Pi(W, p) - mn\right) = \Pi(W, p) - \varepsilon B \frac{mn}{2n^2},$$

where the first inequality holds due to optimality of (W', p') for the rounded instance. Note that $m \leq 2n - 1 < 2n$, and the optimal profit is at least equal to the maximum budget B, so $B \leq \Pi(W, p)$. Thus, $\Pi(W', \tilde{p}) > (1 - \varepsilon)\Pi(W, p)$.

Concerning the computation time to compute the optimal solution (W', p'), observe that the size of the digraph is $\mathcal{O}(n^{6C+1}/\varepsilon^{2C})$. Hence, the computation time to find the longest path is polynomial in terms of n and $1/\varepsilon$. $\qquad \square$

Again, notice that the solution constructed by the FPTAS need not be envy-free.

3   Selling Simple Graphs

In this section we assume that the availability of any item i is one, or in other words, the edge multiplicities c_i are one. We consider three types of graphs, namely paths, trees and grids.

Theorem 4 ([16]). *The profit maximization problem on a simple path can be solved in $\mathcal{O}(n^2)$ time.*

In fact, the result of Theorem 4 is not surprising, since the problem reduces to finding a maximum weight independent set in an interval graph, a problem known to be solvable in polynomial time [14].

3.1   Trees

Guruswami et al. [11] show that the problem with unlimited availability of items is APX-hard even on star graphs. Contrasting this complexity result, we prove that if the availability of each item is exactly one, the profit maximization problem on a tree can be solved in polynomial time. (Again, recall that we do not require the solution to be envy-free.)

Theorem 5. *The profit maximization problem on a simple tree can be solved in $\mathcal{O}(n^5)$ time.*

Proof. Consider the graph $H = (J, E)$ where $(j, k) \in E$ if and only if $I_j \cap I_k \neq \emptyset$, for two customers $j, k \in J$. Since $\{I_j | j \in J\}$ is a collection of simple paths in a tree, graph H is called an *EPT graph* [10]. Since G is a tree and availability of each item is exactly one, the maximum weight independent set in H with vertex weights b_j, $j \in J$, is the optimal set of winners W, and the weight of this independent set is equal to the maximum profit. The vector of optimal prices can be straightforwardly obtained by setting the price of one arbitrary edge from I_j, $j \in W$, to b_j, and setting the prices of all other edges in I_j to 0. The remaining edges in the tree can be priced arbitrarily.

A polynomial time algorithm to compute a maximum weight independent set in an EPT graph was described by Tarjan [17]. The algorithm is a recursive procedure that decomposes the problem on the basis of clique separators. The polynomial running time is a consequence of the fact that the atoms, i.e., the non-decomposable subgraphs of EPT graphs are line graphs. For line graphs, the maximum weight independent set problem is just the maximum weight matching problem, which can be solved in $\mathcal{O}(n^3)$ time by Edmonds algorithm [7]. The total time complexity is bounded by $\mathcal{O}(n^5)$.   □

3.2   Grids

Demaine et al. [6] show that the profit maximization problem where the bundles are arbitrary subsets of items (and with unlimited availability of items) is hard to approximate to within a (semi-)logarithmic factor. If we restrict bundles to be paths in a general graph, Briest and Krysta [5] show that the problem is APX-hard even if several parameters of the problem are constantly bounded. For the even more restricted problem where the bundles are paths in a star, Guruswami et al. [11] also show APX-hardness.

Here we show that if the availability of items is bounded, we can derive an even stronger inapproximability result, even for a very restricted class of graphs and customers' bundles.

Theorem 6. *For all $\varepsilon > 0$, approximating the profit maximization problem on a simple grid to within $n^{1-\varepsilon}$ is NP-hard, even with unit budgets, and when each item is an element of at most two bundles. The same result holds if the solution is required to be envy-free.*

Proof. For the proof we construct an approximation preserving reduction from INDEPENDENT SET. In the latter problem, given a graph $G = (V, E)$, the problem is to find a maximum cardinality subset $S \subseteq V$ such that no two vertices from S are adjacent. It is NP-hard to approximate INDEPENDENT SET within a factor $|V|^{1-\varepsilon}$; see [18].

Let $V = \{v_1, \ldots, v_n\}$ and $E = \{e_1, \ldots, e_m\}$. We construct the instance of the profit maximization problem as follows. We create a grid graph with $(n + 1)$ horizontal layers and $(2m + 2)$ vertical layers. We index the vertices of the grid graph by pairs (i, j) where i is the index of the vertical layer and j is the index of the horizontal layer. Let horizontal layer $j \in \{1, \ldots, n\}$ correspond to vertex $v_j \in V$, and let the edge $((2i, n + 1), (2i + 1, n + 1))$ in the grid correspond to edge $e_i \in E$. Next, for each vertex $v_j \in V$, we introduce a customer in the profit maximization problem with a bundle defined by the following simple path in the grid graph. The path starts at point $(1, j)$ and ends at point $(2m + 2, j)$ following the layer j everywhere except for the edges $((2i, j), (2i + 1, j))$ such

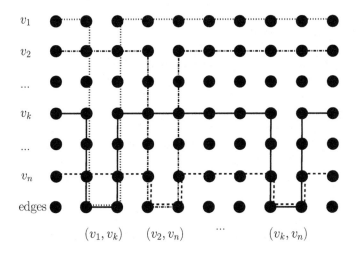

Fig. 1. Grid graph from the reduction

that $v_j \in e_i$. These edges are substituted by vertical detours, passing through edges $((2i, n+1), (2i+1, n+1))$; see Figure 1 for an example. We complete the construction setting the budget of each customer to 1.

We claim that in G there exists an independent set of cardinality K if and only if there exists a solution of the corresponding profit maximization problem with total profit K. By construction, two paths corresponding to adjacent vertices in G must share some edge e in layer $(n+1)$. Since the multiplicity of edge e is 1, only one of these paths can be present in a feasible solution. Hence, the total profit in the profit maximization problem is at most the maximum cardinality independent set in G. Now, consider an independent set S in G and any two vertices in this independent set. By construction, the two corresponding paths in the profit maximization problem are edge disjoint. Therefore, there is a solution of the profit maximization problem where S defines the set of winners and this allocation is feasible with respect to availability of items. For each $v_j \in S$ we set the price of the grid edge $((1, j), (2, j))$ to 1, for each $v_j \notin S$ we set the price of $((1, j), (2, j))$ to 2, and for all other edges of the grid we set the prices to 0. In the constructed solution of the profit maximization problem the total profit equals $|S|$. Thus, the reduction preserves the objective value.

Since the number of customers n in the profit maximization problem exactly equals $|V|$, we derive that the profit maximization problem is hard to approximate to within a factor $n^{1-\varepsilon}$. It remains to notice that the constructed solution is envy-free since the price of the bundle of each non-winning customer equals 2, which is greater than the budget. Therefore, the theorem holds also if we require the solution to be envy-free. □

4 Conclusion

Notice that the currently best known negative result on the tractability of the profit maximization problem on a path is NP-hardness [4,5]. Even though several FPTAS's (including the one of this paper) exist whenever certain parameters are constantly bounded, the best known positive result for the general case is a logarithmic approximation [3,11]. It thus remains an intriguing open problem to obtain a (deterministic) constant approximation algorithm.

Acknowledgements. Thanks to Jason Hartline for several interesting remarks and pointers to several references.

References

1. G. Aggarwal, T. Feder, R. Motwani, and A. Zhu, *Algorithms for multi-product pricing*, Automata, Languages and Programming - ICALP 2004 (J. Díaz, J. Karhumäki, A. Lepistö, and D. Sannella, eds.), Lecture Notes in Computer Science, vol. 3142, Springer, 2004, pp. 72–83.
2. R. K. Ahuja, T. L. Magnanti, and J. B. Orlin, *Network flows*, Prentice Hall, New Jersey, 1993.

3. M.F. Balcan and A. Blum, *Approximation algorithms and online mechanisms for item pricing*, Proc. of the 7th ACM Conference on Electronic Commerce, ACM, 2006, pp. 29–35.

4. H. Bodlaender and E. Penninkx, *An elegant NP-completeness proof for profit maximization on paths*, personal communication, May 2005.

5. P. Briest and P. Krysta, *Single-minded unlimited supply pricing on sparse instances*, Proc. 17th Annual ACM-SIAM Symposium on Discrete Algorithms, ACM-SIAM, 2006, pp. 1093–1102.

6. E. D. Demaine, U. Feige, M.T. Hajiaghayi, and M. R. Salavatipour, *Combination can be hard: Approximability of the unique coverage problem*, Proc. 17th Annual ACM-SIAM Symposium on Discrete Algorithms, ACM-SIAM, 2006, pp. 162–171.

7. J. Edmonds, *Matching and a polyhedron with 0-1 vertices*, Journal of Research of the National Bureau of Standards, B **69** (1965), 125–130.

8. P. W. Glynn, B. Van Roy, and P. Rusmevichientong, *A nonparametric approach to multi-product pricing*, Operations Research **54** (2006), no. 1, 82–98.

9. A. V. Goldberg and J. D. Hartline, *Competitive auctions for multiple digital goods*, Algorithms - ESA 2001 (F. Meyer auf der Heide, ed.), Lecture Notes in Computer Science, vol. 2161, Springer, 2001, pp. 416–427.

10. M. C. Golumbic and R. E. Jamison, *The edge intersection graphs of paths in a tree*, Journal of Combinatorial Theory, Series B **38** (1985), 8–22.

11. V. Guruswami, J. D. Hartline, A. R. Karlin, D. Kempe, C. Kenyon, and F. McSherry, *On profit-maximizing envy-free pricing*, Proc. 16th Annual ACM-SIAM Symposium on Discrete Algorithms, ACM-SIAM, 2005, pp. 1164–1173.

12. J. D. Hartline and V. Koltun, *Near-optimal pricing in near-linear time*, Algorithms and Data Structures - WADS 2005 (F. K. H. A. Dehne, A. López-Ortiz, and J.-R. Sack, eds.), Lecture Notes in Computer Sciences, vol. 3608, Springer, 2005, pp. 422–431.

13. D. Lehman, L. I. O'Callaghan, and Y. Shoham, *Truth revelation in approximately efficient combinatorial auctions*, Journal of the ACM **49** (2002), no. 5, 1–26.

14. R. H. Möhring, *Algorithmic aspects of comparability graphs and interval graphs*, Graphs and Order (I. Rival, ed.), Reidel, Dordrecht, 1985, pp. 41–101.

15. G. L. Nemhauser and L. A. Wolsey, *Integer and combinatorial optimization*, John Wiley & Sons, New York, 1988.

16. M. H. Rothkopf, A. Pekec, and R. M. Harstad, *Computationally manageable combinatorial auctions*, Management Science **44** (1998), no. 8, 1131–1147.

17. R. E. Tarjan, *Decomposition by clique separators*, Discrete Mathematics **55** (1985), 221–232.

18. D. Zuckerman, *Linear degree extractors and the inapproximability of* MAX CLIQUE *and* CHROMATIC NUMBER, Electronic Colloquium on Computational Complexity, Report TR05-100 (2005), `http://www.eccc.uni-trier.de/eccc/`.

Strip Graphs: Recognition and Scheduling*

Magnús M. Halldórsson and Ragnar K. Karlsson

Dept. of Computer Science
University of Iceland
IS-107, Reykjavik, Iceland
{mmh, rkk1}@hi.is

Abstract. We consider the class of strip graphs, a generalization of interval graphs. Intervals are assigned to rows such that two vertices have an edge between them if either their intervals intersect or they belong to the same row. We show that recognition of the class of strip graphs is $\mathcal{NP}$-complete even if all intervals are of length 2. Strip graphs are important to the study of job selection, where we need an equivalence relation to connect multiple intervals that belong to the same job.

The problem we consider is Job Interval Selection (JISP) on m machines. In the single-machine case, this is equivalent to Maximum Independent Set on strip graphs. For m machines, the problem is to choose a maximum number of intervals, one from each job, such that the resulting choices form an m-colorable interval graph. We show the single-machine case to be fixed-parameter tractable in terms of the maximum number of overlapping rows. We also use a concatenation operation on strip graphs to reduce the m-machine case to the 1-machine case. This shows that m-machine JISP is fixed-parameter tractable in the total number of jobs.

1 Introduction

1.1 Strip Graphs

Strip graphs are a generalization of interval graphs. They are defined by an interval graph combined with an equivalence relation on the intervals. For example, we can map different jobs assigned to a machine as an interval graph and allow each equivalence class to be made up of jobs belonging to the same user. In this case, two vertices have an edge between them if either their intervals intersect or they belong to the same equivalence class. We can look at this as the union of two graphs: an interval graph and a graph of equivalence classes, representable as a set of disjoint cliques. Note that we can represent a set of disjoint cliques as an interval graph as well: if we enumerate the cliques, the interval $[i-1, i)$ can be assigned to each vertex of clique i. By using this representation we see that any such graph can be defined by taking the union of two interval graphs, one of which is a set of disjoint cliques. Such a graph is called a *strip graph*, which refers to the *rectangle graph* representation of graphs formed by the union of two

---

* Supported by a grant from RANNÍS — The Icelandic Centre for Research.

interval graphs, defined by Bar-Yehuda *et al.* in [BYHN+06]. In this representation, each vertex is represented on the 2-dimensional plane by an axis parallel rectangle, the vertical side of which is of length 1. The two interval graphs are formed by projecting the sides of the rectangles onto each axis, such that two vertices are adjacent if their corresponding intervals in either of these projections intersect. Throughout this paper, we will assume that all intervals are half-open (open on top).

Gyárfás and West defined a t-track (t-union) graph to be the edgewise union of t interval graphs. Since strip graphs can be represented by two interval graphs, they form a subclass of the 2-union graphs. In [GW95], it was shown that recognizing 2-union graphs is $\mathcal{NP}$-complete. Since we are dealing with only a restricted class of the 2-union graphs, we are interested in seeing if recognition is $\mathcal{NP}$-complete for this subclass as well. To form the sharpest bound possible between $\mathcal{P}$ and $\mathcal{NP}$, we will define a graph as a k-*strip graph* if it is representable as a strip graph such that all strips are of length k and each interval has integral start- and endpoints. Clearly, k-strip graphs are a subclass of strip graphs. We will show that the class of 1-strip graphs is precisely equivalent to the class of line graphs of bipartite graphs and is therefore recognizable in polynomial time, but recognition of strip graphs is $\mathcal{NP}$-complete, even when restricted to k-strip graphs, for any $k \geq 2$.

The reduction is from the problem of determining whether a triangle-free cubic graph is Hamiltonian, which was shown in [WS84] to be $\mathcal{NP}$-complete. We prove in Sect. 3 that the removal of an edge e from any triangle-free cubic graph will form a strip graph if and only if the original graph has a Hamiltonian cycle going through edge e.

1.2 Strip Graph Applications

Interval graphs are often used in scheduling, but they only give information in the time dimension. Modeling more complex relations requires more complex structures. One classic scheduling problem is scheduling the maximum number of jobs in a non-conflicting manner on a single machine, given multiple possible run-times for each job. This problem is commonly referred to as the Job Interval Selection Problem (JISP) and has results going as far back as a 1982 paper by Nakajima and Hakimi [NH82]. By considering each possible run-time for each job as a vertex, and defining equivalence classes of intervals contained in the same job, we can model this problem with a strip graph. While JISP and the generalizations we consider in this paper have been recently studied, most of the research looks at JISP as being a problem on sets of intervals. We take a different route by studying the problem from the perspective of strip graphs and using structural observations of strip graphs to gain new insight into the problems.

Since each equivalence class is a clique, any independent set of this strip graph can only contain one interval from each job. Therefore, the maximum independent set of the strip graph is the maximum number of jobs that can be run on each machine. We show in Sect. 4 that finding the maximum independent set of a strip graph is fixed-parameter tractable in the maximum number of jobs with overlapping windows. This idea of job windows is similar to the one used in Chuzhoy *et al.* [COR01], where they showed that the MIS is computable in pseudo-polynomial time if the size of the job window is small in comparison to the size of the job.

A simple extension to this problem is scheduling the maximum number of jobs in a non-conflicting manner on multiple machines. We will show in Sect. 5 how this is reduced to the single-machine case by generating an instance of single-machine JISP involving concatenated copies of the input graph. We also analyze what effect this has on the running time of our JISP algorithm.

2   Preliminaries

2.1   Definitions and Notation

Given a strip graph G, we define G to be the union of two interval graphs, G_1 and G_2, such that G_1 contains only intervals of length 1 and G_2 is a regular interval graph. For any subgraph $H \subset G$, we will use the standard notation $V(H)$ to refer to the vertex set of H. For any set $V' \subset V(G)$, we will use $G(V')$ to refer to the subgraph of G induced by V'.

Let v be a vertex in a strip graph. We need to know three properties to define v's position. We define, for any vertex v, ρ_v to be an integer such that v is represented by the interval $[\rho_v, \rho_v + 1)$ in G_1. We define s_v and f_v to be the start- and endpoint of v's interval — that is, values such that v is represented by the interval $[s_v, f_v)$ in G_2. When represented as a rectangle graph, we can think of ρ_v as being the "row" containing the rectangle v. We can then define v as being adjacent to another vertex w if either $\rho_v = \rho_w$ — in which case we say v is a *1-neighbor* of w — or $[s_v, f_v) \cap [s_w, f_w) \neq \emptyset$, in which case there are four options, outlined below.

Assume $[s_v, f_v) \cap [s_w, f_w) \neq \emptyset$. We say v is a *left neighbor* of w and w is a *right neighbor* of v if $s_v < s_w < f_v < f_w$. Otherwise, we say v is an *internal neighbor* of w and w is an *external neighbor* of v if $s_w \leq s_v < f_v \leq f_w$. We note that if w and v are identical intervals, then v can be considered both an internal and an external neighbor of w. See Fig. 1.

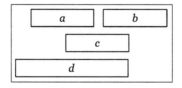

Fig. 1. Vertex a is a 1-neighbor of b, a left-neighbor of c and an internal neighbor of d

3   Hardness of Recognition

3.1   1-Strip Graphs

Remember that a 1-strip graph is defined such that each interval in G_2 is of length 1 and has integral start- and endpoints. The polynomial recognizability of 1-strip graphs follows from the following theorem, mentioned in [HRST99]. The formal proof is quite simple, and the proof is left to the reader. For now, we simply refer to Fig. 2. Since line graphs of bipartite graphs can be recognized in linear time [Leh74], the result follows instantly.

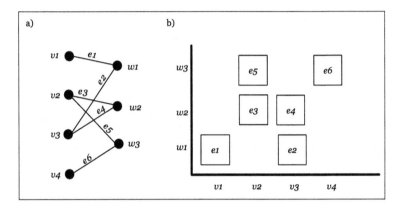

Fig. 2. The 1-strip graph (b) is the line graph of the bipartite graph (a)

Theorem 1. *A graph G is a 1-strip graph if and only if G is the line graph of a bipartite graph.*

3.2   Strip Graphs

West and Shmoys [WS84] showed that the problem of determining whether a triangle-free cubic graph is Hamiltonian is $\mathcal{NP}$-complete. They used this to show that recognizing 2-interval graphs is $\mathcal{NP}$-complete. West also used this to show, with Gyárfás, that recognition of 2-union graphs is $\mathcal{NP}$-complete [GW95]. We now use this problem to show that recognizing strip graphs is $\mathcal{NP}$-complete. What makes our reduction different from the reductions mentioned above is that we have to pay special attention to the lengths of the intervals in the graphs. Neither the reduction from [WS84] nor the reduction from [GW95] applies directly to graphs where one of the intervals is assumed to be of length 1. Additionally, those reductions don't give a boundary between when recognition is in $\mathcal{P}$ and when it is in $\mathcal{NP}$. By additionally considering the length of the intervals of G_2, we can give a restriction that allows recognition in polynomial time, and show that no similar restriction on length is recognizable in polynomial time unless $\mathcal{P} = \mathcal{NP}$.

The following theorem shows that determining whether a triangle-free cubic graph is Hamiltonian reduces to strip graph recognition. Therefore, we can conclude that recognizing strip graphs is $\mathcal{NP}$-complete.

Theorem 2. *A triangle-free cubic graph G is Hamiltonian if and only if there exists an edge e such that $G \setminus \{e\}$ is a strip graph.*

Proof. Assume a graph G is cubic, triangle-free and Hamiltonian. G is a graph on n vertices. Let C be the graph's Hamiltonian cycle, choose an arbitrary starting vertex v_1 and an arbitrary direction for C, and label the remaining vertices in order with the labels $v_2, v_3, \ldots, v_n$. We know that the remaining edges must form a perfect matching, which we denote by M. For any vertex v_i, we denote by $M(v_i)$ the vertex matched to v_i in M. We now remove the edge connecting v_1 to v_n and show that the remaining graph, denoted by G', must be a strip graph.

We begin by forming G'_2. With the removal of (v_n, v_1), C becomes a Hamiltonian path in G', which can be represented with intervals of length 2 by representing each v_i with the interval $[i - 1, i + 1)$. This defines G'_2. The remaining edges, as we saw above, form a perfect matching, so M can be represented with length 1 intervals in G'_1 by representing each v_i and $M(v_i)$ with the interval $[i - 1, i)$. To avoid placing two intervals for each vertex, this operation only happens when v_i is of a lower index than $M(v_i)$. This defines G'_1, and the union of G'_1 and G'_2 clearly forms G'.

Figure 3(b) shows how this transformation works for a triangle-free cubic Hamiltonian graph on 6 vertices after removing the edge (v_6, v_1). The first set of intervals defines G'_2 as a Hamiltonian path, and the second set of intervals defines G'_1 as a perfect matching. Figure 3(c) then shows how this strip graph is represented as a rectangle graph.

For the other direction, assume the removal of some edge (v, w) creates a strip graph, which we again call G'. Note that since G_2 is an interval graph, it must have at least two simplicial vertices — that is, vertices that have a neighborhood covered by at most one clique in G_2. Two of these vertices must be represented by the earliest-ending and latest-starting intervals in G_2. We also know that any vertex in G has a neighborhood covered by one clique in G_1. Therefore, the neighborhoods of the simplicial vertices of G_2 are covered by 2 cliques in G. Since G is triangle-free, these vertices have degree at most 2. Therefore, since v and w are the only two vertices of degree 2, they must be the simplicial (rightmost and leftmost) vertices of G_2.

Let v's neighbor in G'_2 be the vertex z. We know that z's other neighbor in G'_2 cannot be internal unless that neighbor is of degree at most two, so we that neighbor has to be a right neighbor, itself of degree 3. The degree-3 vertices must form a path, which cannot end until it hits a vertex of degree 2, which must then be w. Since w is the vertex with the rightmost endpoint in G'_2, every vertex in G' must appear between v and w. If there exists some vertex p of degree 3 that does not appear on the path between v and w, then it must be an internal neighbor of some vertex on the path (because G'_2 is triangle-free), which means p must be of degree 2, which is a contradiction. We conclude that every degree-3 vertex in

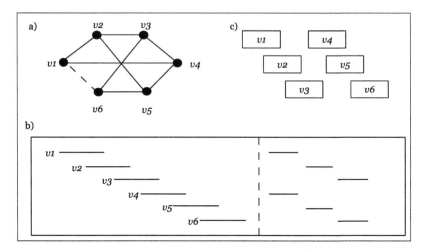

Fig. 3. (a) A cubic triangle-free Hamiltonian graph on 6 vertices. (b) Removing one edge allows us to represent the remaining graph as a strip graph. (c) The rectangle graph representation.

the graph must appear on the path between v and w. That is to say, the path is Hamiltonian.

We note that v_1 was chosen arbitrarily, and therefore any edge of the Hamiltonian cycle can be removed to create G'. For any given vertex v, we know that two of the three edges connected to v are present in any Hamiltonian cycle. Therefore, we have to select two of the edges in v. If the removal of either one creates a strip graph, then G is Hamiltonian. A polynomial-time algorithm to check these two cases in parallel gives a polynomial-time reduction from the Hamiltonian cycle problem on triangle-free cubic graphs to the strip graph recognition problem. We conclude by noting that the proof of the above reduction implies that recognition of k-strip graphs is $\mathcal{NP}$-complete for all $k \geq 2$.

Corollary 3. *Recognition of strip graphs is $\mathcal{NP}$-complete. Additionally, recognition of k-strip graphs is $\mathcal{NP}$-complete for any $k \geq 2$.*

4   Scheduling Applications of Strip Graphs

To see how strip graphs apply to scheduling problems, we first contrast them with t-interval graphs. Consider a set of jobs, each consisting of multiple intervals. These jobs can each be represented by a row in a strip graph. In the t-interval representation, each row would be contracted to a single vertex. This is useful for applications where jobs are split up over multiple intervals, and selecting a job means selecting all of its intervals. In the strip graph representation, on the other hand, each interval is a vertex, which is useful in situations that require access to individual intervals of each job. The classic example of this is the Job

Interval Selection Problem, or JISP, where each job only requires one interval of processing time, and the multiple intervals represent *possible* run-times.

The input to this problem is a set of jobs, called a *request*. Each job is comprised of multiple intervals, each representing a possible run-time for the job. The objective is to run as many jobs as possible such that no two conflict — that is, select at most one interval from each row such that we select the maximum number of intervals possible such that no two overlap. This selection of intervals is referred to as a *schedule* and is preciely equivalent to the Maximum Independent Set (that is, the largest set of mutually non-adjacent vertices, abbreviated MIS) of the corresponding strip graph.

In this section, we will give an exact algorithm for JISP / MIS of strip graphs that runs in exponential time. We will then consider how this algorithm applies to JISP in a multiple-machine environment.

4.1   Maximum Independent Set

The Maximum Independent Set problem is known to be $\mathcal{NP}$-complete for 2-strip graphs [CS96], implying the $\mathcal{NP}$-completeness of the problem on strip graphs. In this section, we will show that finding the maximum independent set is fixed-parameter tractable in the number of overlapping "liveness windows." In the following, assume G is a strip graph on n vertices.

We define $\mathfrak{R}$ to be the partition of G_1 into its connected components. That is, each member of $\mathfrak{R}$ is a row of G. Define $C = \max\{f_v | v \in G\}$. Then the rectangle representation of G is mapped onto an $|\mathfrak{R}| \times C$ grid. For a given subset $R \subset \mathfrak{R}$, we let G^R represent the subgraph of G induced by the rows in R, and for each integer $y \leq C$, we let G_y represent the subgraph of G induced by all vertices ending at or before point y in the interval representation of G_2. That is, for any $R \in \mathfrak{R}$ and $y \leq C$, $G_y^R = G(\{v \in V(G) | v \in R, f_v \leq y\})$.

We store the intermediate results in a matrix M, where $M[R, y]$ is the size of the MIS of the graph G_y^R, for $y \leq C$ an integer, and $R \subset \mathfrak{R}$ a set of rows. At position y, there are two possibilities:

- We can add a vertex v with endpoint $y = f_v$. If this case gives the MIS, then $M[R, f_v]$ is given by $M[R \setminus \{\rho_v\}, s_v] + 1$.
- No vertex with endpoint y is a member of an MIS of G_y^R. If this is the case, then $M[R, y] = M[R, y - 1]$.

We point out that the only interesting values in the above formulation are the start- and endpoints of vertices in G. Since any interval graph can be represented such that all intervals have distinct start- and endpoints, we can assume that this is the case. Therefore, we store all start- and endpoints in an increasing sequence, $y_1, y_2, \ldots, y_{2n}$. Note that since we're only considering start- and endpoints, we can essentially drop the integrality constraints on G_2.

Using this assumption, and the two possibilities above, the dynamic programming relation is given with:

$$M[R, y_i] = \max\{M[R, y_{i-1}], M[R \setminus \{\rho_v\}, s_v] + 1\}$$

where y_i is the endpoint of vertex v (if such a v exists).

To determine the minimum worst-case complexity, we need to determine how many rows need to be checked at each point y_i. For each row r, r is defined to be *live* at all points between the endpoint of its first interval and the endpoint of its last interval. Let $L_t \subset \mathfrak{R}$ be the set of live rows at point t and define $Q = \max_{t \leq C} |L_t|$. Similarly, a row is defined to be *dead* at all points after the endpoint of its last interval. Let $D_t \subset \mathfrak{R}$ be the set of all dead rows at point t. At each start- or endpoint y on the axis, there are two steps to the calculation.

We first compute $M[R \cup D_y, y]$ for all $R \subset L_y$. To understand why we can limit the calculations to this set of rows, consider a row r that is live on the interval $[a, b]$. If $y < a$, then no independent set of G_y includes row r because that row is empty, so r is never included. If, on the other hand, $y > b$, then any independent set of G_y has to consider row r; there is no need to defer the choice to a point further down the axis, because r has no intervals past b. This step is completed in time at most $\mathcal{O}(2^Q)$.

For the second step, consider a vertex v starting at point y. When we reach the endpoint of that vertex, we're going to need to be able to check $M[R' \cup D_{f_v}, y]$ for all $R' \subset L_{f_v}$. To do this, we merely iterate through each subset $R' \subset L_{f_v}$, and set

$$M[R' \cup D_{f_v}, y] = M[(R' \cup D_{f_v}) \cap (L_y \cup D_y), y]$$

this step is completed in time at most $2^{|L_{f_v}|}$. Since each point we consider is the startpoint of at most one vertex, the computation time for step 2 at any point y is at most $\mathcal{O}(2^Q)$.

Since we only consider at most $2 \cdot n$ start-/endpoints of intervals, M can be generated with a total time complexity of $\mathcal{O}(n \cdot 2^Q)$. To generate the MIS from the table M, we merely need to work backwards from C.

Assume the MIS of G is found to be of size k. We then find a vertex v such that $M[L_{f_v} \cup D_{f_v}, f_v] = k$ and $M[(L_{f_v} \cup D_{f_v}) \setminus \{\rho_v\}, s_v] = k - 1$. If we work in decreasing endpoint order, we only need to look at each vertex once, and choose the first one that fits. We add v to the set S and then repeat the procedure for the strip graph $G_{s_v}^{(L_{s_v} \cup D_{s_v}) \setminus \{\rho_v\}}$, which we know has an MIS of size $k - 1$. Using recursion, we get an independent set S of maximum size k in time $\mathcal{O}(n)$. The MIS problem is now clearly fixed-parameter tractable in Q.

Theorem 4. *For a strip graph G, define Q to be the maximum number of live rows overlapping at any point. The above algorithm then finds a maximum independent set of G in time $\mathcal{O}(n \cdot 2^Q)$.*

We conclude with two observations. First, we note that it is easy to generalize this algorithm to the Maximum Weighted Independent Set (MWIS) problem. In that case, each vertex v has a weight w_v, and the objective is to find an independent set such that the sum of the weights of all the vertices in the IS is maximized. The recursive formulation is then given by

$$M[R, y_i] = \max\{M[R, y_{i-1}], M[R \setminus \{\rho_v\}, s_v] + w_v\}$$

where y_i is the endpoint of vertex v (if it exists). The rest of the algorithm then works as described above.

Secondly, we note that a row consisting of only one interval is live at exactly one point. By assuming distinct endpoints for each interval, two rows that each consist of only one interval are never live at the same time. Therefore, our algorithm for MWIS reduces precisely to the classic linear-time DP algorithm for weighted independent set in interval graphs.

5   Multiple Machines

JISP is generalized by assuming that we have m machines running in parallel. The problem then becomes one of finding a maximum-value m-colorable subset of the input graph, which we call an m-schedule. This section will focus on reducing the m machine case to the 1-machine case and analyzing what effect this has on the above algorithm.

Assume we are given as input a strip graph G and asked to calculate the value of JISP for m machines. We define a new graph G^{*m} and show that the value of JISP for 1 machine on G^{*m} is equivalent to the value of JISP for m machines on G. G^{*m} is defined by making m identical copies of G such that the only edges between copies are between all intervals that belong to the same job. This creates a new strip graph, formed by making m identical copies of G's rectangle represenation along the horizontal axis. This transformation is shown in Fig. 4.

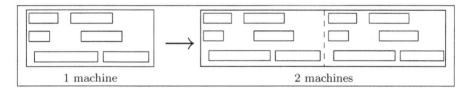

1 machine   2 machines

Fig. 4. A request simulated on 2 machines

Theorem 5. *G has a maximum m-schedule of size k if and only if G^{*m} has a maximum independent set of size k.*

Proof. Assume S is a maximum m-schedule and let $k = |S|$. We enumerate the color classes of S with $S_1, S_2, \ldots, S_m$. Each S_i is an independent set. G^{*m} is a concatenation of m identical requests (which we can refer to as "subrequests"), with edges between intervals belonging to the same job. Therefore, we can choose the vertices of S_1 for the first subrequest of G^{*m}, the vertices of S_2 for the second subrequest, and so on. Call this set of vertices S'. Clearly S' is an independent set of G^{*m}.

Assume now that a larger independent set of G^{*m} exists, which we call T'. Since G^{*m} is formed by taking m copies of G, we can enumerate the copies (in left-to-right order) by $G_1^{*m}, G_2^{*m}, \ldots, G_m^{*m}$. For each $T_i, i \in \{1, \ldots, m\}$, let T_i be the vertices of G corresponding to the vertices of $T' \cap G_i^{*m}$. Clearly, each T_i is an independent set and the union of all the T_i's forms an m-schedule. But this contradicts the assumption that S is a maximum m-schedule.

The following corollary is then obtained by observing that for any G^{*m}, $m \geq 2$, every job is live at the point between the first and second machines. In the terminology of Sect. 4, the running time of the algorithm on m machines is $\mathcal{O}(mn \cdot 2^{|\Re|})$.

Corollary 6. *JISP on m machines, $m \geq 2$, is fixed-parameter tractable in the number of jobs.*

Bibliography

[BYHN+06] Reuven Bar-Yehuda, Magnús M. Halldórsson, Joseph (Seffi) Naor, Hadas Shachnai, and Irina Shapira. Scheduling split intervals. *SIAM Journal of Computing*, 36:1–15, 2006.

[COR01] Julia Chuzhoy, Rafail Ostrovsky, and Yuval Rabani. Approximation algorithms for the job interval selection problem and related scheduling problems. In *FOCS '01: Proceedings of the 42nd IEEE symposium on Foundations of Computer Science*, page 348, Washington, DC, USA, 2001. IEEE Computer Society.

[CS96] Yves Crama and Frits C. R. Spieksma. Scheduling jobs of equal length: complexity, facets and computational results. *Math. Program.*, 72(3):207–227, 1996.

[GW95] András Gyárfás and David B. West. Multitrack interval graphs. *Congressus Numerantium*, 109:109–116, 1995.

[HRST99] Magnús M. Halldórsson, Sridhar Rajagopalan, Hadas Shachnai, and Andrew Tomkins. Scheduling multiple resources. *unpublished manuscript*, 1999.

[Leh74] Philippe G. H. Lehot. An optimal algorithm to detect a line graph and output its root graph. *J. ACM*, 21(4):569–575, 1974.

[NH82] Kazuo Nakajima and S. Louis Hakimi. Complexity results for scheduling tasks with discrete starting times. *Journal of Algorithms*, 3:344–361, 1982.

[WS84] Douglas B. West and David B. Shmoys. Recognizing graphs with fixed interval number is NP-complete. *Discrete Appl. Math.*, 8:295–305, 1984.

Approximating the Traffic Grooming Problem in Tree and Star Networks*

(Extended Abstract)

Michele Flammini[1], Gianpiero Monaco[1], Luca Moscardelli[1],
Mordechai Shalom[2], and Shmuel Zaks[2]

[1] Dipartmento di Informatica, Universita degli Studi dell'Aquila, L'Aquila, Italy
{flammini, gianpiero.monaco, moscardelli}@di.univaq.it
[2] Department of Computer Science, Technion, Haifa, Israel
{cmshalom, zaks}@cs.technion.ac.il

Abstract. We consider the problem of grooming paths in all-optical networks with tree topology so as to minimize the switching cost, measured by the total number of used ADMs. We first present efficient approximation algorithms with approximation factor of $2\ln(\delta \cdot g) + o(\ln(\delta \cdot g))$ for any fixed node degree bound δ and grooming factor g, and $2\ln g + o(\ln g)$ in unbounded degree directed trees, respectively. In the attempt of extending our results to general undirected trees we completely characterize the complexity of the problem in star networks by providing polynomial time optimal algorithms for $g \leq 2$ and proving the intractability of the problem for any fixed $g > 2$. While for general topologies the problem was known to be NP-hard g not constant, the complexity for fixed values of g was still an open question.

Keywords: Optical Networks, Wavelength Division Multiplexing(WDM), Add-Drop Multiplexer(ADM), Traffic Grooming, Tree Networks.

1 Introduction

All-optical networks have been largely investigated in recent years due to the promise of data transmission rates several orders of magnitudes higher than current networks [4,8,19,21]. Major applications are in video conferencing, scientific visualization and real-time medical imaging, high-speed supercomputing and distributed computing [11,19].

The key to high speeds in all-optical networks is to maintain the signal in optical form, thereby avoiding the prohibitive overhead of conversion to and from the electrical form at the intermediate nodes. The high bandwidth of the optical fiber is utilized through *wavelength-division multiplexing*: two signals connecting different source-destination pairs may share a link, provided they are transmitted

---

* This research was partly supported by the EU Project "Graphs and Algorithms in Communication Networks (GRAAL)" - COST Action TIST 293.

F.V. Fomin (Ed.): WG 2006, LNCS 4271, pp. 147–158, 2006.

on carriers having different wavelengths (or colors) of light. The optical spectrum being a scarce resource, given communication patterns in different topologies are often designed so as to minimize the total number of used colors, also as a comparison with the trivial lower bound provided by maximum load, that is the maximum number of connection paths sharing a same physical edge (see [1,16] for a survey of the main related results).

When the various parameters comprising the switching mechanism in these networks became clearer, the focus of studies shifted, and today a large portion of research concentrates with the total hardware cost. This is modelled by considering the basic switching unit of Add-Drop-Multiplexer (ADM), and focusing on the total number of these ADMs. The key point here is that each lightpath uses two ADMs, one at each endpoint. If two adjacent lightpaths are assigned the same wavelength, then they can use the same ADM. An ADM may be shared by at most two lightpaths.

Moreover, in studying the hardware cost, the issue of *grooming* became central. This problem stems from the fact that the network usually supports traffic that is at rates which are lower than the full wavelength capacity, and therefore the network operator has to be able to put together (= groom) low-capacity demands into the high capacity fibers. In graph-theoretic terms, this can viewed as assigning colors to the lightpaths so that at most g of them (g being the *grooming factor*) can share one edge. In terms of ADMs, each lightpath uses two ADMs, one at each endpoint, and in case g lightpaths of the same wavelength enter through the same edge to one node, they can all use the same ADM (thus saving $g - 1$ ADMs). Moreover, all the same colored paths ending at the node through two given incident edges can share the same ADM. The goal is to minimize the total number of ADMs. Note that the above coloring problem is simply the case of $g = 1$.

The problem of minimizing the number of ADMs for the case $g = 1$ was introduced in [17] for ring networks. For such a topology it was shown to be NP-complete in [12] and an approximation algorithm with approximation ratio $3/2$ was presented in [10] and improved in [22,13] to $10/7 + \epsilon$ and $10/7$ respectively. For general topologies [12] described an algorithm with approximation ratio $8/5$ and an improved $(3/2 + \epsilon)$-approximation one was presented in [9].

The notion of traffic grooming ($g > 1$) was introduced in [18] for ring topologies. In such a setting, the ADMs minimization problem was shown to be NP-complete in [7] for rings and for general g, i.e., with g being part of the input instance. The complexity of the problem for fixed values of g before this paper was still an open question. Finally, an algorithm with approximation ratio of $2 \ln g$ for any fixed g in a ring topology has been given in [14].

The all-to-all traffic case, in which there is the same demand between each pair of nodes, has been studied in [3,7] for various values of g, and an optimal construction for the case $g = 2$ in path networks was given in [2].

In a different scenario, the problem of minimizing hardware components in optical networks using grooming in order to exploit large bandwidth has been studied in [6] for ring networks and in [5] for stars networks.

In this paper we extend the results in [14] to trees. Namely, we provide polynomial time algorithms with approximation factors of $2\ln(\Delta \cdot g) + o(\ln(\Delta \cdot g))$ for any fixed node degree bound Δ and $2\ln g + o(\ln g)$ in unbounded degree directed trees, e.g. with all the edges oriented from the root to the leaves or vice versa, respectively. In the attempt of extending our results to general undirected trees, we completely characterize the complexity of the problem, that is for every fixed value of g, in star networks. Such a topology is of its own interest, as it often arises in practice in the interconnection of LANs or MANs with a wide area backbone. In particular we provide polynomial time optimal algorithms for $g \leq 2$ and prove the intractability of the problem for any fixed $g > 2$.

The paper is organized as follows. In the next section, we describe the problem and introduce some useful definitions. The algorithms for tree topologies are presented and analyzed in Section 3. In Section 4 we describe the polynomial algorithm for the star topology for $g \leq 2$ and the NP-completeness proof for $g \geq 3$. Finally, in Section 5 we give some conclusive remarks and discuss some open questions.

Due to space limitations, some proofs are omitted or only sketched in this Extended Abstract.

2 Problem Definition

An instance of the *traffic grooming problem* is a triple (G, P, g) where $G = (V, E)$ is a graph, P is a set of simple paths in G and g is a positive integer, namely the grooming factor.

A coloring (or wavelength assignment) of (G, P) is a function $w : P \mapsto \mathbb{N}^+ = \{1, 2, ...\}$. Given any subset $Q \subseteq P$, let $w(Q) = \{w(p)|p \in Q\}$ be the set of all the colors assigned by w to the paths in Q. Moreover, for a coloring w, a color λ and any $Q \subseteq P$, Q_λ^w is the subset of paths from Q colored λ by w. A proper coloring (or wavelength assignment) w of (G, P, g) is a coloring of P in which for any edge e at most g paths using e are colored with the same color[1].

A coloring w is a 1-coloring of $Q \subseteq P$, if it colors the paths of Q using one color. A set Q is 1-colorable if there exists a proper 1-coloring for it.

As already remarked in the introduction, every colored path $p \in P$ needs at least one ADM at each of its endpoint nodes. Given a grooming factor g, at most g paths incident to a node through the same edge can use the same ADM, and such an ADM can be shared also by the at most g path incident to the same node through another incident edge.

Given a 1-colorable subset $Q \subseteq P$, $\#ADM1COLOR(Q)$ is the minimum number of ADMs needed by the paths in Q (when receiving the same color).

---

[1] This is a generalization of the problem of minimizing the number of ADMs, when traffic grooming is allowed. When restricting to path and ring networks, this means that each of the two edges at a node can have at most g lightpaths of each color; however, when considering a general network, our definition permits splitting of groomed lightpaths at a node, which requires a special treatment by the optical switch. See discussion at the end of Section 5.

More formally, $\#ADM1COLOR(Q) = \sum_{v \in V} \left\lceil \frac{f(v)}{2} \right\rceil$, where $f(v)$ is the number of distinct edges adjacent to v used by the paths in Q.

For a coloring w of P, $\#ADM_\lambda^w$ is the minimum number of ADMs operating at wavelength λ in all the network and $\#ADM^w$ is the minimum total number of ADMs in the network, that is for all the used wavelengths.

The traffic grooming problem is the optimization problem of finding a proper coloring w of (G, P, g) minimizing $\#ADM^w$.

3   Approximated Solutions for Tree Topologies

In this section we present and analyze the approximation algorithm for tree topologies. We first give a general description in terms of a generic parameter k, depending only on g, that will be properly determined in the analysis, depending on the tree topology (bounded degree trees or directed trees). We will always implicitly assume trees rooted at a distinguished arbitrarily chosen node.

3.1   The General Algorithm $GROOMBYSC(k)$

The general algorithm $GROOMBYSC(k)$ described below has three phases. During Phase 1 it computes 1-colorable sets and their corresponding weights. It considers subsets of the paths P of size at most $k \cdot g$. Whenever a 1-colorable set Q is found, it is added to the list of relevant sets, together with a corresponding weight $\#ADM1COLOR(Q)$, that is the minimum number of ADMs needed by Q (when its path receive the same color). In Phase 2 it finds a set cover of P using subsets calculated in Phase 1. It uses the GREEDYSC approximation algorithm for the minimum weight set cover problem presented in [20]. In Phase 3 it transforms the set cover into a partition by eliminating intersections and then colors the paths belonging to each set with the same color, different from the one of the other sets of the partition.

Algorithm $GROOMBYSC(k)$ is a simple modification of the one presented in [14], where Phase 1 is slightly simplified and the notion of weighting is extended. In particular, in order to apply the algorithm of [14] for trees, we extended the notion of weights of the sets to fit to the problem definition (the notion in [14] didn't apply for trees; it was the number of endpoints of the paths, while for trees we need the minimum number of ADMs needed assuming that all paths receive the same color).

The correctness and time complexity of the algorithm follow the arguments of [14]; in particular, it can be shown that its running time is polynomial in $n = |P|$ and $m = |E|$, for any fixed k and g and for all instances (G, P, g), by taking into account the fact that $\#ADM1COLOR(Q)$ can be calculated in polynomial time.

As was pointed out in [14], this algorithm is general enough to be applied to any network topology. Basically, as is often the case in approximation algorithms, it runs in polynomial time, depending on some constraint; the problem is to prove its approximation ratio, and in [14] this is done for ring networks. The novelty and contribution of this paper is in tuning the parameters, and in particular in extending the analysis, so that it applies for tree topologies. Our analysis for undirected bounded degree tree networks and for directed tree networks are presented in the following two subsections.

3.2   Bounded Degree Trees

Now we analyze the algorithm $GROOMBYSC(k)$ applied to trees having their maximum degree equal to Δ.

We will use the shortcut $\#ADM^*$ for $\#ADM^{w^*}$, and given a set $U \subseteq V$ we will denote by $\#ADM^*(U)$ the number of ADMs used by the optimal solution at nodes belonging to U.

Lemma 1. *Given an integer h, there exists a solution $\overline{SC}$ for the instance of the set cover problem determined on Phase 2 of $GROOMBYSC(k)$, with $k = (\Delta - 1)h$, such that $weight(\overline{SC}) \leq \#ADM^* \left(1 + \frac{2g}{h}\right)$.*

Proof. Let $w^*(P) = \{1, 2, ..., W^*\}$ and $1 \leq \lambda \leq W^*$. Consider the set V_λ^* of nodes v having at least one ADM operating at wavelength λ at node v. We divide V_λ^* into sets of at least h and at most $k = (\Delta - 1)h$ nodes in the following way. We consider a node v in the tree such that the number of nodes belonging to V_λ^* in the subtree rooted at v is between h and k. Then we remove such a subtree, put in a separate set all the nodes of the subtree belonging to V_λ^* and repeat the procedure on the remaining tree. Clearly, since the maximum degree of the nodes in the tree is Δ, it is always possible to remove a subtree containing at least h and at most k nodes of V_λ^*, except for the last set (in the following the *residual* set) that may contain less than h nodes.

Let $V_{\lambda,j}$, $j = 1, \ldots, p_\lambda$, be the subsets of nodes obtained in this way, and q_λ the number of them having at least h nodes in V_λ^*. Let $r_\lambda < h$ be the number of nodes in the residual set, if it exists. Notice that if no set with less than h nodes is in the partition, $p_\lambda = q_\lambda$ and $r_\lambda \overset{def}{=} 0$, otherwise $p_\lambda = q_\lambda + 1$. For any $V_{\lambda,j}$ we define $\overline{S}_{\lambda,j}$ to be the set of paths colored λ by w^* with both their endpoints in $V_{\lambda,j}$ or with one endpoint in $V_{\lambda,j}$ and leaving the subtree corresponding to $V_{\lambda,j}$ through the edge going toward the root. In order to upper bound the number of such paths, let us charge each of them to one of its endpoint nodes reached by the path through the incident link toward the root. Since by the feasibility of the coloring at most g paths are then charged to each node in $V_{\lambda,j}$, we have $|\overline{S}_{\lambda,j}| \leq g \cdot k$.

Notice that every p with $w^*(p) = \lambda$ is contained in exactly one set $\overline{S}_{\lambda,j}$. Therefore $\overline{SC}_\lambda \overset{def}{=} \cup_j \{\overline{S}_{\lambda,j}\}$ is a cover of all the paths colored λ by w^*. Considering all colors $1 \leq \lambda \leq W^*$ we conclude that $\overline{SC} \overset{def}{=} \cup_{\lambda=1}^{W^*} \overline{SC}_\lambda$ is a cover of P with sets from S. It remains to show that its weight has the claimed property.

First observe that $\#ADM_\lambda^* \geq hq_\lambda + r_\lambda$. Summing up over all possible values of λ we obtain $\#ADM^* \geq h \sum_\lambda q_\lambda + \sum_\lambda r_\lambda$, which implies:

$$\sum_\lambda q_\lambda \leq \frac{\#ADM^*}{h}. \tag{1}$$

We claim that for every $j \leq q_\lambda$, $weight[\overline{S}_{\lambda,j}] = \#ADM1COLOR(\overline{S}_{\lambda,j}) \leq \#ADM^*(V_{\lambda,j}) + g$. This is because:

- the endpoints of the paths with both endpoints in $\overline{S}_{\lambda,j}$ are in $V_{\lambda,j}$;
- the number of paths having only one endpoint in set $V_{\lambda,j}$ is at most g. This follows from the observation that these paths should use the unique edge going from the subtree containing the nodes in $V_{\lambda,j}$ towards the root of the whole tree. As the set $\overline{S}_{\lambda,j}$ is 1-colorable, the number of these paths is at most g.

For the residual set (which exists only if $r_\lambda > 0$), we have $weight(\overline{S}_{\lambda,q_\lambda+1}) \leq \#ADM^*(V_{\lambda,q_\lambda+1}) + g \cdot q_\lambda$. This is because:

- the endpoints of the paths with both endpoints in $\overline{S}_{\lambda,q_\lambda+1}$ are in $V_{\lambda,q_\lambda+1}$;
- by the same argument as before, the paths having only one endpoint in $V_{\lambda,q_\lambda+1}$ are at most g in number. When $q_\lambda \geq 1$, $g \leq g \cdot q_\lambda$; otherwise $q_\lambda = 0$ meaning that $V_{\lambda,1}$ is the unique set, thus in this case the number of paths having exactly one endpoint in this set is zero.

Summing up for all $1 \leq j \leq q_\lambda + 1$ we get $weight(\overline{SC}_\lambda) \leq \#ADM_\lambda^* + 2g \cdot q_\lambda$. Summing up for all λ and recalling (1) we get

$$weight(\overline{SC}) = \sum_\lambda weight(\overline{SC}_\lambda) \leq \sum_\lambda (\#ADM_\lambda^* + 2g \cdot q_\lambda) =$$
$$= \#ADM^* + 2g \sum_\lambda q_\lambda \leq \#ADM^* + 2g \frac{\#ADM^*}{h} = \left(1 + \frac{2g}{h}\right) \#ADM^*. \qquad \square$$

Now, by choosing $h = g \ln g$, since the greedy algorithm for the Minimum Weight Set Cover Problem is a H_f-approximation algorithm, where f is the maximum cardinality of the sets in the input and $H_f = 1 + \frac{1}{2} + ... + \frac{1}{f}$ the f-th harmonic number, it is possible to show the following theorem.

Theorem 1. *There is a $2 \ln g + \ln \Delta + o(\ln(\Delta \cdot g))$-approximation algorithm for the traffic grooming problem in bounded degree tree networks.*

3.3   Directed Trees

We now analyze the algorithm $GROOMBYSC(k)$ applied to directed trees, i.e. to trees having all the edges directed from the root towards the leaves or vice versa. Notice that the result for this topology is independent of the degree Δ. This stems from the fact that in directed trees lightpaths cannot cross from one subtree to another (in contrast with the case of undirected ones).

By the same arguments used in the proof of Theorem 1, it is possible to show that $GROOMBYSC(k)$ is a $H_{g \cdot k}(1 + \frac{2g}{h})$ approximation algorithm for the traffic grooming problem in directed tree networks, where $k = 3h$ and h is again equal to $g \ln g$. In the following we underline the differences between the analyses on the two tree topologies, using the same notation introduced for bounded degree trees.

The main difference is in the proof of the existence of a set cover of weight at most $\#ADM^* \left(1 + \frac{2g}{h}\right)$. Since the maximum degree of the tree is not bounded, we cannot partition V_λ^* in the same way using node disjoint subtrees. Thus, we divide V_λ^* into sets of at least h and at most $3h$ nodes in the following way. At each step, we consider a node x such that all its subtrees have less than h nodes in V_λ^* and there exists a subset of subtrees having in total at least h and at most $3h$ nodes belonging to V_λ^* satisfying the following property: in the optimal solution no ADM operating with color λ at node x is shared by requests belonging to different components. Notice that this is always possible since without loss of generality we can restrict to optimal solutions having only one ADM at node x with color λ for all the requests coming from each single subtree of x and since by coupling subtrees with requests sharing the same ADM at node x we obtain components having a number of nodes not exceeding $3h$ (if a subset has less than h nodes, adding two subtrees sharing an ADM at x in the optimal solution will yield at most $3h$ nodes). Thus we remove such subtrees inserting all their nodes and eventually node x in a single component; more precisely, node x is inserted in the component if and only if there exists an ADM operating at color λ in the optimal solution at x used by requests belonging to a subtree of the component and other ones crossing the edge going toward the root. Finally, x is removed only if it has not further sons and has been inserted in one component. We keep repeating the same step until a node x with the same property is present in the remaining tree. Clearly, since we remove node x only if it does not disconnect the tree, it is always possible to select a set containing at least h and at most $3h$ nodes of V_λ^*, except for the last (residual) set that may contain less than h nodes.

Since the tree is directed, no request can step through two subtrees of a same parent node, and thus through two components composed by subtrees having the same parent node. Therefore, we can again claim that $weight(\overline{SC}_\lambda) \leq \#ADM_\lambda^* + 2g \cdot q_\lambda$, by conveniently partitioning the ADMs of the optimal solution with special regards for the ADMs at the parent nodes of the subtrees belonging to the components; more precisely, we partition the ADMs at such nodes so that an ADM shared in the optimal solution by requests belonging to a component is counted in the set of the partition corresponding to such a component.

Therefore, it is possible to prove the following theorem.

Theorem 2. *There is a $2 \ln g + o(\ln g)$-approximation algorithm for the traffic grooming problem in directed tree networks.*

4   Star Topology

In this section, we consider star topologies having one *hub* node (node 0) of degree n and n nodes $1, \ldots, n$ of degree 1, connected to the hub node. We first present and analyze optimal algorithms for $g \leq 2$ and then show that the problem is NP-complete for every $g \geq 3$.

4.1   Optimal Algorithms for $g \leq 2$

The case $g = 1$ can be trivially solved by arbitrarily pairing all the possible requests having the hub node as an endpoint (no saving is possible for the other paths). Thus, we restrict to the case $g = 2$ considering instances of the form $(G, P, 2)$ with $G = (V, E)$, where $V = \{0, \ldots, n\}$ and $E = \{\{0, i\} | i = 1, \ldots, n\}$.

Algorithm *GROOMSTAR*. The algorithm has three phases. During Phase 1 it creates a new graph H having the same node set of G and an edge for any request in P between its endpoints. In Phase 2 it finds and removes all the cycles in H and colors the requests corresponding to their edges assigning a different wavelength to each cycle. Finally, in Phase 3 it finds all the remaining paths between nodes having an odd degree in H, and properly colors the corresponding requests assigning a different wavelength to each path.

Running Time. The running time of the algorithm is clearly polynomial. In fact, Phase 1 has a time complexity linear in the size of the instance; in Phase 2 and Phase 3 the problem of finding a cycle or a sequence of edges connecting two nodes is solvable in linear time by visiting the graph, and these problems have to be computed at most $|P|$ times.

Correctness and Optimality. Let x_i, $i = 1, \ldots, n$, be the number of paths of two edges in P having an endpoint in node i; similarly, let y_i be the number of paths of one edge in P having an endpoint in node i.

Lemma 2. *The number of ADMs needed for the traffic groomin problem in a star network is at least* $\left\lceil \frac{\sum_{i=1}^{n} y_i}{2} \right\rceil + \sum_{i=1}^{n} \left\lceil \frac{x_i + y_i}{2} \right\rceil$.

Proof. For each node $1, \ldots, n$ we need at least one ADM for each pair of paths having an endpoint in it, i.e. a total of $\left\lceil \frac{x_i + y_i}{2} \right\rceil$ ADMs. For the hub node 0, since for each edge there exists at most one path $p \in P$ crossing it and having node 0 as endpoint, we again need at least one ADM for each pair of such paths, i.e. a total of $\left\lceil \frac{\sum_{i=1}^{n} y_i}{2} \right\rceil$ ADMs. $\square$

Theorem 3. *Algorithm GROOMSTAR returns an optimal solution using exactly* $\left\lceil \frac{\sum_{i=1}^{n} y_i}{2} \right\rceil + \sum_{i=1}^{n} \left\lceil \frac{x_i + y_i}{2} \right\rceil$ *ADMs.*

Sketch of Proof. During Phase 2, cycles are found and colored. Notice that in each cycle we use one ADM for each pair of paths and eliminating it the parity of the degree of nodes in H is not modified.

In Phase 3, at each step the number of nodes having an odd degree is decreased by 2. Since there are no cycles at the beginning of this phase, it is always possible to find a sequence of edges starting and ending at two nodes with odd degrees and to color the corresponding paths with the same color. Notice that at every node we use one ADM for each pair of paths, except for the nodes having an odd degree. Since such nodes after removing the sequence of edges from H (corresponding to a path between nodes of odd degree) become of even degree, only once such a situation occurs at a node having an initial odd degree. Thus, the claim directly follows by Lemma 2 and by observing that at each node we use a number of ADMs equal to the number of possible pairings of incident paths, plus one in case of odd degree. $\square$

4.2   NP-Completeness for $g > 2$

Now we turn to show that determining a solution using a minimum number of ADMs for any $g > 2$ is an intractable problem. Before proving it formally, we need to show the NP-completeness of the following decision problem.

Edge Partition in δ-regular graphs
INPUT: undirected graph $G = (V, E)$.
QUESTION: is it possible to partition the edge set E in k subsets $E_1, \ldots, E_k$ for some $k > 0$, each inducing a δ-regular subgraph $G_l = (V_l, E_l)$, $1 \le l \le k$, that is with each node having degree exactly δ?

Lemma 3. *For every $\delta \ge 3$ Edge Partition in δ-regular graphs is an NP-complete problem.*

Proof. Due to space limitations, we prove the claim for $\delta = 3$; the extension of the proof to every degree $\delta > 3$ will appear in the full version of the paper. In order to show the NP-completeness, we provide a polynomial time transformation from the *Exact Cover by 3-Sets* problem (known to be NP-complete; see [15]). In this problem we have a universe set $U = \{o_1, \ldots, o_m\}$ of m elements with m multiple of 3, a family $\mathcal{S} = \{S_1, \ldots, S_h\}$ of h subsets of U, each containing exactly 3 elements, and we want to decide if there exist $m/3$ subsets $S_{j_1}, \ldots, S_{j_{m/3}}$ that partition U, i.e. such that $\bigcup_{i=1}^{m/3} S_{j_i} = U$.

Starting from an instance of Exact Cover by 3-Sets, we construct a graph $G = (V, E)$ whose edges can be partitioned in δ-regular graphs if and only such an instance admits a partition.

For any $o_j \in U$, let n_j be the number of subsets in $\mathcal{S}$ containing o_j. Then $G = (V, E)$, where $V = V_1 \cup V_2 \cup V_3 \cup V_4$ and $E = E_1 \cup E_2 \cup E_3 \cup E_4 \cup E_5$, with:
$V_1 = \{u_i \mid i = 1, \ldots, h\}$, $V_2 = \{v_j \mid j = 1, \ldots, m\}$, $V_3 = \{w_{1,j,l}, w_{2,j,l} \mid j = 1, \ldots, m, l = 1, \ldots, n_j\}$, $V_4 = \{z_{1,j,l}, z_{2,j,l} \mid j = 1, \ldots, m, l = 1, \ldots, n_j\}$, and
$E_1 = \{\{u_i, v_j\} \mid o_j \in S_i\}$, $E_2 = \{\{v_j, w_{1,j,l}\}, \{v_j, w_{2,j,l}\} \mid j = 1, \ldots, m, l = 1, \ldots, n_j\}$, $E_3 = \{\{w_{1,j,l}, z_{1,j,l}\}, \{w_{1,j,l}, z_{2,j,l}\}, \{w_{2,j,l}, z_{1,j,l}\}, \{w_{2,j,l}, z_{2,j,l}\} \mid j = 1, \ldots, m, l = 1, \ldots, n_j\}$, $E_4 = \{\{z_{1,j,l}, z_{2,j,l}\} \mid j = 1, \ldots, m, l = 2, \ldots, n_j\}$, $E_5 = \{\{z_{2,j,1}, z_{1,j+1,1}\} \mid j = 1, \ldots, m-1, l = 2, \ldots, n_j\} \cup \{\{z_{2,m,1}, z_{1,1,1}\}\}$.

Informally, in the reduction graph each subset $S_i \in \mathcal{S}$ corresponds to the subgraph induced by vertex u_i and the vertices v_j such that $o_j \in S_i$, which are all connected to u_i. Moreover, at each o_j, for each S_i containing it there is an attached *auxiliary component* induced by v_j and vertices $w_{1,j,l}$, $w_{2,j,l}$, $z_{1,j,l}$ and $z_{2,j,l}$ in $V_3 \cup V_4$ connected among themselves by the edges in $E_3 \cup E_4$ so that they all have degree exactly 3, except v_j. Notice that the first auxiliary component of each v_j is slightly different from the other ones, as $z_{1,j,l}$ and $z_{2,j,l}$ are not adjacent. In fact, the first auxiliary component of v_j for $j < m$ is connected to the first of v_{j+1} by an edge in E_5, and the first of v_m with the one of v_1, still maintaining degree 3 for all nodes $w_{1,j,1}, w_{2,j,1}, z_{1,j,1}$ and $z_{2,j,1}$. The key property here is that, since for a node of degree 3 either all the incident edges must be in a partition inducing a 3-regular graph or none of them, in any partition all the edges of the first auxiliary components must belong to the same partition set.

Assume that the instance of Exact Cover by 3-Sets admits the required partition, and without loss of generality let $S_1, \ldots, S_{m/3}$ be the partition subsets. Then G can be edge partitioned into 3-regular graphs as follows. In the first partition we insert the edges of all the first auxiliary components, including the one connective two consecutive ones, plus the edges incident vertices $u_1, \ldots, u_{m/3}$. In this way, since every $u_i \in V_1$ has degree 3, all the nodes in the auxiliary components have degree 3 and each v_j has 2 incident edges belonging to its first auxiliary component plus exactly one connecting it to $u_1, \ldots, u_{m/3}$, all such induced graphs are 3-regular. The remaining 3-regular components are formed including, for each $u_i \in V_1$ with $i > m/3$, the edges incident to u_i plus the one of a not yet considered auxiliary component of each of the vertices v such that $o_j \in S_i$ (notice that for each v_j they are exactly the ones needed to be joined to the edges incident to vertices $u_i \in V_1$ with $i > m/3$).

On the other hand, if G can be edge partitioned into 3-regular graphs, a partition of the universe U can be determined as follows. Consider the 3-regular subgraph containing all the first auxiliary components of the nodes $v_j \in V_2$ (as previously mentioned they must be all together). Since each $v_j \in V_2$ has exactly two edges belonging to its first auxiliary component, exactly one edge incident to it must be inside the same subgraph. Such an edge cannot belong to another auxiliary component of v_j, as all such edges must be in the same subgraph and this would cause a degree greater than 3 at v_j. Thus the additional edge of v_j must have as the other endpoint a node $u_i \in V_1$. Since this holds for all the nodes $v_j \in V_2$ and for every $u_i \in V_1$ either all its incident edges belong to the subgraph or none of them, vertices $u_i \in V_1$ connected to at least one vertex $v_j \in V_2$ are connected to exactly 3 vertices of V_2. Since each $v_j \in V_2$ is connected to exactly one $u_i \in V_1$, the subsets S_i corresponding to such nodes $u_i \in V_1$ then form a partition of U, hence the claim.   □

We can now show the hardness the grooming problem in star networks for every fixed $g > 2$.

Theorem 4. *For any given $g > 2$, given in input a set of communication requests in a star network, deciding if there exists a solution of the traffic grooming problem using at most a given number h of ADMs is NP-complete.*

Proof. We prove the claim by reduction from Edge Partition in δ-regular graphs with $\delta = g$, shown to be NP-complete in the previous lemma. Given an instance $G = (V, E)$ of such a problem with $V = \{1, \ldots, n\}$, we construct in polynomial time an instance of the grooming problem such that there exists a solution using at most $2|E|/g$ ADMs if and only if the edges of G can be partitioned in δ-regular graphs. The star networks has as node set $V \cup \{0\}$ and as set of communication paths $P = \{\langle i, 0, j \rangle | \{i, j\} \in E\}$.

Assume that E can be edge-partitioned for some $k > 0$ in k subsets $E_1, \ldots, E_k$ inducing g-regular subgraphs $G_\lambda = (V_\lambda, E_\lambda)$, $1 \leq \lambda \leq k$, each with a number of nodes $|V_\lambda| = 2|E_\lambda|/g$.

Since by the star topology each subset of paths $Q_\lambda = \{\langle i, 0, j \rangle | \{i, j\} \in E_\lambda\}$ is 1-colorable, by letting $w(p) = \lambda$ for each $p \in Q_\lambda$, the number of ADMs sufficient for Q_λ is $|V_\lambda|$. Therefore, there exists a solution using an overall number of ADMs equal to $|V_1| + \ldots + |V_k| = 2(|E_1| + \ldots + |E_k|)/g = 2|E|/g$.

On the other side, assume that E cannot be partitioned so as to induce only g-regular subgraphs and consider any solution of the grooming problem using k colors for the paths in P for some $k > 0$. Let $P_\lambda \subseteq P$, $1 \leq \lambda \leq k$, the subset of the paths of P with color λ. Then $P_1, \ldots, P_k$ induce in a natural way a partition $E_1, \ldots, E_k$ of E with corresponding subgraphs $G_1 = (V_1, E_1), \ldots, G_k = (V_k, E_k)$, each having maximum degree upper bounded by g, and thus with a number of nodes $|V_\lambda| \geq 2|E_\lambda|/g$. Moreover, since by hypothesis at least one subgraph is not g-regular, there must exist l, $1 \leq \lambda \leq k$, such that $|V_\lambda| > 2|E_\lambda|/g$. Clearly, the number of ADMs used for each color λ, is at least $|V_\lambda|$, so that the overall number of ADMs used by the solution is at least $|V_1| + \ldots + |V_k| > 2(|E_1| + \ldots + |E_k|)/g = 2|E|/g$, thus proving the claim. $\qquad\square$

5   Discussion and Open Problems

In this paper we addressed the traffic grooming problem with respect to the minimization of ADMs. We presented approximation algorithms for tree networks having bounded degree or directed, with approximation ratios logarithmic in g. Moreover, we provided an optimal polynomial algorithm for the star topology and grooming factor 2 and showed that the problem is NP-complete even for such topologies and every fixed grooming factor at least equal to 3.

The main open problem arising from this paper is the determination of an approximation algorithm for general trees. At the best of our knowledge no approximation algorithm has been proposed even for stars and grooming factor 3. Another interesting research direction is the minimization of ADMs in the setting in which only a limited number of wavelengths is available.

An additional extension concerns the notion of traffic grooming. While grooming in path and ring networks is clear, its extension to a general network can be understood in more than one way. We adopted in this paper a definition that fits well to the graph-theoretic model; however, according to our definition, it is possible that two lightpaths with the same wavelength will be directed to different endpoints at an intermediate node having degree at least 3. Coming back to the hardware switches, one can argue that this splitting is impossible, thus resulting in a more constrained definition of the grooming problem (it is interesting to note that this revised problem is NP-complete even for the case of a star network and $g = 2$). Investigating the possible interpretations of grooming, according to the various hardware components, calls for more future research.

Acknowledgement. We would like to thank David Coudert for helpful comments regarding an earlier version of this paper.

References

1. B. Beauquier, J.-C. Bermond, L. Gargano, P. Hell, S. Perennes, and U. Vaccaro. Graph problems arising from wavelength–routing in all–optical networks. In *Proc. 2nd Workshop on Optics and Computer Science, WOCS'97*, April 1997.

2. J.-C. Bérmond, L. Braud, and D. Coudert. Traffic grooming on the path. In *12 th Colloqium on Structural Information and Communication Complexity, Le Mont Saint-Michel, FRANCE*, May 2005.

3. J.-C. Bermond and D. Coudert. Traffic grooming in unidirectional WDM ring networks using design theory. In *IEEE ICC*, Anchorage, Alaska, May 2003.

4. C. A. Brackett. Dense wavelength division multiplexing networks: principles and applications. *IEEE Journal on Selected Areas in Communications*, 8:948–964, 1990.

5. B. Chen, G. N. Rouskas, and R. Dutta. Traffic grooming in star networks. In *Broadnets*, 2004.

6. B. Chen, G. N. Rouskas, and R. Dutta. Traffic grooming in wdm ring networks with the min-max objective. In *NETWORKING*, pages 174–185, 2004.

7. A. L. Chiu and E. H. Modiano. Traffic grooming algorithms for reducing electronic multiplexing costs in wdm ring networks. *Journal of Lightwave Technology*, 18(1):2–12, January 2000.

8. N. K. Chung, K. Nosu, and G. Winzer. Special issue on dense wdm networks. *IEEE Journal on Selected Areas in Communications*, 8, 1990.

9. G. Călinescu, Ophir Frieder, and Peng-Jun Wan. Minimizing electronic line terminals for automatic ring protection in general wdm optical networks. *IEEE Journal of Selected Area on Communications*, 20(1):183–189, Jan 2002.

10. G. Călinescu and P-J. Wan. Traffic partition in wdm/sonet rings to minimize sonet adms. *Journal of Combinatorial Optimization*, 6(4):425–453, 2002.

11. D. H. C. Du and R. J. Vetter. Distributed computing with high-speed optical networks. In *Proceeding of IEEE Computer*, volume 26, pages 8–18, 1993.

12. T. Eilam, S. Moran, and S. Zaks. Lightpath arrangement in survivable rings to minimize the switching cost. *IEEE Journal of Selected Area on Communications*, 20(1):172–182, Jan 2002.

13. L. Epstein and A. Levin. Better bounds for minimizing sonet adms. In *2nd Workshop on Approximation and Online Algorithms, Bergen, Norway*, September 2004.

14. M. Flammini, L. Moscardelli, M. Shalom, and S. Zaks. Approximating the traffic grooming problem. In *ISAAC*, pages 915–924, 2005.

15. M. Garey and D. S. Johnson. *Computers and Intractability, A Guide to the Theory of NP-Completeness*. Freeman, 1979.

16. L. Gargano and U. Vaccaro. *"Routing in All–Optical Networks: Algorithmic and Graph–Theoretic Problems"in: Numbers, Information and Complexity*. Kluwer Academic, 2000.

17. O. Gerstel, P. Lin, and G. Sasaki. Wavelength assignment in a wdm ring to minimize cost of embedded sonet rings. In *INFOCOM'98, Seventeenth Annual Joint Conference of the IEEE Computer and Communications Societies*, 1998.

18. O. Gerstel, R. Ramaswami, and G. Sasaki. Cost effective traffic grooming in wdm rings. In *INFOCOM'98, Seventeenth Annual Joint Conference of the IEEE Computer and Communications Societies*, 1998.

19. P. E. Green. *Fiber-Optic Communication Networks*. Prentice Hall, 1992.

20. D. S. Johnson. Approximation algorithms for combinatorial problems. *J. Comput. System Sci.*, 9:256–278, 1974.

21. R. Klasing. Methods and problems of wavelength-routing in all-optical networks. In *Proceeding of the MFCS'98 Workshop on Communication, August 24-25, Brno, Czech Republic*, pages 1–9, 1998.

22. M. Shalom and S. Zaks. A $10/7 + \epsilon$ approximation scheme for minimizing the number of adms in sonet rings. In *First Annual International Conference on Broadband Networks, San-José, California, USA*, October 2004.

Bounded Arboricity to Determine the Local Structure of Sparse Graphs*

Gaurav Goel[1,2] and Jens Gustedt[1,3]

[1] INRIA Lorraine, France
[2] IIT Delhi, India
[3] LORIA, France

Abstract. A known approach of detecting dense subgraphs (*communities*) in large sparse graphs involves first computing the *probability vectors* for *short random walks* on the graph, and then using these probability vectors to detect the communities, see Latapy and Pons [2005]. In this paper we focus on the first part of such an approach *i.e.* the computation of the probability vectors for the random walks, and propose a more efficient algorithm for computing these vectors in time complexity that is linear in the size of the output, in case the input graphs are restricted to a family of graphs of bounded arboricity. Such classes of graphs cover a large number of cases of interest, *e.g* all minor closed graph classes (planar graphs, graphs of bounded treewidth etc) and random graphs within the preferential attachment model, see Barabási and Albert [1999]. Our approach is extensible to other models of computation (PRAM, BSP or out-of-core computation) and also w.h.p. stays within the same complexity bounds for Erdős Renyi graphs.

1 Introduction and Overview

Consider a few real world large sparse graphs — the World Wide Web (WWW) graph where the vertices are HTML pages connected by links (edges) pointing from one page to another, the social acquaintance network where the vertices represent people and the edges represent the association between them, the graph representing the citation pattern of scientific publications with the vertices being the publications and the edges being the links to the articles cited in a publication, or for that matter the collaboration graph of movie actors with the vertices representing actors and edges joining actors which have worked together in at least one movie. All of these graphs and most other real world sparse graphs have a unique property — they have a low *arboricity*. Arboricity can be defined as follows:

Definition 1. *For a graph G the* arboricity $A(G)$ *is the smallest integer k for which there exists forests $T_1, \ldots, T_k$ which are subgraphs of G, such that their union is G.*

In fact, besides well-known examples of real world sparse graphs, all minor closed graph families, see *e.g.* Mader [1967], and all random graphs that are generated by the

---

* Research made possible by an internship grant for the first author at INRIA Lorraine, jointly accorded by the Lorraine Region, INRIA and the French ministry of foreign affairs.

F.V. Fomin (Ed.): WG 2006, LNCS 4271, pp. 159–167, 2006.

model discussed by Barabási and Albert [1999], also have a low value for arboricity, namely for each of these classes $\mathcal{C}$ in question there exists a constant $\delta_{\mathcal{C}}$ that bounds this parameter from above.

The method discussed in this paper is much inspired from previous work that explicitly or implicitly uses the property of bounded arboricity graphs to achieve algorithms of linear complexity, see *e.g.* in Cheriton and Tarjan [1976], Kannan et al. [1992], Bodlaender [1996], Gustedt [1998]. It divides the graph into portions based on the degrees of vertices, computes the required quantities only in one of the portions, passes the results to the remaining graph and finally recurses the procedure in this remaining graph.

One of the main tricks of such algorithms is that they may consider the bound on the arboricity $\delta = \delta_{\mathcal{C}}$ as being fixed and that this constant then appears only in the form of some function $f(\delta)$ in the complexity, independent of the input size, and, in our case, also of the output size.

1.1 Problem Definition

The problem of detecting dense subgraphs (*communities*) in large sparse graphs is inherent to many real world domains like social networking or internet computing. A known approach of detecting these communities involves first computing the *probability vectors* for *random walks* on the graph for a *fixed* number d of steps, and then using these probability vectors to detect the communities, see Latapy and Pons [2005]. Their algorithm takes $O(dnm)$ time where n and m are the number of vertices and edges in the graph. We focus on the first part of their approach i.e. computation of the probability vectors for the random walks, and propose a more efficient algorithm (than matrix multiplication) for computing these vectors in time complexity that is linear in the size of the output, in case the input graphs are restricted to a family of graphs for which the arboricity bounded by some constant δ.

The fact that the number d can be considered constant will be used extensively by the algorithms that we will propose. The complete expression of the complexity would be of the form $f(\delta, d) \cdot (N + M)$, for some function f of two parameters. But in this extended abstract we will not have the room for providing all the details. So far, the values for d that have been shown of practical use have been fairly small, usually smaller than 10.

Input: An undirected graph $G = (V, E)$ with average degree of z and a bounded arboricity $A(G) \leq \delta$ for some fixed constant δ. Once a random walk is placed at a particular vertex v, the probability of choosing any of the outgoing edges in the next step is the same p for all edges and the probability of staying at the same vertex in the next step, the *stationary probability* at v, is $q_v = (1 - (p * deg_v))$ where deg_v is the degree of vertex v.

Size of Input: $O(n + m)$ where $n = |V|$ is the number of vertices and $m = |E|$ is the number of edges in the given graph. Let this be denoted by N.

Task: Compute a set of vectors $P_1(v), \ldots, P_d(v)$ for every vertex v which contains the probabilities of reaching all other vertices in exactly $i = 1, \ldots, d$ steps if a random walk is started from v. For computational efficiency, each probability vector is maintained as a list of tuples, each containing a vertex number and the non zero probability of reaching that vertex in i steps from v through a random walk. Vertices which cannot be reached in i steps from v are not present in this list.

It is easy to see that the total size of the probability vectors that we want to compute may be much larger than the input size. This is due to the fact even if the average degree of our graph will be bounded, there may be vertices with a very high degree. All neighbors of such a high degree vertex will see each other with a random walk of distance two, and so their probability vectors will have at least the size of this neighborhood. So we may not expect algorithms to solve our problem in time that is proportional to the size of the input, we have to consider output complexity as well.

Size of Output: The total number of probability vectors obtained are $d \cdot n$ and the length of each one of them may be as large as n. Let the total size of the output be denoted by M. We will always assume that $N \leq M$.

1.2   Our Approach and Result

Our algorithm divides the given graph (of bounded arboricity) into a *core* and a *periphery* based on the average degree of the vertices. Computations are first done on the periphery, then the information is passed on to the core. The core in itself is a graph of bounded arboricity and thus the procedure is recursed on the core. Once the innermost core is reached then the direction of flow of information changes and the information starts flowing from the core to the periphery till it reaches the outermost periphery. The process of exchanging information between core and periphery is repeated a number of times which in our particular case is a function of the length of the random walks d, to collect all probability vectors for those paths that cross several times between core an periphery. This approach will make it possible to compute the probability vectors in time complexity that is linear in the number of non-zero entries of these vectors i.e. in $O(M)$ time which is a major improvement over the existing algorithm.

Moreover, we will always be able to charge all computations to vertices that have bounded degree in the graph under investigation. Thereby it is possible to parallelize our algorithm efficiently for the PRAM (Fortune and Wyllie [1978]), BSP (Valiant [1990]) or PRO (Gebremedhin et al. [2002]) models of parallel or distributed computation. By arguments as given by Dehne et al. [1997] and Gustedt [2003] such a paralellization may then also be extended to out-of-core computations. The later is particularly interesting for practical problems, since the actual hurdle for large scale computations on massif graphs are memory and not time constraints.

It is also possible to extend our results to other classes of graphs, namely graphs that are randomly chosen according to the Erdős-Renyi model G_p. Almost certainly these have a low average degree, too, and with high probability the recursive procedure that we propose will only fail on some very small iterated 'core' graph where all vertices have high degree. Since the problem on such a small core could be solved directly without a major impact on the total complexity, these graphs are tractable by our approach w.h.p, and in particular we obtain linear complexity on average. But due to space restrictions we will no be able to discuss this in detail.

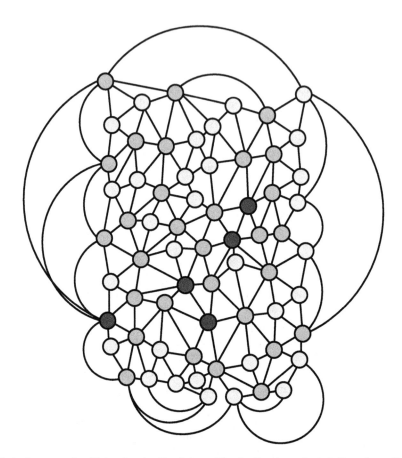

Fig. 1. A planar graph with two levels of periphery. The first level consists of all vertices of degree 5 or less (light colored). In the remaining graph, the *core*, the second level of periphery are then the vertices that have at most degree 5 in that graph (medium colored).

2   Basic Facts

To obtain a lower bound for the arboricity of a graph we may simply compute the quotient of the number of edges and the size of any of its spanning trees.

Definition 2. *Let* $G = (V, E)$ *be a connected graph. The* tree-density *of G is given by*

$$\tilde{A}(G) = \begin{cases} \left\lceil \frac{|E|}{|V|-1} \right\rceil & |V| > 1 \\ |V| & otherwise. \end{cases} \tag{1}$$

If G is not connected $\tilde{A}(G)$ *is the maximum tree-density of its connected components.*

Observation 1. *Let* $G = (V, E)$ *be a graph and* $G' = (V', E') \subseteq G$ *be a subgraph. Then* $\tilde{A}(G') \leq A(G)$.

Proof: Suppose $A(G) \leq \delta$ and let $(F_1, \ldots, F_\delta)$ be a tree-partition of G. Now let $(F_1', \ldots, F_\delta')$ be the forests that correspond to G', i.e., set $F_i' = F_i \cap G'$ for all $i = 1, \ldots, \delta$. Each of the F_i' contains at most $|V'| - 1$ edges and altogether these forest have at most $\delta(|V'| - 1)$ edges.   □

Surprisingly the converse of the above observation is also true, i.e., the tree-density of each subgraph also gives a lower bound on A. The following deep theorem holds

Theorem 1 (Nash-Williams [1961]).

$$\text{Let } G \text{ be a finite graph. Then } A(G) = \max_{G' \subseteq G} \tilde{A}(G').$$

For an easily achievable converse we loose a factor of 2 in the estimation.

Observation 2. Let $G = (V, E)$ be a graph such that each subgraph $G' \subseteq G$ has $\Delta(G') \leq \delta$ then $A(G) \leq \lfloor \delta \rfloor$.

Proof: We proceed by induction on the number of vertices in G. Let v be a vertex of minimum degree in G and $G' = G \setminus \{v\}$. Clearly $deg_G(v) \leq \lfloor \delta \rfloor$. We have $\Delta(G') \leq \delta$ by assumption so by the induction hypotheses there is a partition of G' into forests $(F_1', \ldots, F_{\lfloor \delta \rfloor}')$. Now we may obviously extend $(F_1', \ldots, F_{\lfloor \delta \rfloor}')$ to a partition $(F_1, \ldots, F_{\lfloor \delta \rfloor})$ of G by making v a leaf in each of the forests.   □

Another fact that we will need later concerns not the structure of our graphs but the structure of the probability vectors. The *symmetric* probability distribution which we assume deviates a bit from what is found in literature but it ensures the following useful property for the graph:

Lemma 3. If in G the probability of going from a vertex v_1 to v_m through a m step random walk is ρ, then the probability of going from v_m to v_1 through a m step random walk is also ρ.

Proof: Consider a m step path $v_1, v_2, \ldots, v_m$, not necessarily all distinct. The total probability of this path being chosen during a random walk of length m depends on the number of steps m' in the given path that join two distinct vertices. These steps contribute a factor of $p^{m'}$. The stationary phases of the path contribute by the number of times a particular vertex is repeated on the path. The order in which such a vertex v appears on the path is not important for the probability, it only contributes with a factor of $q_v^{m_v - 1}$, where q_v is the stationary probability of v and m_v is its multiplicity on the path.

So the inverse of the path is chosen with the same probability when starting a walk from v_m. Since this holds for all path of length m the claim holds.   □

3   The Algorithm

Algorithm 1 shows a generic recursive procedure for which in the following we will instantiate the remaining parts to then compute the probability vectors we are interested in. Observe that this algorithm doesnot even request that the graph has an arboricity within some limits; in view of Theorem 1 it could even be used to prove or disprove a given bound of the arboricity. But for the sake of simplicity we will assume that the input graph has an arboricity that is bounded by some fixed constant δ.

Algorithm 1. Recursive shelling to solve problem $\mathcal{P}(G, \mathcal{D})$.

Input: A connected Graph $G = (V, E)$ with eventually some local problem specific data $\mathcal{D}(G)$ attached to vertices and/or edges.

Output: $G = (V, E)$ together with the local information for $\mathcal{P}(G)$ attached to vertices and/or edges.

if *G has degree bounded by 2δ* **then**

direct $\quad \lfloor$ Solve $\mathcal{P}$ directly for G and **return** $\mathcal{P}(G)$;

else

    Let $z = \lceil \Delta_G \rceil$;

split   Let $V_0 \subseteq V$, the *periphery*, be the vertices of degree z or less;

    Let $V_1 = V \setminus V_0$, the *core*;

    $G_i = (V_i, E_i) = V|V_i$, for $i = 0, 1$;

recurse   Recurse on the connected components of G_0 and G_1 to obtain $\mathcal{P}_0^0$ and $\mathcal{P}_1^0$;

    Let $E_2 = E \setminus (E_0 \cup E_1)$;

loop   **foreach** $i = 1, \ldots, k$ **do**

pull   Via E_2, pull $\mathcal{P}_1^{2i-2}$ to G_0 and compute $\mathcal{P}_0^{2i-1}$ and $\mathcal{P}_0^{2i}$;

push   Via E_2, push $\mathcal{P}_0^{2i-1}$ to G_1 to obtain $\mathcal{P}_1^{2i-1}$ and $\mathcal{P}_1^{2i}$;

combine   Combine $\mathcal{P}_0^0, \ldots, \mathcal{P}_0^{2k}, \mathcal{P}_1^0, \ldots, \mathcal{P}_1^{2k}$ into $\mathcal{P}(G)$;

    return $\mathcal{P}(G)$;

Steps split and recurse: Dividing the graph in to a core and a periphery. The main trick of the algorithm is to divide the graph in two parts, low degree vertices form the *periphery* and high degree vertices the *core*. Note that z is bounded by the arboricity $A(G)$. The core G_1 is a subgraph of G and thus $A(G_1) \le A(G) \le \delta$. By definition, recursing on G_0 will immediately run into the case direct.

Recursion returns the probability vectors for the paths that lie entirely inside the core or the periphery. Then, for the final result we have to take into account all paths of length at least d that exist in the entire graph and which cross the boundary between the core and the periphery, *i.e.* that use edges in E_2. In Algorithm 1, $\mathcal{P}_0^i$ and $\mathcal{P}_1^i$ denote the probability vectors of paths restricted to use exactly i edges from E_2.

Step direct: Computations inside degree bounded graphs. Initially each vertex has a probability vector consisting of its neighbors and the probability of reaching them in 1 step of a random walk. All paths of maximal length d starting in a particular vertex v can be enumerated in time $(2\delta)^d$ and the probabilities of all those paths can be maintained in the same complexity. Then, for all target vertices w that can be reached from v in that way, the corresponding probability can be computed. Since $(2\delta)^d$ is considered to be constant, all this computation is proportional to the size of the probability vector.

In total, this means that the computation is linear in the size of the output.

Steps pull: Computations in the periphery. Suppose that we have stored all probabilities for paths that contain exactly $2i - 2$ edges from E_2. Any path that contains $2i - 1$ such edges has at least one endpoint in the periphery. For all vertices in the periphery we may collect all vertices that can be reached with exactly one edge in E_2 and compute the corresponding probabilities.

Now consider paths that contain $2i$ edges from E_2 and that have both endpoints v and v' in the periphery G_0. They can be divided into three parts: paths from v and v' into the core and a path that uses $2i - 2$ edges of E_2 and for which the probability is registered in $\mathcal{P}_1^{2i-1}$.

Step push: Passing of information — periphery to core. With Lemma 3, for every vertex in the periphery look at its d probability vectors, search for probabilities of reaching vertices in the core which has a neighbor in the periphery, and reverse this information, *i.e.*, if a vertex v_j in the periphery has the knowledge that it can reach another vertex v_k in the core (which has a neighbor in the periphery) in m steps of random walk with probability p, then give this information to v_k, telling it that it can reach v_j in m steps of random walk with probability p. This step will generate probability vectors $\mathcal{P}_1^{2i-1}$ in the core.

Similar considerations for the case of paths using an even number of edges in E_2 and that start and end in G_1 lead to the computation of $\mathcal{P}_1^{2i}$. For complexity considerations it is important to note that the corresponding computations can be charged to nodes in the periphery. If a vertex v in the periphery knows that it can reach vertex v' in the core by crossing E_2 exactly once it may use the information in $\mathcal{P}_0^{2i-1}(v)$ to provide information for $\mathcal{P}_1^{2i}(v')$. To avoid duplication, this *stitching* of information needs to be done only by those vertices in the periphery which have a neighbor in the core, *i.e.* the endpoints of edges in E_2 that are in the periphery.

3.1   Correctness and Complexity

It is now easy to see that the sets of paths that are used for the different vectors $\mathcal{P}_{0,1}^{0,\dots,2k}$ are mutual disjoint and thus their probabilities add up to give the total probabilities. Therefore correctness of the algorithm follows immediately.

For the complexity, observe that the **foreach**-loop is executed $O(d)$ times and that all push and pull computations are proportional to modifications on the probability vectors that they produce. Since δ and d are considered to be constant, all calls (without accounting for recursion) are linear in the output they produce. Observe here that not necessarily all vertices of the core are even touched by these updates. They are only touched and new non-zero probabilities are added to their vectors, if the merge of the periphery and core information discovers vertices that have not been reachable inside core or periphery alone.

So, when summing up over all calls the time is proportional to the produced output.

It remains to show that the recursion and its touching of vertices and edges does not worsen the running time. The recursive calls on the periphery G_0 always have depth 1, since by definition G_0 only has vertices of low degree. So the only call that could lead to a deeper recursion (and higher complexity) is the one for the core G_1. V_1 are the vertices in V that have a degree that is above $2\delta \geq 2$ and so we have that $|V_1| \leq |V|/2$. Thereby the total sum of the edges of the graphs that occur during recursion is bounded by $\sum_i 2^{-i}\delta|V| \leq 2\delta|V|$, and thus linear in our setting.

4   Conclusion and Further Work

We proposed an *output sensitive* algorithm for computing probability vectors for short random walks on graphs with bounded arboricity. Thereby it applies to a large number of graph classes which are of theoretical and practical interest. The good complexity behavior of our algorithm also extends to other computational models such as PRAM, BSP or PRO and also to some classes random graphs w.h.p.

The approach we discussed gives an alternative and more efficient way to compute these probability vectors, but it doesnot change the community structure that emerge from the approach discussed by Latapy and Pons [2005]. However, the second phase of using clustering algorithms that was described by them is very time consuming, too, so it might be appropriate to deviate from that approach, even by allowing for algorithms that would give a community structure that is defined slightly differently. This part of the problem is still open. A good approach could be to use Union-Find strategies which have linear complexity under some conditions (see Fiorio and Gustedt [1996]). Any such approach would require an oracle on the basis of which smaller communities would be joined to form bigger ones till a community structure emerges that is satisfactory according to some predefined criteria. We hope to do some work in this direction also.

Bibliography

W. Aiello, F. Chung, and L. Lu. Random evolution in massive graphs. In *Proceedings of the 42nd Annual IEEE Symposium on Foundations of Computer Science*, pages 510–519, 2001. URL citeseer.ifi.unizh.ch/article/aiello01random.html.

Albert-László Barabási and Réka Albert. Emergence of scaling in random networks. *Science*, 286:509–512, 1999.

Hans Leo Bodlaender. A linear time algorithm for finding tree-decompositions of small treewidth. *SIAM J. Comput.*, 25(1305-1317), 1996.

David Cheriton and Robert Endre Tarjan. Finding minimum spanning trees. *SIAM J. Computing*, 5:724–742, 1976.

F. K. H. A. Dehne, W. Dittrich, and D. Hutchinson. Efficient external memory algorithms by simulating coarse-grained parallel algorithms. In *ACM Symposium on Parallel Algorithms and Architectures*, pages 106–115, 1997.

Christophe Fiorio and Jens Gustedt. Two linear time union-find strategies for image processing. *Theoretical Computer Science*, 154:165–181, 1996.

Steven Fortune and James Wyllie. Parallelism in random access machines. In *10th ACM Symposium on Theory of Computing*, pages 114–118, May 1978.

Assefaw Hadish Gebremedhin, Isabelle Guérin Lassous, Jens Gustedt, and Jan Arne Telle. PRO: a model for parallel resource-optimal computation. In *16th Annual International Symposium on High Performance Computing Systems and Applications*, pages 106–113. IEEE, The Institute of Electrical and Electronics Engineers, 2002.

Jens Gustedt. Minimum spanning trees for minor-closed graph classes in parallel. In *Symposium on Theoretical Aspects of Computer Science*, pages 421–431, 1998.

Jens Gustedt. Towards realistic implementations of external memory algorithms using a coarse grained paradigm. In *International Conference on Computational Science and its Applications (ICCSA 2003), part II*, number 2668 in LNCS, pages 269–278. Springer, 2003.

S. Kannan, M. Naor, and S. Rudich. Implicit representation of graphs. *SIAM Journal On Discrete Mathematics*, 5:596–603, November 1992. URL `citeseer.ifi.unizh.ch/kannan92implicit.html`.

Matthieu Latapy and Pascal Pons. Computing communities in large networks using random walks. In *ISCIS'05*, pages 284–293, 2005.

W. Mader. Homomorphieeigenschaften und mittlere Kantendichte von Graphen. *Math. Ann.*, 174:265–268, 1967.

Crispin St. John Alvah Nash-Williams. Edge-disjoint spanning trees of finite graphs. *J. London Math. Soc. (2)*, 36:445–450, 1961.

L. G. Valiant. A bridging model for parallel computation. *Communications of the ACM*, 33(8): 103–111, 1990.

An Implicit Representation of Chordal Comparabilty Graphs in Linear-Time

Andrew R. Curtis, Clemente Izurieta, Benson Joeris,
Scott Lundberg, and Ross M. McConnell

Department of Computer Science
Colorado State University
Fort Collins, CO 80523-1873, U.S.A

Abstract. Ma and Spinrad have shown that every transitive orientation of a chordal comparability graph is the intersection of four linear orders. That is, chordal comparability graphs are comparability graphs of posets of dimension four. Among other uses, this gives an implicit representation of a chordal comparability graph using $O(n)$ integers so that, given two vertices, it can be determined in $O(1)$ time whether they are adjacent, no matter how dense the graph is. We give a linear-time algorithm for finding the four linear orders, improving on their bound of $O(n^2)$.

1 Introduction

A *partial order* or *poset* relation is a transitive antisymmetric relation. In this paper, we consider the graphical representation of a poset using a directed acyclic and transitive graph. When we say the graph is *transitive*, we mean that whenever $x \to y \to z$, $x \to z$. Whether the partial order is reflexive is irrelevant to our goals, so we only consider loopless graphs. The *comparability relation* of a partial order is the set of pairs that are comparable in the partial order. That is, it is the symmetric closure, where, whenever (a, b) is in the partial order, (b, a) is added to it. The comparability relation has a natural representation as an undirected graph that has an edge ab whenever (a, b) and (b, a) are in the comparability relation; it is obtained by ignoring edge directions in the transitive graph that represents the partial order. And given a comparability graph, it is possible to *transitively orient* it in linear time [MS99], that is, to recover a corresponding partial order.

A *chordal graph* is an undirected graph where each cycle of length four or greater has a *chord*, that is, an edge that is not on the cycle but whose endpoints are both on the cycle.

A co-comparability graph or co-chordal graph is one whose complement is a comparability graph or chordal graph, respectively. Many interesting graph classes are defined by intersecting the comparability, co-comparability, chordal and co-chordal graph classes.

An example is an *interval graph*, which is the intersection graph of a set of intervals on the line, that is, the graph that has one vertex for each of the intervals

F.V. Fomin (Ed.): WG 2006, LNCS 4271, pp. 168–178, 2006.

and an edge for each intersecting pair. These are exactly the intersection of the chordal and co-comparability graphs.

A *permutation graph* is defined by a permutation of a linearly ordered set of objects. The vertices are the objects, and the edges are the *non-inversions*, that is, the pairs of objects whose relative order is is the same in the two permutations. These are exactly the intersection of the comparability and co-comparability graphs.

A *split graph* is a graph whose vertices can be partitioned into a clique and an independent set. These are exactly the intersection of the chordal and co-chordal graphs. More information about all graph classes classes mentioned here can be found in [Gol80].

All of these graphs are subclasses of the class of perfect graphs, because comparability graphs and chordal graphs are perfect. Interval graphs can be represented with $O(n)$ integers, numbering the endpoints in left-to-right order and associating each vertex with its endpoint numbers. Adjacency can then be tested in $O(1)$ time by comparing the two pairs of endpoints of the vertices to see if they correspond to intersecting intervals. Similarly, permutation graphs can be represented by numbering the vertices in left-to-right order in two linear orders, and testing adjacency in $O(1)$ time by determining whether the two vertices have the same relative order in both. These are examples of *implicit representations*; for more details see Spinrad's book on the topic of implicit representations of graph classes [Spi03].

A *linear order* is just a special case of a partial order, where the elements are numbered 1 through n, and the relation is the set of ordered pairs $\{(i,j)|i < j\}$. This partial order has $\Theta(n^2)$ elements, but can be represented implicitly by giving the ordering or numbering of the vertices.

It is easy to see that the intersection of two partial orders (the ordered pairs that are common to both) is also a partial order, hence this applies to the intersection of linear orders. In fact, every partial order is the intersection of a set of linear orders [DM41]. A partial order has *dimension k* if there exist k linear orders whose intersection is exactly that partial order. It is easy to see from this that the permutation graphs are just the comparability graphs of two-dimensional partial orders. Two-dimensional partial orders and permutation graphs can be recognized and their representation with two linear orders can be found in linear time [MS99]. In general, k linear orders gives an $O(nk)$ representation, but unfortunately, it is NP-complete to determine whether a partial order has dimension k for $k \geq 3$ [Yan82].

In this paper, we examine *chordal comparability graphs*, that is, the intersection of the class of chordal graphs and the class of comparability graphs. Ma and Spinrad have shown that all chordal comparability graphs are the comparability graphs of partial orders of dimension at most four [MS91, Spi03]. The four linear orders give a way of representing the graph in $O(n)$ space so that for any two vertices, it can be answered in $O(1)$ time whether they are adjacent. Each vertex is labeled with the four position numbers of each vertex in the four linear order, and for two vertices, they are adjacent iff one of them precedes the other in each of the four orders. This type of implicit representation is desirable as a data structure for representing the partial order or its chordal comparability graph, and for organizing algorithmic solutions for combinatorial problems on the graphs.

This bound was shown to be tight by Kierstead, Trotter, and Qin in [KTQ92], who used a non-constructive Ramsey-theoretic proof to show that some chordal comparability graphs actually require four linear orders, but as is typical of Ramsey-theoretic proofs, the upper bound of the smallest one requiring four is an enormous $27^{27} + 1$ vertices. It seems likely that there exist small examples that require four linear orders. Testing a candidate is complicated by the NP-completeness of determining whether a partial order has dimension 3, though we do not know whether that problem remains NP-complete when restricted to chordal comparability graphs. Finding a smallest one, or even a small one, is an open problem.

We should note that, unlike the implicit representations of interval graphs and permutation graphs, this representation does not characterize chordal comparability graphs, as there are posets of dimension four whose comparability graphs are not chordal comparability graphs.

Ma and Spinrad have given a linear-time algorithm for recognizing chordal comparability graphs, but the best bound they give for finding the four linear orders is $O(n^2)$, where n is the number of vertices. In this paper, we improve this latter bound to $O(n + m)$.

2   Preliminaries

In this paper, $G = (V, E)$ denotes a simple, finite graph with vertex-set V and edge-set E. For convenience, we assume that G is connected. If there exists an edge between $v, u \in V$, we say that v and u are adjacent or are neighbors in G. If G is directed, then we use (u, v) to denote an edge from u to v in G. Given a directed graph G, we say that G is *acyclic* or is a *DAG* if G does not contain any directed cycles. The *transitive closure* of a DAG G adds the minimum number of edges to G such that the resulting graph is transitive, ie. $(x, y), (y, z) \in E$ implies that $(x, z) \in E$.

If G is an interval graph, then each vertex in G has a corresponding interval in $1, \ldots, 2n$. Let I_v denote the interval of v. Two vertices v and u are adjacent if their intervals share some common point. We say that v contains u if all points in I_u are also in I_v. Two intervals overlap if they share some point, but neither contains the other, and two intervals are disjoint if they share no common points.

2.1   Union-Find Data Structure

Our algorithm primarily makes use of elementary data structures; additionally, we use the union-find data structure. This data structure maintains a family of disjoint sets under the union operation. In order to identify the sets, each set has a *leader*, which is the representative for all elements in its set. Union-find supports the following operations.

- MakeSet(x): creates a new set containing only x.
- Find(x): returns the leader of the set containing x.
- Union(x, y): unions the sets containing x and y and returns the single set's new leader.

A MakeSet operation on n elements, followed by m union and find operations on them, takes $O(n + m\alpha(m, n))$ time, where α is an extremely slow-growing but unbounded functional inverse of Ackermann's function. Full details can be found in [Tar83].

However, there is a special case of the general union-find data structure developed by Gabow and Tarjan [GT85]. Their data structure requires initializing the structure with an unrooted tree on the n elements, and performing unions in any order that maintains the invariant that each union find class induces a connected subtree of the initializing tree. Given such an initializing tree, it is possible to do n MakeSet operations followed by m union and find operations in $O(n + m)$ time. In our application, we are able to initialize the Gabow-Tarjan structure, and this is critical to obtaining a true linear time bound.

3 Representing a Chordal Comparability Graph with Four Linear Orders

Details of the following properties of chordal graphs are well-known, and can be found in the text by Golumbic [Gol80]. A graph is chordal if and only if it has a *subtree intersection model*, which consists of the following:

1. A tree T that has $O(n)$ vertices;
2. A connected subtree T_v associated with each vertex v such that two vertices x and y are adjacent in G if and only if T_x and T_y contain a common node in T, and that the sum of cardinalities of the vertex sets in the subtrees is $O(m)$.

Such a tree is often called a *clique tree* after one method of generating one by creating one node of T for each maximal clique of G. An example is given in Figure 1.

Following the approach of Ma and Spinrad, we perform an arbitrary depth-first search on the clique tree, labeling the vertices in ascending order of their discovery time. The first and last discovery time i and j of nodes in a subtree T_x defines an interval $I_x = [i, j]$ on the sequence $(1, 2, ..., n)$. It is easy to verify that x and y are adjacent in G if I_x and I_y properly overlap, and that they are nonadjacent if I_x and I_y are disjoint.

Suppose I_y is contained in I_x. Then it is possible that they are adjacent. If this were always the case, then G would be not just a chordal comparability graph, but an interval comparability graph. However, it is also possible that they are not adjacent. In this case, the DFS discovered a vertex in T_x, and sometime during the interval I_x, it left T_x to visit a set of vertices below T_x that contain T_y, before returning upward to T_x to finish traversing it. This shows that it is not necessary for T_x and T_y to intersect for I_y to be a subinterval of I_x.

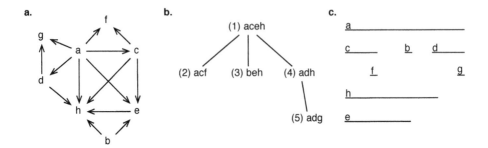

Fig. 1. (a) A transitive orientation of a chordal graph G, (b) a clique-intersection tree of G. T_a consists of vertices $\{1, 2, 4, 5\}$, T_b consists of vertex 3, T_c consists of vertices $\{1, 2\}$, etc. Each T_x is a connected subtree of T, and two vertices x and y are adjacent in G if and only if T_x and T_y contain a common node. (c) The vertices are numbered according to their discovery times during a depth-first search, and the first and last discovery time in T_x defines an interval I_x on the sequence $(1, 2, 3, 4, 5)$. For instance, $I_a = [1, 5]$, $I_b = [3, 3]$, $I_e = [1, 3]$, etc.

We may find a transitive orientation P of the chordal comparability graph $G = (V, E)$ in linear time using the algorithm of McConnell and Spinrad [MS99]. Ma and Spinrad define three disjoint partial orders a transitive orientation P of G, the set R_1 of ordered pairs of the form $\{(x, y) | I_x$ strictly precedes $I_y\}$, and the set R_2 of ordered pairs of of the form $\{(x, y) | I_y \subset I_x$ and x and y are non-neighbors$\}$. Clearly, $P \cup R_1 \cup R_2$ is an orientation of the complete graph, $\{P, R_1, R_2\}$ is a partition of it into three poset relations, and $\{R_1, R_2\}$ is a partition of $\overline{G}$ into two poset relations. Because $R_1 \cup R_2$ is an orientation of $\overline{G}$, we will refer to the members of R_1 and R_2 as *edges*.

In general, the union of two disjoint partial orders is not necessarily a partial order, or even acyclic. However, Ma and Spinrad show that $E_1 = P \cup R_1$, $E_2 = P \cup R_1^T$, where R_1^T denotes the reversal of all edges in R_1, $E_3 = P \cup R_2$ and $E_4 = P \cup R_2^T$ are each acyclic. This shows that P is a four-dimensional partial

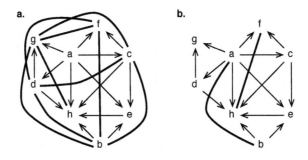

Fig. 2. (a) G from Figure 1 with the type 1 edges drawn with a thicker line and (b) G with the type 2 edges shown

order, as follows. Let L_1, L_2, L_3, and L_4 be arbitrary topological sorts of E_1, E_2, E_3, and E_4, respectively. It must be the case that for $(x, y) \in P$, x precedes y in all four topological sorts, since (x, y) is a directed edge in each of E_1, E_2, E_3, and E_4. Every edge of P is *conserved* in the intersection of L_1 through L_4.

For any edge (u, v) in R_1, $(u, v) \in E_1$ and $(v, u) \in E_2$. Therefore u precedes v in L_1 and follows it in L_2 and (u, v). It follows that neither (u, v) nor (v, u) is in the intersection $L_1 \cap L_2$ of the topological sorts of E_1 and E_2. The act of reversing R_1 in E_1 and E_2 *deletes* the edges of R_1 from the intersection $L_1 \cap L_2$, hence from the intersection of L_1 through L_4.

Similarly, for any edge (x, y) in R_2, x precedes y in L_3 and follows it in L_4, so the act of reversing R_2 in E_3 and E_4 ensures that x precedes y in L_3 and follows it in L_4. Therefore, neither (x, y) nor (y, x) is in the intersection $L_3 \cap L_4$ of the topological sorts of E_3 and E_4. Reversing R_2 in E_3 and E_4 *deletes* R_2 from the intersection.

Together, these observations prove that the intersection of L_1 through L_4 is exactly P: all elements of P are conserved and no elements of R_1, R_2, R_1^T, or R_2^T are conserved. The constructive proof gives the basis of Ma and Spinrad's algorithm, which finds a transitive orientation of P, finds a clique tree, performs a DFS on it to identify R_1 and R_2, and then returns the topological sorts of E_1 through E_4 in $O(n^2)$ time.

On the surface, it seems impossible to improve on this time bound without resorting to an entirely different algorithm, since the topological sorts reference all edges in $P \cup R_1 \cup R_2$, and there are $n(n-1)/2$ of them.

Our approach is similar to Ma and Spinrad's, but we are able to use the properties of partial orders, chordal graphs, and a number of data structure tricks to avoid touching all of the edges in R_1 and R_2 directly, thereby obtaining an $O(n + m)$ bound. Many of the details are nontrivial, especially in the case of L_4.

3.1   Finding L_1 and L_2

We first describe the procedure for finding the topological sort L_1 of $E_1 = P \cup R_1$. To obtain L_1, we perform a depth-first search on E_1, prepending each vertex to L_1 when DFS retreats from it because all of its neighbors have been marked visited. It is well-known that prepending vertices to a list as they finish during DFS results in a topological sort of any DAG [CLRS01]. The edges of P can be handled during the DFS in the standard way with an adjacency-list representation. However, when all neighbors in P of a vertex have been marked visited, it is not necessarily the case that all neighbors in E_1 have been marked visited, since it may have neighbors in R_1. The problem is that $|R_1|$ is not $O(n + m)$, so touching all the members of R_1 would ruin our time bound. To get around this, we create a data structure that supports the following operation in $O(1)$ time:

– **Find next R_1 neighbor:** Given a vertex v, return an unmarked neighbor in R_1 if there is one, or else report that it has no unmarked neighbors in R_1.

To create this data structure, we radix sort all endpoints of intervals using the position of the endpoint as primary sort key, and whether it is a left or right endpoint as a secondary sort key. We then label each right endpoint with a pointer to its *parent*, which is the first left endpoint that follows it. We then remove the right endpoints to obtain a list L of the vertices sorted by left endpoint; each vertex still retains a pointer to its parent in L.

When we run the DFS, we maintain a set of union-find classes on elements of L, using the following invariant:

- Each union-find class starts either at the beginning of the list or at the first element following an unmarked element, and contains all elements either up through the end of the list or through the next unmarked element. Each union-find class is labeled with a pointer to its sole unmarked element, if it has one.

Initially, every element of L is in its own union-find class. Note that at all times, every union-find class except the rightmost one has exactly one unmarked element, and the unmarked element in a class is the rightmost element in the class. In addition, every union-find class is consecutive in L, which allows us to use the path represented by L as the initializing tree for the Gabow-Tarjan data structure. The **find next R_1 neighbor** for a vertex v operation can be implemented by performing a find operation on the parent of v. As v's parent is the first disjoint vertex right of v, its union-find class points to the first unmarked vertex right of v. Therefore, it takes $O(1)$ time to locate the next unmarked vertex right of v, namely, the next R_1 neighbor of v.

When a vertex v is visited, it is marked visited. This requires the following:

- **Mark a vertex v as discovered:** Unless v is the last element of L, merge its union-find class and the union-find class that contains the successor w of v in L. Let the new class point to the unmarked element of w's old class.

To perform the DFS, we make recursive calls on all neighbors of v in P. When the last of these returns, we can find an unmarked neighbor in R_1 in $O(1)$ time executing the **find next R_1 neighbor** operation. Either a recursive call is made on the result, marking it as discovered, or, if v has no remaining neighbors in R_1, v can be marked as finished. Each vertex is marked once as discovered and once as finished, so these marking operations can take place $O(n)$ times during the entire DFS, each at a cost of $O(1)$. Therefore, the inclusion of R_1 along with P in the DFS ends up costing $O(n)$ time, even though R_1 can be $\Theta(n^2)$ in the worst case. The final bound on the DFS to obtain L_1 is $O(n + m)$, where m is the number of edges in G.

By left-right symmetry, a similar algorithm applies to finding E_2.

3.2 Finding L_3

The approach for finding L_3 is similar in spirit to the one for finding L_1 and L_2, except that the charging argument to obtain the time bound is more complicated. We again handle DFS on P using an adjacency-list representation. When a vertex has no more unmarked neighbors in P, it may still have neighbors in R_2. We must define an operation analogous to **find next R_1 neighbor**:

- **Find next R_2 neighbor:** Given a vertex v that has no undiscovered neighbor in P, return an undiscovered neighbor in R_2 if there is one, or else report that it has no unmarked neighbors in R_2.

However, because of additional difficulties posed by R_2 edges, we cannot claim the $O(1)$ bound for this operation that we can for `find next` R_1 `neighbor`. Instead, we use an amortizing argument that shows that all calls to `find next` R_2 `neighbor` made during a DFS take a total of $O(n)$ time.

As with `find next` R_1 `neighbor`, we create a data structure to support the operation by sorting vertices by left endpoint of their interval to obtain a list L, and we maintain union-find classes with one unmarked element in each class except the rightmost class.

All R_2 neighbors have both endpoints in the interior of I_v instead of to the right of I_v. Instead of starting with the first vertex whose interval lies strictly to the right of v's interval, we start with the first interval whose left endpoint is to the right of v's left endpoint. We perform a `find` operation on this vertex to find the first unmarked vertex w that follows it in L. If I_w is to the right of I_v, then we may mark v as *finished*, since I_v has no unmarked vertices in its interior. If $I_v \subset I_w$, then w is the next R_2 neighbor of v in the list, and we can make a recursive call on it, marking it as discovered.

A new problem arises when I_w properly overlaps I_v, since we have now spent $O(1)$ time finding w, but its interval is not contained in w's, so is not an R_2 neighbor of v and we cannot mark it. Fortunately, if I_v and I_w properly overlap, they are neighbors in G. Since v has no unmarked neighbors in P and w is unmarked, it must be the case that (w, v) is an edge in P. We therefore charge the cost of touching w to the edge wv in G. We then continue by performing a `find` operation on the successor of w in L to find the next unmarked vertex. We iterate this operation, each time charging the $O(1)$ cost of finding the next unmarked vertex to an edge of G, halting when we reach the right endpoint of I_v, in which case v can be marked as finished, or else find an unmarked R_2 neighbor z, which we can then make a recursive call to DFS on, marking z discovered. During the recursive call, we retain a pointer to z so that when it returns, we may resume the DFS from v by performing a `find` operation on z.

Each vertex is again marked as discovered once, finished once, and each edge of P directed into a vertex v is charged once. Since $|P| = m$, The additional cost incurred in including R_2 with P in the DFS is $O(n + m)$.

3.3   Finding L_4

Unlike the case of L_1 and L_2, the cases of L_3 and L_4 are not symmetric, so we cannot use the procedure for finding L_3 to find L_4. In the case of L_3, the `find next` R_2 `neighbor` found all unmarked vertices whose left endpoint was interior to I_v. Those that were not R_2 neighbors were neighbors in G, which allowed us to charge the cost of finding them to edges of G.

L_4 needs to be a topological sort of $P \cup R_2^T$. Consider what happens when we reverse R_2 to get R_2^T. The R_2^T neighbors of a vertex v are those non-neighbors

w in G such that $I_v \subset I_w$. Such a neighbor has a left endpoint to the left of I_v's left endpoint and a right endpoint to the right of I_v's right endpoint. Using a Gabow-Tarjan data structure as we did above can identify unmarked neighbors whose left endpoint is to the left of I_v's. The insurmountable problem is that such a vertex may also have a right endpoint to the left of I_v's left endpoint. This means it is a non-neighbor in G. We have spent $O(1)$ time touching it, but we have no edge of G to charge the cost to.

We therefore abandon the union-find approach and instead adopt a strategy that involves partitioning sets into neighbors and non-neighbors in P, and takes advantage of the fact that we already have a topological sort L_3 of $E_3 = P \cup R_2$.

We begin with $L' = L_3^T$. Since every edge of $P \cup R_2$ points from a later vertex to an earlier vertex, so none of the edges of $P \cup R_2$ survive in the intersection of $E_3 \cap L'$ We then modify L' to reverse the relative order of every pair (a, b) such that $(a, b) \in P$ *without affecting the relative order of any pair* $(c, d) \in R_2$. This restores P to the intersection $L_3 \cap L'$ while conserving all edges of R_2^T. We let L_4 be the resulting modification of L'.

Given a subset S of elements of L', let the *subsequence of L' induced by S* denote the result of deleting all elements from L' that are not in S. This is just the ordering of S that is consistent with their relative order in L'.

Let us initially number the vertices in L' in order from 1 to n. By radix sorting all edges of P using vertex of origin as the primary sort key and destination vertex as the secondary sort key, we may obtain, for each vertex v, an adjacency list that is sorted in left-to-right order as the vertices appear in L'. This gives the subsequence of L' defined by neighbors of v. We let L' be a doubly-linked list, so that, given a pointer to a vertex in L', it can be removed from L' in $O(1)$ time.

We now give the algorithm for turning L' into L_4:

Reordering L_4^T to obtain L_4
The algorithm is recursive, so we assume that a subsequence L' of L_4 has been passed into the current recursive call. Let v be the last vertex in L'. As a base case, if $|L'| \le 1$, there is nothing to be done. Otherwise, remove the neighbors of v in P from L', leaving the subsequence L_n of non-neighbors of v in L'. Since the adjacency list of v is in sorted left-to-right order in L', we can put them into a doubly-linked list that gives the subsequence L_a of vertices that are adjacent to v in L'. We recursively reorder L_n and L_a to obtain L_n' and L_a', and return the concatenation $L_n' \cdot v \cdot L_a'$.

The following establishes the correctness:

Lemma 1. $P \subseteq L_4$ and $R_2 \cap L_4 = \emptyset$.

Proof. Since v is a source and all of its edges point to L_a, they all point to the right when L_a is moved ahead of v. Since P is transitive, L_a is not just the neighbors of v, but the set of all nodes reachable from v on a directed path. Therefore, there is no directed edge of P from L_a to L_n. All edges of P that go between L_n and L_a are directed to the right, as they are supposed to in L_4. We conclude that all edges of P that have endpoints in two sets of L_a, $\{v\}$, and L_n point to the right after L_n is moved to the right of v.

Suppose some edge (x, y) of R_2 points to the right after L_a is moved to the right. Then y is a neighbor of v and x is not. Moreover, $(x, y) \in R_2$ implies that $I_y \subset I_x$. Since v is a neighbor of y, I_v's intersects I_y, which means I_v also intersects I_x. If I_v properly overlaps I_x, then v and x are neighbors, contradicting x's membership in L_n. Therefore, $I_v \subset I_x$, and, since v and x are non-neighbors, this implies that $(v, x) \in R_2$. But this contradicts the fact that v is a source in $P \cup R_2$, since it lies at the beginning of L_3. Therefore, edge (x, y) of R_2 cannot exist.

We conclude that all edges of P that go between L_n, $\{v\}$, L_a point to the right after the reordering, and all edges of R_2 that go between these sets point to the left.

None of the elements internal to L_n or L_a have been reordered. By induction on the length of the subsequence passed into a recursive call, the recursive calls on L_n and L_a reorder these sets so that all edges of P point to the right in the final order, and all edges of R_2 point to the left in the final order. The lemma is immediate from this statement.

Since L' is a doubly-linked list, it takes time proportional to the the degree of v to remove L_a out of L' and concatenate it to the front. We charge this cost to edges out of v. At each level of the recursion, a distinct vertex serves in the role of v, so each edge is charged at most once. The running time is therefore $O(n + m)$.

Since $L_1 \cap L_2$ includes all of P and excludes every edge of R_1, and since $L_3 \cap L_4$ includes all of P and excludes every edge of R_2, we get the following:

Theorem 1. *Four linear orders that realize a chordal comparability graph G can be found in $O(n + m)$ time.*

It is worth noting that our algorithm to find L_4 is an example of an *ordered vertex partitioning algorithm*, which proceed by refining a partition by splitting classes into neighbors and non-neighbors of a *pivot vertex*, and maintaining a linear order on the partition classes. Other examples of such algorithms are Lex-BFS [RTL76], which is used for recognizing chordal graphs, and the algorithm for transitively orienting comparability graphs given in [MS99]. The algorithm we give for finding L_4 gives a third variant on this class of graph algorithms, differing significantly from them in the order in which it selects pivot vertices and the linear order it maintains on the partition classes.

References

[CLRS01] T.H. Cormen, C.E. Leiserson, R.L. Rivest, and C. Stein. *Introduction to Algorithms*. McGraw Hill, Boston, 2001.

[DM41] B. Duschnik and E. W. Miller. Partially ordered sets. *Amer. J. Math.*, 63:600–610, 1941.

[Gol80] M. C. Golumbic. *Algorithmic Graph Theory and Perfect Graphs*. Academic Press, New York, 1980.

[GT85]   H. N. Gabow and R. E. Tarjan. A linear-time algorithm for a special case of disjoint set union. *Journal of Computer and System Sciences*, 30:209–221, 1985.

[KTQ92]   H. Kierstead, , W. T. Trotter, and J. Qin. The dimension of cycle-free orders. *Order*, 9:103–110, 1992.

[MS91]   T. Ma and J. P. Spinrad. Cycle-free partial orders and chordal comparability graphs. *Order*, 8:49–61, 1991.

[MS99]   R. M. McConnell and J. P. Spinrad. Modular decomposition and transitive orientation. *Discrete Mathematics*, 201(1-3):189–241, 1999.

[RTL76]   D. Rose, R. E. Tarjan, and G. S. Lueker. Algorithmic aspects of vertex elimination on graphs. *SIAM J. Comput.*, 5:266–283, 1976.

[Spi03]   J. Spinrad. *Efficient Graph Representations*. American Mathematical Society, Providence RI, 2003.

[Tar83]   R. E. Tarjan. *Data structures and network algorithms*. Society for Industrial and Applied Math., Philadelphia, 1983.

[Yan82]   M. Yannakakis. The complexity of the partial order dimension problem. *SIAM J. Algebraic and Discrete Methods*, 3:303–322, 1982.

Partitioned Probe Comparability Graphs
(In Honor of Prof. Dr. J.H. van Lint)

David B. Chandler[1], Maw-Shang Chang[2,*], Ton Kloks[**],
Jiping Liu[3,***], and Sheng-Lung Peng[4]

[1] Institute of Mathematics
Academia Sinica, Taipei 115, Taiwan, R.O.C.
`chandler@math.sinica.edu.tw`
[2] Department of Computer Science and Information Engineering
National Chung Cheng University, Chiayi 621, Taiwan, R.O.C.
`mschang@cs.ccu.edu.tw`
[3] Department of Mathematics and Computer Science
The University of Lethbridge, Alberta, T1K 3M4, Canada
[4] Department of Computer Science and Information Engineering
National Dong Hwa University, Hualien 974, Taiwan, R.O.C.
`lung@csie.ndhu.edu.tw`

Abstract. Given a class of graphs $\mathcal{G}$, a graph G is a probe graph of $\mathcal{G}$ if its vertices can be partitioned into a set $\mathbb{P}$ of probes and an independent set $\mathbb{N}$ of nonprobes such that G can be embedded into a graph of $\mathcal{G}$ by adding edges between certain nonprobes. If the partition of the vertices is a part of the input we call G a partitioned probe graph of $\mathcal{G}$. In this paper we show that there exists a polynomial-time algorithm for the recognition of partitioned probe graphs of comparability graphs. This immediately leads to a polynomial-time algorithm for the recognition of partitioned probe graphs of cocomparability graphs. We then show that a partitioned graph $G = (\mathbb{P} + \mathbb{N}, E)$ is a partitioned probe permutation graph if and only if G is at the same time a partitioned probe graph of comparability and cocomparability graphs.

1 Preliminaries

Given a graph $G = (V, E)$ with a partition of the vertices into a set $\mathbb{P}$ of *probes* and an independent set $\mathbb{N}$ of *nonprobes*, we wish to find an embedding of G into a graph of a certain graph class by adding edges between nonprobes. In this paper we focus on the graph classes of comparability, cocomparability, and permutation graphs.

A graph G is a pair $G = (V, E)$ where the elements of V are called the vertices of G and where E is a family of two-element subsets of V called the edges. We

---

* Partially supported by the National Science Council of Taiwan, grant NSC94-2213-E-194-009.
** Supported by the National Science Council of Taiwan, grant NSC94-2627-B-007-001.
*** Partially supported by the NSERC of Canada.

denote edges of a graph G as (x, y) and we call x and y the endvertices of the edge. Unless stated otherwise, a graph is regarded as undirected. For a vertex x we write $N(x)$ for its set of neighbors in G, and for a subset $W \subseteq V$ we write $N(W) = \cup_{x \in W} N(x) - W$. We write $n = |V|$ for the number of vertices and $m = |E|$ for the number of edges. For a subset A of edges or arcs of a graph we denote by $V(A)$ the set of endvertices incident with elements of A. For a graph $G = (V, E)$ and a subset $S \subseteq V$ of vertices, we write $G[S]$ for the subgraph of G *induced* by S. For a subset $W \subseteq V$ of vertices of a graph $G = (V, E)$ we write $G - W$ for the graph $G[V - W]$, *i.e.*, the subgraph induced by $V - W$. For a vertex x we write $G - x$ rather than $G - \{x\}$. If the graph is directed we use the notation $\overrightarrow{xy}$ to denote the *arc* directed from x to y. Likewise, we use $\overleftarrow{xy}$ to denote the arc directed in the opposite direction. For a subset $\mathcal{E} \subseteq E$ of edges of a graph $G = (V, E)$, let $\overrightarrow{\mathcal{E}} = \{\overrightarrow{xy}, \overrightarrow{yx} \mid (x, y) \in \mathcal{E}\}$. If $\mathcal{E} = E$ we use $\overrightarrow{E} = \overrightarrow{\mathcal{E}}$. We call the elements of $\overrightarrow{E}$ *directed edges* of G. For a set F of directed edges, we write $\hat{F} = \{(x, y) \mid \overrightarrow{xy} \in F \text{ or } \overrightarrow{yx} \in F\}$ for its *symmetric closure*.

Definition 1. *Let* $F \subseteq \overrightarrow{E}$ *be a set of directed edges of a graph* $G = (V, E)$. *Define:*
1. $F^{-1} = \{\overleftarrow{xy} \mid \overrightarrow{xy} \in F\}$, *and*
2. $F^2 = \{\overrightarrow{xz} \mid \overrightarrow{xy} \text{ and } \overrightarrow{yz} \in F \text{ for some } y \in V\}$.

Definition 2. *Let* $\mathcal{E} \subseteq E$ *be a subset of edges of a graph* $G = (V, E)$. *We call F an* orientation *of* $\mathcal{E}$ *if* $F + F^{-1} = \overrightarrow{\mathcal{E}}$ *and* $F \cap F^{-1} = \varnothing$.

If $\mathcal{E} = E$, we call F also an orientation of G.

Definition 3. *A transitive orientation of a graph* $G = (V, E)$ *is an orientation F of E such that* $F^2 \subseteq F$. *A graph G is a* comparability graph *if G has a transitive orientation.*

Given a comparability graph, a transitive orientation of its edges can be obtained in linear time [5,6]. However, checking the transitivity of the orientation needs a verification phase, for which no faster algorithm is known than a fast matrix multiplication [1].

Definition 4 (Golumbic [4]). *Define the binary relation* Γ *on the directed edges* $\overrightarrow{E}$ *of a graph* $G = (V, E)$ *as follows. For* $(x, y), (x, z) \in E$,

$$\overrightarrow{xy} \; \Gamma \; \overrightarrow{xz} \iff \overleftarrow{xy} \; \Gamma \; \overleftarrow{xz} \iff (y, z) \notin E.$$

The relation Γ is reflexive and symmetric and its transitive closure Γ^c is an equivalence relation on $\overrightarrow{E}$. The equivalence classes of Γ^c partition $\overrightarrow{E}$ into the *implication classes* of G. For an implication class A of G, the *symmetric closure* $\hat{A} = A \cup A^{-1}$ is called a *color class* of G.

Golumbic [2,3] gave a simple algorithm to test whether a graph $G = (V, E)$ is a comparability graph and to give it a transitive orientation if it is a comparability graph. The central part of Golumbic's algorithm is to compute a G-*decomposition* of $\overrightarrow{E}$, defined as follows.

Definition 5. *Let* $G = (V, E)$ *be an undirected graph. A* G-*decomposition is a partition* $E = \hat{B}_1 + \cdots + \hat{B}_k$, *where* B_i *is an implication class of* G $[\hat{B}_i + \cdots + \hat{B}_k]$ *for* $i = 1, \ldots, k$.

Golumbic's algorithm follows directly from the following theorem.

Theorem 1 ([2,3,4]). *Let* $G = (V, E)$ *be a graph and let* $E = \hat{B}_1 + \cdots + \hat{B}_k$ *be a* G-*decomposition. The following statements are equivalent.*

(i) G *is a comparability graph;*
(ii) $A \cap A^{-1} = \varnothing$ *for all implication classes* A *of* G;
(iii) $B_i \cap B_i^{-1} = \varnothing$ *for* $i = 1, \ldots, k$.

Furthermore, if these conditions hold, then $B_1 + \cdots + B_k$ *is a transitive orientation of* E.

By Theorem 1, we can test whether a graph $G = (V, E)$ is a comparability graph and give G a transitive orientation through computing a G-decomposition. Golumbic [2,3,4] gave an algorithm to compute a G-decomposition in $O(\Delta \cdot m)$ time where Δ is the maximum degree of a vertex in G.

Definition 6. *Let* $\mathcal{G}$ *be a class of graphs. A graph* $G = (V, E)$ *is a probe graph of* $\mathcal{G}$ *if its vertex set can be partitioned into a set of probes* $\mathbb{P}$ *and an independent set of nonprobes* $\mathbb{N}$, *such that* G *can be embedded in a graph of* $\mathcal{G}$ *by adding edges between certain nonprobes.*

If the partition of the vertices of a graph G into a set of probes $\mathbb{P}$ and a set of nonprobes $\mathbb{N}$ is part of the input then we call G a *partitioned probe graph of* $\mathcal{G}$ if G can be embedded into a graph of $\mathcal{G}$ by adding edges between certain vertices of $\mathbb{N}$. In this paper we denote a partitioned graph as $G = (\mathbb{P} + \mathbb{N}, E)$, and when this notation is used it is to be understood that $\mathbb{N}$ is an independent set. We will refer to the class of (partitioned) probe graphs of the class of (XXX) graphs as (partitioned) probe (XXX) graphs where (XXX) is the name of a graph class.

In Section 2, we give an $O(nm)$-time algorithm for the recognition of partitioned probe comparability graphs, where n and m are the numbers of vertices and edges, respectively. Let $G = (\mathbb{P} + \mathbb{N}, E)$ be a partitioned graph and let G^* be the graph obtained from $\overline{G}$ by deleting all edges between nonprobes. We call G^* the *sandwich conjugate* of G. Since a graph is a cocomparability graph if its complement is a comparability graph, a partitioned graph G is a partitioned probe cocomparability graph if G^* is a partitioned probe comparability graph and hence partitioned probe cocomparability graphs can be recognized in $O(n^3)$ time. A graph is a permutation graph if and only if it is at the same time a comparability graph and a cocomparability graph [7]. An immediate consequence is that a probe permutation graph is both probe comparability and probe cocomparability. For partitioned graphs the converse remains true: G is a partitioned probe permutation graph if and only if G and G^* are both partitioned probe comparability graphs. This result is proved in the last section and leads to an $O(n^3)$-time recognition algorithm for partitioned probe permutation graphs.

2   Recognition of Partitioned Probe Comparability Graphs

In this section we extend the algorithm for recognizing comparability graphs given by Golumbic [2,3,4] to allow recognizing partitioned probe comparability graphs within the same time bound. This algorithm shows that a graph G is a comparability graph by showing that G has a transitive orientation. An orientation of a partitioned probe comparability graph may not be transitive, but the transitive completion may be a transitive orientation of an embedding. The following proposition is clear from the definitions.

Proposition 1. *Let* $G = (\mathbb{P} + \mathbb{N}, E)$ *be a partitioned probe comparability graph with an embedding* H. *Let* $\mathcal{F}$ *be a transitive orientation of* H, $F = \overrightarrow{E} \cap \mathcal{F}$, *and let* (V, F^c) *be the transitive closure of* (V, F). *Then,*

(i) $F^c \subseteq \mathcal{F}$,

(ii) $V(\mathcal{F} - F) \subseteq \mathbb{N}$, *and*

(iii) $(V, \hat{F}^c)$ *is a comparability graph with transitive orientation* F^c.

Observe that F^c is an orientation of the smallest embedding of G such that it can be oriented in agreement with F. We will call F a quasitransitive orientation. Determining whether G is partitioned probe comparability will be equivalent to determining whether it has a quasitransitive orientation.

Definition 7. *Let* $G = (\mathbb{P} + \mathbb{N}, E)$ *be a partitioned graph and let* F *be an orientation of* G. *We call* F *quasitransitive if* $V(F^2 - F) \subseteq \mathbb{N}$.

Theorem 2. *A partitioned graph* $G = (\mathbb{P} + \mathbb{N}, E)$ *is partitioned probe comparability if and only if* G *has a quasitransitive orientation.*

Proof. Using the notation of Proposition 1, if G is partitioned probe comparability with embedding H, then F is clearly a quasitransitive orientation of G. That is, if $\overrightarrow{xy}, \overrightarrow{yz} \in F \subseteq \mathcal{F}$, then $\overrightarrow{xz} \in \mathcal{F}$ and, unless x and z are both nonprobes, $\overrightarrow{xz} \in F$. Now suppose F is a quasitransitive orientation of G. Let $\mathcal{F} = F + F^2$. We prove that G is a partitioned probe comparability graph by showing that $(V, \mathcal{F})$ is transitive, *i.e.*, by showing that $\mathcal{F}^2 \subseteq \mathcal{F}$. Suppose both $\overrightarrow{xy}, \overrightarrow{yz} \in \mathcal{F}$. We show that $\overrightarrow{xz} \in \mathcal{F}$.

If at most one vertex in $\{x, y, z\}$ is a nonprobe, then $\overrightarrow{xy}, \overrightarrow{yz}$, and $\overrightarrow{xz}$ are in F. Now we consider the case that at most one vertex in $\{x, y, z\}$ is a probe. If y is a probe (x and z are nonprobes), then $\overrightarrow{xy}, \overrightarrow{yz} \in F$ and hence $\overrightarrow{xz} \in F^2 \subseteq \mathcal{F}$. If z is a probe (x and y are nonprobes), then $\overrightarrow{xy} \notin F$, but $\overrightarrow{xy} \in F^2$ and there exists a $u \in \mathbb{P}$ such that $\overrightarrow{xu}, \overrightarrow{uy} \in F$. Then $\overrightarrow{xu}, \overrightarrow{uy}, \overrightarrow{yz} \in F \implies \overrightarrow{xz} \in F$. Similarly, if x is a probe (y and z are nonprobes), there exists a $v \in \mathbb{P}$ such that $\overrightarrow{yv}, \overrightarrow{vz} \in F$, $\overrightarrow{yz} \in F^2$. Finally, if x, y, and z are all nonprobes, there exist two vertices $u \neq v$ as above. (If $u = v$, the edge (u, y) would be oriented in two directions.) Then

$$v \in \mathbb{P} \text{ and } \overrightarrow{uy}, \overrightarrow{yv} \in F \implies \overrightarrow{uv} \in F$$

$$v \in \mathbb{P} \text{ and } \overrightarrow{xu}, \overrightarrow{uv} \in F \implies \overrightarrow{xv} \in F$$

$$\overrightarrow{xv}, \overrightarrow{vz} \in F \implies \overrightarrow{xz} \in F^2 \subseteq \mathcal{F}.$$

Thus G is a probe comparability graph and $H = (V, \hat{\mathcal{F}})$ is an embedding of G.   □

We need to modify the Γ relation to use it in partitioned probe comparability recognition. Consider edges $(x, y), (y, z) \in E$ where $x, z \in \mathbb{N}$ and let H be some embedding of G. We do not know *a priori* whether $(x, z) \in E(H)$. Thus we do not know whether $\overrightarrow{xy} \ \Gamma \ \overrightarrow{zy}$ in H before H is constructed. To capture this fact we define a new relation on $\overrightarrow{E}$.

Definition 8. *Let* $G = (\mathbb{P} + \mathbb{N}, E)$ *be a partitioned graph. We define a binary relation* Υ *on* $\overrightarrow{E}$ *as follows. Let* $(x, y), (x, z) \in E$ *be edges of* G. *Then each of* $\overrightarrow{xy} \ \Upsilon \ \overrightarrow{xz}$ *and* $\overrightarrow{yx} \ \Upsilon \ \overrightarrow{zx}$ *if and only if one of the following holds.*
(a) $y = z$, *or*
(b) $(y, z) \notin E$ *and at least one of* y *and* z *is a probe.*

The relation Υ is reflexive and symmetric. Its transitive closure, denoted by Υ^c, defines an equivalence relation on $\overrightarrow{E}$. We call the equivalence classes the *probe implication classes* of G. Let A be a probe implication class of G. We call the symmetric closure of A, *i.e.*, $\hat{A}$, a *probe color class* of G. Next, we define the *probe G-decomposition* as follows:

Definition 9. *Let* $G = (\mathbb{P} + \mathbb{N}, E)$ *be a partitioned graph. A partition* $E = \hat{B}_1 + \cdots + \hat{B}_k$ *is called a* probe G-decomposition *if* B_i *is a probe implication class of* $G[\hat{B}_i + \cdots + \hat{B}_k]$ *for each* $1 \le i \le k$.

The following extension of Theorem 1 is the basis for our algorithm.

Theorem 3 (Probe TRO Theorem). *Let* $G = (\mathbb{P} + \mathbb{N}, E)$ *be a partitioned graph with probe G-decomposition* $E = \hat{B}_1 + \cdots + \hat{B}_k$. *The following statements are equivalent.*
(i) G *is a partitioned probe comparability graph;*
(ii) $A \cap A^{-1} = \varnothing$ *for all probe implication classes* A *of* G;
(iii) $B_i \cap B_i^{-1} = \varnothing$ *for* $i = 1, \ldots, k$.
Furthermore, if these conditions hold, then $F = B_1 + \cdots + B_k$ *is a quasitransitive orientation of* E *and* $H = (V, \hat{\mathcal{F}})$ *is a comparability graph which is an embedding of* G, *where* $\mathcal{F} = F + F^2$.

By the Probe TRO Theorem, it is easy to see that the algorithm for recognizing comparability graphs given in [4] extends to recognizing partitioned probe comparability graphs and assigning quasitransitive orientations. An embedding follows from the quasitransitive orientation. We postpone the proof of the Probe TRO Theorem. Some of the following lemmas are extensions of lemmas given in [2,3,4,8] for proving the TRO Theorem. Note that in the original TRO Theorem, there are four statements. The last statement presents that every circuit has even length. Currently, we do not know how to transform it for the probe graphs. So, we omit this statement.

Arcs $\overrightarrow{xy}$ and $\overrightarrow{uv}$ are in the same probe implication class if and only if they are joined by an Υ-*chain*, *i.e.*, a sequence of edges $(x_i, y_i) \in E$ such that

$$\overrightarrow{xy} = \overrightarrow{x_0 y_0} \ \Upsilon \ \cdots \ \Upsilon \ \overrightarrow{x_k y_k} = \overrightarrow{uv}. \tag{1}$$

Proposition 2. *If $\overrightarrow{xy}$ Υ^c $\overrightarrow{uv}$, then there exists an Υ-chain (1) such that for each i, $1 \le i \le k$, either*
$$x_{i-1} = x_i \text{ and } y_{i-1} \ne y_i, \text{ or}$$
$$x_{i-1} \ne x_i \text{ and } y_{i-1} = y_i.$$
(2)

Such a chain will be called a canonical Υ-chain.

Corollary 1. *Let $G = (\mathbb{P} + \mathbb{N}, E)$ be a partitioned graph and let A be a probe implication class of G. Then $(V(A), \hat{A})$ is a connected undirected graph.*

Lemma 1 (The Probe Triangle Lemma). *Let x, y, and z be three distinct vertices of a partitioned graph $G = (\mathbb{P} + \mathbb{N}, E)$, let $z \in \mathbb{P}$, and let X, Y, and Z be probe implication classes of G with $X \ne Z$ and $Z \ne Y^{-1}$ and having arcs $\overrightarrow{xy} \in Z$, $\overrightarrow{zx} \in Y$, such that $\overrightarrow{zy} \in X$. Then the following four statements hold.*

(i) *If $x \ne u \ne y$, $(x, u) \in E$ and $\overrightarrow{xu}$ Υ $\overrightarrow{xy}$, then $z \ne u$, $(z, u) \in E$ and $\overrightarrow{zu} \in X$.*
(ii) *If $x \ne u \ne y$, $(u, y) \in E$ and $\overrightarrow{uy}$ Υ $\overrightarrow{xy}$, then $z \ne u$, $(z, u) \in E$ and $\overrightarrow{zu} \in Y$.*
(iii) *If $\overrightarrow{pq} \in Z$, then $\overrightarrow{zp} \in Y$ and $\overrightarrow{zq} \in X$.*
(iv) *$z \notin V(Z)$.*

Proof. We first prove (i). Notice that $\overrightarrow{xz} \notin Z$, since $\overrightarrow{zx} \in Y$ and $Z \ne Y^{-1}$. Then $u \ne z$ because $\overrightarrow{xy} \in Z \ne Y^{-1}$ and $\overrightarrow{xu}$ Υ $\overrightarrow{xy}$. Since $z \in \mathbb{P}$, if $(z, u) \notin E$, $\overrightarrow{xz}$ Υ $\overrightarrow{xu}$, a contradiction to the assumption that $\overrightarrow{xz} \notin Z$. Hence $(z, u) \in E$ must hold. Since $\overrightarrow{xu}$ Υ $\overrightarrow{xy}$, we have that $(y, u) \notin E$ and at least one of y and u is a probe. Because $(z, u) \in E$, $(z, y) \in E$, $(y, u) \notin E$, and at least one of y and u is a probe, we have $\overrightarrow{zu}$ Υ $\overrightarrow{zy}$ and $\overrightarrow{zu} \in X$ because $\overrightarrow{zy} \in X$.

The proof of (ii) is similar.

Next, to prove (iii), let $\overrightarrow{pq} \in Z$. By Proposition 2, since $\overrightarrow{xy} \in Z$, there exists a canonical Υ-chain
$$\overrightarrow{xy} = \overrightarrow{x_0 y_0} \ \Upsilon \cdots \Upsilon \ \overrightarrow{x_k y_k} = \overrightarrow{pq}.$$

We claim that $x_i \ne z \ne y_i$, $\overrightarrow{zx_i} \in Y$, and $\overrightarrow{zy_i} \in X$, for $0 \le i \le k$. We prove the claim by induction on i. It holds for $i = 0$ by assumption. Suppose it holds for $i - 1$. If $x_{i-1} = x_i$ and $y_i \ne y_{i-1}$, then $\overrightarrow{zy_i} \in X$ by (i). Otherwise $x_{i-1} \ne x_i$ and $y_i = y_{i-1}$ and hence $\overrightarrow{zx_i} \in Y$ by (ii). In either case, we have $x_i \ne z \ne y_i$, $\overrightarrow{zx_i} \in Y$ and $\overrightarrow{zy_i} \in X$. In particular, $p = x_k \ne z \ne y_k = q$, $\overrightarrow{zp} = \overrightarrow{zx_k} \in Y$, and $\overrightarrow{zq} = \overrightarrow{zy_k} \in X$. We have (iii) and (iv). $\square$

The following is immediate from (iv) of Lemma 1.

Corollary 2. *Let $G = (\mathbb{P} + \mathbb{N}, E)$ be a partitioned graph with a probe implication class Z. Let $x, y, z \in V = \mathbb{P} + \mathbb{N}$, let $(x, y), (x, z), (y, z) \in E$, and let $z \in \mathbb{P}$. If $\overrightarrow{xy} \in Z$, then $z \in V(Z)$ if and only if at least one of $\overrightarrow{xz} \in Z$ or $\overrightarrow{zy} \in Z$.*

Lemma 2. *Let $G = (\mathbb{P} + \mathbb{N}, E)$ be a partitioned graph and let A be a probe implication class of G. Exactly one of the following alternatives holds.*

(i) *either $A = A^{-1}$ or,*
(ii) *$A \cap A^{-1} = \varnothing$, and A and A^{-1} are quasitransitive orientations of the graph $G_A = (V(G), \hat{A})$.*

Proof. If $A \cap A^{-1} \neq \varnothing$, then there exists an arc $\overrightarrow{xy} \in A \cap A^{-1}$, and $\overrightarrow{xy} \, \Upsilon^c \, \overrightarrow{yx}$. For any $\overrightarrow{uv} \in A$, we have $\overrightarrow{uv} \, \Upsilon^c \, \overrightarrow{xy}$ and $\overrightarrow{yx} \, \Upsilon^c \, \overrightarrow{vu}$, implying $\overrightarrow{uv} \, \Upsilon^c \, \overrightarrow{vu}$ and $\overrightarrow{vu} \in A$. Thus $A = A^{-1}$.

On the other hand suppose $A \cap A^{-1} = \varnothing$. We will show that $V(A^2 - A) \subseteq \mathbb{N}$ by showing that for $\overrightarrow{xy}, \overrightarrow{yz} \in A$ where at least one of x and z is a probe of G, $\overrightarrow{xz} \in A$. If $(x, z) \notin E$, then $\overrightarrow{xy} \, \Upsilon \, \overrightarrow{zy}$ and $\overrightarrow{zy} \in A \cap A^{-1} \neq \varnothing$, a contradiction. Thus $(x, z) \in E$ must hold. If $\overrightarrow{xz} \in A$, then we are done. Suppose $\overrightarrow{xz} \notin A$. Consider the case of $x \in \mathbb{P}$. We have $\overrightarrow{yz} \in A$ and $\overrightarrow{yx}, \overrightarrow{xz} \notin A$. By Corollary 2, $x \notin V(A)$, a contradiction. The case of $z \in \mathbb{P}$ is similar. Thus A is a quasitransitive orientation of G_A.

Obviously, quasitransitivity of A implies quasitransitivity of A^{-1} for G_A. $\quad\square$

Lemma 3. *Let* $G = (V, E) = (\mathbb{P} + \mathbb{N}, E)$ *be a partitioned graph, and let* A *be a probe implication class of* G. *If* F *is a quasitransitive orientation of* $G_{\bar{A}} = (V, E - \hat{A})$ *and if* $A \cap A^{-1} = \varnothing$, *then* $F + A$ *is a quasitransitive orientation of* G.

Proof. By the previous lemma, A is a quasitransitive orientation of $G_A = (V, \hat{A})$. Let $\mathcal{F} = F + A$. Clearly $\mathcal{F}$ is an orientation of G and $F \cap A = \varnothing$. If $\mathcal{F}$ is not quasitransitive, then there exist arcs $\overrightarrow{xy}, \overrightarrow{yz} \in \mathcal{F}$, x and z not both nonprobes, such that $\overrightarrow{xz} \notin \mathcal{F}$. If $(x, z) \notin E$ then $\overrightarrow{xy} \, \Upsilon \, \overrightarrow{zy}$, contradicting quasitransitivity of A unless both $\overrightarrow{xy}, \overrightarrow{zy} \in F$. But then $\overrightarrow{xy}$ and $\overrightarrow{yz}$ violate quasitransitivity of $G_{\bar{A}}$.

Suppose then that $\overrightarrow{zx} \in \mathcal{F}$. Two of the three arcs, $\overrightarrow{xy}, \overrightarrow{yz}$, and $\overrightarrow{zx}$, must be in A, or in F. We have a violation of quasitransitivity in G_A, or in $G_{\bar{A}}$, respectively. $\quad\square$

Notice that for every implication class A, $V(A)$ is a partitive (or module) in G [4, pp. 112]. That is, for each vertex $x \in V - V(A)$ either $V(A) \cap N(x) = \varnothing$ or $V(A) \subseteq N(x)$. We define the following generalization which includes probe implication classes.

Definition 10. *Let* $G = (\mathbb{P} + \mathbb{N}, E)$ *be a partitioned graph, and let* $M \subseteq V = \mathbb{P} + \mathbb{N}$. *Let* $C_1, \ldots, C_t$ *be the components of* $G[\mathbb{P}(M)]$. *We call* M *a QT-module if the following conditions are satisfied.*
(a) $\forall_{x \in \mathbb{P} - M}$ *either* $N(x) \cap M = \varnothing$ *or* $M \subseteq N(x)$.
(b) $\forall_{y \in \mathbb{N} - M} \forall_{1 \leq i \leq t}$ *either* $N(y) \cap V(C_i) = \varnothing$ *or* $V(C_i) \subseteq N(y)$.

Lemma 4. *Let* $G = (\mathbb{P} + \mathbb{N}, E)$ *be a partitioned graph and let* A *be a probe implication class of* G. *Then* $V(A)$ *is a QT-module.*

Proof. First suppose there exists an $x \in \mathbb{P} - V(A)$ which is connected to some, but not all, vertices in $V(A)$. Define $R = \{w \in V(A) \mid (x, w) \in E\}$ and $U = V(A) - R$. By Corollary 1, the undirected graph $(V(A), \hat{A})$ is connected. There exist $p \in U$ and $q \in R$ such that $\overrightarrow{pq} \in A$ or $\overleftarrow{pq} \in A$. Since $(p, q), (q, x) \in E$, $(p, x) \notin E$, and $x \in \mathbb{P}$, we obtain

$$\overrightarrow{pq} \, \Upsilon \, \overrightarrow{xq} \quad \text{and} \quad \overleftarrow{pq} \, \Upsilon \, \overleftarrow{xq}, \tag{3}$$

contradicting $x \notin V(A)$. We have shown that condition (a) of Definition 10 is satisfied.

Instead suppose there exist a $y \in N - V(A)$ and a component C_i of $\mathbb{P}(V(A))$ such that y is adjacent to some but not all vertices in C_i. The proof is the same as before, except that now R and U are the neighbors and nonneighbors of y restricted to C_i, and that this time p is a probe in (3) (instead of x). Therefore condition (b) of Definition 10 is also satisfied.   □

Lemma 5. *Let* $G = (\mathbb{P} + N, E)$ *be a partitioned graph, let* M *be a* QT*-module and let* A *be a probe implication class of* G. *Then either*

$$V(A) \subseteq M \quad \text{or} \quad A \cap (M \times M) = \varnothing.$$

Proof. Suppose that $\overrightarrow{xy} \, \Upsilon \, \overrightarrow{zy}$ for some edges $(x, y), (y, z) \in E$ with $x, y \in M$ and $z \in V - M$. If $z \in \mathbb{P}$, Definition 10 gives us that $(x, z) \in E$. If $z \in N$ and $(x, z) \notin E$, then x cannot be a probe, because otherwise x and y are in the same component of $\mathbb{P}(M)$. Either case contradicts $\overrightarrow{xy} \, \Upsilon \, \overrightarrow{zy}$, from which the lemma follows.   □

We are now ready to prove our main theorem.

Theorem 4. *Let* $G = (V = \mathbb{P} + N, E)$ *be a partitioned graph. The following statements are equivalent.*

(i) G *is a partitioned probe comparability graph.*
(ii) $A \cap A^{-1} = \varnothing$ *for each probe implication class* A *of* G.
(iii) *For each probe implication class* A *of* G, *the graphs* $G_A = (V, \hat{A})$ *and* $G_{\bar{A}} = (V, E - \hat{A})$ *are partitioned probe comparability graphs.*

Proof. We will prove the theorem by induction on the number of vertices in G, and on the number k of probe color classes when two graphs have the same number of vertices.

Suppose $k = 1$. Since G has only one probe color class, $E(G_{\bar{A}}) = \varnothing$, and by Lemma 2, all the statements hold or none hold.

For the remainder of the proof we assume that $k > 1$, and that the theorem holds for all partitioned probe graphs on fewer vertices than G, and for all partitioned probe graphs on $V(G)$ for which the number of probe color classes is less than k. We prove that the statements are equivalent for a partitioned probe graph G of k probe color classes.

(i)$\Longrightarrow$(ii).   Let F be a quasitransitive orientation of G and suppose that $\overrightarrow{xy} \in A \cap A^{-1}$. Then there is an Υ-chain from $\overrightarrow{xy}$ to $\overrightarrow{yx}$

$$\overrightarrow{xy} = \overrightarrow{x_1 y_1} \, \Upsilon \, \cdots \, \Upsilon \, \overrightarrow{x_\ell y_\ell} = \overrightarrow{yx}.$$

However, from the definitions of Υ and quasitransitive orientation for G, if (x, y) and (p, q) are two edges of G such that $\overrightarrow{xy} \in F$ and $\overrightarrow{xy} \, \Upsilon \, \overrightarrow{pq}$ then $\overrightarrow{pq} \in F$, which is a contradiction since $\overrightarrow{xy}$ and $\overrightarrow{yx}$ cannot both be in F.

(ii)$\Longrightarrow$(iii).   That G_A is a partitioned probe comparability graph follows from (ii) of Lemma 2. In the following we prove that $G_{\bar{A}}$ is also a partitioned probe comparability graph. We consider two cases:

Case 1. $V(A) = V$. First we show that every probe implication class D of G, with $\hat{D} \neq \hat{A}$, is a subset of some probe implication class of $G_{\bar{A}}$. Let $\Upsilon_{\bar{A}}$ denote the relation Υ for the graph $G_{\bar{A}}$. Suppose $\overrightarrow{xy}, \overrightarrow{xz} \in D$ and $\overrightarrow{xy} \Upsilon \overrightarrow{xz}$. Then $(x, z) \notin E$ and clearly $\overrightarrow{xy} \Upsilon_{\bar{A}} \overrightarrow{xz}$. That is, an Υ-chain between two arcs of D implies an $\Upsilon_{\bar{A}}$-chain between them. Thus D is a subset of some probe implication class of $G_{\bar{A}}$.

We will show that probe implication classes of G that merge in $G_{\bar{A}}$ are all stars in $G_{\bar{A}}$ with a nonprobe as the center, and any two stars which merge together have a common center as the source for all arcs or the sink for all arcs in the classes. Therefore $D \cap D^{-1} = \varnothing$ for every probe implication class D of $G_{\bar{A}}$. Since $G_{\bar{A}}$ has at least one class fewer than G, the result now follows from the induction hypothesis.

Suppose the relation $\Upsilon_{\bar{A}}$ connects $\overrightarrow{xy} \in D_i$ and $\overrightarrow{xz} \in D_j$, where D_i and D_j are two distinct probe implication classes of G. Then $(y, z) \in \hat{A}$. If $x \in \mathbb{P}$, then by Lemma 1 (iv) $x \notin V(A)$, contradicting the assumption $V(A) = V(G)$; thus $x \in \mathbb{N}$, and y and z must be probes. Also note that $D_i \neq D_j^{-1}$, since otherwise $\overrightarrow{zx}, \overrightarrow{xy} \in D_i$, but $\overrightarrow{zy} \notin D_i$. Therefore $\hat{D}_i \cap \hat{D}_j = \varnothing$.

Suppose there exists a vertex $p \neq x$ such that $\overrightarrow{xy} \Upsilon \overrightarrow{py}$. Then $p \in \mathbb{P}$ since $x \in \mathbb{N}$. Since $z \in \mathbb{P}$ and $\overrightarrow{xy}, \overrightarrow{py} \in D_i$, we have $\overrightarrow{pz} \in D_j$ by the Probe Triangle Lemma (iii). Then by (iv) of the same lemma, since $\overrightarrow{py} \in D_i$, $\overrightarrow{pz} \in D_j$, and $\overrightarrow{yz} \in \hat{A}$, we have $p \notin V(A)$, a contradiction. Hence all arcs of the probe implication class of $\overrightarrow{xy}$ must be of the form $\overrightarrow{xq}$. Hence D_i is a star, as claimed. Since $\overrightarrow{xy} \in D_i$ and $\overrightarrow{xz} \in D_j$, we see that D_i and D_j merge at x.

Case 2. $V(A) \subset V$. As in Case 1, every probe implication class of G except A and A^{-1} is contained in some probe implication class of $G_{\bar{A}}$. Therefore $G_{\bar{A}}$ has at least one color class fewer than G. The result follows from the induction hypothesis if $D \cap D^{-1} = \varnothing$ for every implication class of $G_{\bar{A}}$.

We divide the arcs of $\overrightarrow{E}(G_{\bar{A}})$ into two groups: those arcs between two vertices of $M = V(A)$, and those with at least one endvertex not in M. Since M is a QT-module, by Lemma 5, every probe implication class of G is either a subset of $\overrightarrow{E}(G[M])$, or disjoint from $\overrightarrow{E}(G[M])$. First consider $G[M]$. Relation Υ does not connect any arc of $\overrightarrow{E}(G[M])$ with any arc not in $\overrightarrow{E}(G[M])$, and $\Upsilon_{\bar{A}}$ does not either, because the edges of G missing in $G_{\bar{A}}$ do not leave M. Thus, by the same argument as in Case 1, every probe implication class D of $(M, E(G[M]) - A)$ satisfies $D \cap D^{-1} = \varnothing$.

Next we consider those probe implication classes that are disjoint from $\overrightarrow{E}(G[M])$. Let X be a set containing one vertex from each component of $G[\mathbb{P}(M)]$. Since $G[V-M+X]$ is a proper induced subgraph, it satisfies statement (ii) of the theorem. Since the vertex set is not all of G, the induction hypothesis says that $G[V-M+X]$ is a partitioned probe comparability graph having a quasitransitive orientation F. We extend this orientation to an orientation F^* of $(V, E - E(G[M]))$ as follows. Let $v \in V - M$. Since X is an independent set, every arc between v and X in F is from v to X, or every such arc is from X to v. We give every edge between v and M the same orientation.

We show that F^* is again a quasitransitive orientation. Suppose there is a violation of quasitransitivity in F^* involving $(v, x), (v, y) \in E$. If $v \in V - M$, and $x, y \in M$, by the construction of F^*,

$$\vec{vx} \in F^* \iff \vec{vy} \in F^*$$

and there is no violation.

Next, assume $x \in M$ but $v, y \in V - M$, and x and y not both nonprobes. If $x \in \mathbb{P}$, let $z \in X$ be the vertex, possibly the same as x, in the same component of $\mathbb{P}(G[M])$ as x; otherwise, if $x \in \mathbb{N}$, let $z \in X$ be arbitrary. Then

$$(z, y) \in E \iff (x, y) \in E$$

because M is a QT-module. Since (v, x) and (v, z) have the same direction, as well as (x, y) and (z, y) if they are both edges, and since $\{z, y, v\}$ does not contain a violation of quasitransitivity, neither does $\{x, y, v\}$.

Suppose now $v \in M$ and $x, y \in V - M$. Then there exists $z \in X$ such that $(z, x), (z, y) \in E$ and such that (z, x) and (z, y) receive the same orientations as (v, x) and (v, y), respectively. Since F is quasitransitive, $\{x, y, z\}$ does not contain a violation of quasitransitivity; neither therefore does $\{x, y, v\}$.

We conclude that $(V, E - E(G[M]))$ is a partitioned probe comparability graph with quasitransitive orientation F^* and with fewer color classes than G. By the inductive hypothesis $D \cap D^{-1} = \varnothing$ for every implication class D of this graph. Thus we have $D \cap D^{-1} = \varnothing$ for every probe implication class D of $G_{\bar{A}}$, which by induction on the number of classes is a partitioned probe comparability graph.

(iii)$\Longrightarrow$(i).   This implication is due to Lemma 3.   □

The Probe TRO Theorem follows immediately from Theorem 4. The algorithm for the recognition of partitioned probe comparability graphs can be obtained by modifying the algorithm given by Golumbic [4, pp. 130]. Due to the limitation of the length, we omit the detail. We obtain the following theorem.

Theorem 5. *Recognition of partitioned probe comparability graphs and finding a quasitransitive orientation can be done in $O(\Delta \cdot m)$ time and $O(n + m)$ space, where Δ is the maximum degree of a vertex. Moreover, an embedding can also be obtained from a quasitransitive orientation in $O(\Delta \cdot m)$ time.*

Let $G = (\mathbb{P} + \mathbb{N}, E)$ be a partitioned graph. The *sandwich conjugate* G^* of G is the partitioned graph obtained from $\overline{G}$ by removing all edges between vertices of $\mathbb{N}$. Then we have that a partitioned graph is a partitioned probe cocomparability graph if and only if its sandwich conjugate G^* is a partitioned probe comparability graph. Thus we have the following corollary.

Corollary 3. *The recognition of partitioned probe cocomparability graphs can be done in $O(n^3)$ time.*

3   Recognition of Partitioned Probe Permutation Graphs

Definition 11. *Let $\pi \in \mathrm{Sym}(n)$ be a permutation acting on the set of integers $\{1, 2, \ldots, n\}$. We define the* inversion graph $G[\pi]$ *as follows. The graph has vertex $V = \{x_1, \ldots, x_n\}$ and edge set E defined by*

$$(x_i, x_j) \in E \iff (i - j)(\pi^{-1}(i) - \pi^{-1}(j)) < 0.$$

An undirected graph G *is called a* permutation graph *if there exists a permutation* π *such that* $G \cong G[\pi]$.

Pnueli *et al.* showed that a graph G is a permutation graph if and only if both G and $\overline{G}$ are comparability graphs [7]. In the following we extend this statement to cover partitioned probe permutation graphs.

Lemma 6. *Let* $G = (\mathbb{P} + \mathbb{N}, E)$ *be a partitioned graph. Then* G *is a partitioned probe permutation graph if and only if there exists a labeling* L *of the vertices by integers* $1, \ldots, n$, *and a permutation* $\pi \in \mathrm{Sym}(n)$ *such that* $(x, y) \in E$ *if and only if both*

$$\{x, y\} \not\subseteq \mathbb{N} \quad \text{and} \quad (L(x) - L(y)) \left(\pi^{-1}L(x) - \pi^{-1}L(y)\right) < 0. \tag{4}$$

Proof. It is easy to see that if $(x, y) \in E(G[\pi]) - E$ is an edge, then $\{x, y\} \subseteq \mathbb{N}$. Thus $G[\pi]$ is an embedding of G.   □

Theorem 6. *A partitioned probe graph* G *is a probe permutation graph if and only if both* G *and* G^* *are partitioned probe comparability graphs.*

Proof. Let G be a partitioned probe permutation graph and let H be an embedding of G. Both H and $\overline{H}$ are comparability graphs. By definition, $\{x, y\} \subseteq \mathbb{N}$ if either $(x, y) \in E(H) - E(G)$ or $(x, y) \in E(\overline{H}) - E(\overline{G})$. Thus both G and G^* are partitioned probe comparability graphs.

Suppose both G and G^* are partitioned probe comparability graphs. By Theorem 2, both G and G^* have quasitransitive orientations. Let F_1 and F_2 be quasitransitive orientations of G and G^*, respectively. We claim that $(V, F_1 + F_2)$ is an acyclic digraph. If not, let $v_0, \ldots, v_\ell, v_0$ be a cycle of the smallest possible length $\ell > 3$. Since at least one of the two ends of an edge is a probe, without loss of generality let $v_0 \in \mathbb{P}$. Then either $\overleftarrow{v_0 v_2} \in F_1 + F_2$, in which case v_0, v_1, v_2, v_0 is a shorter cycle, or $\overrightarrow{v_0 v_2} \in F_1 + F_2$, in which case $v_0, v_2, v_3, \ldots, v_0$ is a shorter cycle, contradicting minimality in each case.

If $\ell = 3$, then at least two of the three vertices visited by the cycle are probes and at least two of the edges of the cycle are in the same F_i, $1 \le i \le 2$, implying that F_i is not quasitransitive. Thus $(V, F_1 + F_2)$ is acyclic. Similarly $(V, F_1^{-1} + F_2)$ is acyclic. In the following we construct a permutation π such that $G[\pi]$ is an embedding of G. Define two labelings L and L' as follows.

1. Label the vertices in the order determined by a topological sort of vertices of $(V, F_1 + F_2)$, that is, $L(x) = i$ if x is the ith vertex of this sort.
2. Label the vertices according to the order determined by a topological sort of vertices of $(V, F_1^{-1} + F_2)$, that is, that is, $L'(x) = i$ if x is the ith vertex of the new sort.

Then $\pi(i) = L \circ L'^{-1}(i)$, for $i = 1, \ldots, n$.

Notice that $L(y) > L(x)$ if and only if $\overrightarrow{xy} \in F_1 + F_2$. Similarly $L'(y) > L'(x)$ if and only if $\overrightarrow{xy} \in F_1^{-1} + F_2$. Since it is the edges of E which have their orientations reversed between Steps I and II and $\{x, y\} \not\subseteq \mathbb{N}$ if $(x, y) \in E$, we have

$$(x, y) \in E \iff \{x, y\} \not\subseteq \mathbb{N} \text{ and } (L(x) - L(y))(L'(x) - L'(y)) < 0$$

which is exactly what we get by substituting $\pi = L \circ L'^{-1}$ into (4).   □

By Theorem 5 and Theorem 6 we obtain:

Theorem 7. *A partitioned probe permutation graph can be recognized in* $O(n^3)$ *time.*

Acknowledgments

We would like to thank the anonymous referees for their helpful suggestions and comments.

References

1. Coppersmith, D. and S. Winograd, Matrix multiplication via arithmetic progressions, *Proceedings 19*[th] *ACM Syposium on Theory of Computing* (1987), pp. 1–6.
2. Golumbic, M. C., The complexity of comparability graph recognition and coloring, *Computing* **18** (1977), pp. 199–208.
3. Golumbic, M. C., Comparability graphs and a new matroid, *J. Combin. Theory Ser. B* **22** (1977), pp. 68–90.
4. Golumbic, M. C., *Algorithmic Graph Theory and Perfect Graphs*, Academic Press, New York, 1980.
5. Habib, M., F. de Montgolfier, and C. Paul, A simple linear-time modular decomposition algorithm for graphs, using order extensions, *Proceedings SWAT'04*, LNCS 3111 (2004), pp. 187–198.
6. McConnell, R. M. and J. P. Spinrad, Modular decomposition and transitive orientation, *Discrete Mathematics* **201** (1999), pp. 189–241.
7. Pnueli, A., A. Lempel, and S. Even, Transitive orientation of graphs and identification of permutation graphs, *Canad. J. Math.* **23** (1971), pp. 160–175.
8. Simon, K. and P. Trunz, A cleanup on transitive orientation, *Proceedings ORDAL'94*, LNCS 831 (1994), pp. 59–85.

Computing Graph Polynomials on Graphs of Bounded Clique-Width

J.A. Makowsky[1], Udi Rotics[2], Ilya Averbouch[1], and Benny Godlin[1,3]

[1] Department of Computer Science
Technion–Israel Institute of Technology, Haifa, Israel
janos@cs.technion.ac.il
[2] School of Computer Science and Mathematics,
Netanya Academic College, Netanya, Israel
rotics@mars.netanya.ac.il
[3] IBM Research and Development Laboratory, Haifa, Israel

Abstract. We discuss the complexity of computing various graph polynomials of graphs of fixed clique-width. We show that the chromatic polynomial, the matching polynomial and the two-variable interlace polynomial of a graph G of clique-width at most k with n vertices can be computed in time $O(n^{f(k)})$, where $f(k) \leq 3$ for the inerlace polynomial, $f(k) \leq 2k + 1$ for the matching polynomial and $f(k) \leq 3 \cdot 2^{k+2}$ for the chromatic polynomial.

1 Introduction

In this paper[1] we deal with the complexity of computing various graph polynomials of a simple graph of clique-width at most k. Our discussion focuses on the univariate **characteristic** polynomial $P(G, \lambda)$, the **matching** polynomial $m(G, \lambda)$, the **chromatic** polynomial $\chi(G, \lambda)$, the multivariate **Tutte** polynomial $T(G, X, Y)$ and the **interlace** polynomial $q(G, XY)$, cf. [Bol99], [GR01] and [ABS04b]. All these polynomials are not only of graph theoretic interest, but all of them have been motivated by or found applications to problems in chemistry, physics and biology. We give the necessary technical definitions in Section 3.

Without restrictions on the graph, computing the characteristic polynomial is in **P**, whereas all the other polynomials are ♯**P** hard to compute. Clique-width is a parameter of graphs similar, but more flexible, than its related notion of tree-width, [CO00], [CMR01]. Tree-width plays an important rôle in parametrized complexity theory, as shown forcefully in the monograph of Downey and Fellows [DF99].

We assume the reader is familiar with the basic of complexity theory as given in [Pap94], and with the notion of tree-width of a graph. General background on tree-width may be found in [Die96].

---

[1] We report here some of the results obtained in the first author's seminar 238900 on graph polynomials, held in 2005 at the CS Department of the Technion.

F.V. Fomin (Ed.): WG 2006, LNCS 4271, pp. 191–204, 2006.
© Springer-Verlag Berlin Heidelberg 2006

We distinguish between the following upper bounds for the running time of algorithms with input size n and a parameter k of the input:

Fixed parameter exponential time (FPEXP). Runtime less than $2^{n^{c_1(k)}}$ with $c_1(k) \geq 1$.

Fixed parameter subexponential time (FPSUBEXP).
Runtime less than $2^{c_2(k) \cdot n^{1-\epsilon(k)}}$.

Fixed parameter polynomial time (FPPT). Runtime less than $n^{c_3(k)}$.

Fixed parameter tractable (FPT). Runtime less than $c_4(k) \cdot n^d$.

Polynomial time (P). Runtime less than $O(n^d)$.

Here d is independent of n and k, whereas the other constants may depend on k. In this paper k will be either the tree-width or the clique-width of the input graph G.

Our main results are summarized in the following theorem.

Theorem 1. *There are algorithms such that, for a graph G with n vertices of clique-width at most k, given together with its k-expression,*

(i) *the interlace polynomial $q(G, X, Y)$ can be computed in time $O(n^d)$, where d is constant independent of n and k. Hence it is in* **FPT** *for k, cf. Proposition 3.*

(ii) *the matching polynomial $m(G, \lambda)$ can be computed in time $O(n^{f(k)})$, where f is function independent of n and linear in k. Hence it is in* **FPPT** *for k. cf. Theorem 4.*

(iii) *the chromatic polynomial $\chi(G, \lambda)$ can be computed in time $O(n^{f(k)})$, where f is function independent of n and simply exponential in k. Hence it is in* **FPPT** *for k. cf. Theorem 6.*

We summarize the known and new results in table 1.

Table 1. Overview of complexity

Polynomial	tree-width k	clique-width k	
$P(G, \lambda)$	in **P**	in **P**	
$\chi(G, \lambda)$	in **FPT** for k [And98, Nob98]	in **FPPT** for k	NEW
$m(G, \lambda)$	in **FPT** for k [Mak04]	in **FPPT** for k	NEW
$q(G, X, Y)$	in **FPT** for k NEW	in **FPT** for k	NEW
$T(G, X, Y)$	in **FPT** for k [And98, Nob98]	in **FPSUBEXP** for k [GHN05]	

Note that in the case of the matching polynomial f is linear in k, whereas for the chromatic polynomial it is exponential in k. For arbitrary graphs only the characteristic polynomial is in **P**. The others are all $\sharp$**P**-hard, [Val79], [ABS04b], [JVW90].

Relevance of the results. We report here on ongoing research into the complexity of computing graph polynomials on graphs of bounded clique-width. There are two ways of approaching the problem: Either one establishes that one can apply [Mak04, Theorem 6.6], which can be far from obvious, or use explicit methods. For the interlace polynomial we indeed use the first apporach. For counting problems the explicit approach was applied to the chromatic number in [KR03], and to ♯**SAT** in [FMR06]. For the Tutte polynomial it was applied in [GHN06]. To the best of our knowledge no other work on the complexity of graph polynomials on graph classes of bounded clique-width was published. Our explicit results show remarkable differences concerning the various polynomials. Whether this is inherent or just due to the absence of more sophisticated algorithms remains open. We suspect that the differences are inherent.

2   Clique-Width of Graphs

The notion of clique-width was introduced in [CER93] and developed more systematically in [CO00]. k-graphs are graphs with vertices labeled with possibly one out of k many labels. The class of graphs $CW(k)$ of clique-width at most k is defined inductively. Singleton k-graphs are in $CW(k)$. $CW(k)$ is closed under disjoint union $\oplus$, relabeling of the label i by the label j, denoted by $\rho_{i \to j}$, and, for $i \neq j$, edge creation $\eta_{i,j}$ where all the vertices labeled i are connected to all the vertices labeled j. The clique-width of a graph G is the minimum k such that $G \in CW(k)$. A k-expression t is a term built using the terms for singleton k-graphs, disjoint union, relabeling and edge creation. We denote by $val(t)$ the k-graph described by t. Courcelle and Olariu in [CO00] showed that the clique-width of graphs of tree-width at most k are in $CW(2^{k+1}+1)$. Therefore, any class of graphs of bounded tree-width is automatically of bounded clique-width. The best known bound is due to [CR05]. Given a graph G, computing its tree-width is **NP**-hard, [ACP87]. The same was recently shown to be true for clique-width [FRRS05]. Deciding whether G has tree-width at most k is in **FPT** for k. For clique-width this is not known to be true. However, S. Oum and P. Seymour, [OS05], showed

Theorem 1 (S. Oum and P. Seymour, 2004). *There is polynomial time algorithm, which for fixed k, decides whether the graph has clique-width at least $k + 1$ or else outputs a $k_1 = (2^{3k+2} - 1)$-expression for the graph showing that its clique-width is at most k_1.*

In general, it seems that finding an explicit bound for the clique-width is a more complicated task than finding a bound for the tree-width.

3   Graph Polynomials

Let $\mathcal{G}$ be the class of graphs $G = (V, E)$ without loops and multiple edges. Let $\mathcal{R}$ be a ring and $\bar{X}$ be a (not necessarily finite) set of indeterminates. A *graph polynomial* is a function

$$p : \mathcal{G} \to \mathcal{R}[\bar{X}]$$

such that for isomorphic graphs $G_1 \simeq G_2$ we have $p(G_1) = p(G_2)$.

There are plenty of graph polynomials which have been discussed in the literature, although no systematic treatment on graph polynomials in general is available. To put our results into perspective we discuss briefly several classical graph polynomials, the *chromatic polynomial* $\chi(G, \lambda)$, the *characteristic polynomial* $P(G, \lambda)$, the *acyclic generating matching polynomials* $m(G, \lambda)$ and $g(G, \lambda)$ and the *Tutte polynomial* $T(G, X, Y)$.

The chromatic polynomial. Let $\chi(G, \lambda)$ denote the number of proper vertex colorings of G with at most λ many colors. G. Birkhoff, [Bir12], observed in 1912 that $\chi(G, \lambda)$ is, for a fixed graph G, a polynomial in λ, which is now called the *chromatic polynomial of G*. The chromatic polynomial is the oldest graph polynomial to appear in the literature, and since then a substantial body of knowledge about the chromatic polynomial of graphs and its applications has been accumulated. The recent book by F.M. Dong, K.M. Koh and K.L. Teo [DKT05] gives an excellent and extensive survey. One of the surprising facts is a theorem of R.P. Stanley, [Sta73], which states that $\chi(G, -1)$ is the number of acyclic orientations of G.

The Tutte polynomial. Interesting generalizations of the chromatic polynomial were introduced by H. Whitney in 1932 and Tutte in 1947. The most prominent among them is now called the *Tutte polynomial* $T(G, X, Y)$ which is a two variable polynomial from which the chromatic polynomial can be obtained via a simple substitution and multiplication with a prefactor. For a modern exposition the reader is referred to [Bol99, chapter X], [Wel93]. For this paper we do not need a definition of the Tutte polynomial.

Other univariate graph polynomials were introduced after 1955, often first motivated by problems from chemistry and physics.

The characteristic polynomial of a graph G, denoted by $P(G, \lambda)$ is the characteristic polynomial of the adjacency matrix M_G of the graph G, $P(G, \lambda) = \det(\lambda \cdot \mathbf{1} - M_G)$ and is completely determined by the eigenvalues of M_G, which are all real, as the matrix is symmetric.

The matching polynomials. The *acyclic polynomial of G* is the polynomial $m(G, \lambda) = \sum_k (-1)^k \cdot m_k(G) \cdot \lambda^{n-2k}$, where the coefficients $m_k(G)$ count k-matchings. A chemical point of view of these polynomials is given in [CDS95] and [Tri92], where also algorithmic aspects are touched. A close relative of the acyclic polynomial is the *generating matching polynomial of a graph G* $g(G, \lambda) = \sum_k m_k(G) \lambda^k$ where $m(G, \lambda) = \lambda^n g(G, (-\lambda^{-2}))$. An excellent survey on these two matching polynomials may be found in [LP86, Chapter 8.5]. We shall refer to both as *matching polynomials*.

The interlace polynomials. The interlace polynomials were introduced in [ABS00, ABS04a, ABS04b]. and further studied in [AvdH04]. The most general version is a two-variable version from [ABS04b] given by

$$q(G, X, Y) = \sum_{S \subseteq V(G)} (X-1)^{r_2(G[S])} (Y-1)^{|V|-r_2(G[S])}$$

where $G[S]$ is the subgraph induced by S and $r_2(G)$ is the matrix rank over $\mathbb{Z}_2$ of the adjacency matrix of G. The polynomial introduced in [ABS00] can be written as

$$q_N(G, Y) = q(G, 2, Y) = \sum_{S \subseteq V(G)} (Y-1)^{|V|-r_2(G[S])}$$

The evaluation $q(G, 1, 2)$ counts the number of independent sets of a graph, [ABS04b].

Relationship between the polynomials. It is well known that the chromatic polynomial can be obtained from the Tutte polynomial by a simple substitution and multiplication with a polynomial time computable graph invariant. It is open whether such a reduction exists between the remaining polynomials (not including the trivial relationships with the characteristic polynomial). But for each two polynomials one can find pairs of graphs G_1, G_2 for which one polynomial gives the same value and for the other it gives different values.

4   Complexity of Computing the Graph Polynomials

The general situation. It is natural to ask how difficult it is to compute the various graph polynomials. We look at two versions of this problem:

Coefficents: Given a graph G, compute all the coefficients of the polynomial $p(G, \lambda)$.

Evaluation: Given a graph G, evaluate the polynomial $p(G, \lambda)$ for a fixed $\lambda = \lambda_0 > 0$ in the underlying ring $\mathcal{R}$.

If the ring $\mathcal{R}$ is the ring of integers $\mathbb{Z}$ these problems are well defined in the Turing model of computations. For arbitrary rings it is best to work in the unit-cost model of computation as described in, say, [BCSS98].

Clearly, if we can evaluate a polynomial efficiently, also its coefficents can be computed efficiently. Evaluating the polynomial $\chi(G, \lambda)$ for integers $\lambda \geq 3$ is $\sharp\mathbf{P}$-complete, [Val79]. For the best known algorithms, see [FK03]. Even if restricted to planar graphs, counting the number of 3-colorings or the number of acyclic orientations, i.e. evaluating the polynomial $\chi(G, \lambda)$ at $\lambda = 3$ and $\lambda = -1$, is known to be $\sharp\mathbf{P}$-hard [VW92]. This also makes evaluating the Tutte polynomial $\sharp\mathbf{P}$-hard. The same is true for the acyclic polynomial due to its connection to counting matchings, and for the interlace polynomial, due to its counting of independent sets, cf. [Val79]. However, it remains open, whether evaluating $q_N(G, Y)$ is $\sharp\mathbf{P}$-hard, cf. [ABS04b]. In their remarkable paper, [JVW90], F. Jaeger, X. Vertigan and D. Welsh, have characterized completely the points (a, b) in the complex plane, where evaluating the Tutte polynomial $T(G, a, b)$ is difficult for arbitrary graphs.

Computing the coefficients for univariate polynomials of degree d can be done in polynomial time from $d + 1$ evaluations at different points. Computing the coefficients for multivariate polynomials is in general more complicated. The the coefficients of the characteristic polynomial are computable in polynomial time using classical algorithms for the determinant of a matrix.

5   Bounded Tree-Width

J. Oxley and D. Welsh [OW92] also noted that the Tutte polynomial for series parallel graphs, which are graphs of tree-width at most 2, can be computed in polynomial time. This was extended to arbitrary fixed tree-width k independently by A. Andrzejak [And98] and S. Noble [Nob98], and therefore also holds for the chromatic polynomial. Actually, they showed that computing the Tutte polynomial is in **FPT** on graph classes of tree-width at most k with computation time roughly $f(k)n^3$ where $f(k)$ is simply exponential in k and the tree-decomposition of the graph is given in advance.

The same result also follows from a more general method presented in [Mak04], which also covers the acyclic polynomial and a wide range of other graph polynomials where summations is restricted to families of subsets of edges which are definable in Monadic Second Order Logic, including, using Proposition 3 below, the interlace polynomial. However, even when the tree decomposition is given in advance, $f(k)$ here would be at least doubly exponential.

For the matching polynomial we have improved this.

Theorem 2. *If the graph G is given together with its k-tree decomposition then the matching polynomial can be computed in time $O(2^{3k} \cdot n^3)$ where the remaining constants are independent of k and n.*

Proof (Sketch). As in [CO00], we translate a k-tree decomposition of the input graph into a k-expression, but we note that the $\eta_{i,j}$'s are only applied to sets with label i and j which are smaller than k. Then we proceed as in the proof of Theorem 4 below.   □

Similar results for polygraphs, which are special case of graph classes of bounded path-width, were obtained already in [BGMP86]. For the Tutte polynomial, and hence for the chromatic polynomial, similar bounds can be obtained from [And98, Nob98].

6   Bounded Clique-Width: The Interlace Polynomial

In the same paper [Mak04], in Theorem 6.6 it is shown that, in combination with the work of P. Seymour and S. Oum [OS05], graph polynomials, where summations are restricted to families of subsets of vertices which are definable in Monadic Second Order Logic, are in **FPT** for graph classes of clique-width at most k. To apply this to the interlace polynomial it suffices to show

Proposition 3. *The interlace polynomial can be written as*

$$\sum_{A:\phi(A,B,C)} \prod_{u:u\in B} U \prod_{v:v\in C} V$$

where $\phi(A,B,C)$ is a formula with free monadic variables A,B,C in **MSOL** *with a parity quantifier. In other words, it is a* **C$_2$MSOL**-*definable graph polynomial.*

Proof (Sketch). The interlace polynomial is defined as

$$q(G,X,Y) = \sum_{S\subseteq V(G)} (X-1)^{r_2(G[S])}(Y-1)^{|V|-r_2(G[S])}$$

First we substitute $X - 1 = U$ and $Y - 1 = V$. To get the formula ϕ, we formalize vectors of the adjacency matrix by unary predicates. This can be done in **MSOL**. To compute the rank in $GF(2)$ we use the fact that in $GF(2)$ a sum of 1's vanishes iff it consists of an even number of summands. For this part we need a counting quantifier **C$_2$**. □

A very similar argument can be found in [CO06].

Theorem 1 (i) now follows from [Mak04, OS05]. However, this method does not apply to the chromatic polynomial, the Tutte polynomial and the matching polynomials.

7 Bounded Clique-Width: The Matching Polynomial

A more explicit version of Theorem 1(ii), combined with Theorem 1 is the following:

Theorem 4. *The matching polynomial $m(G,\lambda)$ of a graph G with n vertices of clique-width at most k can be computed in polynomial time. More precisely, in time $O(n^{\alpha(k)})$, with $\alpha(k) = O(2k+1)$ if a k-expression for the graph G is given. Hence it is in* **FPPT**. *Otherwise, $\alpha(k)$ is simply exponential in k.*

For the proof we first restrict our attention to k-expression which are *irredundant*. Then we introduce a set of auxiliary polynomials, the *constrained matching polynomials*. We use the dynamic programming approach to compute the auxiliary polynomials. Finally, we piece everything together.

Irredundant k-expressions. Let $K = 1, 2, ..., k$ be a set of k labels, and $T(K)$ be the set of all the possible k-expressions using labels of K. A k-expression is called *irredundant* if for every of its subexpression of the form $\eta_{i,j}(t')$ no vertex labeled i is adjacent to vertex labeled j in $val(t')$. According to [CO00], every graph built inductively from a k-expression $t \in T(K)$, can also be built from an irredundant k-expression $t' \in T(K)$. To proceed, let G be a graph, and t be its irredundant k-expression. We want to compute its matching polynomial $m(G,\lambda)$. We denote by $tree(t)$ the parse-tree of t. We shall use a set of auxiliary polynomials.

Constrained matching polynomials $cm_F(H, \lambda)$. Let $H(V, E)$ be a labeled graph with labels $1, 2, ..., k$. We denote by P_i the set of vertices $v \in V$ labeled i. Let $F = (f_1, f_2, ..., f_k)$ be a vector of k non-negative integers $0 \leq f_i \leq n$, where $n = |V|$. We will denote by $cm_F^l(H)$ the number of l-matchings, which leave unused exactly f_i vertices labeled i, for every label $1 \leq i \leq k$. The **constrained matching polynomial** is defined by:

$$cm_F(H, \lambda) = \sum_{l=0}^{\frac{n}{2}} cm_F^l(H)\lambda^l$$

Obviously, the size of suitable matchings is uniquely defined by F, so actually $cm_F(H, \lambda)$ is a monomial. Our set of auxiliary functions is: $\{cm_F(H, \lambda)\}$.

The algorithm

Singletons labeled i: If $F = (0, 0, ...0, 1, 0, ..., 0)$ with a single 1 at the i-th position, then $cm_F(H, \lambda) = 1$, otherwise $cm_F(H, \lambda) = 0$.

$\oplus$: Assume $H = H_1 \oplus H_2$. Since the graphs H_1 and H_2 are disjoint, it is easy to check that

$$cm_F(H) = \sum_{F_1} cm_{F_1}(H_1) \cdot cm_{(F-F_1)}(H_2) \tag{1}$$

$F - F_1$ is vector subtraction. If $F - F_1$ has a negative coordinate we put $cm_{(F-F_1)}(H_2) = 0$.

$\rho_{i \to j}$: Assume $H = \rho_{i \to j}(H_1), i \neq j$. Then we have:

$$cm_F(H) = \sum_{F':(F,F') \models \varphi} cm_{F'}(H_1) \tag{2}$$

where

$$\varphi = \left(\forall_{l:1 \leq l \leq k} : f_l = \begin{cases} f_i' + f_j' & if \ \ l = j \\ 0 & if \ \ l = i \\ f_l' & otherwise \end{cases} \right) \tag{3}$$

$\eta_{i,j}$: Assume $H = \eta_{i,j}(H_1), i \neq j$. Since t is an irredundant k-expression, this operation adds new edges between all the vertices labeled i and all the vertices labeled j. Any matching of H consists of some matching of H_1 and (optionally) some new edges. If such a matching includes q new edges, then it chooses two sets A and B and a bijection between A and B, where A consists of q of f_i free vertices labeled i, and B consists of q of f_j free vertices labeled j. There are $q!$ many such bijections. Using the definition of cm_F and counting the suitable matchings of the left and the right side of the equation we get:

$$cm_F(H) = \sum_{q, F' \models \varphi} \lambda^q \cdot cm_{F'}(H_1) \cdot \binom{f_i'}{q} \cdot \binom{f_j'}{q} \cdot q! \tag{4}$$

where

$$\varphi = \left(\forall_{l:1 \leq l \leq k} : f_l = \begin{cases} f'_i - q \ if \ \ l = i \\ f'_j - q \ if \ \ l = j \\ f'_l \qquad otherwise \end{cases} \right) \tag{5}$$

For the inductive computation of $m(G, \lambda)$ we observe that for different F_1 and F_2 the sets of matchings counted by $cm_{F_1}(H, \lambda)$ and $cm_{F_2}(H, \lambda)$ are disjoint. Furthermore, every matching of H is counted by some $cm_F(H, \lambda)$. Hence, the matching polynomial of graph G can be obtained by summation over F:

$$m(G, \lambda) = \sum_F cm_F(H, \lambda) \tag{6}$$

Complexity analysis

  (i) We initialize up to (n^k) auxiliary monomials by 0 or 1.
 (ii) For the disjoint union $\oplus$ we sum over all possible F_1, which is $O(n^k)$, for all the (n^k) auxiliary monomials. This gives $O(n^{2k})$ steps.
(iii) For ρ we sum over F_1, but the only free value is the j'th field of the vector $(O(n))$. This give, for (n^k) auxiliary monomials, $O(n^{k+1})$ steps.
 (iv) For η we sum over q, which uniquely defines F'. We get, for (n^k) auxiliary monomials, $O(n^{k+1})$ steps.
  (v) Finally, we sum of all the auxiliary monomials, which gives $O(n^k)$ steps.

The number of times each operation is applied depends on the term t. $\oplus$ can appear at most n times in t because every time the number of vertices in $val(t)$ grows at least by 1. η cannot be applied more than n^2 times, because it adds edges every time. ρ can appear at most $n * k$ times for sequences of recolorings without repetitions. Summation of all the monomials is performed once. Hence we get that the total time complexity of the algorithm is bounded by

$$O(n \cdot (n^k) + n \cdot (n^{2k}) + n^2 \cdot (n^{k+1}) + n \cdot k \cdot (n^{k+1}) + (n^k)) = O(n^{2k+1}).$$

8   Bounded Clique-Width: The Chromatic Polynomial

For clique-width at most 2, which are the cographs, cf. [CO00], we have, using an observation of N. Biggs, [Big93, Chapter 9]:

Theorem 5. *The chromatic polynomial of cographs with n vertices can be computed in polynomial time.*

D. Kobler and U. Rotics [KR03] showed that the **chromatic number** of a graph with n vertices of clique-width at most k given as a k-expression can be computed in polynomial time $O(n^{\alpha(k)})$, with $\alpha(k) = O(2^k)$.

A more explicit version of Theorem 1(iii) combined with Theorem 1 is the following:

Theorem 6. *The chromatic polynomial $\chi(G, \lambda)$ of a graph G with n vertices of clique-width at most k can be computed in polynomial time. More precisely, in time $O(n^{\alpha(k)})$, with $\alpha(k) = O(2^k)$ if a k-expression for the graph G is given. Otherwise, $\alpha(k)$ is doubly exponential in k.*

Overview of the proof. The value of the chromatic polynomial $\chi(G, \lambda)$ gives the number of different colorings of a graph G using *at most* λ colors. We denote by $\overline{\chi}(G, \lambda)$ the number of different colorings of a graph G using *exactly* λ colors. We say that two colorings c and d of G are *isomorphic* if the set of sets of vertices induced by the colors in c is equal to the set of sets of vertices induced by the colors in d. We denote by $\overline{\psi}(G, \lambda)$, the number of non-isomorphic colorings of a graph G using exactly λ colors.

The input to our algorithm is a graph G together with a k-expression t defining G. We denote by K the set of labels $\{1, ..., k\}$ in the k-expression. The k-expression contains operations $\oplus, \eta_{i,j}$ and $\rho_{i \to j}$. The algorithms will traverse $tree(t)$ from bottom to top. It constructs at each step the labeled graph H corresponding to a subtree of $tree(t)$ scanned so far, and keeps track of the labels and the number of their corresponding colorings.

If we wanted to compute $\chi(G, \lambda)$ directly using the k-expression t, the only difficult case would be $\eta_{i,j}$. However, we shall compute a different quantity, $num\text{-}cols(N, H)$, defined below. It turns out that the inductive computation of $num - cols(N, H)$ will be easy in the case of $\eta_{i,j}$ and $\rho_{i \to j}$ but a bit more involved in the case of the disjoint union $\oplus$.

Introducing $num\text{-}cols(N, H)$. Let c be a coloring of a labeled graph H. The type of a color i of c is defined as the set of labels $B \subseteq K$, such that label l belongs to B if and only if there is a vertex of H having label l and color i. The type of the coloring c is defined by counting for each $B \subseteq K$, the number of colors of type B in c. In other words, the type of the coloring c is defined as an array N of size 2^k, such that for every set of labels $B \subseteq K$, $N[B]$ contains the number of colors in c of type B.

For an array N we denote by $num\text{-}cols(N, H)$, the number of non-isomorphic colorings of H which are of type N, and by $total(N)$ the sum of all the values of N. The information we keep is for each array N and for each H the value of $num\text{-}cols(N, H)$. Using this notation it is easy to see that

$$\overline{\psi}(G, \lambda) = \sum_{N : total(N) = \lambda} num\text{-}cols(N, G) \tag{7}$$

Computing $num\text{-}cols(N, H)$ inductively. Computing $num\text{-}cols(N, H)$, when H is a singlton is straightforward. When $H = \eta_{i,j}(H_1)$ or when $H = \rho_{i \to j}(H_1)$ it is easy to see how to compute $num\text{-}cols(N, H)$, from $num\text{-}cols(N, H_1)$. We now consider the more complicated case when $H = H_1 \oplus H_2$.

Let d and e be colorings of H_1 and H_2, respectively using disjoint sets of colors. From these two colorings we can obtain colorings of $H_1 \oplus H_2$, by merging some of the colors of d and e into the same color. The colors merged are idenitifed by a set of pairs, denoted by M, where the pair (i, j) belongs to M if and only if color i of d is merged into color j of e. The set of all possible merges for colorings d and e is senoted by $Merges(d, e)$. The coloring of H obtained from colorings d and e using set of merges M is denoted by $coloring(d, e, M)$.

For an array N we denote by $Colorings(d, e, N)$ the set of all non-isomorphic colorings of H which of type N and are obtained from the colorings d and e using a set of merges in $Merges(d, e)$.

Lemma 7. *For every four colorings d, d', e, e' such that colorings d and d' are of the same type and the colorings e and e' are of the same type and for every array N, the number of non-isomorphic colorings in $Colorings(d, e, N)$ is equal to the number of non-isomorphic colorings in $Colorings(d', e', N)$.*

Note that Lemma 7 holds also when d and d' are colorings of different graphs and e and e' are colorings of different graphs.

Let d and e be any two colorings of H_1 and H_2 of type N_1 and N_2 respectively. We define $f(N_1, N_2, N) = |Colorings(d, e, N)|$. By Lemma 7 the value of $f(N_1, N_2, N)$ does not depend on the specific colorings d and e and on the graphs H_1 and H_2 colored by the colorings d and e, respectively. Thus, the value $f(N_1, N_2, N)$ for each triple of arrays N_1, N_2 and N can be calculated in a table before the algorithm starts. Using this notation we have

$$num\text{-}cols(N, H) = \sum_{N_1, N_2} num\text{-}cols(N_1, H_1) \cdot num\text{-}cols(N_2, H_2) \cdot f(N_1, N_2, N) \quad (8)$$

Formula 8 gives us the inductive step in the calculation $num\text{-}cols(N, H)$. Using Formula 7 we now can compute $\bar{\psi}(G, \lambda)$. Finally we have

$$\chi(G, \lambda) = \sum_{\alpha=1}^{\lambda} \bar{\psi}(G, \alpha) \cdot \alpha!.$$

Complexity analysis. The number of different arrays is at most $(\lambda + 1)^{2^k}$, since for each array N and each $B \subseteq K$ the number of possible different values of $N[B]$ is at most $\lambda + 1$.

To analyze the complexity of the algorithm we consider the time taken by the $\oplus$ operations, since the other operations take less time. It can be shown that the routine which calculates all the values of $f(N_1, N_2, N)$ before the algorithm starts takes at most $O(2^{k+1} * (\lambda + 1)^{3*2^k})$ time which can be considered as $O(\lambda^{3*2^k})$, since k is assumed to be a fixed constant.

For each array N, the value of $num\text{-}cols(N, H)$, evaluated by Formula 8 is achieved by considering all pairs of arrays N_1, N_2 and for each such pair evaluating two multiplications. Thus, the total time for calculating $num\text{-}cols(N, H)$ is at most $O(\lambda^{2*2^k})$. Repeating this calculation for every array N we obtain that total time taken in calculating $F(H)$ on behalf of one $\oplus$ operation is at most $O(\lambda^{3*2^k})$. Since there are at most n $\oplus$ operations the total complexity of the algorithm is at most $O(n * (\lambda^{3*2^k}))$. Putting $n = \lambda$, evaluating sufficiently often, and interpolating now gives Theorem 6.

By Stanley's Theorem, [Sta73], $\chi(G, -1)$ is the number of acyclic orientations of G. Hence we have:

Corollary 8. *The number of acyclic orientations of a graph of clique-width at most k can be computed in polynomial time.*

Comparison with the Tutte polynomial. O. Gimenez, P. Hliněný and M. Noy [GHN06] showed that the Tutte polynomial on graphs of clique-width at most k given as a k-expression can be computed in time $2^{O(n^{1-\epsilon})}$ with $\epsilon = \frac{1}{k+2}$. This does not exclude that computing the Tutte polynomial could be $\sharp\mathbf{P}$-hard on graphs of bounded clique-width. But, together with Theorem 6, it excludes showing this via a reduction to the chromatic polynomial.

9   Conclusions and Open Problems

We have done a first step into analyzing the complexity of prominent graph polynomials on graphs of bounded clique width. All our polynomials are fixed parameter tractable (are in **FPT**) on graphs of tree-width at most k. However, on graphs of clique-width at most k their complexity seems to differ. It is a challenge for future research to establish whether this is really so.

Comparing the complexities. The complexity of computing the chromatic polynomial for graphs G of clique-width at most k, given the k-expression of G, is determined by the dimension of N, which is at least as numerous as the power set of the labels, i.e. 2^k.

   In contrast to this, in the case of the matching polynomial, the complexity is determined by the dimension of F, which is equal to k.

   For the Tutte polynomial we could proceed like for the matching polynomial, but the analogue to the vector F then contains values for each label and for each connected component of the spanning subgraphs. However, the number of components may equal the number of vertices of the graph. It is not clear how to avoid this and how to get an F the size of which depends only on k.

Open problems
 (i) Find explicit constants in the case of the interlace polynomial.
 (ii) Is computing the matching polynomial or the chromatic polynomial from k-expressions for graphs of clique-width at most k in **FPT** for k?
 (iii) Can the chromatic polynomial be computed in time comparable to the time needed for the matching polynomial?
 (iv) Is computing the Tutte polynomial of graphs of bounded clique-width in **FPPT**, or even in **FPT** for k?
 (v) How can one generalize the underlying method for other graph polynomials. Although the ideas of [KR03] also helped in computing the matching polynomial, it is not yet clear, how to formulate a general principle.
 (vi) Can one show, using the methods of [DF99], that computing the matching polynomial, chromatic polynomial or Tutte polynomial is not in **FPT** for graphs of clique-width at most k, for some k?

Acknowledgements. We would like to thank Y. Altschuler, B. Dubrov and A. Matsliach for their active participation in our seminar. We are indebted to B. Courcelle for suggesting Proposition 3 and to P. Hliněný and M. Noy for making [GHN06] avaible. We would like to thank various anonymous referees for valuable comments and suggestions.

References

[ABS00]   R. Arratia, B. Bollobas, and G.B. Sorkin. The interlace polynomial: a new graph polynomial. In *Proceedings of the 11th Annual ACM-SIAM Symposium on Discrete Mathematics*, pages 237–245, 2000.

[ABS04a]   R. Arratia, B. Bollobas, and G.B. Sorkin. The interlace polynomial: a new graph polynomial. *Journal of Combinatorial Theory, Series B*, 92:199–233, 2004.

[ABS04b]   R. Arratia, B. Bollobas, and G.B. Sorkin. A two-variable interlace polynomial. *Combinatorica*, 24.4:567–584, 2004.

[ACP87]   S. Arnborg, D.G. Corneil, and A. Proskurowski. Complexity of finding embedding in a k–tree. *SIAM. J. Algebraic Discrete Methods*, 8:277–284, 1987.

[And98]   A. Andrzejak. An algorithm for the Tutte polynomials of graphs of bounded treewidth. *Discrete Mathematics*, 190:39–54, 1998.

[AvdH04]   M. Aigner and H. van der Holst. Interlace polynomials. *Linear Algebra and Applications*, 377:11–30, 2004.

[BCSS98]   L. Blum, F. Cucker, M. Shub, and S. Smale. *Complexity and real computation*. Springer-Verlag, New York, 1998.

[BGMP86]   D. Babić, A. Graovac, B. Mohar, and T. Pisanski. The matching polynomial of a polygraph. *Discrete Applied Mathematics*, 15:11–24, 1986.

[Big93]   N. Biggs. *Algebraic Graph Theory, 2nd edition*. Cambridge University Press, 1993.

[Bir12]   G.D. Birkhoff. A determinant formula for the number of ways of coloring a map. *Annals of Mathematics*, 14:42–46, 1912.

[Bol99]   B. Bollobás. *Modern Graph Theory*. Springer, 1999.

[CDS95]   D.M. Cvetković, M. Doob, and H. Sachs. *Spectra of graphs*. Johann Ambrosius Barth, 3 edition, 1995.

[CER93]   B. Courcelle, J. Engelfriet, and G. Rozenberg. Handle-rewriting hypergraph grammars. *J. Comput. System Sci.*, 46:218–270, 1993.

[CMR01]   B. Courcelle, J.A. Makowsky, and U. Rotics. On the fixed parameter complexity of graph enumeration problems definable in monadic second order logic. *Discrete Applied Mathematics*, 108(1-2):23–52, 2001.

[CO00]   B. Courcelle and S. Olariu. Upper bounds to the clique–width of graphs. *Discrete Applied Mathematics*, 101:77–114, 2000.

[CO06]   B. Courcelle and S. Oum. Vertex-minors, monadic second-order logic, and a conjecture by Seese. *Journal of Combinatorial Theory, Series B*, xx:xx–xx, 2006.

[CR05]   D. G. Corneil and U. Rotics. On the relationship between clique-width and treewidth. *SIAM J. Comput.*, 34(4):825–847, 2005.

[DF99]   R.G. Downey and M.F Fellows. *Parametrized Complexity*. Springer, 1999.

[Die96]   R. Diestel. *Graph Theory*. Graduate Texts in Mathematics. Springer, 1996.

[DKT05]   F.M. Dong, K.M. Koh, and K.L. Teo. *Chromatic polynomials and chromaticity of graphs*. World Scientific, 2005.

[FK03]   M. Fürer and S. P. Kasiviswanathan. Algorithms for counting 2-SAT solutions and colorings with applications. *Electronic Colloquium on Computational Complexity*, 1:R 33, 2003.

[FMR06]   E. Fischer, J.A. Makowsky, and E.V. Ravve. Counting truth assignments of formulas of bounded tree width and clique-width. *Discrete Applied Mathematics*, xx:xx–xx, 2006.

[FRRS05]   M.R. Fellows, F.A. Rosamond, U. Rotics, and S. Szeider. Proving NP-hardness for clique width. *ECCC*, xx:xx–yy, 2005.

[GHN05]   O. Giménez, P. Hliněný, and M . Noy. Computing the Tutte polynomial on graphs of bounded clique-width. In *Graph Theoretic Concepts in Computer Science, WG 2005*, volume 3787 of *Lecture Notes in Computer Science*, pages 59–68, 2005.

[GHN06]   O. Giménez, P. Hliněný, and M . Noy. Computing the Tutte polynomial on graphs of bounded clique-width. *XXX*, xx:xx–yy, 2006.

[GR01]   C. Godsil and G. Royle. *Algebraic Graph Theory*. Graduate Texts in Mathematics. Springer, 2001.

[JVW90]   F. Jaeger, D.L. Vertigan, and D.J.A. Welsh. On the computational complexity of the Jones and Tutte polynomials. *Math. Proc. Camb. Phil. Soc.*, 108:35–53, 1990.

[KR03]   D. Kobler and U. Rotics. Edge dominating set and colorings on graphs with fixed clique-width. *Discrete Applied Mathematics*, 126:197–221, 2003.

[LP86]   L. Lovasz and M. Plummer. *Matching Theory*. North Holland, 1986.

[Mak04]   J.A. Makowsky. Algorithmic uses of the Feferman-Vaught theorem. *Annals of Pure and Applied Logic*, 126:1–3, 2004.

[Nob98]   S.D. Noble. Evaluating the Tutte polynomial for graphs of bounded tree-width. *Combinatorics, Probability and Computing*, 7:307–321, 1998.

[OS05]   S. Oum and P. Seymour. Approximating clique-width and branch-width. *Journal of Combinatorial Theory, Ser. B*, xx(x):xx–yy, 2005.

[OW92]   J.G. Oxley and D.J.A. Welsh. Tutte polynomials computable in polynomial time. *Discrete Mathematics*, 109:185–192, 1992.

[Pap94]   C. Papadimitriou. *Computational Complexity*. Addison Wesley, 1994.

[Sta73]   R. P. Stanley. Acyclic orientations of graphs. *Discrete Mathematics*, 5:171–178, 1973.

[Tri92]   N. Trinajstić. *Chemical graph theory*. CRC Press, 2 edition, 1992.

[Val79]   L.G. Valiant. The complexity of enumeration and reliability problems. *SIAM Journal on Computing*, 8(3):410–421, 1979.

[VW92]   D.L. Vertigan and D.J.A. Welsh. The computational complexity of the Tutte plane: The bipartite case. *Combinatorics, Probability, and Computing*, 1:181–187, 1992.

[Wel93]   D.J.A. Welsh. *Complexity: Knots, Colourings and Counting*, volume 186 of *London Mathematical Society Lecture Notes Series*. Cambridge University Press, 1993.

Generation of Graphs with Bounded Branchwidth[*]

Christophe Paul[1], Andrzej Proskurowski[2], and Jan Arne Telle[3]

[1] CNRS - LIRMM, Montpellier, France
paul@lirmm.fr
[2] Department of Computer and Information Science, University of Oregon, USA
andrzej@cs.uoregon.edu
[3] Department of Informatics, University of Bergen, Norway (Research conducted
while on sabbatical at LIRMM)
telle@ii.uib.no

Abstract. Branchwidth is a connectivity parameter of graphs closely related to treewidth. Graphs of treewidth at most k can be generated algorithmically as the subgraphs of k-trees. n this paper, we investigate the family of edge-maximal graphs of branchwidth k, that we call k-branches. The k-branches are, just as the k-trees, a subclass of the chordal graphs where all minimal separators have size k. However, a striking difference arises when considering subgraph-minimal members of the family. Whereas K_{k+1} is the only subgraph-minimal k-tree, we show that for any $k \geq 7$ a minimal k-branch having q maximal cliques exists for any value of $q \notin \{3, 5\}$, except for $k = 8, q = 2$. We characterize subgraph-minimal k-branches for all values of k. Our investigation leads to a generation algorithm, that adds one or two new maximal cliques in each step, producing exactly the k-branches.

1 Introduction

Branchwidth and treewidth are mutually related connectivity parameters of graphs: whenever one of these parameters is bounded by some fixed constant for a graph, then so is the other [17]. Since many graph problems that are NP-hard in general can be solved in linear time when restricted to such classes of graphs both treewidth and branchwidth have played a large role in many investigations in algorithmic graph theory. Tree-decompositions have traditionally been the choice when solving NP-hard graph problems by dynamic programming to give FPT algorithms when parameterized by treewidth, see e.g. [2,16] for overviews. Recently it is the branchwidth parameter that has been in the focus of several algorithmic research results. For example, several papers [7,5,8,9,6] show that for graphs of bounded genus the base of the exponent in the running time of these FPT algorithms could be improved by the dynamic programming following instead a branch-decomposition of optimal branchwidth. Also, a strong

---

[*] This extended abstract does not contain all proofs. For a full version, please refer to [14] or contact one of the authors.

F.V. Fomin (Ed.): WG 2006, LNCS 4271, pp. 205–216, 2006.
© Springer-Verlag Berlin Heidelberg 2006

heuristic algorithm for the travelling salesman problem [4] has been developed based on branch-decompositions and an exact (exponential-time) algorithm has been given to compute branchwidth [10]. Given these recent developments in favor of branchwidth one may wonder why treewidth has historically been preferred over branchwidth? Mainly, this is because of the equivalent definition of 'G has treewidth $\leq k$' by 'G is a partial k-tree'. This alternative definition is intuitively appealing since the k-trees are the graphs generated by the following very simple algorithm: 'Start with K_{k+1}; Repeatedly choose a k-clique C and add a new vertex adjacent to vertices in C'. Can we define branchwidth in an analogous algorithmic way? This is the question that has inspired our research and in this paper we give an affirmative answer.

We start by investigating the family of edge-maximal graphs of branchwidth k, that we call $\underline{k\text{-branches}}$. The k-branches are chordal, as can be easily deduced from earlier work on branchwidth [11,10]. In Section 2 we report on related work [13] where we have given a characterization of k-branches. In Section 3 we consider subgraph-minimal k-branches. They form the starting graphs of our algorithm generating k-branches, just as the minimal k-tree K_{k+1} is the starting graph of the generation algorithm for k-trees. K_n has branchwidth $\lceil 2n/3 \rceil$ for any $n \geq 3$ and $K_{\lfloor 3(k-1)/2 \rfloor + 1}$ is one of the minimal k-branches. However, for $k \geq 7$ we find that there is a minimal k-branch on q maximal cliques for any $q \notin \{3,5\}$, except for the pathological case $k = 8, q = 2$. We show that the minimal k-branches have clique trees that are caterpillars and give a characterization of the family of minimal k-branches for all values of k. Our investigation culminates in Section 4 with a non-deterministic generation algorithm, that adds one or two new maximal cliques in each step, yielding as output exactly the graphs that are k-branches, and whose spanning subgraphs (i.e., partial graphs) are exactly the graphs of branchwidth at most k. Our results lead to a better understanding of the branchwidth parameter by defining graphs of branchwidth k through the algorithmic concept of partial k-branches. The algorithm will generate a random graph of branchwidth k together its branch-decomposition and can be used to provide test instances for optimization codes based on branch-decomposition.

2   Definitions and Earlier Results

A *branch-decomposition* (T, μ) of a graph G is a tree T with nodes of degree one and three only, together with a bijection μ from the edge-set of G to the set of degree-one nodes (leaves) of T. For an edge e of T let T_1 and T_2 be the two subtrees resulting from $T \setminus \{e\}$, let G_1 and G_2 be the graphs induced by the edges of G mapped by μ to leaves of T_1 and T_2 respectively, and let $mid(e) = V(G_1) \cap V(G_2)$. The width of (T, μ) is the size of the largest $mid(e)$ thus defined. For a graph G its *branchwidth* $bw(G)$ is the smallest width of any branch-decomposition of G. [1]

---

[1] The connected graphs of branchwidth 1 are the stars, and constitute a somewhat pathological case. To simplify certain statements we therefore restrict attention to graphs having branchwidth $k \geq 2$.

A tree-decomposition $(T, \mathcal{X})$ of a graph G is an arrangement of the vertex subsets $\mathcal{X}$ of G, called bags, as nodes of the tree T such that for any two adjacent vertices in G there is some bag containing them both, and for each vertex of G the bags containing it induce a connected subtree. For a subtree T' of T the induced tree-decomposition $(T', \mathcal{X}')$ is the result of removing from $(T, \mathcal{X})$ all nodes of $V(T) \setminus V(T')$ and their corresponding bags.

Definition 1. *A k-troika (A, B, C) of a set X are 3 subsets of X such that $|A| \leq k$, $|B| \leq k$, $|C| \leq k$, and $A \cup B = A \cup C = C \cup B = X$. (A, B, C) respects $S_1, S_2, ..., S_q$ if any $S_i, 1 \leq i \leq q$ is contained in at least one of A, B or C.*

A necessary condition for a graph to be a k-branch is that it is a chordal graph where all minimal separators have size k, with the property that every maximal clique has a k-troika respecting the minimal separators contained in it [13]. This motivates the following definition.

Definition 2. *Let G be a chordal graph with C_G its set of maximal cliques and S_G its set of minimal separators. A tree-decomposition $(T, \mathcal{X})$ of G is called k-full if the following conditions hold: 1) The set of bags $\mathcal{X}$ is in 1-1 correspondence with $C_G \cup S_G$ (we call the nodes with bags in C_G the maxclique nodes and the nodes with bags in S_G the minsep nodes.) 2) The bags of the minsep nodes all have cardinality k. 3) There is an edge ij in the tree T iff $X_i \in S_G, X_j \in C_G$ and $X_i \subseteq X_j$. 4) Every maxclique bag X_j has a k-troika respecting its neighbor minsep bags.*

Note that if G has a k-full tree-decomposition then it is unique. We need additional constraints on k-full tree-decompositions to characterize exactly the k-branches.

Definition 3. *A mergeable subtree of a k-full tree-decomposition $(T, \mathcal{X})$ of a graph G is a subtree T' of T that: contains at least one edge, has leaves that are maxclique nodes, and satisfies:*

1. $|\{v : v \in X \text{ where } X \text{ a node in } T'\}| \leq \lfloor 3k/2 \rfloor$
2. *Either the subtree T' has at most one node that in T has a neighbor in $V(T) \setminus V(T')$ or else T' is a path X, B, Y with X, B, Y and all their neighbors in T inducing a path A, X, B, Y, C satisfying $B \setminus (A \cup C) = \emptyset$.*

Lemma 1. *Let $A - X - B - Y - C$ be a path in T for some k-full tree-decomposition $(T, \mathcal{X})$ with X and Y maxclique nodes. $X \cup Y$ has a k-troika respecting A, C if and only if $|X \cup Y| \leq \lfloor 3k/2 \rfloor$ and $B \setminus (A \cup C) = \emptyset$.*

Proof. If $|X \cup Y| > \lfloor 3k/2 \rfloor$ then $X \cup Y$ does not have a k-troika. Let $P = B \setminus (A \cup C)$. Note that we have $A \cap C \subseteq B$ and since $|A| = |C| = k$ we have $|A \cap C| = 2k - |(X \cup Y) \setminus P|$. But then $|X \cup Y| + |A \cap C| = 2k + |P|$ and this means that by Theorem 2 of [15] (also by results of [11]) $X \cup Y$ has a k-troika respecting A, C if and only if $P = \emptyset$.

Lemma 1 is implicit in [13], and implies that for mergeable subtree T' we can add edges to G to make a clique of $\{v : v \in X$ where X a node in $T'\}$ without increasing branchwidth of G.

Definition 4. *A k-full tree-decomposition $(T, \mathcal{X})$ of a graph G is a k-skeleton of G if G has at least $\lfloor 3(k-1)/2 \rfloor + 1$ vertices and T does not have a mergeable subtree.*

Theorem 1. *[13] G is a k-branch $\Leftrightarrow$ G has a k-skeleton*

3   Minimal k-Branches

We characterize the subgraph-minimal k-branches on q maximal cliques, by describing the structure of the minimal k-skeletons, as defined below. We divide the caracterization into two Theorems, one for the cases when $k \leq 6$ or $q \leq 5$ and the other for the cases $k \geq 7, q \geq 6$.

Definition 5. *A k-branch G is a minimal k-branch if no strict subgraph of G is a k-branch. Let the set of minimal k-skeletons be $MS(k) = \{(T, \mathcal{X}) : (T, \mathcal{X})$ is a k-skeleton but for no proper subtree T' of T is the induced tree-decomposition $(T', \mathcal{X}')$ a k-skeleton$\}$. Let $MS(k, q)$ be the set of minimal k-skeletons on q maxclique nodes.*

If G is a minimal k-branch then for its k-skeleton $(T_G, \mathcal{X})$ we have $(T_G, \mathcal{X}) \in MS(k)$. However, the graph represented by a minimal k-skeleton may have some cliques that are too big for it to be a minimal k-branch. For example, if $(T, \mathcal{X})$ is the tree T having a single maxclique node on 6 vertices then we have $(T, \mathcal{X}) \in MS(4)$ since it is a minimal 4-skeleton but the graph K_6 that it represents is not a minimal 4-branch since it contains the 4-branch K_5 as a subgraph. Since our algorithm in Section 4 builds k-skeletons, rather than graphs, we focus in the following on the minimal k-skeletons.

Lemma 2. *In a minimal k-skeleton $(T, \mathcal{X})$, the tree T does not contain a maxclique leaf X with path $X - A - Y$ and both A and Y having degree 2.*

Lemma 3. *In a minimal k-skeleton $(T, \mathcal{X})$, any minsep node S of degree larger than 2 must have degree 3 with exactly one of its neighbors being a maxclique leaf and the other two having degree 2.*

Lemma 4. *In a minimal k-skeleton $(T, \mathcal{X})$, any maxclique node X of degree 3 has all 3 minsep neighbors A_1, A_2, A_3 of degree 2 and at least one of them has a maxclique leaf as neighbor.*

See Figure 1 for an illustration of the following Theorem, which characterizes the minimal k-skeletons on q maximal cliques when $k \leq 6$ or $q \leq 5$.

Theorem 2.   *1. For $k \geqslant 2$, $MS(k, 1)$ contains $K_{\lfloor 3(k-1)/2 \rfloor + 1}$ and if k even then also $K_{\lfloor 3(k-1)/2 \rfloor + 2}$.*

2. For $k \leqslant 6$ and $k = 8$, $MS(k,2) = \emptyset$. For $k = 7$ and $k \geq 9$, $MS(k,2)$ is nonempty and consists of the trees with two maxclique nodes of size $x + k$ and $y + k$ (with k common vertices) for any $x \leq y$ satisfying

$$3 - (k \bmod 2) \leq x \leq y \leq \lceil k/2 \rceil - 2 \qquad and \qquad x + y \geq \lfloor k/2 \rfloor + 1 \quad (1)$$

3. For any k, $MS(k,3) = \emptyset$.
4. For $k \leqslant 4$ and $k = 6$, $MS(k,4) = \emptyset$. For $k = 5$ and $k \geq 7$ $MS(k,4)$ is nonempty, and consists of the k-full tree-decompositions $(T, \mathcal{X})$ on $q = 4$ maxclique nodes X_1, X_2, X_3, Y with Y a node of degree 3 in T such that
   (a) $|Y|, |X_i| \leqslant \lfloor \frac{3(k-1)}{2} \rfloor$ for any $i \in [1,3]$;
   (b) for any $i \in [1,3]$, $|X_i \cup Y| \leqslant \lfloor \frac{3k}{2} \rfloor$;
   (c) $\lfloor \frac{3k}{2} \rfloor + 1 \leqslant |X_i \cup Y \cup X_j|$ with $1 \leq i < j \leq 3$
5. For any k, $MS(k,5) = \emptyset$
6. For any $k \leqslant 6$ and $q \geqslant 6$, $MS(k,q) = \emptyset$.

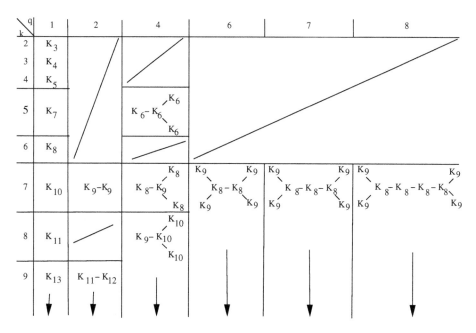

Fig. 1. Examples of minimal k-skeletons $(T, \mathcal{X})$ on q maxclique nodes, for $k \leq 9, q \leq 8$. Downward arrows indicate that minimal k-sketelons, with trees isomorphic to those depicted, exist also for larger k. Only the maxclique nodes are drawn. The minsep nodes have size k and appear on each edge of the trees. Only for the case $q = 8$ does the intersection of minsep nodes matter. For $q = 8$, if $A - X - B - Y - C$ the path with X and Y the maxclique nodes of degree two then minsep nodes A, B, C must satisfy $B \setminus (A \cup C) \neq \emptyset$.

From this characterization of minimal k-skeletons we can deduce the characterization of minimal k-branches. For lack of space we only sketch how to do this for the case $q = 2$. Note that if two distinct pairs $x \leq y$ and $x' \leq y'$ both satisfy Equations (1) then the graph associated with the first pair is a subgraph of the graph associated with the second pair if and only if $x \leq x'$ and $y \leq y'$. Thus, the minimal k-branches on $q = 2$ maximal cliques correspond with such smallest pairs $x \leq y$.

To describe the minimal k-skeletons for $k \geq 7$ having $q \geq 6$ the following definition of the adjacencies in a special caterpillar T will be useful (see also Figure 2).

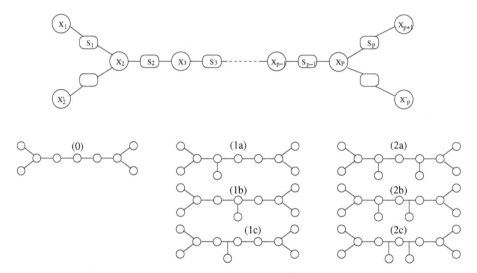

Fig. 2. On top is a special caterpillar with names of nodes as in Definition 6, in Theorem 3 and in Algorithm Stage 1. Below are the 7 non-isomorphic special caterpillars with $p = 6$, with maxclique nodes drawn as circles and minsep nodes not drawn explicitly but present on any edge between two adjacent maxclique nodes. Thus, 1c, 2b, 2c have minsep nodes of degree 3.

Definition 6. *A tree T is a special caterpillar if T consists of a body which is a path $X_1, S_1, X_2, S_2, \dots X_p, S_p, X_{p+1}$ alternating between maxclique and minsep nodes for some $p \geq 3$ with added hairs of length one or two (a hair of length one being a new maxclique node added as neighbor of a minsep node of the body, and a hair of length two being two new adjacent maxclique-minsep nodes with the minsep node added as neighbor of a maxclique node of the body) satisfying the following conditions:*

1. *at most one hair for each node of the body*
2. *no hair on any of $X_1, S_1, S_2, S_{p-1}, S_p, X_{p+1}$*
3. *hair X_2' on X_2 and hair X_p' on X_p, but no hair on minseps $X_2' \cap X_2, X_p' \cap X_p$*
4. *if hair on S_i then no hair on X_i and no hair on X_{i+1}*
5. *if hair on X_i then not hairs on both of X_{i-1} and X_{i+1}*

Theorem 3. $(T, \mathcal{X})$ *is a minimal k-skeleton for some $k \geq 7$ on at least $q \geq 6$ maxclique nodes $\Leftrightarrow$ $(T, \mathcal{X})$ is a k-full tree-decomposition with T a special caterpillar whose bags satisfy (bag names as in Definition 6 and Figure 2):*

1. *either $|X_1 \cup X_2 \cup X_3| \leq 3k/2$ or $|X_2' \cup X_2 \cup X_3| \leq 3k/2$ and also either $|X_{p+1} \cup X_p \cup X_{p-1}| \leq 3k/2$ or $|X_p' \cup X_p \cup X_{p-1}| \leq 3k/2$*
2. *$|X_1 \cup X_2 \cup X_2'| > 3k/2$ and $|X_{p+1} \cup X_p \cup X_p'| > 3k/2$*
3. *For maxcliques X, Y with a common neighbor, $|X| \leq \lfloor 3(k-1)/2 \rfloor$ and $|X \cup Y| \leq 3k/2$*
4. *If S_i has a hair then $S_i \setminus (S_{i-1} \cup S_{i+1}) = \emptyset$*
5. *If X_i has a hair then either i) no hair on X_{i-1} and $S_{i-1} \setminus (S_{i-2} \cup S_i) = \emptyset$ or ii) no hair on X_{i+1} and $S_i \setminus (S_{i-1} \cup S_{i+1}) = \emptyset$*
6. *If no hair on neither of X_i, S_i, X_{i+1} then $S_i \setminus (S_{i-1} \cup S_{i+1}) \neq \emptyset$*

Proof. $\Leftarrow$: We first show that the k-full tree-decomposition $(T, \mathcal{X})$ is a k-skeleton, by showing that T does not have a mergeable subtree as in Definition 3. Any subtree T' having at most one node that in T has a neighbor in $V(T) \setminus V(T')$ is by condition 2 not mergeable since we would have $|\{v : v \in X \text{ and } X \text{ a maxclique node in } T'\}| > \lfloor 3k/2 \rfloor$. Any subtree T' which is a path X, B, Y with X, B, Y and all their neighbors in T inducing a path A, X, B, Y, C will by condition 6 satisfy $B \setminus (A \cup C) \neq \emptyset$ and is thus not mergeable. Thus $(T, \mathcal{X})$ is a k-skeleton and it remains to show that it is a minimal k-skeleton. We prove by contradiction, that for any proper subtree T' of T the induced tree-decomposition $(T', \mathcal{X}')$ is not a k-skeleton. Unless the graph G' that $(T', \mathcal{X}')$ represents has at least $\lfloor 3(k-1)/2 \rfloor + 1$ vertices, $(T', \mathcal{X}')$ is not a k-skeleton. By condition 3 this means that T' must contain at least 2 maxclique nodes. We show that in any such T' there is a mergeable subtree T''. There are 5 special cases of subtrees T' to consider:

1. Suppose maxclique bags of T' are X_1, X_2, X_2', X_3 or $X_{p-1}, X_p, X_p', X_{p+1}$. In both cases the 3 maxclique bags satisfying the size constraint in condition 1 form the mergeable subtree T''.
2. T' contains a leaf X having a minsep neighbor S of degree 2 that itself has neighbor Y. By condition 3, X, S, Y makes up the mergeable subtree T''.
3. T' contains two maxclique leaves X, Y with a common minsep neighbor S. Again by condition 3 X, S, Y is the mergeable subtree.
4. Suppose in T there was a hair on minsep S_i and that T' does not contain this hair but does contain $S_{i-1}, X_i, S_i, X_{i+1}, S_{i+1}$. In this case the mergeable subtree T'' is X_i, S_i, X_{i+1} by condition 4 and Lemma 1.
5. Suppose T has a hair on maxclique X_i and that T' does not contain this hair but that T' does contain $X_{i-1}, S_{i-1}, X_i, S_i, X_{i+1}$. Since T is a special caterpillar, neither S_{i-1} nor S_i has a hair. Thus, condition 5 and Lemma 1 guarantee that either the subtree X_{i-1}, S_{i-1}, X_i or the subtree X_i, S_i, X_{i+1} is mergeable.

$\Rightarrow$: Using Lemmas 2, 3 and 4. we can describe all trees $T \in MS(k)$. There are two trees containing respectively 1 and 2 maxclique nodes except for $k = 8$ (see

Theorem 2). For the remaining trees we note that Lemmas 2, 3 and 4 together imply that for any maxclique leaf X in T with parent A we have either: A of degree 3 with the other two neighbors of A having degree 2 and not being leaves (call these leaves of type i); or A of degree 2 with parent Y of degree 3 having 3 neighbors of degree 2 with 1, 2 or 3, respectively, of these being neighbors of a leaf (leaves of type ii.1, ii.2, ii.3 respectively.) Moreover, all nodes of degree 3 in T (which is the maximum) have at least one neighbor that is a leaf or neighbor of a leaf. Thus we can use the 4 types (i, ii.1, ii.2, ii.3) as building-blocks for any tree $T \in MS(k)$. If we use a building-block of type ii.3) then there is only a unique tree possible, with 4 maxclique nodes, covered already in Theorem 2. Building blocks of type i) and ii.1) contain one leaf and two nodes needing new neighbors, while type ii.2) contains two leaves and one node needing a new neighbor. Thus, when using building-blocks of types i), ii.1) or ii.2) we must always have exactly two building-blocks of type ii.2), that will correspond to two ends of the body of a caterpillar having hairs of length 1 (type ii.1) or 2 (type i). A minsep node of degree 3 cannot be adjacent to a maxclique node of degree 3, because the maxclique hair of this minsep could then have been dropped and we would still have an induced k-skeleton. Likewise, no three consecutive maxclique nodes of the body all have a hair since then the middle hair could have been dropped and we would still have an induced k-skeleton. Thus, T is a special caterpillar.

To end the proof, it suffices to note that conditions 1-6 of the Theorem hold, since otherwise T would either have had a mergeable subtree or it would have been a k-skeleton but not minimal. For example, if condition 6 did not hold for some i then by Lemma 1 $X_i - S_i - X_{i+1}$ would have been a mergeable subtree. For space reasons we do not give the details of all cases.

4   An Algorithm That Generates k-Branches

In this section we give an algorithm generating each possible k-skeleton, which by Theorem 1 will correspond to generation of the k-branches.

Definition 7. *We get an* extended k-skeleton *by taking a k-skeleton $(T, \mathcal{X})$ and adding zero or more minsep leaves with bag-size k as neighbors of maxclique nodes of T while ensuring that each maxclique node still has a k-troika respecting its minsep neighbors. When starting with a minimal k-skeleton $(T, \mathcal{X})$ the result is an* extended minimal k-skeleton*. We define $ES(k)$ to be the set of extended k-skeletons and $EMS(k)$ to be the set of extended minimal k-skeletons.*

Recall that $MS(k)$ are the minimal k-skeletons, and note that by definition $MS(k) \subseteq EMS(k) \subseteq ES(k)$. The algorithm is organised in 3 stages with the outputs of the previous stage forming the inputs to the next stage. STAGE 1 generates $MS(k)$, STAGE 2 generates $EMS(k)$ and STAGE 3 generates $ES(k)$. Note that the extended k-skeletons $ES(k)$ have the dual property that we get a k-skeleton both if we remove all minsep leaves and also if we add a new legal maxclique leaf to each minsep leaf. For our generation algorithm this implies that generating k-branches is equivalent to generating extended k-skeletons where all leaves are maxclique nodes. The reason we generate extended k-skeletons, and not only the k-skeletons, is to be able to enforce that all eventual minsep neighbors of a maxclique node are added as soon as the maxclique node is added. This to easily satisfy the constraint that a maxclique node have a k-troika respecting its minsep neighbors.

Description of STAGE 1*:* Generation of the minimal k-skeletons $MS(k)$.
See Algorithm 1. The minimal k-skeletons on 1, 2, 4, or 6 maxclique nodes are generated by the special rules 1clique, 2clique, 4clique or 6clique respectively. The special caterpillar T in a minimal k-skeleton $(T, \mathcal{X})$ on 6 maxclique nodes is unique, $p = 6$ in Definition 6

For the minimal k-skeletons $(T, \mathcal{X})$ on more than 6 maxcliques we enter a Repeat-loop that will generate the special caterpillar T from left to right by adding in each iteration one or two new maxclique nodes to the current right end of its body. The Repeat-loop is prefixed and postfixed by special operations Start and End that add the building-blocks that in the proof of Theorem 3 are called type ii.2). Note that throughout the code the names of parameters denoting maxclique and minsep nodes are in accordance with Definition 6 and Figure 2, with certain exceptions. In particular, in the prefix operation Start$(X_1, X_2, X_2', X_3, Hair, S_3)$ the parameter $Hair$ is a maxclique leaf hair of length 2 added to maxclique X_3.

In the repeat-loop we maintain the loop invariant that S_i will be the minimal separator at the current right end of the body at which construction of the caterpillar will continue. Throughout STAGE 1 we make the assumption that when adding a new maxclique node X adjacent to some minsep node S then for any other neighbor Y of S the pair X, Y must satisfy condition 3 of Theorem 3. To not clutter the code we do not explicitly state these conditions. When adding new maxclique nodes, both here and in *Stage 3*, the syntax for the operation is ADD$(oldsep, newclique, newsep1, newsep2)$, where the two latter parameters may be missing. The *newclique* node is added as a neighbor of *oldsep* and the *newsep* nodes are added as neighbors of *newclique*. Thus, to extend the rightmost end of the body by a path S_i, X_{i+1}, S_{i+1} we use in rule I and rule II the operation ADD(S_i, X_{i+1}, S_{i+1}). In rule III we are additionally adding a hair of length two consisting of minsep B and maxclique $Hair$ to the new maxclique node X_{i+1} and express this by the two operations ADD$(S_i, X_{i+1}, B, S_{i+1})$ and ADD$(B, Hair)$. In rule IV we are additionally adding a hair W of length one to minsep node S_i and express this by the additional operation ADD(S_i, W). The boolean values HasHair and NeedsPair govern which of rule I to rule IV can be applied while ensuring that the conditions for minimal k-branches are fullfilled. HasHair is True iff the rightmost maxclique node X_i of the current body has a hair H attached to it. NeedsPair is True iff HasHair is True and the next-to-last maxclique node X_{i-1} would not be mergeable with X_i even if we had removed the hair H (in which case the next maxclique node X_{i+1} must satisfy that X_{i+1} and X_i would be mergeable if we had removed H.)

Algorithm 1. STAGE 1: Generate any $(T, \mathcal{X}) \in MS(k)$ by choosing 1,2,3,4 or 5

1: $(T, \mathcal{X}) := 1clique(X)$ s.t. Theorem 2, case $q = 1$ holds ;

2: $(T, \mathcal{X}) := 2clique(X, Y)$ s.t. Theorem 2, case $q = 2$ holds ;

3: $(T, \mathcal{X}) := 4clique(X, Y, Z, W)$ s.t. Theorem 2, case $q = 4$ holds ;

4: $(T, \mathcal{X}) := 6clique(X_1, X_2, X'_2, X_3, X'_3, X_4)$ s.t. T is the unique special caterpillar with $p = 3$ and $q = 6$ and conditions 1,2,3 in Theorem 3 hold ;

5: **begin**

  first choose a or b while ensuring that condition 1 and 2 of Theorem 3 hold;

  a: Start$(X_1, X_2, X'_2, X_3, S_3)$, $i := 3$, HasHair:= 0, NeedsPair:= 0;

  b: Start$(X_1, X_2, X'_2, X_3, Hair, S_3)$, $i := 3$, HasHair:= 1, NeedsPair:= 1;

  repeat

    if *HasHair and NeedsPair* **then** choose **rule I**;

    else if *HasHair and* **not** *NeedsPair* **then** choose **rule I,II or III**;

    else choose **rule II, III or IV**;

    rule I: ADD(S_i, X_{i+1}, S_{i+1}) s.t. $S_i \setminus (S_{i-1} \cup S_{i+1}) = \emptyset$;

    rule II: ADD(S_i, X_{i+1}, S_{i+1}) s.t. $S_i \setminus (S_{i-1} \cup S_{i+1}) \neq \emptyset$;

    rule III: ADD$(S_i, X_{i+1}, B, S_{i+1})$ and ADD$(B, Hair)$;

    rule IV: ADD(S_i, X_{i+1}, S_{i+1}) and ADD(S_i, W) s.t. $S_i \setminus (S_{i-1} \cup S_{i+1}) = \emptyset$;

    if *rule III was chosen* **then** HasHair:= 1 and

    NeedsPair:= $(S_i \setminus (S_{i-1} \cup S_{i+1}) \neq \emptyset)$;

    else HasHair:= 0 and NeedsPair:= 0;

    $i := i + 1$;

  until *body of caterpillar is finished and NeedsPair= 0* ;

  End$(S_i, X_i, X'_i, X_{i+1})$ s.t. Thm 3 (cond. 1 and 2) holds with $i = p$;

end

Description of STAGE 2: Generation of the set $EMS(k)$

For space reasons we do not show a separate algorithm in the text. The input to STAGE 2 is a minimal k-skeleton $(T, \mathcal{X}) \in MS(k)$ as generated by STAGE 1. STAGE 2 is a repeat-loop that can be exited at any time and which in each iteration adds one new minsep leaf S as neighbor of some maxclique node X of $(T, \mathcal{X})$, according to Definition 7. We must ensure that X will still have a k-troika respecting its minsep neighbors. If X already had one neighbor A then condition $OK(X, A, S)$ must hold, if it had two neighbors A, B then condition $OK(X, A, B, S)$ must hold, while if it had three neighbors then a new neighbor cannot be added. These conditions are used also in STAGE 3 and defined by: '$OK(X, A, B)$ is True iff $|X| + |A \cap B| \leq 2k$' and '$OK(X, A, B, C)$ is True iff $|A \cup B| = |A \cup C| = |B \cup C| = |X|$'.

Lemma 5. $EMS(k) = \{(T, \mathcal{X}) : \exists$ *sequence of choices in* STAGE 1 *and in* STAGE 2 *s.t.* STAGE 2 *gives as output* $(T, \mathcal{X})\}$

Description of STAGE 3: Generate the set $ES(k)$.

See Algorithm 2. As in STAGE 1 the rule adding a new maxclique node X adjacent to an existing minsep node A with new promise leaves B and C will have the syntax ADD(A, X, B, C). In case we have one or zero promise leaves the syntax is ADD(A, X, B) and ADD(A, X). The shorthand ADD$(A, X, ...)$ can be replaced by any of the 3 rules. Similarly, the shorthand $\text{OK}(X, A, ...)$ appearing right after some $\text{ADD}(A, X, ...)$ has the intepretation that any third and fourth parameters B and C of the ADD also becomes a third and fourth parameter of the OK.

Algorithm 2. STAGE 3: Takes as input some $(T, X) \in EMS(k)$ produced by STAGE 2 and builds on this to produce as output an extended k-skeleton in $ES(k)$

---

repeat
> Choose a minsep node A of T;
> **if** A a leaf with parent W having a single other neighbor S **then**
>> choose 1, 2, 3, 4 or 5;
>> 1: $\text{ADD}(A, New)$ s.t. $|W \cup New| > \lfloor 3k/2 \rfloor$;
>> 2: $\text{ADD}(A, New, B)$ s.t. $\text{OK}(New, A, B)$ and $|W \cup New| + |B \cap S| > 2k$;
>> 3: $\text{ADD}(A, New1)$ and $\text{ADD}(A, New2)$ s.t. $|New1 \cup New2| > \lfloor 3k/2 \rfloor$;
>> 4: $\text{ADD}(A, New1, B, ...)$ and $\text{ADD}(A, New2, ...)$ s.t. $\text{OK}(New1, A, B, ...)$ and $\text{OK}(New2, A, ...)$;
>> 5: $\text{ADD}(A, New, B, C)$ s.t. $\text{OK}(New, A, B, C)$;
> **else**
>> choose 6 or 7;
>> 6: $\text{ADD}(A, New, B, ...)$ s.t. OK;
>> 7: $\text{ADD}(A, New)$ s.t. $|Y \cup New| > \lfloor 3k/2 \rfloor$ for $\forall Y$ maxclique leaf with parent A;

until *done* ;
Output extended k-skeleton $(T, \mathcal{X})$, which represents a k-branch iff it has no minsep leaves;

Theorem 4. $ES(k) = \{(T, \mathcal{X}) : \exists \text{ sequence of choices of rules in the 3 stages s.t. output is } (T, \mathcal{X})\}$

5   Concluding Remarks

The results of this paper lead to a new understanding of the branchwidth parameter by defining graphs of branchwidth k through the algorithmic concept of partial k-branches. The given algorithm will generate the k-skeleton of a random edge-maximal graph of branchwidth k. This algorithm can be used to provide test instances for optimization codes based on branch-decomposition.

References

1. H.L. Bodlaender, T. Kloks and D. Kratsch. Treewidth and pathwidth of permutation graphs. *SIAM J.Computing*, 25:1305-1317, 1996.
2. H.L. Bodlaender. Treewidth: Algorithmic techniques and results. In *22nd International Symposium on Mathematical Foundations of Computer Science (MFCS)*. Vol. 1295 of *Lecture Notes in Computer Science*, p. 19–36, 1997.
3. H.L. Bodlaender and D.M. Thilikos. Graphs with branchwidth at most three. *Journal of Algorithms*, 32:167–194, 1999.

4. W. Cook and P.D. Seymour. Tour merging via branch-decompositions. *Journal on Computing*, 15:233–248, 2003.
5. E. Demaine, F. Fomin, M. Hajiaghayi, and D.M. Thilikos. Fixed-parameter algorithms for (k,r)-center in planar graphs and map graphs. In *30th Int. Colloquium on Automata, Languages, and Programming (ICALP)*. Vol. 2719 of *Lecture Notes in Computer Science*, p. 829–844, 2003.
6. F. Dorn, E. Penninkx, H.L. Bodlaender and F.V. Fomin. Efficient Exact Algorithms on Planar Graphs: Exploiting Sphere Cut Branch Decompositions. In *13th European Symposium on Algorithm (ESA)*. Vol. 3669 of *Lecture Notes in Computer Science*, p. 95-106, 2005.
7. F. Fomin and D.M. Thilikos. Dominating sets in planar graphs: Branch-width and exponential speedup. In *14th Annual ACM-SIAM Symposium on Discrete Algorithms (SODA)*, p. 168–177, 2003.
8. F. Fomin and D.M. Thilikos. A simple and fast approach for solving problems on planar graphs. In *22nd Annual Symposium on Theoretical Aspect of Computer Science (STACS)* Vol. 2996 of *Lecture Notes in Computer Science*, p. 56-67, 2004.
9. F. Fomin and D. Thilikos. Fast parameterized algorithms for graphs on surfaces: Linear kernel and exponential speedup. In *31st International Colloquium on Automata, Languages, and Programming (ICALP)*, Vol. 3142 of *Lecture Notes in Computer Science*, p. 581-592, 2004.
10. F. Fomin, F. Mazoit and I. Todinca. Computing branchwidth via efficient triangulation and blocks. In *31st Workshop on Graph Theoretic Concepts in Computer Science (WG)*, Vol. 3787 of *Lecture Notes in Computer Science*, p. 374-384, 2005.
11. T. Kloks, J. Kratochvil, and H. Müller. New branchwidth territories. *Discrete Applied Mathematics.* 145:266-275, 2005.
12. J. Kleinberg and E. Tardos. Algorithm design. *Addison-Wesley*, 2005.
13. C. Paul and J.A. Telle. Edge-maximal graphs of branchwidth k. In *International Conference on Graph Theory - ICGT*. Vol. 23 *Electronic Notes in Discrete Mathematics*, 363-368, 2005.
14. C. Paul and A. Proskurowski and J.A. Telle. Algorithm generation of graphs of branchwidht $\leqslant k$. LIRMM Technical report number RR-05047. 2005.
15. C. Paul and J.A. Telle. New tools and simpler algorithms for branchwidth. In *13th European Symposium on Algorithm (ESA)*. Vol. 3669 of *Lecture Notes in Computer Science*, p. 379-390, 2005.
16. B. Reed. Treewidth and tangles, a new measure of connectivity and some applications. In *Surveys in Combinatorics*. Vol. 241 of *London Mathematical Society Lecture Note Series* Cambridge University Press, 1997.
17. N. Robertson and P.D. Seymour. Graph minors X: Obstructions to tree-decomposition. *Journal on Combinatorial Theory Series B*, 52:153–190, 1991.
18. D. Rose. On simple characterization of k-trees. *Discrete Mathematics*, 7:317–322, 1974.

Minimal Proper Interval Completions*

Ivan Rapaport[1], Karol Suchan[2,3], and Ioan Todinca[2]

[1] Departamento de Ingeniería Matemática and Centro de Modelamiento Matemático,
Universidad de Chile, Santiago, Chile
irapapor@dim.uchile.cl
[2] LIFO, Université d'Orléans, 45067 Orléans Cedex 2, France,
{Karol.Suchan, Ioan.Todinca}@univ-orleans.fr
[3] Department of Discrete Mathematics, Faculty of Applied Mathematics,
AGH - University of Science and Technology, Cracow, Poland

Abstract. Given an arbitrary graph $G = (V, E)$ and a proper interval graph $H = (V, F)$ with $E \subseteq F$ we say that H is a *proper interval completion* of G. The graph H is called a *minimal proper interval completion* of G if, for any sandwich graph $H' = (V, F')$ with $E \subseteq F' \subset F$, H' is not a proper interval graph. In this paper we give a $\mathcal{O}(n + m)$ time algorithm computing a minimal proper interval completion of an arbitrary graph. The output is a proper interval model of the completion.

1 Introduction

Various well-known graph parameters, like *treewidth*, *minimum fill-in*, *pathwidth* or *bandwidth* are defined in terms of graph embeddings. The general framework consists in taking an arbitrary graph $G = (V, E)$ and adding edges to G in order to obtain a graph $H = (V, E \cup E')$ belonging to a specified class $\mathcal{H}$. For example, if H is chordal then it is called a *triangulation* of G. The *treewidth* can be defined as $\min(\omega(H)) - 1$, where the minimum is taken over all triangulations of G (here $\omega(H)$ denotes the maximum cliquesize of H). If instead of minimizing the cliquesize of H we minimize $|E'|$, the number of added edges, we define the *minimum fill-in* of G.

If $H = (V, E \cup E')$ is an interval (resp. a proper interval) graph, we say that H is an interval completion (resp. proper interval completion) of G. Recall that an interval graph is a *proper interval graph* if it has an interval model such that no interval is properly contained into another. The *pathwidth* of G can be defined as $\min(\omega(H)) - 1$, where the minimum is taken over all interval completions of G. The minimum number of edges that we need to add for obtaining an interval completion is called the *profile* of the graph.

Proper interval graph completions have been discussed in [11]. Independently, Kaplan et al. [11] and Cai [3] show that the problem of computing the minimum number of edges $|E'|$ such that $H = (V, E \cup E')$ becomes a proper interval

---

* Partially supported by Programs Conicyt "Anillo en Redes" (I.R.) and Ecos-Conicyt (I.R., I.T).

graph is fixed parameter tractable. The problem is adressed as the "proper interval graph completion problem", motivated by applications to genetics. The *bandwidth* of a graph is usually expressed as follows. Consider an ordering (also called layout) $\sigma = (v_1, \ldots, v_n)$ of the vertices of G. The width of the layout is $\max\{|i - j| \mid v_i, v_j \text{ adjacent in } G\}$. The bandwidth of G is the minimum width over all layouts of G. It has been proved in [10] that the bandwidth of G is also equal to $\min(\omega(H)) - 1$, the minimum being taken over all proper interval completions of G (see also Section 2 for the relationship between layouts and proper interval completions). The bandwidth problem for graphs, motivated by the bandwidth minimization problem for matrices, is one of the few graph problems NP-hard even for the class of trees [13]. Computing the bandwidth is also $W[t]$-hard for all t, thus unlikely to be fixed parameter tractable.

For each of the parameters cited above, the problem of computing the parameter is NP-hard. Obviously, for all of them, the optimal solution can be found among the *minimal* embeddings. We say that $H = (V, E \cup E')$ is a *minimal triangulation* (*minimal interval completion, minimal proper interval completion*) if no proper subgraph of H is a triangulation (interval completion, proper interval completion) of G.

Computing minimal triangulations is a standard technique used in heuristics for the treewidth or the minimum fill-in problem. The deep understanding of minimal triangulations lead to many theoretical and practical results for the treewidth and the minimum fill-in. We believe that, similarily, the study of other types of minimal completions might bring new powerfull tools for the corresponding problems.

Related work. Much research has been devoted to the minimal triangulation problem. Rose, Tarjan and Lueker propose the first algorithm solving the problem in $O(nm)$ time [16]. Several authors give different approaches for the same problem, with the same running time. Only recently this $O(nm)$ (in the worst case $O(n^3)$) time complexity has been improved by the algorithms of Kratsch and Spinrad ([12], running in $\mathcal{O}(n^{2.69})$ time) and Heggernes, Telle and Villanger ([9], running in $\mathcal{O}(n^\alpha \log n)$ time where $\mathcal{O}(n^\alpha)$ is the time needed for the multiplication of two $n \times n$ matrices). The later algorithm is the fastest up to now for the minimal triangulation problem.

A first polynomial algorithm solving the minimal interval completion problem has been given in [8]. Heggernes and Mancini [7] gave a linear time algorithm for computing a minimal embedding into split graphs.

Our result. We study the minimal proper interval completion problem. Our main result is a linear time algorithm computing a minimal proper interval completion of an arbitrary graph. One of the main tools is a special ordering of the proper interval graph, called bicompatible ordering [14]. Its role is similar to the simplicial elimination schemes for chordal graph. We define a family of orderings such that the associated proper interval graph is a minimal proper interval completion. Eventually, we give a linear-time algorithm (based on a

BFS) computing such an ordering. The ordering can be efficiently transformed into a proper interval model.

2   Definitions and Basic Results

Let $G = (V, E)$ be a finite, undirected and simple graph. Moreover we only consider connected graphs — in the disconnected case each connected component can be treated separately. Denote $n = |V|$, $m = |E|$. If $G = (V, E)$ is a subgraph of $G' = (V', E')$ (i.e. $V \subseteq V'$ and $E \subseteq E'$) we wrire $G \subseteq G'$. The *neighborhood* of a vertex v in G is $N_G(v) = \{u \mid \{u, v\} \in E\}$. Similarly, for a set $A \subseteq V$, $N_G(A) = \bigcup_{v \in A} N_G(v) \setminus A$. As usual, the subscript is sometimes omitted.

A graph G is an *interval* graph if continuous intervals can be assigned to each vertex of G such that two vertices are neighbors if and only if their intervals intersect. The family of intervals is called the *interval model* of the graph. A graph G is interval if and only is has a *clique path* CP, i.e. a path whose vertex set is the set of all maximal cliques of G, such that for each vertex v of G, the subgraph of CP induced by the maximal cliques containing v is connected. Taking this induced interval for each vertex of G yields an interval model. If an interval graph G has an interval model where no interval is properly contained in another, then the graph is called a *proper interval* graph.

Proper interval graphs can also be characterized as unit interval graphs (all intervals have equal length) or claw-free interval graphs (interval graphs without induced $K_{1,3}$). See e.g. [4] for more details. For our purpose, we use their caracterisation in terms of *bicompatible orderings*. A *perfect elimination ordering* of a graph $G = (V, E)$ is an ordering $\sigma = (v_1, v_2, \ldots, v_n)$ of V such that, for each vertex v_i, its neighbours appearing after v_i in σ induce a clique in the graph G.

Definition 1 ([14]). *Let* $G = (V, E)$ *be a graph and* $\sigma = (v_1, v_2, \ldots, v_n)$ *be an ordering of its vertices. If both* σ *and the reverse of* σ *is a perfect elimination ordering, then* σ *is called* bicompatible.

Theorem 1 ([14]). *H is a proper interval graph if and only if there exists a bicompatible ordering of its vertices.*

The following statement can be considered as an equivalent definition for bicompatible orderings. In our work we rather use this characterization.

Lemma 1 (Characterization of bicompatible orderings [14]). *Let* $H = (V, F)$ *be a proper interval graph. Then* $\sigma = (v_1, v_2 \ldots, v_n)$ *is a bicompatible ordering of H if and only if* $\{v_i, v_l\} \in F$ *implies that* $\{v_j, v_k\} \in F$ *for all* i, j, k, l, $1 \leq i \leq j < k \leq l \leq n$.

Definition 2. *A tuple of disjoint subsets of V,* $P = (P_1, \ldots, P_k)$ *whose union is exactly V is called an* ordered partition *of V. A refinement of P is an ordered partition P' obtained by replacing each set P_i by an ordered partition of P_i. We write* $P' \preccurlyeq P$.

Definition 3. *Given an ordered partition $P = (P_1, \ldots, P_k)$, any tuple $P' = (P_1, \ldots, P_j)$, with $0 \leq j \leq k$, is called a* prefix *of P. We use $V(P')$ to denote $\bigcup \{P_i \mid 1 \leq i \leq j\}$.*

In the particular case where $P = (P_1)$, we simply write P_1. Moreover if P_1 is formed by a single vertex x, we write x instead of $\{x\}$. Given two tuples $P' = (P_1, \ldots, P_k)$, $P'' = (P_{k+1}, \ldots, P_{k+l})$ we write $P' \bullet P''$ to denote their concatenation $P = (P_1, \ldots, P_k, P_{k+1}, \ldots, P_{k+l})$.

Let $\sigma = (v_1, \ldots, v_n)$ be any ordering of V. Notice that an ordering is a special case of an ordered partition.

Definition 4. *Let $G = (V, E)$ be an arbitrary graph and $\sigma = (v_1, \ldots, v_n)$ be an ordering of V. The graph $G(\sigma) = (V, F)$ is defined by*

$$F = \{\{v_j, v_k\} \mid \text{ there are } i, l \text{ such that } 1 \leq i \leq j < k \leq l \leq n \text{ and } \{v_i, v_l\} \in E\}.$$

Lemma 2. *$G(\sigma)$ is a proper interval graph.*

Proof. It is a direct consequence of Lemma 1 and Theorem 1. $\qquad\square$

Remark 1. Let $\sigma = (v_1, v_2 \ldots, v_n)$ be a bicompatible ordering of a proper interval graph $G = (V, E)$. Let $(v_{l_1}, \ldots, v_{l_j})$, where l_c is monotonically increasing, be the list of vertices v_l such that $N(v_l) \setminus N(\{v_1, \ldots, v_{l-1}\}) \neq \emptyset$. Let v_{r_c} be the last neighbor of v_{l_c} in σ, for $1 \leq c \leq j$. For each $c, 1 \leq c \leq j$ let $K_c = [v_{l_c} : v_{r_c}]$ be the set of vertices appearing between v_{l_c} and v_{r_c} in σ. The tuple $(K_1, \ldots, K_j)$ forms a clique path of $G(\sigma)$.

Theorem 2. *Let $G = (V, E)$ be an arbitrary graph and $H = (V, F)$ be a minimal proper interval completion of G. Then there is an ordering σ such that $H = G(\sigma)$.*

Proof. By Theorem 1, there is an ordering σ of V bicompatible for H. As a straight consequence of Definition 4 and Lemma 1, $E(G(\sigma)) \subseteq E(H)$. By Lemma 2, $G(\sigma)$ is also a proper interval graph. Thus, by minimality of H, we deduce that $E(G(\sigma)) = E(H)$. $\qquad\square$

Definition 5. *An ordering σ is called* nice *if $G(\sigma)$ is a minimal proper interval completion of G. Any prefix of a nice ordering is also called* nice.

3 Nice Orderings and Nice Prefixes

3.1 Choosing a First Vertex

A *module* is a set of vertices M such that for any $x, y \in M$, $N(x) \setminus M = N(y) \setminus M$. A *clique module* is a module inducing a clique. A *minimal separator* S is a set of vertices such that there exist two connected components of $G - S$ with vertex sets C and D satisfying $N(C) = N(D) = S$.

Definition 6 ([1]). *A* moplex *is a maximal clique module M such that $N(M)$ is a minimal separator of G. A vertex $v \in M$ of G is called* moplexian.

Proposition 1. *Let M be a moplex of G and $v \in M$. There exist a nice ordering σ starting with v such that the neighborhood of v in $G(\sigma)$ is exactly the neighborhood of v in G. Moreover, for any minimal interval completion H' of G such that $N_G(v) = N_{H'}(v)$, there exists an ordering σ', starting with v and such that $H' = G(\sigma')$.*

Proof. Let M be a moplex such that $v \in M$ (actually this moplex is unique) and let H be the graph obtained from G by completing $V \setminus M$ into a clique. We first show that H is a proper interval graph. Let $S = N(M)$. By definition of a moplex and by construction of H, the graph H is formed by two cliques, namely $M \cup S$ and $V \setminus M$. Their intersection is exactly S. Clearly H is an interval graph. Moreover it has no independent set of size greater that 2, in particular it has no induced $K_{1,3}$. Hence H is interval and claw-free, so H is a proper interval graph (see [4]). In particular there is a minimal proper interval completion of G contained in H.

Consider any minimal proper interval completion H' of G such that $N_G(v) = N_{H'}(v)$ (H' exists by the previous remark). By Theorem 1, there exists an ordering σ' such that $H' = G(\sigma')$. If all vertices appearing before v in σ' are elements of the module M, we can permute v and the first element of σ' without changing the graph $G(\sigma')$. Similarly, if all vertices appearing after v are in M, we reverse σ' and then permute v and the first vertex. In both cases v becomes the first vertex of σ'.

It remains to consider the case when there are two vertices $a, b \notin M$, such that $a < v < b$ in σ'. There is a path from a to b in G, such that all vertices of the path are in $V \setminus M$. Consequently there are two consecutive vertices of the path, say a' and b', such that $a' < v < b'$ in the ordering σ'. Thus $\{v, a'\}$ and $\{v, b'\}$ are edges of H'. Since $a', b' \notin M$, by construction of H' we must have $a', b' \in S$. Recall that S is a minimal separator, thus there are two connected components C and D of $G - S$ such that $N(C) = N(D) = S$. At least one of them, say C, is different from M. Let μ be a path from a' to b' in $G[C \cup \{a', b'\}]$, not using the edge $\{a', b'\}$. Like above, there are two consecutive vertices a'' and b'' of μ with $a'' < v < b''$ in σ. Hence v is adjacent in H' to both a'' and b''. At least one of a'', b'' is in C, contradicting the fact that H' has no edges between v and $V \setminus (M \cup S)$. $\square$

A moplexian vertex always exists an can be found efficiently.

Theorem 3 ([1]). *Every graph has a moplexian vertex. Such a vertex can be found in $O(n + m)$ time. More precisely, the algorithm LexBFS ends on a moplexian vertex.*

3.2   A Family of Nice Orderings

Definition 7. *Let ρ be a non-empty prefix of a vertex ordering. We denote by $\mathrm{First}(\rho)$ the first vertex in ρ having a neighbor in $V \setminus V(\rho)$. We define the strong neighborhood (denoted $\mathrm{N_S}(\rho)$), weak neighborhood $(\mathrm{N_W}(\rho))$ and non-neighborhood $\overline{\mathrm{N}}(\rho)$ as follows:*

- $N_S(\rho) = N(\text{First}(\rho)) \setminus V(\rho)$,
- $N_W(\rho) = N(V(\rho)) \setminus N_S(\rho)$,
- $\overline{N}(\rho) = V \setminus (V(\rho) \cup N_S(\rho) \cup N_W(\rho))$.

Definition 8. *We say that an ordering σ respects a prefix ρ if σ is a refinement of $\rho \bullet (N_S(\rho), N_W(\rho), \overline{N}(\rho))$.*

Our goal is to show that if ρ is a nice prefix starting with a moplexian vertex, then there is a nice ordering respecting it. This is a first step towards the extension of a nice prefix by adding a new vertex. Also note that a BFS ordering respects all its prefixes. Actually our construction of a nice ordering will be based on a BFS starting from a moplexian vertex.

Lemma 3. *Let σ and σ' be two orderings with a common prefix ρ and such that $G(\sigma') \subseteq G(\sigma)$. Suppose that σ is a refinement of $\rho \bullet (N_S(\rho), N_W(\rho) \cup \overline{N}(\rho))$. Then σ' is also a refinement of $\rho \bullet (N_S(\rho), N_W(\rho) \cup \overline{N}(\rho))$.*

Proof. Assume that both sets $N_S(\rho)$ and $N_W(\rho) \cup \overline{N}(\rho)$ are not empty, otherwise the conclusion is true for any σ' starting with ρ. Let v_f denote $\text{First}(\rho)$.

By contradiction suppose that there are two vertices $a \in N_S(\rho)$ and $b \in N_W(\rho) \cup \overline{N}(\rho)$ such that $v_f < b < a$ in the ordering σ'. Therefore $\{v_f, b\}$ is an edge of $G(\sigma')$. If $G(\sigma)$ contained the edge $\{v_f, b\}$, then there are two adjacent vertices v' and a' of G such that $v' \le v_f < b \le a'$ in the ordering σ. By definition of $v_f = \text{First}(\rho)$ we must have $v' = v_f$. Therefore $a' \in N_S(\rho)$, contradicting the fact that $N_S(\rho)$ appears before b in σ.

We conclude that the edge $\{v_f, b\}$ appears in $G(\sigma')$ but not in $G(\sigma)$. $\square$

Lemma 4. *Let σ and σ' be two orderings with a common prefix ρ and such that $G(\sigma') \subseteq G(\sigma)$. Assume that σ respects ρ and let $u \in N_W(\rho)$, $w \in \overline{N}(\rho)$. Then u appears before w in σ'.*

Proof. By contradiction, suppose that w appears before u in σ'. Let $u' \in V(\rho)$ be a neighbor of u. The edge $\{w, u'\}$ is present in $G(\sigma')$, since w is between u' and u in σ'. On the other hand, σ respects ρ, so w appears after $\rho \bullet (N_S(\rho), N_W(\rho))$. No element of ρ is adjacent in G to a vertex appearing after w in σ. By construction of $G(\sigma)$, this graph does not contain the edge $\{w, u'\}$. $\square$

Lemmas 3 and 4 directly imply the following:

Proposition 2. *Let σ and σ' be two orderings with a common prefix ρ and such that $G(\sigma') \subseteq G(\sigma)$. If σ respects ρ, then σ' also respects ρ.*

Lemma 5. *Let ρ be a non-empty prefix. Let $u, w \in N_S(\rho)$. Let σ be an ordering that respects $\rho \bullet u$. Let σ' be an ordering, with ρ as a prefix, in which w appears before u. If there is $w' \in (N(w) \cap \overline{N}(\rho)) \setminus (N(u) \cap \overline{N}(\rho))$, then the graph $G(\sigma')$ contains an edge not appearing in $G(\sigma)$.*

Proof. If w' is between $\text{First}(\rho)$ and u in σ', then by Definition 4 $\{u, w'\}$ is present in $G(\sigma')$. Else, u is between w and w' in σ' and the same holds. On the other hand, σ respects $\rho \bullet u$, so w' appears after $\rho \bullet (u, N_S(\rho \bullet u), N_W(\rho \bullet u))$. No element of $\rho \bullet u$ is adjacent in G to a vertex appearing after w' in σ. By Definition 4, $\{u, w'\}$ is not an edge of $G(\sigma)$. $\square$

Lemma 6. *Let $\sigma = (v_1, \ldots, v_n)$ be an ordering of V. Let σ' be obtained from σ by permuting vertices strictly between v_i, v_k in σ. Then every edge in the symmetric difference $E(G(\sigma')) \bar{\cup} E(G(\sigma))$ is incident to a vertex v_j between v_i and v_k in σ.*

Proof. The proof is a straightforward consequence of the construction of $G(\sigma)$ and $G(\sigma')$. □

3.3 Nice Orderings: A Sufficient Condition

Our main combinatorial result is that nice orderings can be obtained from a BFS ordering starting with a moplexian vertex, with an additional tie-break rule.

Theorem 4. *Let $G = (V, E)$ be a graph. Let $\sigma = (v_1, \ldots, v_n)$ be an ordering of V such that v_1 is a moplexian vertex and for each $1 < i < n$:*

1. *σ respects ρ, where $\rho = (v_1, \ldots, v_{i-1})$,*
2. *v_i is such that $N(v_i) \cap \overline{N}(\rho)$ is inclusion-minimal over all vertices in $N_S(\rho)$.*

Then σ is a nice ordering.

Proof. Suppppose that σ is not a nice ordering and let σ' be an ordering such that $G(\sigma')$ is a strict subgraph of $G(\sigma)$. Take σ' in order to maximize the common prefix of σ and σ'. Let $\rho = (v_1, \ldots, v_p)$ be this maximum common prefix. By construction of σ, all the edges of $G(\sigma)$ incident to v_1 are also edges of G. By Proposition 1, σ' starts with v. Consequently ρ has at least one vertex.

Let $u = v_{p+1}$ be the vertex of index $p + 1$ in σ and w be the vertex of index $p + 1$ in σ'.

By Proposition 2, σ' respects ρ.

Let σ'' be the ordering obtained from σ' by exchanging u and w. We claim that $G(\sigma'') = G(\sigma')$. By Lemma 6, any edge that might differ from $G(\sigma'')$ to $G(\sigma')$ is adjacent to a vertex between u and w, let I denote this interval. Since σ and σ' respect ρ, we have that $u, w \in N_S(\rho)$, hence $I \subseteq N_S(\rho)$. As a consequence of Lemma 5 and by the condition 2 of the theorem, $N(x) \cap \overline{N}(\rho) = N(u) \cap \overline{N}(\rho)$ for every $x \in I$. Let z be the last vertex of σ' contained in $N(u) \cap \overline{N}(\rho)$, if such a vertex exists. In particular z is also the last vertex of σ'' in $N(u) \cap \overline{N}(\rho)$.

Consider any $y \in V(I)$. Both in $G(\sigma')$ and $G(\sigma'')$, y is adjacent to all vertices of $V(\rho)$ appearing after $\text{First}(\rho)$ and has no neighbor appearing strictly before $\text{First}(\rho)$. Since $y \in N_S(\rho)$, the vertices of $\overline{N}(\rho)$ adjacent to y in $G(\sigma')$ are precisely the ones appearing before z – or this neighborhood is empty if z does not exist. The same holds for $G(\sigma'')$. Eventually, $N_S(\rho) \cup N_W(\rho)$ induces a clique both in $G(\sigma')$ and $G(\sigma'')$. Indeed the last vertex b of $N_S(\rho) \cup N_W(\rho)$ in σ' (resp. σ'') is adjacent in G to some vertex a of ρ. Since σ' and σ'' respect ρ, all the vertices of $N_S(\rho) \cup N_W(\rho)$ are in between a and b, so they form a clique. That proves that $G(\sigma') = G(\sigma'')$.

We have proved that σ'' and σ have $\rho \bullet u$ as common prefix, and $G(\sigma'') \subseteq G(\sigma)$. This contradicts the choice of σ'. □

Function IntervalModel

Input: $\sigma = (v_1, \ldots, v_n)$ - a BFS ordering of a simple connected graph G;
Output: an interval model of the graph $G(\sigma)$;
Data structures:

> r is the biggest index of a neighbor of a vertex considered so far.
> c is a counter for numbering the maximal cliques of $G(\sigma)$.
> $[v_{l_c} : v_{r_c}]$, $1 \le c \le j$ are maximal cliques of $G(\sigma)$. (see Remark 1)
> v_{l_c}, v_{r_c}, $1 \le c \le j$ are marks on the leftmost
> > and rightmost vertices of the maximal clique c.
> Q is a queue containing the numbers of maximal cliques
> > that the current vertex belongs to.

begin
$r := 1$
$c := 1$
for $i := 1$ **to** n **do**
> **if** $\max\{q \mid v_q \in N_G(v_i)\} > r$ **then**
> > $r := \max\{q \mid v_q \in N_G(v_i)\}$
> > mark v_i as v_{l_c}
> > mark v_r as v_{r_c}
> > increment c

for $i := 1$ **to** n **do**
> **if** v_i is marked as v_{l_c} **then**
> > add c at the end of the queue Q
> assign to v_i the interval $[First(Q) : Last(Q)]$
> **if** v_i is marked as v_{r_c} **then**
> > remove c from the beginning of the queue Q

$im :=$ the interval model
FixIntervalModel(σ, im)
end

Fig. 1. Algorithm Interval Model

4 The Algorithm

The algorithm is based on a BFS, see Figure 2. It creates an ordering σ of the vertices like in Theorem 4 and then returns a proper minimal model of the minimal proper interval completion $G(\sigma)$.

Theorem 5. *There is a linear time algorithm that, given an arbitrary graph G, computes a proper interval model of a minimal proper interval completion of G.*

Proof. The ordering produced by the algorithm respects the conditions of Theorem 4. Indeed, it is sufficient to notice that the function **ChooseNextVertex** chooses a vertex in $N_S(\rho)$ for the current prefix ρ, and moreover this vertex v is of minimum $d_{\overline{N}}(v)$. Since $d_{\overline{N}}(v)$ is the cardinality of $N(v) \cap \overline{N}(\rho)$, the latter is inclusion-minimal among all the elements of $N_S(\rho)$.

Let us discuss a linear time implementation of the algorithm. The choice of the first vertex can be done in linear time by Proposition 1. The main difficulty

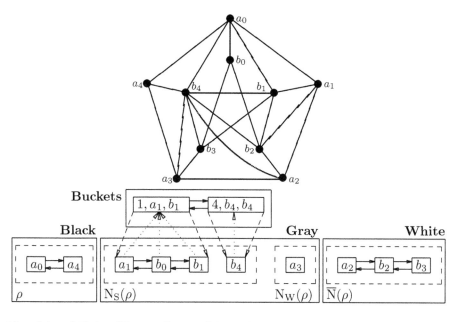

Algorithm MinimalProperIntervalCompletion

Input: a simple graph $G = (V, E)$;
Output: the proper interval model of a minimal proper interval completion of G;
Data structures:

> $mark$: each vertex is marked white (unprocessed), grey (being processed) or black (processed).
>
> Q is the queue of processed vertices.
>
> ρ is the current prefix (an ordering on the black vertices).
>
> $d_{\overline{N}}(v)$ is the number of white neighbours of the vertex v.

Function ChooseNextVertex : chooses a vertex v in the queue Q such that $v \in N_S(\rho)$ and $d_{\overline{N}}(v)$ is minimum for this property.
Function IntervalModel : computes the interval model.
begin
compute a moplexian vertex v_1 and put $\rho := (v_1)$
mark all vertices as white, mark v_1 as black
init Q with the neighbours of v_1 in G and mark these vertices as grey
compute $d_{\overline{N}}(x)$ for all vertices x
for $i := 2$ to n **do**
  $v_i :=$ **ChooseNextVertex**()
  mark v_i as black, $\rho := \rho \bullet v_i$
  compute the set N_i of white neighbours of v_i
  mark the elements of N_i as grey and add them to Q
  for each $y \in N_i$ **for** each $z \in N(y)$ **do** $d_{\overline{N}}(z) := d_{\overline{N}}(z) - 1$
IntervalModel(ρ)
end

Fig. 2. Algorithm Minimal Proper Interval Completion and data structure

is that the function **ChooseNextVertex** must work in constant time. For this purpose, the queue Q will actually be a queue of sets $(N_{j_1}, N_{j_2}, \ldots, N_{j_k})$, where N_{j_p} is the set of neighbours of v_{j_p} added to the queue when processing v_{j_p} (empty sets are not enqued). Hence $N_S(\rho)$ is the first set in the queue, and vertices are dequed from it.

In order to choose the vertex $v \in N_S(\rho) = N_{j_1}$ of minimum $d_{\overline{N}}(v)$ in constant time, we need to sort $N_S(\rho)$ by increasing $d_{\overline{N}}()$. Notice that the value of some $d_{\overline{N}}(z)$ might change during the algorithm, and we whish to update the value in constant time. We use a bucket sort with a special data structure (see [5,6] for a detailed description of the data structure, the authors use it for partition refinement algorithms). The buckets are kept as a doubly chained list (instead of the usual array). Each bucket has its value (the $d_{\overline{N}}(u)$ for the elements of the bucket), and points towards the previous and next non-empty buckets, according to their values. The vertices of a bucket are kept in a doubly chained list, and each vertex points towards the bucket to which it belongs. An example is given in Figure 2, where the bucket of value 1 contains a_1, b_0, b_1 and the bucket 4 contains b_4.

When a set N_i becomes the first set of Q, we apply a classical bucket sort on the vertices of N_i. This sort costs $\mathcal{O}(|N_i| + \max\{d_{\overline{N}}(u) \mid u \in N_i\})$. Then we construct our data structure for the buckets, within the same running time. During the whole algorithm, this initialization of the buckets costs $\mathcal{O}(n + m)$, due to the fact that the sets N_i are pairwise disjoint.

During the algorithm, we decrement the value $d_{\overline{N}}(z)$ for some vertices z (see the two last **for** loops). If z is in the set $N_S(\rho)$, we must uptade the buckets in constant time. Let B be the bucket containing z and B' be the previous bucket in the list of buckets. If the bucket B' corresponds to the value $d_{\overline{N}}(z) - 1$ (before decrementing it), we simply move z from B to B', and possibly remove B if it becomes empty. Otherwise, B' corresponds to a value strictly smaller than $d_{\overline{N}}(z) - 1$, we create a new bucket B'', of value $d_{\overline{N}}(z) - 1$, and add it to the list of buc'kets between B' and B. Thanks to our data structure, this operation can be done in linear time. Note that the total number of iterations of the two last **for** loops is at most $n + m$. Indeed, each vertex y becomes grey exactly once, thus each edge $\{y, z\}$ is visited at most twice.

The function **IntervalModel** (see Figure 1) constructs a clique path of $G(\sigma)$ like in Remark 1 and computes an interval model based on this clique path in linear time. Unfortunately the interval model obtained from the clique path is not directly a proper interval model, thus we have to mend it into a proper one. This can be done by standard techniques, see also the full version of the paper [15]. □

5   Conclusions and Perspectives

We presented a polynomial time algorithm computing a minimal proper interval completion of an arbitrary graph.

There are two very natural questions related to minimal proper interval completions that we leave open. The first would be to characterize all minimal proper interval completions, for example by describing all the orderings σ such that $G(\sigma)$ is a minimal proper interval completion of G. We point out that our algorithm cannot obtain any such ordering of the input graph. Indeed, if we consider the graph $K_{1,4}$, our algorithm chooses a simplicial vertex and completes the rest into a clique. A different minimal proper interval completion of the $K_{1,4}$ can be obtained by adding a matching to the independent set. For this particular example, we are able to construct all nice orderings, by a slightly different (and slower) technique. Roughly speaking, we can use a minimal separator S to split the graph into two parts (by partitioning the components of $G - S$ in two). We compute an ordering starting with the vertices of the minimal separator for one of the parts, then reverse it and use it as prefix to order the second part. It is tempting to ask whether this technique provides all possible nice completions.

The second question consists in exctracting a minimal proper interval completion from some non-minimal proper interval completion H of G. The naive technique would consist in checking, for each edge $e \in E(H) \setminus E(G)$, if $H - e$ is a proper interval graph. Although this ideea works for minimal triangulations and minimal split completions, in our case we have examples showing that it does not always yield a minimal proper interval completion.

References

1. A. BERRY, J. P. BORDAT, *Separability Generalizes Dirac's Theorem*. Discrete Applied Mathematics, 84(1-3): 43-53, 1998.
2. H. L. BODLAENDER, *A Linear-Time Algorithm for Finding Tree-Decompositions of Small Treewidth*. SIAM Journal on Computing, 25(6):1305-1317, 1996.
3. L. CAI, *Fixed-Parameter Tractability of Graph Modification Problems for Hereditary Properties*. Information Processing Letters, 58(4):171-176, 1996.
4. M. C. GOLUMBIC, *Algorithmic Graph Theory and Perfect Graphs*. Academic Press, 1980.
5. M. HABIB, C. PAUL, L. VIENNOT, *Partition Refinement Techniques: An Interesting Algorithmic Tool Kit*. International Journal of Foundations of Computer Science, 10(2): 147-170, 1999.
6. M. HABIB, R. M. MCCONNELL, C. PAUL, L. VIENNOT, *Lex-BFS and partition refinement, with applications to transitive orientation, interval graph recognition and consecutive ones testing*. Theoretical Computer Science, 234(1-2): 59-84, 2000.
7. P. HEGGERNES, F. MANCINI, *Minimal Split Completions of Graphs*. Proceedings of LATIN 2006, Lecture Notes in Computer Science, 3887:592-604, 2006.
8. P. HEGGERNES, K. SUCHAN, I. TODINCA,Y. VILLANGER, *Minimal Interval Completions*. Proceedings of the 13th Annual European Symposium on Algorithms - ESA 2005, Lecture Notes in Computer Science, 3669:403-414, 2005.
9. P. HEGGERNES, J. A. TELLE, Y. VILLANGER, *Computing minimal triangulations in time $O(n^{\alpha} log n) = o(n^{2.376})$*. Proceedings of the 16th Annual ACM-SIAM Symposium on Discrete Algorithms - SODA 2005, SIAM, 907-916, 2005.
10. H. KAPLAN, R. SHAMIR, *Pathwidth, Bandwidth, and Completion Problems to Proper Interval Graphs with Small Cliques*. SIAM Journal on Computing, 25(3): 540-561, 1996.

11. H. KAPLAN, R. SHAMIR, R. E. TARJAN, *Tractability of Parameterized Completion Problems on Chordal, Strongly Chordal, and Proper Interval Graphs.* SIAM Journal on Computing, 28(5): 1906-1922, 1999.

12. D. KRATSCH, J. SPINRAD, *Minimal fill in $\mathcal{O}(n^{2.69})$ time.* To appear in Discrete Applied Mathematics.

13. B. MONIEN, *The bandwidth minimization problem for caterpillars with hair length 3 in NP-complete.* SIAM Journal on Algebraic and Discrete Methods, 7:505-512, 1986.

14. B. S. PANDA, S. K. DAS, *A linear time recognition algorithm for proper interval graphs.* Information Processing Letters, 87(3): 153-161, 2003.

15. I. RAPPAPORT, K. SUCHAN, I. TODINCA, *Minimal proper interval completions.* Technical Report RR-2006-02, LIFO - University of Orléans, 2006. http://www.univ-orleans.fr/SCIENCES/LIFO/prodsci/rapports/RR2006.htm.en.

16. D. ROSE, R.E. TARJAN, AND G. LUEKER, *Algorithmic aspects of vertex elimination on graphs.* SIAM J. Comput., 5:146–160, 1976.

Monotony Properties of Connected Visible Graph Searching[*]

Pierre Fraigniaud and Nicolas Nisse

CNRS
Laboratoire de Recherche en Informatique
Université Paris-Sud
91405 Orsay, France
{pierre, nisse}@lri.fr

Abstract. Search games are attractive for their correspondence with classical width parameters. For instance, the *invisible* search number (a.k.a. *node* search number) of a graph is equal to its pathwidth plus 1, and the *visible* search number of a graph is equal to its treewidth plus 1. The *connected* variants of these games ask for search strategies that are connected, i.e., at every step of the strategy, the searched part of the graph induces a connected subgraph. We focus on *monotone* search strategies, i.e., strategies for which every node is searched exactly once. It is known that the monotone connected visible search number of an n-node graph is at most $O(\log n)$ times its visible search number. First, we prove that this logarithmic bound is tight. Precisely, we prove that there is an infinite family of graphs for which the ratio monotone connected visible search number over visible search number is $\Omega(\log n)$. Second, we prove that, as opposed to the non-connected variant of visible graph searching, "recontamination helps" for connected visible search. Precisely, we describe an infinite family of graphs for which any monotone connected visible search strategy for any graph in this family requires strictly more searchers than the connected visible search number of the graph.

Keywords: Graph Searching, Treewidth, Pathwidth.

1 Introduction

Introduced in [5,12], graph searching is a game between two players on a graph: one is playing the *fugitive* while the other is playing the *searchers*. They play alternatively. At each step: a searcher is placed at a node, or a searcher is removed from a node; then the fugitive can move from its current node u to any node v in the graph under the constraint that there is a path from u to v that does not cross any node occupied by a searcher. The fugitive is caught when a searcher

---

[*] Both authors received additional supports from the project "PairAPair" of the ACI Masses de Données, from the project "Fragile" of the ACI Sécurité Informatique, and from the project "Grand Large" of INRIA.

F.V. Fomin (Ed.): WG 2006, LNCS 4271, pp. 229–240, 2006.
© Springer-Verlag Berlin Heidelberg 2006

is placed at the node it occupies . The goal is to find, for every graph G, the minimum k such that there is a *winning* search strategy with k searchers, i.e., a strategy using k searchers that captures any fugitive in G. This minimum k is called the *search number* of the graph (see [3] for a survey on graph searching).

Two main variants of the game have been considered: visible and invisible search. In visible search [6,14], the fugitive is visible to the searchers, and they can thus adapt their search strategy according to the current position of the fugitive. The corresponding search number is called the *visible search* number, denoted by vs. In invisible search [4], the fugitive is not visible to the searchers, and thus they have to perform a blind strategy to capture the fugitive. The corresponding search number is traditionally named the *node search* number. In this paper however, we call it the *invisible search* number for it measures the ability of a team of searchers to capture an invisible fugitive. The invisible search number is denoted by is.

The importance of the search games comes from the correspondence between search numbers and standard width parameters [13], providing different interpretations of these parameters, and hence different ways of handling them. Precisely, it is known that, for any graph G:

- $\mathtt{is}(G) = \mathtt{pw}(G) + 1$ where $\mathtt{pw}(G)$ denotes the *pathwidth* of G (cf. [7]), and
- $\mathtt{vs}(G) = \mathtt{tw}(G) + 1$ where $\mathtt{tw}(G)$ denotes the *treewidth* of G (cf. [6,14]).

Monotony plays a crucial role in graph searching (cf., [11]). A search strategy is *monotone* if once a node has been cleared (a node is cleared at a step of the strategy if the fugitive cannot access to this node at this step), the fugitive cannot ever have access to this node during the rest of the search. Since a monotone search strategy finds the fugitive in a linear number of steps, it gives a polynomially checkable certificate to the decision problem corresponding to a monotone game. Hence the importance of monotony. Proving that visible and invisible search are both monotone games were two major achievements within the theory of graph searching. Precisely, [4,10] proved that if $\mathtt{is}(G) \leq k$ then there exists a winning monotone invisible search strategy using at most k searchers in G. Similarly, [14] proved that if $\mathtt{vs}(G) \leq k$ then there exists a winning monotone visible search strategy using at most k searchers in G.

Connectedness also plays an important role in graph searching, as far as practical applications are concerned (e.g., network security [1], speleological rescue [5], etc). A search strategy in a graph G is *connected* if, at any step of the strategy, the clear part of the graph (i.e., the part of the graph where the fugitive cannot stand) forms a connected subgraph of G. The minimum k for which there is a winning connected search strategy in G using at most k searchers is called the *connected search* number of G. Considering invisible or visible search defines two parameters denoted by $\mathtt{cis}(G)$ and $\mathtt{cvs}(G)$, respectively. The connectivity constraints generally implies a higher number of searchers for capturing the fugitive. The ratio connected search number over search number can however be bounded. Precisely, it is known (see [8]) that for any n-node graph G, we have

$$\mathtt{cis}(G)/\mathtt{is}(G) \leq \log n + 1 \quad \text{and} \quad \mathtt{cvs}(G)/\mathtt{vs}(G) \leq \log n + 1. \tag{1}$$

For trees, the bound for invisible search can be improved to $\mathtt{cis}(T)/\mathtt{is}(T) \leq 2$ (cf. [2]), and this bound is tight. For visible search, it trivially holds that $\mathtt{cvs}(T) = \mathtt{vs}(T)$ for any tree T.

As for standard (i.e., non-connected) search, monotony is a crucial property for connected search strategies, and it is natural to ask whether monotony holds for connected search games the same way it holds for standard search games. The answer is known to be no for invisible search. Precisely, [15] proves that there is a graph G such that any monotone connected invisible search strategy for G requires more searchers than $\mathtt{cis}(G)$. The impact of this result is important because it is a priori difficult to design non-monotone search strategies, and therefore the connected search problem seems significantly harder than the non-connected one. In particular, it is not known whether the decision problem corresponding to connected search is in NP. The good news though is that [1] proves that monotony holds for trees, i.e., for any tree T there is a winning monotone connected invisible search strategy using $\mathtt{cis}(T)$ searchers.

All these results are summarized in Table 1.

Table 1. An overview of connected graph searching

	search in arbitrary graphs	connected search in trees		connected search in arbitrary graphs	
	monotone	monotone	ratio	monotone	ratio
invisible fugitive	yes [4,10]	yes [1]	≤ 2 [2]	no [15]	$\leq \log n + 1$ [8]
visible fugitive	yes [14]	yes [trivial]	1 [trivial]	no [this paper]	$O(\log n)$ [8] $\Omega(\log n)$ [this paper]

Our results. First, we prove that the bound on the right hand side of Equation 1 is asymptotically tight when restricted to a monotone search strategy. That is, we prove that there is an infinite family of graphs such that, for any n-node graph G in this family, the number of searchers of any winning monotone connected visible search strategy for G is at least $\Omega(\mathtt{vs}(G) \log n)$.

Second, we prove that, as for the connected invisible search game, the connected visible search game is not monotone. Precisely, we describe an infinite family of graphs with arbitrarily large connected visible search number for which any monotone connected visible search strategy for any graph G in this family requires strictly more than $\mathtt{cvs}(G)$ searchers.

Due to space limitation, several proofs are omitted. They can however be found in [9].

2   The Lower Bound

It is known (cf., [8]) that for any connected n-node graph G, there exists a winning monotone connected invisible search using at most $\mathtt{tw}(G)(\log n + 1)$ searchers. Thus there exists a winning monotone connected visible search using at most $\mathtt{tw}(G)(\log n+1)$ searchers. Since $\mathtt{vs}(G) = \mathtt{tw}(G)+1$, it follows that there exists a winning monotone connected visible search using at most $\mathtt{vs}(G)(\log n+1)$ searchers. We prove that this bound is asymptotically tight.

Theorem 1. *For any n_0, there is $n \geq n_0$ and an n-node graph G such that any winning monotone connected visible search for G uses at least $\Omega(\mathtt{vs}(G) \cdot \log n)$ searchers.*

Proof. We construct an infinite family of connected graphs such that any winning monotone connected visible search for any n-node graph G in this family uses at least $c \, \mathtt{vs}(G) \log n$ searchers for some constant $c > 0$. For this purpose, we construct an infinite family $\{G_i, i \geq 1\}$ of connected graphs as follows.

We define the *scale* of length $k > 0$ to be the graph of $2k$ vertices $u_1, \ldots, u_k$, $v_1, \ldots, v_k$ where the u_i's are called *top* nodes, and the v_i's are called *bottom* nodes. There is an edge between u_i and u_{i+1} for all $i = 1, \ldots, k-1$; there is an edge between v_i and v_{i+1} for all $i = 1, \ldots, k-1$; and there is an edge between u_i and v_j for all i, j such that $|i - j| \leq 1$. The *center* of a scale of even length $2k$ is the subgraph induced by the four nodes $u_k, u_{k+1}, v_k, v_{k+1}$. The *extremities* of a scale of length k are the four nodes u_1, v_1, and u_k, v_k, respectively called the left and right extremities.

G_1 is defined as the scale of length $k = 10$, plus one node r_1 called the *root*, and connected to the two extremities u_1 and u_k of the scale. For any $i \geq 1$, the *base* of G_i is a subgraph of G_i that is a scale of even length, and the *kernel* of G_i is the center of its base. For instance, the base of G_1 is the scale of length 10, and the kernel of G_1 is the set $\{u_5, v_5, u_6, v_6\}$, where v_5 and v_6 are the bottom nodes of the kernel of G_1.

Given G_i for $i \geq 1$, we construct G_{i+1} as follows (cf. Fig. 1). Let S_i be the base of G_i (i.e., a scale of even length $2k$), and let r_i be the root of G_i. First, take a copy H of G_i. Let $u_k, u_{k+1}, v_k, v_{k+1}$ be the four nodes of the kernel of H (i.e., the center of the base of H). This kernel is replaced by a scale of length 6, that is: the edges $\{u_k, u_{k+1}\}$, $\{u_k, v_{k+1}\}$, $\{v_k, v_{k+1}\}$, and $\{v_k, u_{k+1}\}$ are removed, u_k and v_k are identified to the left extremities of the length-6 scale, and u_{k+1} and v_{k+1} are identified to the right extremities of the length-6 scale. This operation results in a scale S_{i+1} of length $2k + 4$, that becomes the base of G_{i+1}. Next, we take two copies H_1 and H_2 of G_i, and connect the two copies of r_i to the root of H, that becomes the root r_{i+1} of G_{i+1}. Finally, a complete set of connections are added between the two nodes u_k and v_k of H, and the two bottom nodes of the kernel of H_1, and a complete set of connections are added between the two nodes u_{k+1} and v_{k+1} of H, and the two bottom nodes of the kernel of H_2.

We have $|V(G_{i+1})| = 1 + 2|V(G_i)| + (|V(G_i)| - 1 + 8) = 3|V(G_i)| + 8$. Thus $|V(G_i)| = 25 \cdot 3^{i-1} - 4$.

To summarize, we have the base of G_i consisting of a scale of length $2k$ for $k = 2i+3$, with top nodes $u_1, \ldots, u_{2k}$, and bottom nodes $v_1, \ldots, v_{2k}$. The kernel of G_i is the center $\{u_k, v_k, u_{k+1}, v_{k+1}\}$ of this base. Thus the bottom nodes of this kernel are the two nodes v_k and v_{k+1}. The two nodes u_1 and u_{2k} are the top extremities of the base of G_i.

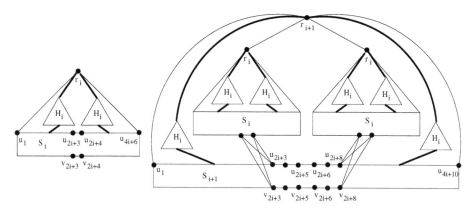

Fig. 1. Recursive construction of G_{i+1} (right) from G_i (left). The dotted lines represent sets of connections.

Claim. For any $i \geq 1$, $\mathtt{tw}(G_i) \leq 4$.

The proof of this claim can be found in [9].

Since for any graph G, $\mathtt{vs}(G) = \mathtt{tw}(G) + 1$, a consequence of Claim 2 is that $\mathtt{vs}(G_i) \leq 5$ for all $i \geq 1$. Before going further in the proof of Theorem 1, we need to present another vision of the graphs G_i. From the definition of G_{i+1}, one can check that it consists of two copies of G_j, for $j = 1, \ldots, i$, connected to a scale of length $2k = 4i + 10$ (cf. Fig. 2). This holds even for $i = 0$ by defining G_0 as the empty graph. More precisely the copies of the G_j's are placed back-to-back in order $G_1, G_2, \ldots, G_i, G_i, \ldots, G_2, G_1$. For every j, the root r_j of any of the two copies of G_j is connected to the root r_{i+1} of G_{i+1}. The two bottom nodes in the kernel of the first copy of G_j are connected to the nodes u_{2j+3} and v_{2j+3} of the base of G_{i+1}, and the two bottom nodes in the kernel of the second copy of G_j are connected to the nodes $u_{2k-(2j+2)}$ and $v_{2k-(2j+2)}$ of the base of G_{i+1}. Finally, the two extremities u_1 and u_{4i+10} of the base scale of G_{i+1} are connected to r_{i+1}. This vision of the graphs G_i's enables us to prove the following.

Claim. For any $i \geq 1$, any winning monotone connected search strategy for G_i whose two first steps consist in placing a searcher at each node v_k and v_{k+1} of the kernel of G_i uses at least $2i + 4$ searchers.

Proof. The proof is by induction on $i \geq 1$. In fact we prove that any monotone connected search strategy starting from v_k and v_{k+1} in G_i has at least $2i + 4$ searchers placed in G_i at the step before it clears the root r_i of G_i. One can easily check that the result holds for G_1, that is any monotone connected search strategy starting from v_5 and v_6 in G_1 has at least 6 searchers placed in G_1 before it clears the root r_1. Let $i \geq 1$ and let us assume that the result holds for any $1 \leq j \leq i$. Let S be a winning monotone connected search strategy for G_{i+1} starting from the two nodes v_k and v_{k+1} of the kernel of G_{i+1}. Consider G_{i+1} as depicted in Fig. 2. To access r_{i+1} from v_k and v_{k+1} in a monotone connected way, S must clear the root r_j of one of the two copies of some G_j for $1 \leq j \leq i$,

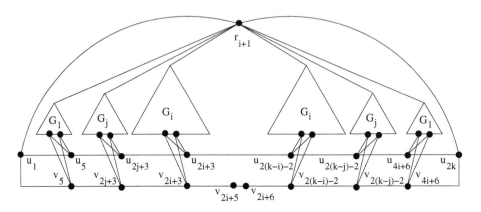

Fig. 2. Alternative definition of G_{i+1}

or one of the two extremities u_1 or u_{2k} of the base of G_{i+1}. Let R be the set of nodes composed of all the roots of the G_j's composing G_{i+1}, plus the two extremities u_1 and u_{2k}. R contains $2i + 2$ nodes. Let v be the first node in R that is cleared by S. We consider two cases.

The first case assumes that v is one of the two extremities of the base of G_{i+1}. By symmetry of G_{i+1}, one can assume, w.l.o.g., that $v = u_1$. Consider every G_j that is connected to nodes of the base between u_1 and u_k. Recall that the two bottom nodes in the kernel of G_j are connected to the nodes u_{2j+3} and v_{2j+3} of the base. There are two vertex-disjoint paths between the root r_j of the considered G_j to any of the nodes u_{2j+3} and v_{2j+3} of the base. Therefore, if less than two nodes in $V(G_j) \cup \{u_{2j+3}, v_{2j+3}\}$ are occupied by searchers, then one searcher must occupy either u_{2j+3} or v_{2j+3} because otherwise the search will not be connected. Indeed, u_{2j+3} and v_{2j+3} could be contaminated by r_j. Moreover, if one searcher only occupies u_{2j+3} or v_{2j+3}, then another searcher must occupy either u_{2j+4} or v_{2j+4} because otherwise the search will not be connected. As a consequence, for any $1 \leq j \leq i$, at least two nodes of $V(G_j) \cup \{u_{2j+3}, v_{2j+3}, u_{2j+4}, v_{2j+4}\}$ are occupied by searchers. Moreover, two searchers must occupy nodes in $\{u_j, k \leq j \leq 2k\} \cup \{v_j, k \leq j \leq 2k\}$ to avoid recontamination of v_k and v_{k+1} from u_{2k}. Finally, at least four searchers are occupying nodes in $\{u_1, v_1, u_2, v_2, u_3, v_3, u_4, v_4\}$ to connect u_1 with the clear part of G_{i+1}. This yields a total of at least $2i + 6$ searchers in the graph when u_1 is cleared, hence S uses at least $2(i + 1) + 4$ searchers in G_{i+1}.

The second case assumes that the first node $v \in R$ that is cleared by S is the root of some G_j, $1 \leq j \leq i$. We prove that S uses at least $2(i + 1) + 4$ searchers in G_{i+1}. The proof of this case can be found in [9]. This completes the induction step, and thus the proof of the claim.   ◇

Let G be connected graph, and let $e = \{u, v\} \in E(G)$. We define the *symmeric* graph of G with respect to e as the graph obtained from two copies of G linked by a set of complete connections between the four nodes resulting from the two copies of $\{u, v\}$. The symmetric of G with respect to $e = \{u, v\}$ is denoted by $G^*_{u,v}$. The K_4 connecting the two copies of G in $G^*_{u,v}$ is called the *center* of $G^*_{u,v}$.

Claim. Let G be a connected graph, and let $\{u, v\} \in E(G)$. Let k be the minimum number of searchers required to clear $G^*_{u,v}$ by a monotone connected visible search strategy. There exists a monotone connected visible search strategy for G using at most k searchers, and whose two first steps consist in placing a searcher at u and a searcher at v.

The proof of this claim can be found in [9].

For any $i \geq 1$, let $\mathcal{G}_i$ be the symmetric of G_i with respect to $\{v_k, v_{k+1}\}$ where v_k and v_{k+1} are the two bottom nodes of the kernel of G_i. We have $|V(\mathcal{G}_i)| = n_i = 2(25 \cdot 3^{i-1} - 4)$. We have $\mathtt{tw}(\mathcal{G}_i) \leq \max\{\mathtt{tw}(G_i), 3\}$ by connecting a bag containing $\{v_k, v_{k+1}\}$ in the tree-decomposition of the first copy of G_i with a bag containing $\{v_k, v_{k+1}\}$ in the tree-decomposition of the second copy of G_i by a path of length two containing a 4-node bag in the middle with two copies of v_k and two copies of v_{k+1}. Hence, from Claim 2, $\mathtt{tw}(\mathcal{G}_i) \leq 4$, and thus $\mathtt{vs}(\mathcal{G}_i) \leq 5$. On the other hand, by combining Claim 2 with Claim 2, we get that any winning monotone connected visible search strategy for $\mathcal{G}_i$ uses at least $2i + 4$ searchers. Therefore, any winning monotone connected visible search strategy for $\mathcal{G}_i$ uses at least $2 \log_3(\frac{\frac{n_i}{2}+4}{25}) + 6$ searchers. $\qquad\square$

3   Monotony

In this section, we prove that the connected visible search game does not satisfy the monotony property.

Theorem 2. *For any $k \geq 4$, there exists a graph G such that $\mathtt{cvs}(G) = 4k + 1$ and any winning monotone connected visible search strategy uses at least $4k + 2$ searchers.*

Proof. The proof is constructive. For the construction of the graphs mentioned in the statement of the theorem, we reuse the family $\{G_i, i \geq 1\}$ introduced for proving Theorem 1. The intuition of the proof is the following. Consider the graph $I^{(k)}$ depicted in Figure 3. We will show that the symmetric of this graph with respect from $\{u, v\}$ cannot be cleared optimally by a monotone search strategy. In this figure, the graphs E and F are two copies of a graph G_i. Roughly, the placements of these graphs force the strategy to clear them from nodes D and B. We show that it is not possible to do that with the minimal number of searchers in a monotone way.

Claim. There exists a connected visible search strategy for G_i, using at most 5 searchers, and starting from r_i (i.e., the first step of the search consists in placing a searcher at r_i, and the strategy clears the graph by expanding from r_i).

The proof of this claim can be found in [9].

Let P_n be the n-node path. Let $P_{k,n}$ be the graph obtained by replacing every vertex of P_n by a complete graph on k vertices, and replacing every edge of P_n by a perfect matching between the complete graphs corresponding to the two extremities of the edge. A graph $P_{k,n}$ is called a *clique-path*.

Claim. For any $n \geq 1$ and any $k \geq 1$:

- There exists a connected visible search strategy for $P_{k,n}$ using at most $k+1$ searchers, and starting from any vertex of the clique corresponding to an extremity of P_n.
- If $n \geq k+1$, then any monotone connected visible search strategy for $P_{k,n}$, using at most k searchers, and starting from any vertex of the clique corresponding to an extremity of P_n cannot clear any vertex of the clique at the other extremity of $P_{k,n}$.

The proof is straightforward and is thus omitted.

For $k \geq 1$, let $I^{(k)}$ be the graph represented in Fig. 3. This representation uses the following coding:

- A black point represents a vertex.
- A circle represents a clique with the indicated number of vertices.
- A thin line between two vertices represents an edge.
- A thin line between a vertex x an a clique represents an edge between x and a vertex of the clique;
- A double line between two cliques represents a perfect matching between them if they are of same size, or between the smallest one and a sub-clique of the largest one if they are of different size.
- a double dotted line between two cliques of same size s represents a path of cliques of size s linked by perfect matchings.
- The graphs K_A, K_B, K_C and K_D are pairwise disjoint k-cliques, all subgraphs of the clique K of size $4k+1$, and extremities of clique-paths.
- The subgraphs E and F are isomorphic to $G_{\lceil 3k/2 \rceil}$ (the marked nodes are the root, and the two bottom nodes of the kernel of $G_{\lceil 3k/2 \rceil}$).

Claim. For any $k \geq 1$, there exists a connected visible search strategy for $I^{(k)}$, starting from u and v, and using at most $4k+1$ searchers.

Proof. The following (non-monotone) strategy uses $4k+1$ searchers. Place searchers at u and v, and use $k+1$ searchers to clear the clique-path leading to A. Let P be a shortest path from A to B going through the central clique K. Place a searcher at every vertex of P, using $2k+3$ searchers (in addition to the k searchers occupying nodes in A). If the fugitive is in the subgraph E, then, from Claim 3, one can use 5 searchers to clear E starting from its root. Thus we assume that the fugitive is not in E. Remove all searchers but $k+1$ searchers occupying A and B, thus E remains isolated. (Note that the strategy is not monotone because of this step). Use the $3k$ remaining searchers to clear the clique-path between B and C (cf. point 1 of Claim 3). After this step, k searchers

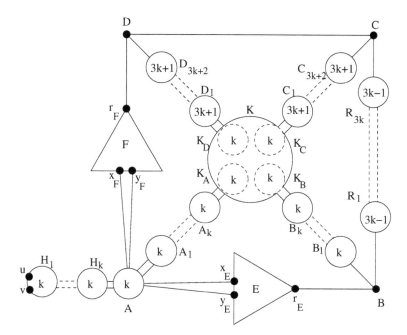

Fig. 3. The graph $I^{(k)}$

occupy vertices of A, one searcher occupies B and one searcher occupies C. Place a searcher at D. If the fugitive is in the subgraph F, then, from Claim 3, one can use 5 searchers to clear F starting from its root. Thus we assume that the fugitive is not in F. Use the k searchers at A, plus one extra searcher, to clear the clique-path between A and K_A. At this step, k searchers occupy vertices of K_A, and three searchers occupy B, C, and D. Let us place k searchers at K_D. If the fugitive is in one of the cliques D_i, then remove all searchers but those occupying K_D and D, and use the k searchers at K_D and the $3k$ remaining searchers to clear the clique-path between K_D and D. Thus we assume that the fugitive is not in one of the cliques D_i. Place k searchers at K_C. If the fugitive is in one of the cliques C_i, then remove all searchers but those occupying K_C and C, and use the k searchers at K_C and the $3k$ remaining searchers to clear the clique-path between K_C and C. Thus we assume that the fugitive is not in one of the cliques C_i. Use the searcher at B, and the k remaining searchers to clear the clique-path from B to K_B. At this point, $4k$ searchers occupy vertices of K_A, K_B, K_C and K_D. Use the remaining searcher to clear the last vertex of K. The fugitive is caught, which concludes the strategy.   ◇

Note that, since $I^{(k)}$ contains a $4k + 1$-clique, the strategy above is optimal.

Claim. For any $k \geq 4$, any winning monotone connected visible search strategy for $I^{(k)}$ starting from u and v uses at least $4k + 3$ searchers.

Proof. The proof is inspired from the non-monotony proof for connected invisible search in [15]. In the following, we say that two paths P and P' between a vertex

v and a clique are vertex-disjoint if $P \cap P' \in \{v\}$. Let us consider a winning monotone connected visible search strategy S for $I^{(k)}$, starting from u and v.

Let us first assume that the root r_E of E is cleared before vertex B is cleared. Let s be the step at which r_E is cleared in S. Let P be a clear path between u and r_E, and let P' be the subpath of P from u to a vertex in A. Since there are k vertex-disjoint paths between B and P', all passing through the clique K of $I^{(k)}$, k searchers have to guard these paths until step s to avoid recontamination. Moreover, from Claim 2, $G_{\lceil 3k/2 \rceil}$ cannot be cleared by a monotone connected visible search strategy starting from x_E and y_E using less than $3k + 4$ searchers. Thus if r_E is cleared before B then S needs at least $4k + 4$ searchers. Similarly, one can prove that if r_F is cleared before D then S needs at least $4k + 4$ searchers.

Thus, for S to use less searchers, B must be cleared before r_E, and D must be cleared before r_F. Thus, there is a vertex in K_A that is cleared before any of the vertices B, C, and D. Let x be the first vertex of K_A to be cleared by S, say at step s'. (Note that, while none of the vertices B, C and D are cleared, they belong to the same component of the contaminated part, and thus the fact that the fugitive is visible does not help to clear any of these vertices).

Let P_0 be a clear path between u and x at step s'. Let P_1 (resp., P_2) be the subpath of P_0 that goes from u to A (resp., from A_1 to x).

Let us assume that, among B, C, and D, D is the first vertex to be cleared in S. Let $s'' > s'$ be the step when D is cleared. Let P'_1 and P'_2 be two vertex-disjoint paths from r_E to two distinct nodes of P_1. Let P'_3 and P'_4 be two vertex-disjoint paths from r_F to two distinct nodes of P_1 that are as well pairwise distinct from the two extremities of P'_1 and P'_2. Finally, let $P'_5, \ldots, P'_{k+4}$ be k vertex-disjoint paths from B to k distinct nodes of P_2. Since $k \geq 4$, these $k + 4$ paths can be chosen pairwise vertex-disjoint, and disjoint from any clique D_i. Thus, for any $1 \leq i \leq k + 4$, and for any step in $[s', s'']$, there must be a distinct searcher occupying a vertex of P'_i to avoid recontamination of P_0 from r_E, r_F, or B. Point 2 of Claim 3 says that, starting from a vertex of D_1, clearing a vertex of D_{3k+2} in a monotone connected visible way requires at least $3k + 2$ searchers. Hence the total number of searchers used by S is at least $4k + 6$.

Thus, for S to use less searchers, D should not be the first vertex among B, C, and D to be cleared. Similarly, one can prove that for S to use less searchers, C should not be the first vertex among B, C, and D to be cleared.

Thus, for S to use less searchers, B must be, among B, C, and D, the first node to be cleared by S. Let $s'' > s'$ be the step when B is cleared. At this step, there is a clear path from x to B, through the cliques B_i — recall that we are assuming that B is cleared before r_E. (Note that while C and D are not cleared, both these vertices belong to the same component of the contaminated part, and thus the fact that the fugitive is visible does not help to clear these vertices).

Let P_3 be a clear path from x to B at step s''.

We now consider the two cases depending on whether D is cleared before C, or the other way around.

The first case assumes that D is cleared before C by S. Let $s''' > s''$ be the first step when a searcher is placed at D. Let P'_1 and P'_2 be two vertex-disjoint paths from r_F to two distinct nodes of P_1, and let $P'_3, \ldots, P'_{k+2}$ be k vertex-disjoint paths from C to k disjoint nodes of P_2. Since $k \geq 2$, the $k+2$ paths $P'_1, \ldots, P'_{k+2}$ can be taken pairwise vertex-disjoint, and disjoint from any D_i clique. Thus, for any $1 \leq i \leq k+2$, and for any step in $[s', s''']$, there must be a searcher at a vertex of P'_i to avoid recontamination of P_0 from r_F or C. Point 2 of Claim 3 says that, starting from a vertex of D_1, clearing a vertex of D_{3k+2} in a monotone connected visible way requires at least $3k + 2$ searchers. Hence the total number of searchers used by S is at least $4k + 4$.

The second case assumes that C is cleared before D by S. Node C can be reached in two different manners: either along the clique-path from C_1 to C_{3k+2}, or along the clique-path from R_1 to R_{3k}. We consider these sub-cases separately. We prove that in both cases, the total number of searchers used by S is at least $4k + 3$. The proof of these cases can be found in [9].

Therefore, the monotone connected visible strategy S for $I^{(k)}$ uses at least $4k + 3$ searchers.   ◇

Let $k \geq 4$. Let $G = I_{u,v}^{(k)*}$ be the symmetric of $I^{(k)}$ with respect to the edge $\{u, v\}$. From Claim 3, there exists a connected visible search strategy for $I^{(k)}$, starting from u and v, and using at most $4k + 1$ searchers. Therefore $\mathsf{cvs}(G) \leq 4k + 1$. On the other hand, Claim 3 states that any winning monotone connected visible search strategy for $I^{(k)}$ starting from u and v uses at least $4k + 3$ searchers. By Claim 2, this implies that any winning monotone connected visible search strategy for G uses at least $4k + 3$ searchers, that is strictly more than $\mathsf{cvs}(G)$. This completes the proof of the theorem.   □

The graphs used in the proof of Theorem 2 have a connected visible search number equal to $4k + 1$ for $k \geq 4$, thus at least 17. We can however design examples with smaller search number.

4 Conclusion

In this paper, we first prove that the connectedness requirement for monotone visible search leads to a logarithmic factor in the number of searchers needed. Our second result is that the connected visible search is not monotone. A quick glance at Table 1 indicates that our results combined with the previous results in this field let only one problem to be solved, as far as connected search is concerned. Namely: is the bound on the left hand side of Equation 1, i.e., $\mathsf{cis}(G)/\mathsf{is}(G) \leq O(\log n)$, tight? In [2], the authors express their belief that, for any graph G, $\mathsf{cis}(G)/\mathsf{is}(G) \leq 2$. That is, the worst case for connected invisible search is actually reached for trees. Up to now, no one was able to prove or disprove this belief.

We also want to rise the question of minimality for counter examples to monotony of connected search games. Precisely, what is the minimum k such that there is a graph G with $\mathsf{cvs}(G) = k$ for which any winning monotone connected visible search strategy uses more than k searchers. Trivially, $k \geq 3$. Moreover, we prove that $k \leq 4$ [9]. The same question seems far more complex in the context of invisible search (i.e., node search). Indeed, the minimum value that is known for this setting is... $k = 281$ (cf. [15]). Is it possible to design counter examples with smaller connected search numbers?

Finally, what is the complexity of the decision problems "$\mathtt{cis}(G) \leq k?$" and "$\mathtt{cvs}(G) \leq k?$". Both are known to be NP-hard, but are they in NP?

References

1. L. Barrière, P. Flocchini, P. Fraigniaud, and N. Santoro. Capture of an intruder by mobile agents. In 14th ACM Symp. on Parallel Algorithms and Architectures (SPAA), pages 200-209, 2002.
2. L. Barrière, P. Fraigniaud, N. Santoro, and D. Thilikos. Connected and Internal Graph Searching. In 29th Workshop on Graph Theoretic Concepts in Computer Science (WG), Springer-Verlag, LNCS 2880, pages 34–45, 2003.
3. D. Bienstock. Graph searching, path-width, tree-width and related problems (a survey). DIMACS Series in Discrete Mathematics and Theoretical Computer Science 5, pages 33-49, 1991.
4. D. Bienstock and P. Seymour. Monotonicity in graph searching. Journal of Algorithms 12, pages 239-245, 1991.
5. R. Breisch. An intuitive approach to speleotopology. Southwestern Cavers VI(5), pages 72-78, 1967.
6. N. D. Dendris, L. M. Kirousis, and D. M. Thilikos. Fugitive search games on graphs and related parameters. Theoretical Computer Science vol. 172, No. 1, pages 233-254, 1997.
7. J. A. Ellis, I.H. Sudborough, J.S. Turner. The Vertex Separation and Search Number of a Graph Information and computation 113, pages 50-79, 1994.
8. P. Fraigniaud and N. Nisse. Connected Treewidth and Connected Graph Searching. Proceedings of Latin American Theoretical Informatics Symposium (LATIN), LNCS 3887, pages 479-490, 2006.
9. P. Fraigniaud and N. Nisse. Monotony Properties of Connected Visible Graph Searching. Technical Report LRI-1456, University Paris-Sud, France, Jul. 2006.
10. A. LaPaugh. Recontamination does not help to search a graph. Journal of the ACM 40(2), pages 224-245, 1993.
11. N. Megiddo, S. Hakimi, M. Garey, D. Johnson and C. Papadimitriou. The complexity of searching a graph. Journal of the ACM 35(1), pages 18-44, 1988.
12. T. Parson. Pursuit-evasion in a graph. Theory and Applications of Graphs, Lecture Notes in Mathematics, Springer-Verlag, pages 426-441, 1976.
13. N. Robertson and P. D. Seymour. Graph minors II, Algorithmic Aspects of Tree-Width. Journal of Algorithms 7, pages 309-322, 1986.
14. P. Seymour and R. Thomas. Graph searching and a min-max theorem for tree-width, J. Combin. Theory Ser. B, 58, pages 22-33, 1993.
15. B. Yang, D. Dyer, and B. Alspach. Sweeping Graphs with Large Clique Number. In 5th International Symposium on Algorithms and Computation (ISAAC), Springer, LNCS 3341, pages 908-920, 2004.

Finding Intersection Models of Weakly Chordal Graphs

Martin Charles Golumbic[1], Marina Lipshteyn[1], and Michal Stern[1,2]

[1] Caesarea Rothschild Institute, University of Haifa, Haifa, Israel
[2] The Academic College of Tel-Aviv - Jaffa, Tel-Aviv, Israel

Abstract. We first present new structural properties of a two-pair in various graphs. A two-pair is used for characterizing weakly chordal graphs. Based on these properties, we prove the main theorem: a graph G is a weakly chordal $(K_{2,3}, \overline{P_6}, \overline{4P_2}, \overline{P_2 \cup P_4}, H_1, H_2, H_3)$-free graph if and only if G is an edge intersection graph of subtrees on a tree with maximum degree 4. This characterizes the so called $[4, 4, 2]$ graphs. The proof of the theorem constructively finds the representation. Thus, we obtain a algorithm to construct an edge intersection model of subtrees on a tree with maximum degree 4 for such a given graph. This is a recognition algorithm for $[4, 4, 2]$ graphs.

1 Introduction

The chordality of a graph plays a fundamental role in graph theory. The class of chordal graphs is widely investigated. One of the reasons is that the class has a natural intersection model and hence a concise tree representation. The tree representation can be constructed in linear time (see e.g. [6,11]), and the tree is called a clique tree since each node of the tree corresponds to a maximal clique of the chordal graph.

In many real world applications, the intersection representation of a graph is more important than the graph itself. In [8], [9], the intersection representations of a graph on a tree is generally defined as follows. An (h, s, t)-representation consists of a collection of subtrees of a tree, such that (i) the maximum degree of T is at most h, (ii) every subtree has maximum degree at most s, and (iii) there is an edge between two vertices in the graph if and only if the corresponding subtrees in T have at least t vertices in common. Notation of ∞ here means that no restriction is imposed. The class of graphs that have an (h, s, t)-representation is denoted by $[h, s, t]$. It is well known ([2,3,12]) that the chordal graphs correspond to $[\infty, \infty, 1]$, which was strengthened in [10] and [9], respectively, to be equivalent to $[3, 3, 1]$ and $[3, 3, 2]$. Interval graphs are also $[2, 2, 1]$ graphs. There are other papers that study $[h, s, t]$ graphs, for specific values of h, s and t, although without using this notion.

The class of weakly chordal graphs is also well studied and has number of known applications. Our main motivation in this paper is to find an $[h, s, t]$ class of graphs that corresponds to weakly chordal graphs. Our general long

F.V. Fomin (Ed.): WG 2006, LNCS 4271, pp. 241–255, 2006.

term motivation is to find an (h, s, t)-representations of holes and anti-holes. In this paper we prove that weakly chordal graphs with a finite set of forbidden subgraphs corresponds to $[4, 4, 2]$ graphs.

Our result bridges between characterization of a graph based on its structural properties and characterization of a graph based on its intersection model. Significantly important is that we present new structural properties of a two-pair in various graphs. In particular, a two-pair is used for a characterization of weakly chordal graphs.

We first prove the new structural properties of two-pairs in Section 3. The main theorem of this paper is proved in Section 4: A graph G is a weakly chordal $(K_{2,3}, \overline{P_6}, \overline{4P_2}, \overline{P_2 \cup P_4}, H_1, H_2, H_3)$-free graph if and only if the graph G has a $(4, 4, 2)$-representation. The proof of this theoretical result is based on the structural properties given in Section 3 and on forbidden structures of $[4, 4, 2]$ graphs shown in Section 4.1. Moreover, the proof constructively finds a $(4, 4, 2)$-representation. Thus, in addition, in Section 4.2, we provide an algorithm to construct an edge intersection model of subtrees on a tree for a given graph.

2 Preliminaries

All standard definitions can be found in [1,4,6,11].

Consider an undirected graph $G = (V, E)$. A sequence $[v_1, \ldots, v_k]$ of distinct vertices is a *path* in G if (v_1, v_2), (v_2, v_3), $\ldots$, $(v_{k-1}, v_k) \in E$. These edges are called the edges of the path. The length of the path is the number $k - 1$ of its edges. A closed path $[v_1, \ldots, v_k, v_1]$ is called a *cycle* if in addition $(v_k, v_1) \in E$. A *chord* of a cycle $[v_1, \ldots, v_k, v_1]$ is an edge between two vertices of the cycle that is not an edge of the cycle. A cycle is *chordless* if it contains no chords. Trivially, a triangle has no chord, so we refer to a chordless cycle in this work as having length strictly greater than 3. We denote by C_k the chordless cycle on k vertices, and we always assume $k > 3$. An undirected graph G is a *chordal (triangulated)* graph, if every cycle in G of length strictly greater than 3 possesses a chord, i.e., there is no chordless cycle C_k, $k \geq 4$, in G.

A graph G is a *weakly chordal* graph if neither G nor its complement $\overline{G}$ have an induced subgraph C_k, $k \geq 5$. Weakly chordal graphs satisfy the hereditary property, i.e., any induced subgraph of a weakly chordal graph is also weakly chordal.

A subset S of vertices of a connected graph G is called a *separator* if G_{V-S} is not connected. A separator S is called an (a, b)-*separator* if a and b are in different connected components of G_{V-S}. The set S is a *minimal* (a, b)-*separator* if S is an (a, b)-separator and no proper subset of S is an (a, b)-separator. Finally, a separator S is a *minimal separator* if there is some pair $\{a, b\}$ such that S is a minimal (a, b)-separator.

A *two-pair* in a graph G is a pair of vertices $\{x, y\}$, such that every chordless path between x and y contains exactly two edges.

Clearly, the common neighborhood of a two-pair $\{x, y\}$ is a minimal (x, y)-separator, which we denote by $Sep(x, y)$.

Theorem 1. [7] *A graph G is weakly chordal if and only if every induced sub-graph of G either has a two-pair or is a clique.*

We denote by $\langle \mathcal{S}, T \rangle$ an (h, s, t)-representation of a graph, where $\mathcal{S}$ is a collection of subtrees on a host tree T. Every vertex x in G corresponds to a subtree S_x in $\langle \mathcal{S}, T \rangle$, and we define $\mathcal{E}_x$ to be the set of edges of S_x in T. Every subset $X \subseteq V$ in G corresponds to a collection of subtrees $\mathcal{S}_X$ and we define $\mathcal{E}_X$ to be the set of edges of $\mathcal{S}_X$ in T.

The chordless cycle C_n is an $[\infty, \infty, 2]$ graph. The representation of C_n is called a pie. A *pie* is a star subgraph of T with n edges (a_0, b), (a_1, b), ..., (a_{n-1}, b), such that each "slice" $(a_i, b) \cup (a_{i+1}, b)$, for $i = 0, 1, \ldots, n - 1$, is contained in a different member of $\mathcal{S}$. (Addition is assumed to be modulo n). The vertex b is called the *center of the pie*. The following Theorem shows that this is essentially the only representation for C_n, which generalizes a result of [5] for $[\infty, 2, 2]$.

Theorem 2. *If an $[\infty, \infty, 2]$ graph G contains a chordless cycle $C = (x_0, x_1, \ldots, x_{n-1}, x_0)$ $(n \geq 4)$, then T contains a pie on these n vertices.*

Proof. Let S_i be the subtree in T corresponding to x_i. Choose an edge $e_i \in S_i \cap S_{i+1}$ and let c_i be an endpoint of e_i.

First we prove that for all i and j, S_i and S_j share a common vertex. Suppose S_i and S_j do not share a vertex for some i and j, so in particular $|i - j| \geq 2$. Let P be a path in T from c_i to c_j. On one hand, since T is a tree, P is contained in the subtree $S_{i+1} \cup \cdots \cup S_j$ and, since S_i and S_j do not share a vertex, there exists an edge $e \in (P - S_i - S_j) \cap S_k$ for some $i < k < j$. On the other hand, P is contained in the subtree $S_{j+1} \cup \cdots \cup S_i$, but the edge e cannot be in any S_l for $j + 1 \leq l < i$. Contradiction!

Finally, since subtrees in a tree have Helly number 2, there is a vertex $b \in S_i$ for all $0 \leq i < n$. Choosing vertices a_i, such that $(a_i, b) \in S_i \cup S_{i+1}$ we obtain a pie which realizes C. $\qquad\square$

3   New Structural Properties of a Two-Pair

In this section, we show new interesting structural properties of a two-pair in specific families of graphs. Our main motivation to investigate these properties is that two-pairs characterize weakly chordal graphs.

Given a two-pair $\{x, y\}$ in G, let $Z = \{z \in V | z \notin Sep(x, y),\ z \neq x, y,$ and z has two non-adjacent neighbors in $Sep(x, y)\ \}$. We denote by Z_x the vertices in Z that are adjacent to x, and Z_y to be the vertices in Z that are adjacent to y. The vertices in the set $\{x\} \cup \{y\} \cup Sep(x, y)$ are called the *core* vertices. Therefore, the vertices of a graph G can be partitioned into core and non-core vertices. The vertices in the set $Z \cup \{x\} \cup \{y\} \cup Sep(x, y)$ are called *essential*. Therefore, the vertices of a graph G can be partitioned into essential and non-essential vertices. Clearly, a core vertex is also an essential vertex. Recall that $N(v) = \{u | (v, u) \in E(G)\}$ and $N[v] = \{v\} \cup N(v)$. We denote by $N[X] = \{N[v] | v \in X\}$ for $X \subseteq V$. We denote by $N'[v] = \{v\} \cup (N(v) \cap Sep(x, y))$.

Property 3. *Let $\{x, y\}$ be a two-pair in a $K_{2,3}$-free graph G. Then there are no three independent vertices in $Sep(x, y)$, since otherwise they form an induced $K_{2,3}$ together with the vertices x and y.*

Property 4. *Let $\{x, y\}$ be a two-pair in a $\overline{P_2 \cup P_4}$-free graph G. Then there is no chordless path of four vertices in $Sep(x, y)$, since otherwise they form an induced $\overline{P_2 \cup P_4}$ together with the vertices x and y.*

Property 5. *Let $\{x, y\}$ be a two-pair in a $K_{2,3}$-free graph G and let $z \in Z$, then z must be adjacent to either x or y. Moreover, Z_x and Z_y are disjoint sets and are not connected by an edge in G.*

Lemma 6. *Let $\{x, y\}$ be a two-pair in a $(K_{2,3}, \overline{P_2 \cup P_4}, \overline{P_6})$-free graph G: (i) Z_x and Z_y are cliques in G, (ii) Every vertex in Z_x and every vertex in Z_y is adjacent to every vertex in $Sep(x, y)$, (iii) Each vertex in $Z_x \cup x$ has the same essential neighbors in G, and each vertex in $Z_y \cup y$ has the same essential neighbors in G.*

Let $\{x, y\}$ be a two-pair. We now define the sets $\mathcal{C}_1, \mathcal{C}_2, \mathcal{C}_3, \mathcal{C}_4, \mathcal{C}_5$ with respect to $Sep(x, y)$, as illustrated in Figure 1, and whose structure will be shown below.

(I) If $Sep(x, y)$ is a clique, then $\mathcal{C}_5 = Sep(x, y)$ and $\mathcal{C}_i = \emptyset, 1 \leq i \leq 4$.
   Otherwise, $Sep(x, y)$ is not a clique and there exist two non-adjacent vertices $s_1, s_2 \in Sep(x, y)$. Let $\mathcal{C}_1 = N'[s_1] \setminus N'[s_2]$ and let $\mathcal{C}_2 = N'[s_2] \setminus N'[s_1]$. Note that $\mathcal{C}_1$ and $\mathcal{C}_2$ are non-empty since $s_1 \in \mathcal{C}_1$ and $s_2 \in \mathcal{C}_2$.
(II) If $N'[s_1] \cap N'[s_2] = \emptyset$, then $\mathcal{C}_3 = \mathcal{C}_4 = \mathcal{C}_5 = \emptyset$.
(III) If $N'[s_1] \cap N'[s_2]$ is a clique, then $\mathcal{C}_3 = \mathcal{C}_4 = \emptyset$ and $\mathcal{C}_5 = N'[s_1] \cap N'[s_2]$.
(IV) If $N'[s_1] \cap N'[s_2]$ is not a clique, then there exist two non-adjacent vertices $s_3, s_4 \in N'[s_1] \cap N'[s_2]$. In this case, $\mathcal{C}_3 = N'[s_3] \setminus N'[s_4]$, $\mathcal{C}_4 = N'[s_4] \setminus N'[s_3]$ and $\mathcal{C}_5 = N'[s_3] \cap N'[s_4]$.

We now prove the following claims on the structural properties of the sets $\mathcal{C}_1, \ldots, \mathcal{C}_5$ in G.

Lemma 7. *Let $\{x, y\}$ be a two-pair in G:*
(i) If G is a $K_{2,3}$-free graph, then $Sep(x, y) = \mathcal{C}_1 \cup \mathcal{C}_2 \cup \mathcal{C}_3 \cup \mathcal{C}_4 \cup \mathcal{C}_5$,
(ii) If G is a $(K_{2,3}, \overline{4P_2})$-free graph, then $\mathcal{C}_1, \ldots, \mathcal{C}_5$ are disjoint cliques,
(iii) If G is a $(K_{2,3}, \overline{4P_2}, \overline{P_2 \cup P_4})$-free graph, then the adjacency of $\mathcal{C}_1, \ldots, \mathcal{C}_5$ are illustrated in Figure 1, where an edge $(\mathcal{C}_i, \mathcal{C}_j)$ illustrates that every vertex in $\mathcal{C}_i$ is adjacent to every vertex in $\mathcal{C}_j$.

Let $\mathcal{C}_1, \ldots, \mathcal{C}_5$ as defined above. We now define the collection $A = \{\{x\} \cup \mathcal{C}_1 \cup \mathcal{C}_3, \{x\} \cup \mathcal{C}_2 \cup \mathcal{C}_4, \{y\} \cup \mathcal{C}_1 \cup \mathcal{C}_4, \{y\} \cup \mathcal{C}_2 \cup \mathcal{C}_3\}$ and the collection $B = \{\{y\} \cup \mathcal{C}_1 \cup \mathcal{C}_3, \{y\} \cup \mathcal{C}_2 \cup \mathcal{C}_4, \{x\} \cup \mathcal{C}_1 \cup \mathcal{C}_4, \{x\} \cup \mathcal{C}_2 \cup \mathcal{C}_3\}$.

We define $A \sqcup B$ as the union of A and B, such that no element of $A \sqcup B$ is contained in another element of A or B, i.e., the "containment maximal" elements of $A \cup B$.

Claim 8. *Let $\{x, y\}$ be a two-pair in a $(K_{2,3}, \overline{P_2 \cup P_4}, \overline{P_6}, \overline{4P_2})$-free graph G. If one of $\mathcal{C}_1, \ldots, \mathcal{C}_4$ is an empty set, then $|A \sqcup B| \leq 4$.*

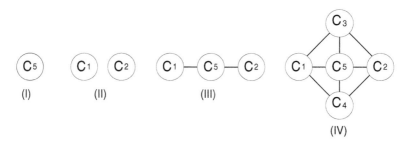

Fig. 1. Adjacency of the sets $\mathcal{C}_1, \ldots, \mathcal{C}_5$ in the following cases: (I) $Sep(x, y)$ is a clique, (II) $N'[s_1] \cap N'[s_2] = \emptyset$, (III) $N'[s_1] \cap N'[s_2]$ is a clique, (IV) $N'[s_1] \cap N'[s_2]$ is not a clique ($\mathcal{C}_5$ is possibly empty)

Claim 9. *Let $\{x, y\}$ be a two-pair in a weakly chordal $(K_{2,3}, \overline{P_2 \cup P_4}, \overline{P_6}, \overline{4P_2})$-free graph G. Let D be a connected component of non-essential vertices, such that $N[D] \subseteq \{x\} \cup \mathcal{C}_i \cup \mathcal{C}_j$, where there exists an edge between $\mathcal{C}_i$ and $\mathcal{C}_j$ for $1 \le i \ne j \le 4$. Then there exists a clique K of size at most 2 in D, such that $N[K] \subseteq \{x\} \cup \mathcal{C}_i \cup \mathcal{C}_j$.*

Claim 10. *Let $\{x, y\}$ be a two-pair in a $(K_{2,3}, \overline{P_2 \cup P_4}, \overline{P_6}, \overline{4P_2}, H_1, H_2, H_3)$-free graph G. Let D_1 be a connected component of non-essential vertices in G, such that $N[D_1]$ is contained in or equal to an element in A (or B) and $\mathcal{C}_1$, $\mathcal{C}_2$, $\mathcal{C}_3$, $\mathcal{C}_4 \ne \emptyset$. If there exists a connected component $D_2 \ne D_1$ of non-essential vertices in G, then $N[D_2]$ is not contained in and not equal to an element in B (or A).*

4   The Main Theorem and Algorithm

In this section we present the main result based on the structural properties proved in Section 3.

4.1   Forbidden Structures of $[4, 4, 2]$ Graphs

Theorem 11. *Let G be a $[4, 4, 2]$ graph. Then G is weakly chordal and does not contain the induced subgraphs $K_{2,3}$, $\overline{P_6}$, $\overline{4P_2}$ $\overline{P_2 \cup P_4}$, $\overline{C_6}$, H_1, H_2 and H_3 that are shown in Figure 2.*

Proof. According to Theorem 2, a $[4, 4, 2]$ graph G contains no C_n, $n > 4$. Since $C_5 = \overline{C_5}$, the graph G does not contain $\overline{C_5}$. Moreover, G contains no $\overline{C_6}$ and no $\overline{P_6}$ and therefore G contains no $\overline{C_n}$, $n > 6$. Hence, G is weakly chordal.

According to Theorem 2, all the graphs in Figure 2 must contain a pie on the cycle $C = (a, b, c, d, a)$ in any $(4, 4, 2)$ representation, and the center of the pie has degree 4. Let e_1, e_2, e_3, e_4 be the edges of the pie, where e_1 is contained in $S_a \cap S_b$, e_2 is contained in $S_b \cap S_c$, e_3 is contained in $S_c \cap S_d$ and e_4 is contained in $S_d \cap S_a$.

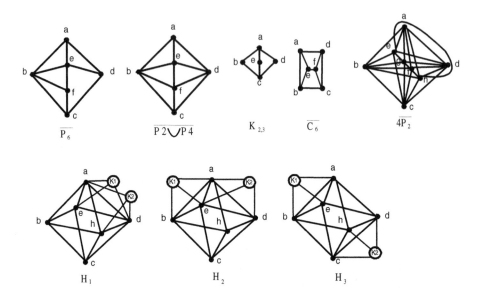

Fig. 2. Forbidden subgraphs, where K_1 and K_2 are cliques of size at most 2

In any $(4, 4, 2)$-representation of $K_{2,3}$, the subtree S_e cannot contain any of the edges e_1, e_2, e_3, e_4, but must contain a common edge with S_a and S_c. Contradiction!

In any $(4, 4, 2)$-representation of $\overline{P_6}$ and $\overline{P_2 \cup P_4}$, the subtree S_e contains the edges e_1, e_4 and does not contain the edges e_2, e_3. The subtree S_f contains the edge e_2 (and possibly e_3), but does not contain the edges e_1, e_4. Therefore, $S_e \cap S_f = \emptyset$. Contradiction!

In any $(4, 4, 2)$-representation of $\overline{C_6}$, the subtree S_e contains the edge e_1 and does not contain the edges e_2, e_3, e_4. The subtree S_f contains the edge e_3 and does not contain the edges e_1, e_2, e_4. Therefore, $S_e \cap S_f = \emptyset$. Contradiction!

In any $(4, 4, 2)$-representation of $\overline{4P_2}$, either S_e contains e_1, e_3 and S_f contains e_2, e_4, or S_e contains e_2, e_4 and S_f contains e_1, e_3. In both cases, each one of the subtrees S_g and S_h must have at least three edges among e_1, e_2, e_3, e_4. Since $S_g \cap S_h = \emptyset$, this is a contradiction.

In any $(4, 4, 2)$-representation of H_1, the collection of subtrees that correspond to connected components K_1 and K_2 do not contain e_1, e_2, e_3. Thus both S_e and S_h must contain the edge e_4. This is a contradiction, since $S_e \cap S_h = \emptyset$.

In any $(4, 4, 2)$-representation of H_2, the collection of subtrees that correspond to connected component K_1 contains e_1 and the collection of subtrees that correspond to connected component K_2 contains e_4. Then S_e must contain e_1 and e_4. Furthermore, S_e must contain one among e_3 and e_2. Thus, S_h contains at most one edge among e_2, e_3. This is a contradiction, since $S_h \cap S_a \neq \emptyset$.

In any $(4, 4, 2)$-representation of H_3, the collection of subtrees that correspond to connected components K_1 do not contain e_2, e_3, e_4 and the subtrees that correspond to connected components K_2 do not contain e_1, e_2, e_4. Therefore, S_e must contain e_1 and S_h must contain e_3. Since $S_e \cap S_h = \emptyset$, S_e cannot contain

e_3 and therefore must contain e_2 and e_4. Then S_h cannot contain e_1, e_2, e_4. This is a contradiction, since $S_h \cap S_a \neq \emptyset$.

Thus, G does not contain the induced subgraphs shown in Figure 2.   □

4.2   The Algorithm

Construct $(4, 4, 2)$**-representation algorithm**
input: a $(K_{2,3}, \overline{P_6}, \overline{4P_2}, \overline{P_2 \cup P_4}, H_1, H_2, H_3)$-free weakly chordal graph G
output: a $(4, 4, 2)$-representation $\langle \mathcal{S}, T \rangle$ of G

begin
1 $\langle \mathcal{S}, T \rangle \leftarrow$ **create-initial-star**(G);
2 **while** $h > 4$ *(h is the maximum degree of T)* **do**
   $\lfloor$ $\langle \mathcal{S}, T \rangle \leftarrow$ **Degree-reduce**$(G, \langle \mathcal{S}, T \rangle)$;

end

Create-initial-star procedure
input: a graph G with $m = |E(G)|$ and $v = |V(G)|$
output: an $(m, n - 1, 2)$-representation $\langle \mathcal{S}, T \rangle$ of G, where T is a star

begin
$T \leftarrow$ star with central vertex b and m edges $(v_1, b), \dots, (v_m, b)$;
$l \leftarrow 1$;
foreach $(x_i, x_j) \in E(G)$ **do**
   $\lfloor$ add the edge (v_l, b) to S_i and S_j;
   $\quad l \leftarrow l + 1$;

end

Definition 12. *Let $\langle \mathcal{S}, T \rangle$ be an $(h, s, 2)$-representation of a graph G. Core or essential subtree in $\langle \mathcal{S}, T \rangle$ corresponds to a core or essential vertex in G, respectively.*

Let $\{x, y\}$ be a two-pair in the graph G. For an edge $e \in \mathcal{E}_x \cup \mathcal{E}_y \cup \mathcal{E}_{Sep(x,y)}$, we define the collection of subtrees $\widehat{\mathcal{S}}(e) = \{S \in \mathcal{S} | S$ is a non-core subtree and S contains the edge $e\}$.

An edge e, such that $e \notin \mathcal{E}_{Sep(x,y)}$ or e is contained in a non-essential subtree, is called a flexible edge.

Let u be a vertex in T. We denote the induced subgraph G_U, where $U \subseteq V$ is the set of vertices in G that corresponds to the collection of subtrees that contain the vertex u in T.

Theorem 13. *Let G be a weakly chordal $(K_{2,3}, \overline{P_6}, \overline{4P_2}, \overline{P_2 \cup P_4}, H_1, H_2, H_3)$-free graph, then the **Construct (4,4,2)-representation algorithm** finds a $(4, 4, 2)$-representation of G.*

Proof. At the first step of the algorithm, we perform the **Create-initial-star** procedure, which clearly finds an initial $(h, s, 2)$-representation of G, where

Degree-reduce procedure

input : G, an $(h, s, 2)$-representation $\langle \mathcal{S}, T \rangle$ of G, $h > 4$

output : an $(h, s, 2)$-representation $< \mathcal{S}', T' >$ of G with fewer vertices of degree h

begin

1 find a vertex u of degree h;

2 $\langle \mathcal{S}, T \rangle \leftarrow$ **PreprocessingA**$(\langle \mathcal{S}, T \rangle, u)$;

//every subtree in $\mathcal{S}$ contains either 0 or at least 2 edges incident to u;

3 find the induced subgraph G_U;

4 **if** G_U *is a clique* **then**

   $\quad \langle \mathcal{S}', T' \rangle \leftarrow$ **TransformationB**$(\langle \mathcal{S}, T \rangle, u)$;

   end procedure;

//G_U is not a clique:

5 find a two-pair $\{x, y\}$ and $Sep(x, y)$;

6 **if** $G_{Sep(x,y)}$ *is a clique* **then**

   $\quad ColoringC \leftarrow$ **ColoringC**$(\langle \mathcal{S}, T \rangle, u, \{x, y\}, Sep(x, y)$);

   $\quad \langle \mathcal{S}', T' \rangle \leftarrow$ **TransformationC**$(\langle \mathcal{S}, T \rangle, u, Sep(x, y), ColoringC)$;

   end procedure;

//$G_{Sep(x,y)}$ is not a clique:

7 **foreach** *star edge* $e \in \mathcal{E}_x \cup \mathcal{E}_y \cup \mathcal{E}_{Sep(x,y)}$ **do**

   $\quad$ find the collection $\widehat{\mathcal{S}}(e)$;

   $\quad ColoringD \leftarrow$ **ColoringD**$(\langle \mathcal{S}, T \rangle, u, \widehat{\mathcal{S}}(e)$);

   $\quad$ **if** $\exists$ *a flexible red edge* $e' \neq e$ **then**

   $\quad\quad \langle \mathcal{S}', T' \rangle \leftarrow$ **TransformationD**$(\langle \mathcal{S}, T \rangle, u, e, ColoringD)$;

   $\quad\quad$ **end procedure**;

   $\quad$ **else**

   $\quad\quad$ uncolor $\langle \mathcal{S}, T \rangle$;

//$\forall e \in \mathcal{E}_x \cup \mathcal{E}_y \cup \mathcal{E}_{Sep(x,y)}$ in the corresponding $ColoringD$ there is no flexible red edge:

8 $\langle \mathcal{S}', T' \rangle \leftarrow$ **TransformationE**$(G, \langle \mathcal{S}, T \rangle, u, \{x, y\}, Sep(x, y)$);

end

See Appendix A for the procedures.

$h = |E(G)|$ and $s \leq |V(G)| - 1$. If $h \leq 4$, then the Theorem holds trivially. Otherwise, $h > 4$.

While there exists a vertex in T of degree greater than 4, i.e., $h > 4$, we perform the **Degree-reduce** procedure, which by Lemma 14 obtains an $(h, s, 2)$-representation of G with fewer vertices of degree h. If the input representation of **Degree-reduce** procedure has only one vertex of degree h, then the output is an $(h', s', 2)$-representation of G with maximum degree h', where $h' < h$. The algorithm stops when a $(4, 4, 2)$-representation is obtained.

The following Lemma is the main justification for the correctness of algorithm.

Lemma 14. *Let* $\langle \mathcal{S}, T \rangle$ *be an* $(h, s, 2)$-*representation of a weakly chordal* $(K_{2,3}, \overline{P_6}, \overline{4P_2}, \overline{P_2 \cup P_4}, H_1, H_2, H_3)$-*free graph* G, *where* T *has a maximal degree* $h > 4$.

*If $\langle S,T \rangle$ is an input to **Degree-reduce** procedure, then the output is an $(h, s, 2)$-representation of G with fewer vertices of degree h.*

Proof. **Step 1** Since $\langle S,T \rangle$ is an $(h, s, 2)$-representation, there exists a vertex u of degree h, which is found in this step.

Step 2
Claim. The output of **PreprocessingA**($\langle S,T \rangle$, u) is an $(h, s, 2)$-representation of G, such that every subtree in S either uses no edges with the endpoint u or uses at least two edges with the endpoint u.

Step 3. We find the induced subgraph G_U, where each vertex in G_U corresponds to a subtree that contains the vertex u in T. By the hereditary property, the induced subgraph G_U is also a weakly chordal ($K_{2,3}$, $\overline{P_6}$, $\overline{4P_2}$, $\overline{P_2 \cup P_4}$, H_1,H_2, H_3)-free graph.

Step 4. According to Theorem 1, the subgraph G_U is either a clique or has a two-pair. If G_U is a clique, then at Step 4 we call **TransformationB**($\langle S,T \rangle$, u) procedure and **Degree-reduce procedure** ends.

Claim. If G_U is a clique, then the output of **TransformationB** procedure ($\langle S,T \rangle$, u) is an $(h, s, 2)$-representation of G with fewer vertices of degree h than in $\langle S,T \rangle$. Otherwise, at Steps 5-9, G_U is not a clique and therefore has a two-pair.

Step 5. We find a two-pair $\{x, y\}$ and the set $Sep(x, y)$.

Step 6. If $G_{Sep(x,y)}$ is a clique, then first we call **ColoringC**($\langle S,T \rangle$, u, $\{x, y\}$, $Sep(x, y)$) and then we call to **TransformationC**($\langle S,T \rangle$, u, $Sep(x, y)$, *Coloring C*) and **Degree-reduce procedure** ends.

Claim. Let $G_{Sep(x,y)}$ be a clique. Then in the output of **ColoringC** procedure ($\langle S,T \rangle$, u, $\{x, y\}$, $Sep(x, y)$) every subtree not in $S_{Sep(x,y)}$ has either red edges or blue edges. The output of **TransformationC** procedure ($\langle S,T \rangle$, u, $Sep(x, y)$, *ColoringC*) is an $(h, s, 2)$-representation of G with fewer vertices of degree h than in $\langle S,T \rangle$.
    Otherwise, at Steps 7-8, $G_{Sep(x,y)}$ is not a clique.

Step 7. For every edge $e \in \mathcal{E}_x \cup \mathcal{E}_y \cup \mathcal{E}_{Sep(x,y)}$, we call to **ColoringD** ($\langle S,T \rangle$, u, $\widehat{S}(e)$) procedure until there exists a flexible red edge e'. Then we perform **TransformationD procedure** ($\langle S,T \rangle$, u, e, *ColoringD*) and **Degree-reduce procedure** ends.

Claim. In the output of **ColoringD procedure**($\langle S,T \rangle$, u, $\widehat{S}(e)$):
(i) Every red edge is contained in a non-core subtree
(ii) If $e \in \mathcal{E}_x$, then S_x has at least one red edge
(iii) At most one of S_x and S_y has a red edge
(iv) Every subtree in $S_{Sep(x,y)}$ has at least one uncolored star edge
(v) For every non-core subtree, either all its star edges are uncolored or are red
(vi) At most one of the collections S_{Z_x} and S_{Z_y} has a subtree with a red edge.

Claim. Let G be a $(K_{2,3}, \overline{P_2 \cup P_4}, \overline{P_6})$-free graph and $G_{Sep(x,y)}$ is not a clique. If there exists a flexible red edge $e' \neq e$, then the output of **TransformationD procedure**$(\langle \mathcal{S}, T \rangle, u, e, ColoringD)$ is an $(h, s, 2)$-representation of G with fewer vertices of degree h than in $\langle \mathcal{S}, T \rangle$.

Otherwise, at Step 8, for every edge $e \in \mathcal{E}_x \cup \mathcal{E}_y \cup \mathcal{E}_{Sep(x,y)}$, in the corresponding coloring with respect to $\widehat{\mathcal{S}}(e)$, there is no flexible red edge. In addition, G_U has the following structural property:

Claim. If Step 8 of the **Degree-reduce procedure** is performed, then a vertex is in G_U if and only if the vertex is essential.

Step 8. In **TransformationE procedure** $(G, \langle \mathcal{S}, T \rangle, u, \{x, y\}, Sep(x, y))$, we first find cliques $\mathcal{C}_1, \ldots, \mathcal{C}_5$ of $Sep(x, y)$. Then we call to **ColoringE procedure** $(\langle \mathcal{S}, T \rangle, u, \{x, y\}, \mathcal{C}_1, \ldots, \mathcal{C}_4)$ and to **FindColorPivot procedure** $(ColoringE, \{x, y\}, \mathcal{C}_1, \ldots, \mathcal{C}_4, G)$.

Definition 15. *Let $C(S)$ be the color set of the edges of a subtree S in $\langle \mathcal{S}, T \rangle$.*

Claim. In the output representation of **ColoringE procedure**$(\langle \mathcal{S}, T \rangle, u, \{x, y\}, \mathcal{C}_1, \ldots, \mathcal{C}_4)$: (i) Every non-star edge with an endpoint $\{v_1, \ldots, v_h\}$ is colored, (ii) For every non-essential subtree, all its non-star edges with an endpoint among $\{v_1, \ldots, v_h\}$ are colored with the same set of colors $C(S)$.

Claim. Let G be a $(K_{2,3}, \overline{P_2 \cup P_4}, \overline{P_6}, \overline{4P_2}, H_1, H_2, H_3)$-free graph. In the output representation of **FindColorPivot procedure** $(ColoringE, \{x, y\}, \mathcal{C}_1, \ldots, \mathcal{C}_4, G)$:
(i) For every colored subtree S, $C(S)$ is contained in or equal to an element in $A \sqcup B$.
(ii) If $\mathcal{C}_1, \mathcal{C}_2, \mathcal{C}_3, \mathcal{C}_4 \neq \emptyset$, then for every colored non-essential subtree S, $C(S)$ is contained in or equal to an element in the set H found by **FindColorPivot** procedure.

At the end of **TransformationE procedure**$(G, \langle \mathcal{S}, T \rangle, u, \{x, y\}, Sep(x, y))$ we consider the coloring and the pivot found and construct a new representation, such that u has degree at most 4.

Claim. In the output of **TransformationE procedure** the degree of u is at most 4.

We have shown that in all cases the output is an $(h, s, 2)$-representation with fewer number of vertices of degree h than in $\langle \mathcal{S}, T \rangle$.   □

4.3   The Main Theorem

The following Theorem summarizes the main contribution of the paper:

Theorem 16. *A graph G is a weakly chordal $(K_{2,3}, \overline{P_6}, \overline{4P_2}, \overline{P_2 \cup P_4}, H_1, H_2, H_3)$-free graph if and only if the graph G has a $(4, 4, 2)$-representation.*

Proof. Follows from Theorem 11 and Theorem 13.   □

Combining Theorem 16 and Theorem 13, we conclude that the **Construct (4,4,2)-representation algorithm** correctly recognizes $[4, 4, 2]$ graphs.

5   Future Work

From the main theorem we obtain the following implication. We are currently working on the algorithm based on the following Theorem.

Theorem 17. *Let G be a $(K_{2,3}, \overline{P_6}, \overline{4P_2}, \overline{P_2 \cup P_4})$-free graph. G is weakly chordal if and only if G is an $[8,8,2]$ graph and $(C_5, C_6, C_7, C_8, \overline{C_n}, n \geq 6)$-free graph.*

We are looking for other forbidden structures on graphs that will lead to characterization of various $[h, s, t]$ families, e.g. hole-free and anti-hole free graphs. It will be interesting to find more intersection models for other well known structured families of graphs, e.g. for weakly chordal graphs.

References

1. C. Berge, *Graphs and Hypergraphs*, North-Holland, Amsterdam, 1973.
2. P. Buneman, *A characterization of rigid circuit graphs*, Discrete Math., 9 (1974), 205-212.
3. F. Gavril, *The intersection graphs of subtrees in trees are exactly the chordal graphs*, J. Comb. Theory Ser. B (1974), 47-56.
4. M.C. Golumbic, *Algorithmic Graph Theory and Perfect Graphs*, Second Edition, Annals of Discrete Math. 57, Elsevier (2004).
5. M.C. Golumbic, R.E. Jamison, *The edge intersection graphs of paths in a tree*, Journal of Combinatorial Theory, Series B 38 (1985), 8-22.
6. M.C. Golumbic, A.N. Trenk, *Tolerance Graphs*, Chapter 11, Cambridge University Press (2004).
7. R.B. Hayward, C.T. Hoàng, F. Maffray, *Optimizing weakly triangulated graphs*, Graphs and Combinatorcs 5, 339-349, 1989.
8. R.E. Jamison, H.M. Mulder, *Tolerance intersection graphs on binary trees with constant tolerance 3*, Discrete Math. 215 (2000), 115-131.
9. R.E. Jamison, H.M. Mulder, *Constant tolerance intersection graphs of subtrees of a tree*, Discrete Math. 290 (2005), 27-46.
10. F.R. McMorris, E. Scheinerman, *Connectivity threshold for random chordal graphs*, Graphs and Combin. 7 (1991), 177-181.
11. J. Spinrad, *Efficient Graph Representations*, Providence, R.I.: American Mathematical Society (2003).
12. J.R. Walter, *Representations of rigid cycle graphs*, PhD Thesis, Wayne State Univ., 1972.

Appendix

PreprocessingA procedure
input: $\langle \mathcal{S}, T \rangle$, u
output: $\langle \mathcal{S}, T \rangle$ let $v_1, \ldots, v_h$ be the neighbors of u in T;
foreach *star edge* (v_i, u) **do**

> find $\mathcal{S}(v_i, u)$, which is the collection of subtrees in $\mathcal{S}$ that contains only the edge (v_i, u) among $(v_1, u), \ldots, (v_h, u)$;
> **if** $\mathcal{S}(v_i, u) \neq \emptyset$ **then**
>
> > split the edge (v_i, u) into two edges, by adding a dummy vertex w such that:
> > **foreach** *subtree* $S \in \mathcal{S}(v_i, u)$ **do**
> >
> > > replace (v_i, u) by the edge (v_i, w) in S (thus making w the endpoint of S);
> >
> > **foreach** *subtree* $S \notin \mathcal{S}(v_i, u)$ *and* (v_i, u) *is contained in* S **do**
> >
> > > replace (v_i, u) by the two edges (v_i, w) and (w, u) in S.

rename the neighbors of u to be $v_1, \ldots, v_h$ of u;

TransformationB procedure
input: $\langle \mathcal{S}, T \rangle$, u
output: $\langle \mathcal{S}', T' \rangle$
split the edge (v_1, u) into two edges by adding a dummy vertex w, remove the edge (v_2, u), add the edge (v_2, w) such that:
foreach *subtree* $S \in \mathcal{S}_U$ *that contains the edge* $(v_i, u), i = 1, 2$, **do**

> replace (v_i, u) by the two edges (v_i, w) and (w, u) in S;

ColoringC procedure
input: $\langle \mathcal{S}, T \rangle$, u, $\{x, y\}$, $Sep(x, y)$
output: $ColoringC$
find $\mathcal{E}_x$ and $\mathcal{E}_y$;
color the edges of $\mathcal{E}_x$ red and color the edges of $\mathcal{E}_y$ blue;
repeat

> **if** *uncolored edge* $(v_i, u) \notin \mathcal{E}_{Sep(x,y)}$ *is contained in a subtree with a red edge* **then**
>
> > color (v_i, u) red;
>
> **if** *uncolored edge* $(v_i, u) \notin \mathcal{E}_{Sep(x,y)}$ *is contained in a subtree with a blue edge* **then**
>
> > color (v_i, u) blue;

until *no further coloring is possible*;

TransformationC procedure
input: $\langle \mathcal{S}, T \rangle$, u, $Sep(x, y)$, $ColoringC$
output: $\langle \mathcal{S}', T' \rangle$
split the vertex u into two vertices u' and u'', and add the edge (u', u'') such that:

foreach *star edge* (v_i, u) **do**
   switch (v_i, u) **do**
      case *red*
          replace (v_i, u) by (v_i, u') in every subtree containing it;
      case *blue*
          replace (v_i, u) by (v_i, u'') in every subtree containing it;
      case *uncolored*
         if (v_i, u) *is contained in a subtree with a red edge* **then**
             replace (v_i, u) by (v_i, u') in every subtree containing it;
         else
             replace (v_i, u) by (v_i, u'') in every subtree containing it;

add the edge (u', u'') to every subtree in $\mathcal{S}_{S\top\sqrt{(\S,\dagger)}}$;

ColoringD procedure
input: $\langle \mathcal{S}, T \rangle$, u, $\widehat{\mathcal{S}}(e)$
output: *ColoringD*
foreach *edge* (v_i, u) *that is contained in a subtree in* $\widehat{\mathcal{S}}(e)$ **do**
    color (v_i, u) red;
repeat
   if *uncolored edge* (v_i, u) *is contained in a non-core subtree* S *with a red edge*
   then
       color (v_i, u) red;
until *no further coloring is possible*;

TransformationD procedure
input: $\langle \mathcal{S}, T \rangle$, u, e, *ColoringD*
output: $< \mathcal{S}', T' >$
split the edge $e = (v_j, u)$ into two edges by adding a dummy vertex w such that:

replace (v_j, u) by (v_j, w) in every non-essential subtree containing it;
replace (v_j, u) by $(v_j, w), (w, u)$ in every essential subtree containing it;
foreach *flexible red edge* $(v_i, u), i \neq j$ **do**
    replace (v_i, u) by (v_i, w) in every non-essential subtree containing it;
    replace (v_i, u) by $(v_i, w), (w, u)$ in every essential subtree containing it;

TransformationE procedure
input: G, $\langle S,T \rangle$, u, $\{x,y\}$, $Sep(x,y)$
output: $< S',T' >$
$\{C_1, C_2, C_3, C_4, C_5\} \leftarrow$ **FindSepCliques**$(G, Sep(x,y)\ )$;
$ColoringE \leftarrow$ **ColoringE**$(\langle S,T \rangle, u, \{x,y\}, C_1, \ldots, C_4)$;
$H \leftarrow$ **FindColorPivot**$(ColoringE, \{x,y\}, C_1, \ldots, C_4, G)$;
add four new edges $(u_1, u), (u_2, u), (u_3, u), (u_4, u)$ such that:
foreach *subtree* $S \in C_5$ **do**
  └ add the four edges $(u_1, u), (u_2, u), (u_3, u), (u_4, u)$ to S

foreach $h_i \in H$ **do**
  └ add the edge (u_i, u) to all the subtrees corresponding to a vertex in h_i;

foreach *star edge* (v_i, u) **do**
  │ **foreach** *non-star edge* $e = (v_i, q), q \neq u$ **do**
  │ │ **if** e *is colored with more than one color* **then**
  │ │ │ find h_i that contains the color set of e;
  │ │ └ pivot $\leftarrow i$
  │ │ **else**
  │ │ └ pivot $\leftarrow$ the minimal i, such that h_i contains the only color of e;
  │ │ split the star edge on the path $[u_{pivot}, u]$ into two edges by adding a dummy vertex w;
  │ │ add a dummy pendant edge (w', w);
  │ │ replace e by $(w, w'), (w', q)$ in every essential subtree containing it;
  │ └ replace e by (w', q) in every non-essential subtree containing it;
  └ remove the star edge (v_i, u);

FindSepCliques procedure
input: G, $Sep(x,y)$
output: C_1, C_2, C_3, C_4, C_5
choose two non-adjacent vertices $s_1, s_2 \in Sep(x,y)$;
$C_1 \leftarrow \{N'(s_1)\} - \{N'(s_2)\}$;
$C_2 \leftarrow \{N'(s_2)\} - \{N'(s_1)\}$;
if $N'(s_1) \cap N'(s_2) = \emptyset$ **then**
  └ $C_3 = C_4 = C_5 = \emptyset$;
else
  │ **if** $N'(s_1) \cap N'(s_2)$ *is a clique* **then**
  │ │ $C_3 \leftarrow \emptyset$;
  │ │ $C_4 \leftarrow \emptyset$;
  │ │ $C_5 \leftarrow N'(s_1) \cap N'(s_2)$;
  │ **else**
  │ │ choose non-adjacent s_3 and s_4 in $N'(s_1) \cap N'(s_2)$;
  │ │ $C_3 \leftarrow \{N'(s_3)\} - \{N'(s_4)\}$;
  │ │ $C_4 \leftarrow \{N'(s_4)\} - \{N'(s_3)\}$;
  │ └ $C_5 \leftarrow \{N'(s_3)\} \cap \{N'(s_4)\}$

ColoringE procedure
input: $\langle S,T \rangle$, u, $\{x,y\}$, $\mathcal{C}_1, \ldots, \mathcal{C}_4$
output: $ColoringE$
define six colors, such that each color corresponds to one of $x, y, \mathcal{C}_1, \mathcal{C}_2, \mathcal{C}_3, \mathcal{C}_4$;
foreach *non-star edge* $(v_i, a), a \neq u$ *that is contained in a subtree corresponding to*
a vertex among $\{x, y, \mathcal{C}_1, \mathcal{C}_2, \mathcal{C}_3, \mathcal{C}_4\}$ **do**
$\quad$ color (v_i, a) with the corresponding color;

repeat
$\quad$ **if** *a non-star edge* $(v_i, a), a \neq u$ *is contained in a non-essential subtree* S *with*
$\quad$ *an edge colored* c **then**
$\quad\quad$ color edge (v_i, a) with color c;

until *no further coloring is possible*;
//some edges will receive multiple colors;

FindColorPivot procedure
input: $ColoringE$, $\{x, y\}$, $\mathcal{C}_1, \ldots, \mathcal{C}_4$, G
output: H
$A \leftarrow \{h_1 = \{x, \mathcal{C}_1, \mathcal{C}_3\}, h_2 = \{x, \mathcal{C}_2, \mathcal{C}_4\}, h_3 = \{y, \mathcal{C}_1, \mathcal{C}_4\}, h_4 = \{y, \mathcal{C}_2, \mathcal{C}_3\}\}$;
$B \leftarrow \{h_1 = \{y, \mathcal{C}_1, \mathcal{C}_3\}, h_2 = \{y, \mathcal{C}_2, \mathcal{C}_4\}, h_3 = \{x, \mathcal{C}_1, \mathcal{C}_4\}, h_4 = \{x, \mathcal{C}_2, \mathcal{C}_3\}\}$;
if $\mathcal{C}_1, \mathcal{C}_2, \mathcal{C}_3, \mathcal{C}_4 \neq \emptyset$ **then**
$\quad$ **if** $\exists$ *a non-essential subtree whose color set has three colors and forms an element*
$\quad$ *of* A **then**
$\quad\quad$ $H = A$;
$\quad$ **else**
$\quad\quad$ $H = B$;
else
$\quad$ //each triple in A or in B that contains an empty set is substituted by a pair;
$\quad$ $H \leftarrow A \cup B$;
$\quad$ **while** $\exists h, h' \in H$ *such that* h *is contained in* h' **do**
$\quad\quad$ $H \leftarrow H - h$;

A Fully Dynamic Algorithm for the Recognition of P_4-Sparse Graphs

Stavros D. Nikolopoulos[1], Leonidas Palios[1], and Charis Papadopoulos[2],[*]

[1] Department of Computer Science, University of Ioannina
P.O. Box 1186, GR-45110 Ioannina, Greece
{stavros, palios}@cs.uoi.gr
[2] Department of Informatics, University of Bergen
P.B. 7800, N-5020 Bergen, Norway
charis@ii.uib.no

Abstract. We consider the dynamic recognition problem for the class of P_4-sparse graphs: the objective is to handle edge/vertex additions and deletions, to recognize if each such modification yields a P_4-sparse graph, and if yes, to update a representation of the graph. Our approach relies on maintaining the modular decomposition tree of the graph, which we use for solving the recognition problem. We establish conditions for each modification to yield a P_4-sparse graph and obtain a fully dynamic recognition algorithm which handles edge modifications in $O(1)$ time and vertex modifications in $O(d)$ time for a vertex of degree d. Thus, our algorithm implies an optimal edges-only dynamic algorithm and a new optimal incremental algorithm for P_4-sparse graphs. Moreover, by maintaining the children of each node of the modular decomposition tree in a binomial heap, we can handle vertex deletions in $O(\log n)$ time, at the expense of needing $O(\log n)$ time for each edge modification and $O(d \log n)$ time for the addition of a vertex adjacent to d vertices.

Keywords: fully dynamic algorithms, P_4-sparse graphs, modular decomposition, recognition.

1 Introduction

A *dynamic graph* algorithm for a class Π of graphs is an algorithm that handles a series of on-line modifications (i.e., insertions or deletions of vertices or edges) on a graph in Π; if the modification yields a graph in Π, the algorithm performs it (updating an internal representation), otherwise it outputs false and does nothing. Such algorithms are categorized depending on the modifications they support: an *incremental* (*decremental*) algorithm supports only vertex insertions (deletions); an *additions-only* (*deletions-only*) algorithm supports only edge additions (deletions); an *edges-only fully dynamic* algorithm supports both

---

[*] Work by Charis Papadopoulos was carried while he was a graduate student at the Department of Computer Science, University of Ioannina.

F.V. Fomin (Ed.): WG 2006, LNCS 4271, pp. 256–268, 2006.

edge additions and edge deletions; a *fully dynamic* algorithm supports all edge as well as all vertex modifications.

Several authors have studied the dynamic recognition problem for graphs of specific families. Incremental recognition algorithms have been proposed by Hsu [12] for interval graphs and by Deng *et al.* [8] for connected proper interval graphs. Ibarra [13] has given an edges-only fully dynamic algorithm for chordal graphs which handles each edge operation in $O(n)$ time and an edges-only fully dynamic algorithm for split graphs which handles each edge operation in $O(1)$ time. More recently, Hell *et al.* [10] have given a fully dynamic algorithm for recognizing proper interval graphs which works in $O(d + \log n)$ time per modification, where d is the degree of a vertex in case of a vertex modification; Shamir and Sharan [19] have developed a fully dynamic algorithm for the recognition of cographs, threshold graphs, and trivially perfect graphs, which handles edge modifications in $O(1)$ time and vertex modifications in $O(d)$ time; finally, Crespelle and Paul have presented fully dynamic algorithms for directed cographs [5] and permutation graphs [6] which require $O(d)$ time if d arcs are involved, and $O(n)$ time, respectively. For the class of P_4-sparse graphs, an incremental algorithm for recognizing a P_4-sparse graph has been proposed by Jamison and Olariu [15] which handles the insertion of a vertex of degree d in $O(d)$ time.

Researchers have also considered the problem of the dynamic maintenance of the modular decomposition tree of a graph: Muller and Spinrad [18] have given an incremental algorithm, which handles each vertex insertion in $O(n)$ time; for cographs, Corneil *et al.* [3] have given an optimal incremental algorithm, which handles the insertion of a vertex of degree d in $O(d)$ time.

Our work in this paper focuses on P_4-*sparse graphs*; these are the graphs in which every set of five vertices induces at most one chordless path on four vertices [11]. They are perfect and also perfectly orderable [11], and properly contain the cographs, the P_4-reducible graphs, etc. (see [1,15,16]). The P_4-sparse graphs have received considerable attention in recent years and they find applications in applied mathematics and computer science (e.g., communications, transportation, clustering, scheduling, computational semantics) in problems on graphs featuring "local density"; local density is often associated with the absence of P_4s and the P_4-sparse graphs are unlikely to have many P_4s.

In this paper, we describe a fully dynamic algorithm for the class of P_4-sparse graphs. Our algorithm maintains the modular decomposition tree of the graph; it checks whether the requested edge/vertex operations yield a P_4-sparse graph, and if yes, it updates the modular decomposition tree. Edge operations are handled in $O(1)$ time while vertex operations are handled in $O(d)$ time. As a result, we obtain an optimal edges-only dynamic algorithm and a new optimal incremental algorithm for P_4-sparse graphs. Moreover, in order to improve the time complexity of the vertex deletion operation, we can maintain the children of each node of the modular decomposition tree in a binomial heap [2]. Then, we can handle vertex deletions in $O(\log n)$ time; the drawback is that then the time required for each edge modification becomes $O(\log n)$ and for the addition of a vertex adjacent to d vertices becomes $O(d \log n)$.

2   Theoretical Framework

Let G be a simple graph; we denote by $V(G)$ and $E(G)$, the vertex and edge set of G. The subgraph of G induced by a set $S \subseteq V(G)$ is denoted by $G[S]$. If a vertex u is adjacent to a vertex v, we say that u *sees* v, otherwise, we say that it *misses* v; more generally, a vertex set A sees (misses, resp.) a vertex set B, if every vertex in A sees (misses, resp.) every vertex in B.

Modular Decomposition and P_4-sparse Graphs.   A subset M of vertices of a graph G is a *module* of G, if every vertex outside M is either adjacent to all vertices in M or to none of them. The emptyset, the singletons, and the vertex set $V(G)$ are *trivial* modules and whenever G has only trivial modules it is called a *prime* (or *indecomposable*) *graph*. A module M of G is called a *strong module* if, for any module M' of G, either $M' \cap M = \emptyset$ or one module is included into the other. Furthermore, a module in G is also a module in $\overline{G}$.

The *modular decomposition* of a graph G is a linear-space representation of all the partitions of $V(G)$ where each partition class is a module. The *modular decomposition tree* $T(G)$ of the graph G (or *md-tree* for short) is a unique (up to isomorphism) labeled tree associated with the modular decomposition of G in which the leaves of $T(G)$ are the vertices of G and the set of leaves associated with the subtree rooted at an internal node induces a strong module of G (Figure 1). Thus, the md-tree $T(G)$ represents all the strong modules of G. It is known that for every graph G the md-tree $T(G)$ can be constructed in linear time [4,7,17].

Let t be an internal node of the md-tree $T(G)$ of a graph G. We denote by $M(t)$ the module corresponding to t which consists of the set of vertices of G associated with the subtree of $T(G)$ rooted at node t. The node t is labeled by either P (for parallel module) if the subgraph $G[M(t)]$ is disconnected, S (for series module) if the complement of $G[M(t)]$ is disconnected, or N (for neighborhood module) otherwise. Let $u_1, u_2, \ldots, u_p$ be the children of the node t of $T(G)$. We denote by $G(t)$ the *representative graph* of the module $M(t)$ defined as follows: $V(G(t)) = \{u_1, u_2, \ldots, u_p\}$ and $u_i u_j \in E(G(t))$ if there exists edge $v_k v_\ell \in E(G)$ such that $v_k \in M(u_i)$ and $v_\ell \in M(u_j)$; by the definition of a module, if a vertex of $M(t_i)$ is adjacent to a vertex of $M(t_j)$ then every vertex of $M(t_i)$ is adjacent to every vertex of $M(t_j)$. Thus, $G(t)$ is isomorphic to the graph induced by a

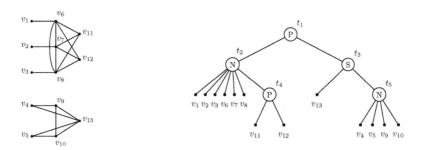

Fig. 1. A disconnected P_4-sparse graph on 13 vertices and its md-tree

subset of $M(t)$ consisting of a single vertex from each maximal strong submodule of $M(t)$ in the modular decomposition of G. Depending on whether an internal node t of $T(G)$ is a P-, S-, or N-node, the following holds:

- ○ if t is a P-node, $G(t)$ is an edgeless graph;
- ○ if t is an S-node, $G(t)$ is complete graph;
- ○ if t is an N-node, $G(t)$ is a prime graph.

In particular, for the class of P_4-sparse graphs, Giakoumakis and Vanherpe [9] showed that:

Lemma 1. *Let $T(G)$ be the modular decomposition tree of a graph G. Then, G is P_4-sparse iff for every N-node t of $T(G)$, $G(t)$ is a prime spider with a spider-partition (S, K, R) and no vertex of $S \cup K$ is an internal node in $T(G)$.*

A graph G is called a *spider* if the vertex set $V(G)$ of the graph G admits a partition into sets S, K, and R such that:

C1: $|S| = |K| \geq 2$, the set S is an independent set, and the set K is a clique;
C2: each vertex in R is adjacent to all the vertices in K and to no vertex in S;
C3: there exists a bijection $f : S \longrightarrow K$ such that for each vertex $v \in S$, either
   (i) $N(v) \cap K = \{f(v)\}$ or (ii) $N(v) \cap K = K - \{f(v)\}$.

The triple (S, K, R) is called the *spider-partition*. A graph G is a *prime spider* if G is a spider with $|R| \leq 1$. If the condition of case C3(i) holds, then the spider G is called a *thin spider*, whereas if the condition of case C3(ii) holds then G is a *thick spider*; note that the complement of a thin spider is a thick spider and vice versa. A prime spider with $|S| = |K| = 2$ is simultaneously thin and thick.

3 The Fully-Dynamic Algorithm

As mentioned, our algorithm maintains the modular decomposition tree $T(G)$ of the P_4-sparse graph.

3.1 Adding an Edge

Let uv be the edge to be added and let $G' = G \cup \{uv\}$. For the two vertices $u, v \in G$ we denote by t_{uv} the least common ancestor of u and v in $T(G)$. Since u, v are non-adjacent in G, node t_{uv} is either a P-node or an N-node. Let t_u and t_v be the children of t_{uv} such that $M(t_u)$ and $M(t_v)$ contain the vertices u and v, respectively. Note that if $|M(t_u)| = 1$ ($|M(t_v)| = 1$, resp.) then $t_u = u$ ($t_v = v$, resp.). Without loss of generality, we make the following assumption:

Assumption 1. *We assume that $|M(t_v)| \geq |M(t_u)|$.*

Then, the following 3 lemmata cover all possible cases that may arise.

Lemma 2. *Let $|M(t_u)| \geq 2$. Then G' is a P_4-sparse graph if and only if t_{uv} is a P-node and $|M(t_u)| = |M(t_v)| = 2$.*

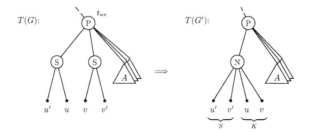

Fig. 2. Illustrating Lemma 2 and the corresponding updates of the md-tree

Lemma 3. *Let $|M(t_u)| = 1$ (i.e., $M(t_u) = \{u\}$) and suppose that t_{uv} is a P-node. Then G' is a P_4-sparse graph if and only if one of the following (mutually exclusive) cases holds:*

(i) vertex v sees all the vertices in $M(t_v)$;

(ii) vertex v misses exactly one vertex $y \in M(t_v)$ such that y sees only one vertex $x \in M(t_v)$, and only the vertex x sees every vertex in $M(t_v)$;

(iii) vertex v misses $\ell > 1$ vertices in $M(t_v)$ such that $G(t_v)$ is a thin spider (S, K, R) with $|S| = |K| = \ell$, $R = \{r\}$ and the vertex v belongs to the set $M(r)$ and sees all the vertices of $M(r)$.

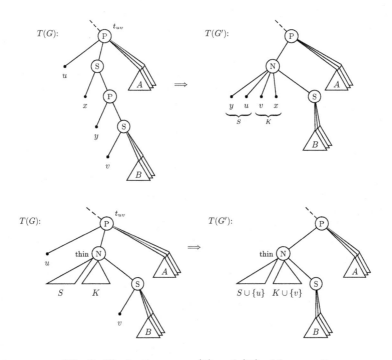

Fig. 3. Illustrating cases (ii) and (iii) of Lemma 3

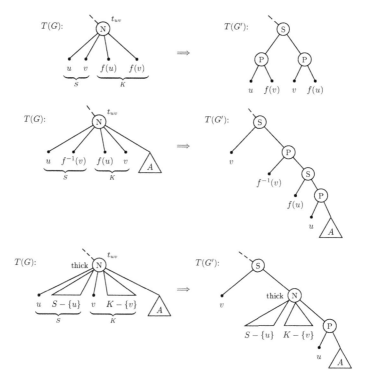

Fig. 4. Illustrating the cases of Lemma 4

Lemma 4. *Let* $|M(t_u)| = 1$ *(i.e.,* $M(t_u) = \{u\}$*) and suppose that* t_{uv} *is an N-node such that* (S, K, R) *is the spider partition of* $G(t_{uv})$*. Then* G' *is a* P_4*-sparse graph if and only if either* $S = \{u, v\}$ *and* $R = \emptyset$ *or* $u \in S$, $v \in K$*, and* $G(t_{uv})$ *is a thick spider.*

3.2   Removing an Edge

Since the P_4-sparse graphs have the complement-invariant property, we take advantage of the following theorem by Shamir and Sharan [19]:

Theorem 1. *[19] Let* Π *be a complement-invariant graph property. Let Alg be a dynamic algorithm for* Π*-recognition, which supports either edge additions only or edge deletions only, and is based on modular decomposition. Then Alg can be extended to support both operations with the same time complexity.*

3.3   Adding a Vertex

Let G be a P_4-sparse graph and a vertex $x \notin V(G)$ which is adjacent to d vertices in $V(G)$, where $d \in \{0, 1, \ldots, |V(G)|\}$. In this section, we show how to recognize if the graph G' with vertex set $V(G) \cup \{x\}$ is P_4-sparse, and if so, we show how to obtain the md-tree $T(G')$ of G' from the md-tree $T(G)$ in $O(d)$ time.

Let us classify the internal nodes of the md-tree $T(G)$ into the following three categories: an internal node t is *x-fully-adjacent*, *x-partly-adjacent*, *x-non-adjacent* iff x is adjacent to all, some but not all, and none, respectively, of the vertices in the module $M(t)$. The above classification is extended to leaf-nodes: a leaf-node a is *x-fully-adjacent* or *x-non-adjacent* iff x is adjacent or non-adjacent, respectively, to a. Because the x-fully-adjacent nodes form a forest of subtrees of $T(G)$ whose total number of leaves is d and because every internal node in $T(G)$ (and in these subtrees) has at least two children, we have:

Observation 1. *The number of x-fully-adjacent (internal and leaf) nodes of $T(G)$ is less than $2d$, where d is the number of vertices of G adjacent to x.*

In turn, for the x-partly-adjacent nodes, the fact that the module of an S-node induces a connected graph, the module of a P-node induces a graph whose complement is connected, and the module of an N-node induces a graph which is connected and whose complement is also connected implies:

P1: if an internal node t of the md-tree $T(G)$ is x-partly-adjacent, then all its ancestors in $T(G)$ are x-partly-adjacent;

P2: for every x-partly-adjacent P-node t_P of $T(G)$, the subgraph of G induced by the module $M(t_P)$ contains two non-adjacent vertices a, b such that a is adjacent and b is not adjacent to x;

P3: for every x-partly-adjacent S-node t_S of $T(G)$, the subgraph of G induced by the module $M(t_S)$ contains an edge ab such that a is adjacent and b is not adjacent to x;

P4: for every x-partly-adjacent N-node t_N of $T(G)$, the subgraph of G induced by the module $M(t_N)$ contains both an edge ab such that a is adjacent and b is not adjacent to x and a pair of non-adjacent vertices a', b' such that a' is adjacent and b' is not adjacent to x.

Additionally, the following very important property holds:

Theorem 2. *For any two x-partly-adjacent nodes of $T(G)$, the graph G' is P_4-sparse only if one of them is an ancestor of the other.*

Let $\rho_x = t_0 t_1 \cdots t_k$ denote the path in $T(G)$ containing all the x-partly-adjacent nodes (Theorem 2) where t_0 is the root of $T(G)$ and t_k is the x-partly-adjacent node farthest away from the root. Then, Theorem 2 implies that for each node t_i, $0 \leq i < k$, each of t_i's children, other than t_{i+1}, is either x-fully-adjacent or x-non-adjacent; for the node t_k, each of t_k's children is either x-fully-adjacent or x-non-adjacent and there is at least one child of each kind. Additionally, for the x-partly-adjacent N-nodes, the following holds:

Lemma 5. *Let t be an x-partly-adjacent N-node of $T(G)$ whose corresponding spider partition of $M(t)$ is (S, K, R), and suppose that the vertex x is adjacent to a vertex in $S \cup K$. Then, the graph G' is P_4-sparse only if x sees $S \cup K$, or sees K and misses S.*

Let us consider the partition of the vertex set $M(t_0) - M(t_k) \subset V(G)$ into the following four sets:

$$V_P = \bigcup_{\substack{t_i \text{ is a P-node} \\ 0 \leq i < k}} (M(t_i) - M(t_{i+1})), \qquad V_{N_S} = \bigcup_{\substack{t_i \text{ is an N-node} \\ 0 \leq i < k}} S(t_i),$$

$$V_S = \bigcup_{\substack{t_i \text{ is an S-node} \\ 0 \leq i < k}} (M(t_i) - M(t_{i+1})), \qquad V_{N_K} = \bigcup_{\substack{t_i \text{ is an N-node} \\ 0 \leq i < k}} K(t_i),$$

where for an N-node t_i, $S(t_i)$ and $K(t_i)$ are the independent set and the clique of the spider induced by the module $M(t_i)$. Then, every vertex in V_P (in V_S, resp.) is non-adjacent (adjacent, resp.) to the vertices in $M(t_k)$ since their least common ancestor t_i in $T(G)$ is a P-node (S-node, resp.), while the structural properties of a spider imply that every vertex in $K(t_j)$ ($S(t_j)$, resp.) for an N-node t_j is adjacent (non-adjacent, resp.) to the vertices in $M(t_k)$.

Our vertex-addition procedure relies on the following lemmata:

Lemma 6. *Suppose that the x-partly-adjacent nodes of the md-tree $T(G)$ lie on a path $t_0 t_1 \cdots t_k$, where t_0 is the root of $T(G)$. If t_k is a P-node then G' is P_4-sparse if and only if one of the following four (mutually exclusive) cases holds:*

(i) *Vertex x sees V_S and V_{N_K}, and misses V_P and V_{N_S}.*
(ii) *Vertex x sees V_S, V_{N_K}, and exactly one vertex, say, y, in V_P, and misses V_{N_S} where*
    (ii.1) vertex y is a child of node t_{k-2} (which is a P-node),
    (ii.2) t_{k-1} is an S-node with two children, the node t_k and one vertex, say, u (which is adjacent to x), and
    (ii.3) vertex x sees all the vertices in $M(t_k)$ except for a single vertex, say, b, which is a child of t_k.
(iii) *Vertex x sees V_{N_K}, all but one vertex, say, z, in V_S, and misses V_P and V_{N_S} where*
    (iii.1) vertex z is a child of node t_{k-1} (which is an S-node), and
    (iii.2) node t_k has two children a and b, which are leaf-nodes such that a is adjacent and b is non-adjacent to x.
(iv) *The node t_{k-1} is an N-node corresponding to a thick spider with independent set $S(t_{k-1})$, vertex x sees V_S, V_{N_K}, $S(t_{k-1})$, and all but one vertex, say, b, in $M(t_k)$, and misses V_P and $V_{N_S} - S(t_{k-1})$.*

The case where t_k is an S-node is precisely the complement version of Lemma 6: we need to exchange P- and S-nodes, thin and thick spiders, their cliques and independent sets, and what x sees/misses in the conditions of Lemma 6.

Lemma 7. *Suppose that the x-partly-adjacent nodes of the md-tree $T(G)$ lie on a path $t_0 t_1 \cdots t_k$, where t_0 is the root of $T(G)$. If t_k is an S-node then G' is P_4-sparse if and only if one of the following four (mutually exclusive) cases holds:*

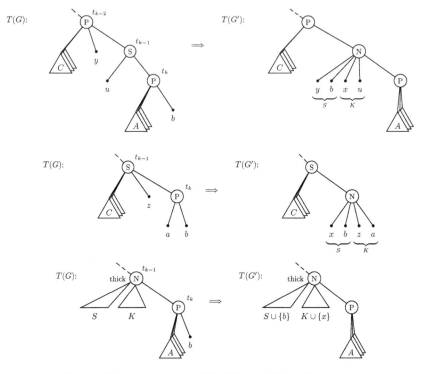

Fig. 5. Illustrating cases (ii), (iii), and (iv) of Lemma 6

(i)   Vertex x sees V_S and V_{N_K}, and misses V_P and V_{N_S}.

(ii) Vertex x sees V_{N_K}, all but one vertex, say, y, in V_S, and misses V_P and V_{N_S} where
  (ii.1) vertex y is a child of node t_{k-2} (which is an S-node),
  (ii.2) node t_{k-1} is a P-node with two children, the node t_k and one vertex, say, u (which is non-adjacent to x), and
  (ii.3) vertex x sees only a single vertex of $M(t_k)$, which is a child of t_k.

(iii) Vertex x sees V_S, V_{N_K}, and exactly one vertex, say, z, in V_P, and misses V_{N_S} where
  (iii.1) vertex z is a child of node t_{k-1} (which is a P-node), and
  (iii.2) node t_k has two children a, b, which are leaf-nodes such that a is adjacent and b is non-adjacent to x.

(iv) The node t_{k-1} is an N-node corresponding to a thin spider with clique $K(t_{k-1})$, vertex x misses V_P, V_{N_S}, $K(t_{k-1})$, and all but one vertex, say, b, in $M(t_k)$, and sees V_S and $V_{N_K} - K(t_{k-1})$.

Lemma 8. *Suppose that the x-partly-adjacent nodes of the md-tree $T(G)$ lie on a path $t_0 t_1 \cdots t_k$, where t_0 is the root of $T(G)$. If t_k is an N-node and the partition of the spider $G(t_k)$ is (S, K, R), then G' is P_4-sparse if and only if one of the following three (mutually exclusive) cases holds:*

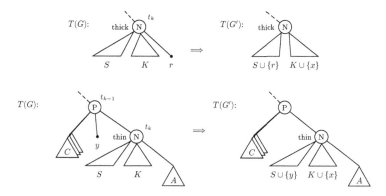

Fig. 6. Illustrating cases (i) and (ii.2) of Lemma 8

(i) *Vertex x sees $S \cup K$ (and misses $M(r)$ where $R = \{r\}$): vertex x sees V_S and V_{N_K}, and misses V_P and V_{N_S}, the spider corresponding to t_k is a thick spider, and the node r is a leaf.*

(ii) *Vertex x sees K (and misses S): one of the following three cases holds:*

   (ii.1) *vertex x sees V_S and V_{K_N}, and misses V_P and V_{N_S};*

   (ii.2) *vertex x sees V_S, V_{N_K}, and exactly one vertex, say, y, in V_P, and misses V_{N_S} where y is a child of t_{k-1}, the spider corresponding to t_k is thin, and all the vertices in $M(r)$ (if $R = \{r\}$) see x;*

   (ii.3) *vertex x sees V_{N_K}, all but one vertex, say, y, in V_S, and misses V_P and V_{N_S} where y is a child of t_{k-1}, the spider corresponding to t_k is thick, and all the verticts in $M(r)$ (if $R = \{r\}$) miss x.*

(iii) *Vertex x misses $S \cup K$ (and sees $M(r)$ where $R = \{r\}$): vertex x sees V_S and V_{N_K}, and misses V_P and V_{N_S}, the spider corresponding to t_k is a thin spider, and the node r is a leaf.*

Since in each case of Lemmata 6–8, x sees V_{N_K} and all but at most one of the elements of V_S, (i.e., all the x-partly-adjacent N-nodes and all but at most one x-partly-adjacent S-nodes belong to *Partial*), and since the parent of a P-node cannot be a P-node, we can show the following:

Observation 2. *For each node $t \in$ Partial at distance at least 4 from the root of the tree $T(G)$, if none of t's parent, grandparent, great-grandparent, and great-great-grandparent belongs to Partial, then the graph G' is not P_4-sparse.*

This implies that for G' to be P_4-sparse, node t_k of $T(G)$ is at depth at most $4d$.

The procedure that handles the addition of vertex x finds the node t_k and takes advantage of Lemmata 6–8. It starts from the leaves of the md-tree $T(G)$ which correspond to the neighbors of x and moving in a bottom-up fashion constructs the set A of internal nodes of $T(G)$ having at least one x-fully-adjacent child. Then, it splits A obtaining the set *Full* of x-fully-adjacent nodes of $T(G)$ and a subset *Partial* of x-partly-adjacent nodes, from which it determines t_k (vertex t' of Step 3). In detail, the procedure works as follows:

Procedure VERTEX_ADD(vertex x)

1. $A \leftarrow \emptyset$;
   construct a queue Q whose elements are pointers to each of the leaf-nodes of $T(G)$ which correspond to the neighbors of x;
   while the queue Q is not empty **do**
     remove from Q an element (i.e., a pointer to a node, say, t, of $T(G)$);
     increment the *counter*-field of the parent $p(t)$ of t by 1 and let its new value by *val*;
     if $val = 1$
     then insert in A a pointer to $p(t)$;
     if val = number of $p(t)$'s children
     then insert in Q a pointer to $p(t)$; {*t is x-fully-adjacent*}

2. $Full \leftarrow$ set of pointers to each of the leaf-nodes of $T(G)$ which correspond to the neighbors of x;
   $Partial \leftarrow \emptyset$;
   for each element a of the set A **do**
     let t be the node of $T(G)$ pointed to by a;
     if the value of t's *counter*-field is equal to the number of t's children
     then insert a in $Full$; {*t is x-fully-adjacent*}
     else insert a in $Partial$; {*t is x-partly-adjacent*}
     set t's *counter*-field equal to 0; {*reset the value of counter-field*}

3. $t' \leftarrow$ a node of largest depth in $T(G)$ among the nodes pointed to by the elements in $Partial$;
   if the depth of t' in $T(G)$ exceeds $4d$
   or there exists a node in $T(G)$ pointed to by an element in $Partial$ which is not an ancestor of t'
   or none of the cases of Lemmata 6, 7, and 8 applies to t'
   then output *false* (i.e., G' is not P_4-sparse); **return**;
   modify $T(G)$ depending on the case of Lemma 6, 7, or 8 which applies to t';

The correctness of the algorithm follows from Theorem 2, Observation 2, and from the following facts: (i) the set of nodes of the tree $T(G)$ pointed to by the elements of the set $Full$ is precisely the set of x-fully-adjacent nodes; (ii) the set of nodes of the tree $T(G)$ pointed to by the elements of the set $Partial$ are the x-partly-adjacent nodes of $T(G)$ with at least one x-fully-adjacent child (note that $t_k \in Partial$); (iii) the node t' found in Step 3 is precisely the x-partly-adjacent node t_k farthest away from the root.

3.4   Deleting a Vertex

Let $v \in V(G)$ be a vertex with d incident edges in G which has to be deleted. Clearly, the graph G' which results after the deletion of v is a P_4-sparse graph as it is an induced subgraph of G. Hence we focus on properly updating the md-tree $T(G)$ so that we obtain the md-tree $T(G')$.

Let us first consider the case where the parent-node $p(v)$ of v in $T(G)$ is an N-node t such that the spider partition of $G(t)$ is (S, K, R). We have:

(i) $v \in S$: First suppose that $S = \{v, v'\}$, $K = \{k, k'\}$, and let v be adjacent to k: then, the spider is replaced by an S-node with children the vertex k' and a P-node; if $R = \emptyset$, then this P-node has as children the vertices v' and k, else if $R = \{r\}$, it has as children the vertex v' and an S-node with children the vertex k and the node r. Now, suppose that $|S| = |K| \geq 3$ and let $f(v) = k \in K$. If the spider is thin then: if $R = \emptyset$, then after the removal of v, k is removed from K and is linked at the pointer for R; if $R = \{r\}$, then k is removed from K and if r is an S-node then k is linked as a child, otherwise the place of r is taken by an S-node with k and r as children. If the spider is thick, then after the removal of v, vertex k sees all the remaining vertices in $M(t)$; thus, the N-node t is replaced by an S-node with children the vertex k and the node t after we have removed the vertices v, k.

(ii) $v \in K$: Since the complement of a thin spider is a thick spider (and vice versa) with the clique and independent sets swapped (and if $R = \{r\}$, the P- and S-nodes in the subtree rooted at r swapped as well), this is the complement version of the previous case and takes the same time to handle.

(iii) $R = \{v\}$: In this case, v is deleted, and we obtain a spider with $R = \emptyset$.

Next, we consider the case where the parent-node $p(v)$ of v in $T(G)$ is a P- or S-node; if $p(v)$ has more than 2 children, it suffices to simply delete v. However, caution is needed if $p(v)$ has only two children, in which case the sibling u of v needs to be linked to the grandparent $p(p(v))$ of v; furthermore, if u and $p(p(v))$ are both P- or S-nodes, then the children of u are placed as children of $p(p(v))$. Finally, in either of the remaining two cases when the parent-node $p(v)$ has 2 children, i.e., if the sibling u of v is an N-node or if the grandparent $p(p(v))$ is an N-node, then u is linked as a child of $p(p(v))$.

3.5   Time Complexity

Lemmata 2–8 and Figures 2–6 show that handling each modification requires only local checks and changes. In order to perform them efficiently, we store in each node of the md-tree $T(G)$ its type (P, S, or N), the number of its children, as well as ways to access its parent and its children, and an auxiliary field *counter* (initialized to 0). If each P- or S-node stores pointers to its parent and to a list of its children, then edge additions (and deletions, by Theorem 1) are handled in $O(1)$ time, whereas vertex additions/deletions are handled in $O(d)$ time. Alternatively, the children of a P- or S-node may be stored in a binomial min-heap [2] in which the pointer to the parent of these children in $T(G)$ is stored at the minimum element of the heap. Then, edge additions (and deletions) and vertex deletions take $O(\log n)$ time, whereas vertex additions take $O(d \log n)$ time. Our results are summarized in the following theorem.

Theorem 3. *We have described a fully dynamic algorithm for recognizing P_4-sparse graphs and maintaining their modular decomposition tree. Edge modifications can be handled in $O(1)$ time while vertex modifications can be handled in $O(d)$ time; alternatively, edge modifications and vertex deletions can be handled in $O(\log n)$ time and vertex additions in $O(d \log n)$ time.*

References

1. A. Brandstädt, V.B. Le, and J. Spinrad, *Graph Classes – a Survey*, SIAM Monographs in Discrete Mathematics and Applications, SIAM, Philadelphia, 1999.
2. T.H. Cormen, C.E. Leiserson, R.L. Rivest, and C. Stein, *Introduction to Algorithms* (2nd edition), MIT Press, Inc., 2001.
3. D.G. Corneil, Y. Perl, and L.K. Stewart, A linear recognition algorithm for cographs, *SIAM J. Comput.* **14** (1985) 926–984.
4. A. Cournier and M. Habib, A new linear algorithm for modular decomposition, *Proc. 19th Int'l Colloquium on Trees in Algebra and Programming (CAAP'94)*, LNCS **787** (1994) 68–84.
5. C. Crespelle and C. Paul, Fully-dynamic recognition algorithm and certificate for directed cographs, *Proc. 30th Int'l Workshop on Graph-Theoretic Concepts in Computer Science (WG'04)*, LNCS **3353** (2004) 93–104.
6. C. Crespelle and C. Paul, Fully-dynamic algorithm for modular decomposition and recognition of permutation graphs, *Proc. 31st Int'l Workshop on Graph-Theoretic Concepts in Computer Science (WG'05)*, LNCS **3787** (2005) 38–48.
7. E. Dalhaus, J. Gustedt, and R.M. McConnell, Efficient and practical algorithms for sequential modular decomposition, *J. Algorithms* **41** (2001) 360–387.
8. X. Deng, P. Hell, and J. Huang, Linear time representation algorithms for proper circular arc graphs and proper interval graphs, *SIAM J. Comput.* **25** (1996) 390–403.
9. V. Giakoumakis and J.-M. Vanherpe, On extended P_4-reducible and P_4-sparse graphs, *Theoret. Comput. Sci.* **180** (1997) 269–286.
10. P. Hell, R. Shamir, and R. Sharan, A fully dynamic algorithm for recognizing and representing proper interval graphs, *SIAM J. Comput.* **31** (2002) 289–305.
11. C. Hoàng, Perfect graphs, Ph.D. Thesis, McGill University, Montreal, Canada, 1985.
12. W.-L. Hsu, On-line recognition of interval graphs in $O(m + n \log n)$ time, *Combinatorics and Computer Science* 1995, LNCS **1120** (1996) 27–38.
13. L. Ibarra, Fully dynamic algorithms for chordal graphs, *Proc. 10th Annual ACM-SIAM Symp. on Discrete Algorithms (SODA'99)*, (1999) 923–924.
14. L. Ibarra, A fully dynamic algorithm for recognizing interval graphs using the clique-separator graph, Technical Report, DCS-263-IR, University of Victoria, 2001.
15. B. Jamison and S. Olariu, Recognizing P_4-sparse graphs in linear time, *SIAM J. Comput.* **21** (1992) 381–406.
16. B. Jamison and S. Olariu, A tree representation for P_4-sparse graphs, *Discrete Appl. Math.* **35** (1992) 115–129.
17. R.M. McConnell and J. Spinrad, Modular decomposition and transitive orientation, *Discrete Math.* **201** (1999) 189–241.
18. J.H. Muller and J. Spinrad, Incremental modular decomposition, *J. ACM* **36** (1989) 1–19.
19. R. Shamir and R. Sharan, A fully dynamic algorithm for modular decomposition and recognition of cographs, *Discrete Appl. Math.* **136** (2004) 329–340.
20. J. Spinrad, P_4-trees and substitution decomposition, *Discrete Appl. Math.* **39** (1992) 263–291.

Clique Graph Recognition Is NP-Complete*

L. Alcón[1], L. Faria[2], C.M.H. de Figueiredo[3], and M. Gutierrez[1]

[1] Departamento de Matemática, UNLP, Argentina
[2] Departamento de Matemática, FFP, UERJ, Brazil
[3] Instituto de Matemática and COPPE, UFRJ, Brazil

Abstract. A *complete set* of a graph G is a subset of V inducing a complete subgraph. A *clique* is a maximal complete set. Denote by $\mathcal{C}(G)$ the *clique family* of G. The *clique graph* of G, denoted by $K(G)$, is the intersection graph of $\mathcal{C}(G)$. Say that G is *a clique graph* if there exists a graph H such that $G = K(H)$. The clique graph recognition problem asks whether a given graph is a clique graph. A sufficient condition was given by Hamelink in 1968, and a characterization was proposed by Roberts and Spencer in 1971. We prove that the clique graph recognition problem is NP-complete.

1 Introduction

We consider finite, simple and undirected graphs. V and E denote the vertex set and the edge set of the graph G, respectively. A *complete set* of G is a subset of V inducing a complete subgraph. A *clique* is a maximal complete set.

If G is a graph, $\mathcal{C}(G)$ denotes the *clique family* of G. The *clique graph* of G, denoted by $K(G)$, is the intersection graph of $\mathcal{C}(G)$. Say that G is *a clique graph* if there exists a graph H such that $G = K(H)$. Not every graph is a clique graph. The Clique graph recognition problem can be formulated as follows.

CLIQUE GRAPH
INSTANCE: A graph $G = (V, E)$.
QUESTION: Is there a graph H such that $G = K(H)$?

A sufficient condition for a graph to be a clique graph was given in [6], and characterizations of clique graphs are given in [9] and more recently in [1]. However the time complexity of the problem of recognizing clique graphs is still open [4,8,10].

Given a *set family* $\mathcal{F} = (F_i)_{i \in I}$, the sets F_i are called *members* of the family. $F \in \mathcal{F}$ means that F is a member of $\mathcal{F}$. The family is *pairwise intersecting* if the intersection of any two members is not the empty set. The *intersection* or *total intersection* of $\mathcal{F}$ is the set $\cap \mathcal{F} = \cap_{i \in I} F_i$. The family $\mathcal{F}$ has the *Helly property*, if any pairwise intersecting subfamily has nonempty total intersection.

The edge with end vertices u and v is represented by uv. We say that the complete set C *covers* the edge uv when u and v belong to C. A *complete edge cover* of a graph G is a family of complete sets of G covering all edges of G.

---

* Dedicated to Alberto Santos Dumont, aviation pioneer, on the 100th anniversary of the flight of his *14 Bis* in Paris in October 1906.

F.V. Fomin (Ed.): WG 2006, LNCS 4271, pp. 269–277, 2006.
© Springer-Verlag Berlin Heidelberg 2006

The following Theorem is a well known characterization of Clique Graphs.

Theorem 1 (Roberts and Spencer [9]). *G is a clique graph if and only if there exists a complete edge cover of G satisfying the Helly property.*

A *triangle* is a complete set with exactly 3 vertices. The set of triangles of G is denoted $T(G)$. Let $\mathcal{F}$ be a complete edge cover of G and T a triangle, $\mathcal{F}_T$ is the subfamily of $\mathcal{F}$ formed by all the members containing at least two vertices of T.

Next lemma is a characterization of a complete edge cover satisfying the Helly property, in what follows *RS-family*, which will be used in the proof of our main theorem.

Lemma 1 (Alcón and Gutierrez [2]). *Let $\mathcal{F}$ be a complete edge cover of G. The following conditions are equivalent:*
i) $\mathcal{F}$ has the Helly property.
ii) For every $T \in T(G)$, the subfamily $\mathcal{F}_T$ has the Helly property.
iii) For every $T \in T(G)$, the subfamily $\mathcal{F}_T$ has nonempty intersection, this means $\cap \mathcal{F}_T \neq \emptyset$.

As noted by Roberts and Spencer [9], Theorem 1 yields a polynomial certificate of G being a clique graph. First, for the polynomial size of the edge cover certificate, note that if $\mathcal{F}$ has the Helly property, then every subfamily $\mathcal{F}'$ of $\mathcal{F}$ has the Helly property as well. In addition, we prove that if G admits a complete edge cover $\mathcal{F}$, then G admits a complete edge cover $\mathcal{F}'$ of size at most $|E|$ which is our considered certificate: just greedily scan the edges of E, select for $\mathcal{F}'$ one complete set of $\mathcal{F}$ covering the first edge, and for each edge e not yet covered by $\mathcal{F}'$, select for $\mathcal{F}'$ one complete set of $\mathcal{F}$ covering e. Clearly this greedy procedure labels each selected set with a corresponding scanned edge of E, yielding a subfamily $\mathcal{F}'$ of size at most $|E|$. Second, for the polynomial verification of the certificate, a result of Berge [3] says that a family of sets has the Helly property, if and only if for any triple of elements, the subfamily of sets containing at least two out of these three elements has non-empty intersection. Actually, by Lemma 1, it is enough to consider the triples of vertices a, b, c of G defining a triangle T. We consider the members of $\mathcal{F}'_T$ and check for every vertex v of V if v belongs to $\cap \mathcal{F}'_T$. This produces an $O(n^4 m)$ algorithm that checks if a complete edge cover $\mathcal{F}'$ of size $O(m)$ is Helly. Thus CLIQUE GRAPH belongs to NP.

In this paper we prove that CLIQUE GRAPH is NP-complete by a reduction from the following version of the 3–satisfiability problem with at most 3 occurrences per variable.

Let $U = \{u_i, 1 \leq i \leq n\}$ be a set of boolean variables. A literal is either a variable u_i or its complement $\overline{u_i}$. A clause over U is a set of literals of L. Let $C = \{c_j, 1 \leq j \leq m\}$ be a collection of clauses over U. We say that variable u_i occurs in clause c_j (and then in C) if u_i or $\overline{u_i} \in c_j$. We say that variable u_i occurs in clause c_j as literal u_i (or that literal u_i occurs in c_j) if $u_i \in c_j$, and as literal $\overline{u_i}$ (or that literal $\overline{u_i}$ occurs in c_j) if $\overline{u_i} \in c_j$.

3SAT$_{\overline{3}}$

INSTANCE: $I = (U, C)$, where $U = \{u_i, 1 \le i \le n\}$ is a set of boolean variables, and $C = \{c_j, 1 \le j \le m\}$ a set of clauses over U such that each clause has two or three variables, each variable occurs two or three times in C, each variable occurs never twice in the same clause. If variable u_i occurs twice in C, then it is once as literal u_i and once as literal $\overline{u_i}$. If variable u_i occurs three times in C, then it is once as literal u_i and twice as literal $\overline{u_i}$.

QUESTION: Is there a truth assignment for U such that each clause in C has at least one true literal?

It is a known result that 3SAT$_{\overline{3}}$ is an NP-complete problem [5,7].

In order to reduce 3SAT$_{\overline{3}}$ to CLIQUE GRAPH we need to construct in polynomial time a particular instance G of CLIQUE GRAPH from a generic instance $I = (U, C)$ of 3SAT$_{\overline{3}}$, in such a way that C is satisfiable if and only if G is a clique graph.

In Section 2 we describe the construction of instance G of CLIQUE GRAPH from instance $I = (U, C)$ of 3SAT$_{\overline{3}}$. In Section 3, we state and prove the main theorem; and in Section 4 give some conclusions and propose new related problems.

2   Construction of G from $I = (U, C)$

Let $I = (U, C)$ be any instance of 3SAT$_{\overline{3}}$,

For each variable u_i let j_i be the subindex of the unique clause where variable u_i occurs as literal u_i; and $\overline{J}_i = \{j \mid$ literal $\overline{u_i}$ occurs in $c_j\}$.

For each clause c_j with $\mid c_j \mid = 3$, let $I_j = \{i \mid$ variable u_i occurs in $c_j\}$; and for each clause c_j with $\mid c_j \mid = 2$, let $I_j = \{i \mid$ variable u_i occurs in $c_j\} \cup \{n+1\}$. Notice that in any case $\mid I_j \mid = 3$. Given $I_j = \{i_1, i_2, i_3\}$, with $i_1 < i_2 < i_3$, let $i_1^* = i_2$, $i_2^* = i_3$ and $i_3^* = i_1$.

From instance $I = (U, C)$, we construct a graph $G = (V, E)$. Please refer to Figures 1 and 2. The vertex set V is the union:

$$V = \bigcup_{1 \le i \le n} \left[\{a_{j_i}^i, c_{j_i}^i, d_{j_i}^i, e_{j_i}^i, f_{j_i}^i, g_{j_i}^i, h_{j_i}^i\} \bigcup_{j \in \overline{J}_i} \{a_j^i, c_j^i, d_j^i, e_j^i, f_j^i, g_j^i, h_j^i, z_j^i, v_j^i, w_j^i\} \right] \bigcup$$

$$\bigcup_{1 \le j \le m, |c_j| = 2} \left[\{a_j^{n+1}, c_j^{n+1}, d_j^{n+1}, e_j^{n+1}, f_j^{n+1}, g_j^{n+1}, h_j^{n+1}\} \right].$$

Since $\mid \overline{J}_i \mid \le 2$, $\mid V \mid$ is bounded by $(n+1) \times 7 + n \times 2 \times 10 = 27 \times n + 7$.

The edge set E contains:

For each $j, 1 \le j \le m$, the edges of the complete graph induced by the vertex set $K_{12}(j) = \{a_j^i, d_j^i, g_j^i, h_j^i \mid i \in I_j\}$; the edges of the sets $\{c_j^i d_j^i \mid i \in I_j, i \ne n+1\}$ and $\{c_j^i a_j^i, c_j^i a_j^{i^*}, e_j^i d_j^i, e_j^i h_j^i, f_j^i g_j^i, f_j^i a_j^{i^*} \mid i \in I_j\}$.

And for each $i, 1 \le i \le n$, for each $j \in \overline{J}_i$, the edges of the complete graph induced by the vertex set $K_5(j, i) = \{h_{j_i}^i, g_{j_i}^i, v_j^i, h_j^i, g_j^i\}$; and the edges of the set $\{h_{j_i}^i w_j^i, w_j^i h_j^i, g_{j_i}^i z_j^i, z_j^i g_j^i, a_{j_i}^i v_j^i, v_j^i a_j^i\}$.

Notice that for each variable u_i, graph G contains as induced subgraph the graph depicted in Figure 1; and for each clause c_j, graph G contains as induced

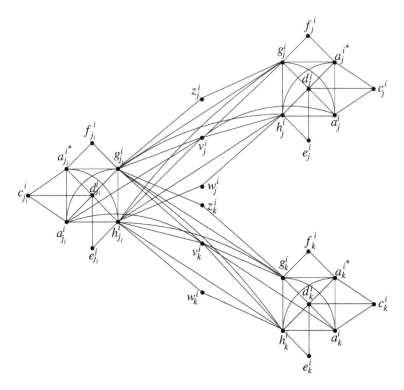

Fig. 1. Truth Setting component T_i for variable u_i with $\overline{J}_i = \{j, k\}$

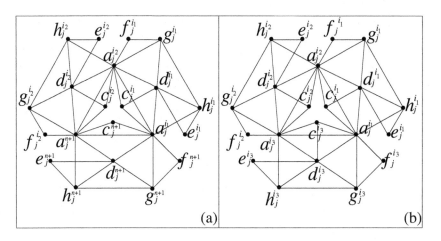

Fig. 2. Satisfaction Testing component S_j for clause c_j with: (a) 2 literals corresponding to variables u_{i_1} and u_{i_2}; and (b) 3 literals corresponding to variables u_{i_1}, u_{i_2} and u_{i_3}. Vertices $\{a_j^i, d_j^i, g_j^i, h_j^i \mid i \in I_j\}$ induce a complete graph but for simplicity some edges are not drawn.

subgraph the graph depicted either in Figure 2 (a) or (b), where some edges have been omitted. We obtain the whole graph G by superposing these subgraphs.

For the convenience of the reader we offer an example in Figure 5 of graph G obtained from the instance $I = (U, C)$, $U = \{u_1, u_2, u_3\}$, $C = \{\{u_1, u_3\}, \{\overline{u_1}, u_2, \overline{u_3}\}, \{\overline{u_1}, \overline{u_2}\}\}$.

2.1   About Graph G

The following two lemmata present properties of graph G that we will use in the proof of the main theorem. Notice that any RS-family of a graph contains the triangles of the graph that are cliques, in particular the triangles with a vertex of degree 2.

Lemma 2. (*Two Cover Lemma*) *Let $\mathcal{F}$ be an RS-family of the graph G. For each $j, 1 \leq j \leq m$, and for each $i \in I_j, i \neq n + 1$, exactly one of the triangles $\{a_j^i, a_j^{i^*}, c_j^i\}$, $\{a_j^i, c_j^i, d_j^i\}$ belongs to $\mathcal{F}$.*

Proof. Please refer to Figure 1. There are three possible complete sets of G covering a_j^i, c_j^i: $\{a_j^i, a_j^{i^*}, c_j^i, d_j^i\}$, $\{a_j^i, a_j^{i^*}, c_j^i\}$, $\{a_j^i, c_j^i, d_j^i\}$.

Note that both triangles $\{d_j^i, e_j^i, h_j^i\}$ and $\{f_j^i, a_j^{i^*}, g_j^i\}$ belong to $\mathcal{F}$, as e_j^i and f_j^i are vertices of degree 2 in G.

Suppose $\{a_j^i, a_j^{i^*}, c_j^i, d_j^i\} \in \mathcal{F}$. The Helly property implies $\{a_j^i, g_j^i, v_j^i\}$ is not contained in a complete set of $\mathcal{F}$, which implies $\{a_j^i, h_j^i, v_j^i\} \in \mathcal{F}$, in order to cover edge $a_j^i v_j^i$; but then $\{a_j^i, a_j^{i^*}, c_j^i, d_j^i\}$, $\{e_j^i, h_j^i, d_j^i\}$ and $\{a_j^i, h_j^i, v_j^i\}$ violate the Helly property. Then $\{a_j^i, a_j^{i^*}, c_j^i, d_j^i\} \notin \mathcal{F}$.

Assume $\{a_j^i, a_j^{i^*}, c_j^i\}$ and $\{a_j^i, c_j^i, d_j^i\}$ belong to $\mathcal{F}$. Since $\{a_j^i, a_j^{i^*}, c_j^i\} \in \mathcal{F}$, then $\{a_j^i, v_j^i, g_j^i\}$ is not contained in a complete of $\mathcal{F}$, thus $\{a_j^i, v_j^i, h_j^i\} \in \mathcal{F}$, but in this case $\{a_j^i, v_j^i, h_j^i\}$, $\{a_j^i, c_j^i, d_j^i\}$ and $\{e_j^i, h_j^i, d_j^i\}$ violate the Helly property.   □

Lemma 3. (*Literal Communication Lemma*) *Let $\mathcal{F}$ be an RS-family of the graph G. For each $i, 1 \leq i \leq n$, and for each $j \in \overline{J}_i$, if $\{a_j^i, c_j^i, d_j^i\} \in \mathcal{F}$ then $\{a_{j_i}^i, a_{j_i}^{i^*}, c_{j_i}^i\} \in \mathcal{F}$.*

Proof. Please refer to Figure 1. Since $\{a_j^i, c_j^i, d_j^i\} \in \mathcal{F}$ and $\{e_j^i, h_j^i, d_j^i\} \in \mathcal{F}$, we have that $\{a_j^i, h_j^i, v_j^i\}$ and $\{a_j^i, h_j^i, g_j^i, v_j^i\}$ do not belong to $\mathcal{F}$, because this violates the Helly property. Thus $\{a_j^i, g_j^i, v_j^i\} \in \mathcal{F}$, in order to cover edge $a_j^i v_j^i$.

Triangles $\{a_j^i, g_j^i, v_j^i\}$, $\{g_{j_i}^i, g_j^i, z_j^i\}$ belong to $\mathcal{F}$ implies $\{a_{j_i}^i, g_{j_i}^i, v_j^i\}$ and $\{a_{j_i}^i, h_{j_i}^i, g_{j_i}^i, v_j^i\}$ do not belong to $\mathcal{F}$. Thus $\{a_{j_i}^i, h_{j_i}^i, v_j^i\} \in \mathcal{F}$.

Since $\{a_{j_i}^i, h_{j_i}^i, v_j^i\}$ and $\{d_{j_i}^i, h_{j_i}^i, e_{j_i}^i\}$ belong to $\mathcal{F}$, then $\{a_{j_i}^i, c_{j_i}^i, d_{j_i}^i\} \notin \mathcal{F}$. By the <u>Two Cover Lemma</u> $\{a_{j_i}^i, a_{j_i}^{i^*}, c_{j_i}^i\} \in \mathcal{F}$.   □

These two lemmata are the basis of the proof of the main theorem. It follows that given any RS-family of G, and any variable u_i, by looking if one triangle of the satisfaction testing subgraph S_{j_i} belongs or not to the RS-family it is possible to know whether one triangle of the satisfaction testing subgraph S_j, with $j \in \overline{J}_i$, belongs or not to the RS-family. The two possible cases are shown in Figures 3 and 4.

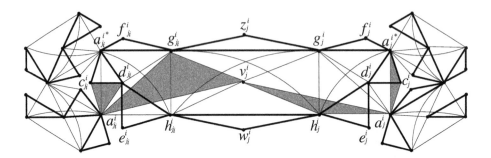

Fig. 3. Truth value of u_i equals to True

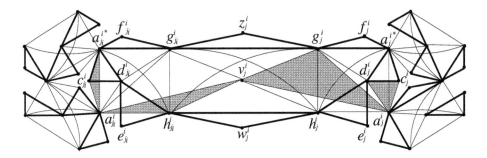

Fig. 4. Truth value of u_i equals to False

3   Main Theorem

Theorem 2. CLIQUE GRAPH *is NP-complete.*

Proof. As shown in the Introduction, CLIQUE GRAPH belongs to NP.

Let G be the graph obtained by Section 2 process from an instance $I = (U, C)$ of 3SAT$_{\overline{3}}$. Suppose G is a clique graph, we will exhibit a truth assignment for U such that C is satisfied.

Let $\mathcal{F}$ be an RS-family for G. Let $u_i \in U$ be a variable. Set u_i equal to true if and only if $\{a^i_{j_i}, c^i_{j_i}, d^i_{j_i}\} \in \mathcal{F}$.

To see that this truth assignment for U satisfies C consider a clause c_j.

The Helly property on $\mathcal{F}$ implies there exists $i \in I_j$ such that the triangle $\{a^i_j, a^{i^*}_j, c^i_j\}$ is not a member of $\mathcal{F}$. Notice that $i \neq n + 1$.

By the Two Cover Lemma, $\{a^i_j, a^{i^*}_j, c^i_j\} \notin \mathcal{F}$ implies $\{a^i_j, c^i_j, d^i_j\} \in \mathcal{F}$.

If $j = j_i$ then variable u_i is true and clause c_j is satisfied.

If $j \neq j_i$, then $j \in \overline{J}_i$, by Literal Communication Lemma, $\{a^i_j, c^i_j, d^i_j\} \in \mathcal{F}$ implies $\{a^i_{j_i}, a^{i^*}_{j_i}, c^i_{j_i}\} \in \mathcal{F}$, thus by Two Cover Lemma, $\{a^i_{j_i}, c^i_{j_i}, d^i_{j_i}\} \notin \mathcal{F}$. It follows that u_i is false, then c_j is satisfied.

Conversely, given a truth assignment of U that satisfies C, we exhibit a complete edge cover $\mathcal{F}$ of G.

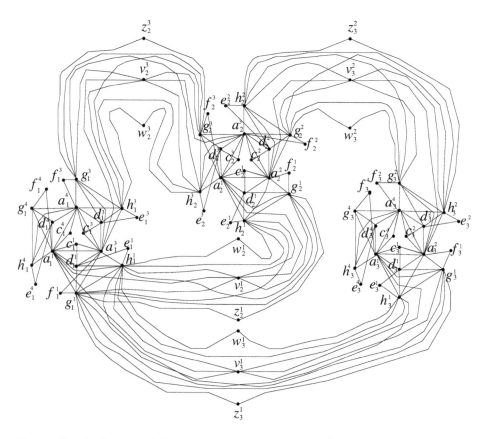

Fig. 5. Graph G obtained from the $3\text{SAT}_{\overline{3}}$ instance $U = \{u_1, u_2, u_3\}$, $C = \{\{u_1, u_3\}, \{\overline{u_1}, u_2, \overline{u_3}\}, \{\overline{u_1}, \overline{u_2}\}\}$.

For each $j, 1 \leq j \leq m$, complete set $K_{12}(j) = \{a_j^i, d_j^i, g_j^i, h_j^i \mid i \in I_j\}$;

For each $j, 1 \leq j \leq m$, for each $i \in I_j$, the triangles $\{f_j^i, a_j^{i^*}, g_j^i\}$, $\{e_j^i, d_j^i, h_j^i\}$.

For each $j, 1 \leq j \leq m$, for each $i \in I_j, i \neq n+1$, $\{c_j^i, a_j^{i^*}, d_j^i\}$; and for $i = n+1$, $\{c_j^{n+1}, a_j^{n+1^*}, a_j^{n+1}\}$.

For each $i, 1 \leq i \leq n$, for each $j \in \overline{J}_i$, the complete set $K_5(j, i) = \{h_{j_i}^i, g_{j_i}^i, v_j^i, h_j^i, g_j^i\}$;

For each $i, 1 \leq i \leq n$, for each $j \in \overline{J}_i$, $\{z_j^i, g_{j_i}^i, g_j^i\}$, $\{w_j^i, h_{j_i}^i, h_j^i\}$.

For each $i, 1 \leq i \leq n$, such that variable u_i is true, $\{c_{j_i}^i, d_{j_i}^i, a_{j_i}^i\}$; and for each $j \in \overline{J}_i$, $\{a_{j_i}^i, g_{j_i}^i, v_j^i\}$, $\{v_j^i, h_j^i, a_j^i\}$, $\{a_j^i, a_j^{i^*}, c_j^i\}$.

For each $i, 1 \leq i \leq n$, such that variable u_i is false, $\{c_{j_i}^i, a_{j_i}^i, a_{j_i}^{i^*}\}$; and for each $j \in \overline{J}_i$, $\{a_{j_i}^i, h_{j_i}^i, v_j^i\}$, $\{v_j^i, g_j^i, a_j^i\}$, $\{a_j^i, d_j^i, c_j^i\}$.

The proof is completed by showing that the complete edge cover $\mathcal{F}$ of G has the Helly property. By Lemma 1, it is enough to show that for each triangle $T \in T(G)$, $\cap \mathcal{F}_T \neq \emptyset$.

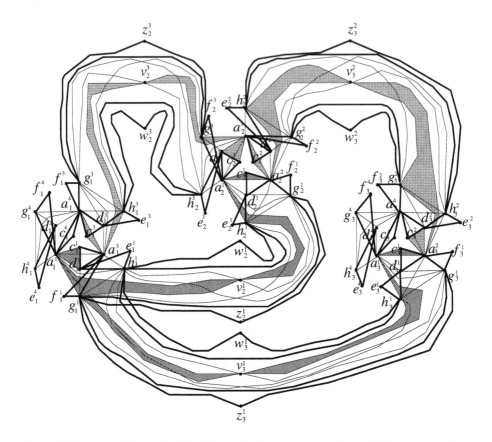

Fig. 6. RS-cover $\mathcal{F}$ for graph G of Figure 5, obtained from the satisfying truth assignment where u_1 is true, and u_2 and u_3 are false. Bold edges depict forced triangles of $\mathcal{F}$, dashed connected regions depict triangles of $\mathcal{F}$ which depend on the truth assignment for $I = (U, C)$, complete sets $K_5(j, i)$ and $K_{12}(j)$ are not depicted in order to make simpler the drawing.

If a triangle T contains an edge e for which any complete set of $\mathcal{F}$ covering e contains also T, then $\cap \mathcal{F}_T \neq \emptyset$. We call such a triangle an *easy triangle* and use this tool in order to accomplish the proof. We classify the triangles of G into types according to either they are, or they are not contained in a $K_{12}(j)$ or in a $K_5(j, i)$. Details are omitted in the extended abstract.   □

Figure 6 exhibits the RS-cover $\mathcal{F}$ defined from Theorem 2 for graph G of Figure 5.

4   Final Remarks

We have proved that deciding whether a given graph is a clique graph is an NP-complete problem. From the same proof, it follows that the problem remains NP-complete even for bounded degree graphs and for graphs with bounded clique size. However the problem is polynomial when restricted to graphs with maximum degree less than 5 and also when restricted to graphs with clique size less than 4 [10]. This fact suggests the search of the best bounds both for the maximum degree and for the clique size for which the problem is polynomial. Notice that the problem of recognizing clique graphs restricted to Planar Graphs of maximum clique size 4 was left open in [2]. It seems not to be trivial.

Acknowledgement. This research was partially supported by CAPES, CNPq, FAPERJ, PROCIÊNCIA-UERJ Project, CNPq–Project PROSUL- Proc. no. 490333/04-4.

References

1. Alcón, L., Gutierrez, M.: A new characterization of Clique Graphs. Matemática Contemporânea **25** (2003) 1–7
2. Alcón, L., Gutierrez, M.: Cliques and Extended Triangles. A necessary condition to be Clique Planar Graph. Discrete Applied Mathematics **141/1-3** (2004) 3–17
3. Berge, C.: "Hypergraphes". Gauthier-Villars Paris (1987)
4. Brandstädt, A., Le, V. B., Spinrad, J. P.: Graph Classes: A survey. SIAM Monographs on Discrete Mathematics and Applications (1999)
5. Garey, M. R., Johnson, D. S.: Computers and Intractability: a guide to the theory of NP-completeness. W. H. Freeman (1979)
6. Hamelink, R. C.: A partial characterization of clique graphs. Journal of Combinatorial Theory B **5** (1968), 192–197
7. Papadimitriou, C. M.: Computational Complexity. Addison-Wesley (1994)
8. Prisner, E.: A common generalization of Line Graphs and Clique Graphs. Journal of Graph Theory **18** (1994) 301–313
9. Roberts, F. S., Spencer, J. H.: A characterization of clique graphs. Journal of Combinatorial Theory B **10** (1971) 102–108
10. Szwarcfiter, J. L.: A survey on Clique Graphs. Recent Advances in Algorithmic Combinatorics, C. Linhares-Sales and B. Reed eds., Springer-Verlag (2002)

Homogeneity vs. Adjacency: Generalising Some Graph Decomposition Algorithms

B.-M. Bui Xuan[1], M. Habib[2], V. Limouzy[2], and F. de Montgolfier[2]

[1] LIRMM, Université Montpellier 2, France
buixuan@lirmm.fr
[2] LIAFA, Université Paris 7, France
{habib, limouzy, fm}@liafa.jussieu.fr

Abstract. In this paper, a new general decomposition theory inspired from modular graph decomposition is presented. Our main result shows that, within this general theory, most of the nice algorithmic tools developed for modular decomposition are still efficient.

This theory not only unifies the usual modular decomposition generalisations such as modular decomposition of directed graphs and of 2-structures, but also decomposition by star cutsets.

1 Introduction

Several combinatorial algorithms are based on partition refinement techniques [16]. Graph algorithms make an intensive use of vertex splitting, the action of partitioning classes between neighbours and non-neighbours of a vertex. For instance, all known linear-time modular decomposition algorithms [4,6,8,12,14] use this technique.

In bioinformatics also, the distinction of a set by an element, so-called splitter, seems to play an important role, as for example in the nice computation of the set of common intervals of two permutations by T. Uno and M. Yagiura [17].

We investigate the abstract notion of splitters and subsequently propose a formalism based on the concept of homogeneity. Our aim is a better understanding of the existing modular decomposition algorithms by characterising the algebraic properties on which they are based. Our main result is that most of the nice algorithmic tools developed to compute a representation for modular decomposition [4,6,8,12,14] are still efficient within this general theory.

This theory not only unifies the usual modular decomposition generalisations such as modular decomposition of directed graphs [13] and of of 2-structures [9], but also allows to handle star cutsets.

The paper is structured as follows: we first detail the new combinatorial decomposition theory, then present a general algorithmic framework, and close the paper with interesting outcomes.

2 Homogeneity, a New Viewpoint

Throughout this section X is a finite set, and $\mathcal{P}(X)$ the family of all subsets of X. Two sets A and B *overlap* if $A \cap B$, $A \setminus B$ and $B \setminus A$ are all nonempty. It

F.V. Fomin (Ed.): WG 2006, LNCS 4271, pp. 278–288, 2006.

is denoted $A \infty B$. A *reflectless* triple is $(x, y, z) \subseteq X^3$ with $x \neq y$ and $x \neq z$. Reflectless triples will be denoted by $(x|yz)$ instead of (x, y, z) since the first element does not play the same role. Let H be a relation over the reflectless triples of X. Given $s \in X$, we define H_s as the binary relation on $X \setminus \{s\}$ such that $H_s(x, y)$ iff $H(s|xy)$.

Definition 1. *H is a* homogeneous relation *on X if, for all $s \in X$, H_s is an equivalence relation on $X \setminus \{s\}$ (i.e. it fulfills the Symmetry and Reflexivity and Transitivity properties). The equivalence classes of H_s are called the s-classes and denoted $H_s^1...H_s^k$.*

Definition 2. *Let H be a homogeneous relation on X. $M \subseteq X$ is a* module *of H if $\forall m, m' \in M$, $\forall s \in X \setminus M$, $H(s|mm')$.*

If $\neg H(s|mm')$ we say that s distinguishes m from m', or is a splitter *of $\{m, m'\}$. A module M is* trivial *if $|M| \leq 1$ or $M = X$. The family of modules of H is denoted by $\mathcal{M}_H$, or $\mathcal{M}$ if not ambiguous. H is* prime *if $\mathcal{M}_H$ is reduced to the trivial modules.*

Remark 1. From the definition it is obvious that, given a module M, if $\neg H(s|xy)$ for some $x, y \in M$ then $s \in M$.

Homogeneity and distinction can be applied to graphs. Indeed, there is a natural homogeneous relation associated to graphs as follow:

Definition 3. *The* standard homogeneous relation $H(G)$ *of a directed graph $G = (X, E)$ is defined such that, for all $s, x, y \in X$, $H(s|xy)$ is true if and only if the following two conditions hold:*
1. either both x and y or none of them are in-neighbours of s, and
2. either both x and y or none of them are out-neighbours of s.

In other word, $H(s|xy)$ tells if s "sees" x and y in the same way. Of course this definition also holds for undirected graphs, tournaments, and can also be extended to 2-structures [9].

Proposition 1. *For a graph G, the modules of its standard homogeneous relation $H(G)$ are the modules of G in the usual sense [11,15].*

Proposition 2. *For all $A, B \in \mathcal{M}$ if $A \infty B$ then $(A \cap B) \in \mathcal{M}$ and $(A \cup B) \in \mathcal{M}$.*

This property is called here *closure under intersection and union*. It is easy to check, and can be used to prove:

Proposition 3 (Lattice structure). *Let H be a homogeneous relation on X and $\mathcal{M}'_H = \mathcal{M}_H \cup \{\emptyset\}$. Then, $(\mathcal{M}'_H, \subseteq)$ is a lattice.*

This lattice is a sublattice of the boolean lattice (hypercube) on X. Moreover, if we consider $A \in \mathcal{M}$ such that $|A| \geq 1$, $\mathcal{M}(A) = \{M \in \mathcal{M}_H \text{ and } M \supseteq A\}$ then $(\mathcal{M}(A), \subseteq)$ is a distributive lattice. Let us now define some useful types of homogeneous relations.

Definition 4. *A homogeneous relation H is said to be*

- **Graphical** *if $\forall \ x, y, z \in X$, $H(x|yz) \ \wedge \ H(y|xz) \ \Rightarrow \ H(z|xy)$*
- **Quotiental** *if $\forall \ s, t, x, y \in X$, $H(x|st) \ \wedge \ H(y|st) \ \wedge \ H(t|xy) \Rightarrow H(s|xy)$*
- **Digraphical** *if $\forall s, t, x, y \in X, H(x|st) \wedge H(y|st) \wedge H(t|sx) \wedge H(t|sy) \Rightarrow H(s|xy)$*

Notice that if H is **Quotiental** then for each module M and for all $x, y \in M$ and $s, t \in X \setminus M$, $H(x|st) \ \Leftrightarrow \ H(y|st)$. Indeed, for the *Quotiental* relations, elements in a module M uniformly perceive a set A not intersecting M: if one element of M distinguishes A then so do all. This allows to shrink M into a single element, the quotient by M, or to pick a *representative element* from the module. This is here called the *Quotient* property.

Given $A \subseteq X$ one can define the induced relation $H[A]$ as H restricted to reflectless triples of A^3. If A is a module we have the following nice property:

Proposition 4 (Restriction). *Let H be a homogeneous relation, M a module and $M' \subseteq M$. Then, $M' \in \mathcal{M}_{H[M]} \ \Leftrightarrow \ M' \in \mathcal{M}_H$.*

Recursiveness can therefore be used when dealing with modules. Notice that the proposition is not always true if M is not a module. The *Quotient* and *Restriction* properties were used first with modular decomposition of graphs and are useful for algorithmics [15].

3   Submodularity of Homogeneous Relations

Definition 5. *A set function $\mu : \mathcal{P}(X) \to \mathbb{R}$ is submodular if and only if for all $A, B \subseteq X$ $\mu(A) + \mu(B) \geq \mu(A \cup B) + \mu(A \cap B)$ (see e.g. [10]).*

Theorem 1 (Submodularity). *Let H be a homogeneous relation on X. Let $s(E)$ be the function counting the number of splitters of a nonempty subset E of X, and such that $s(\emptyset) = -|X|$. s is submodular.*

Proof. It suffices to prove $s(A) + s(B) \geq s(A \cup B) + s(A \cap B)$ for all overlapping $A, B \subseteq X$. So let $A, B \subseteq X$ be two overlapping sets. For convenience $\mathcal{S}_A$ denotes the set of all splitters of A. We note $X = \{X_1, \ldots, X_k\}$ if $\{X_1, \ldots, X_k\}$ is a partition of X. Obviously, $\mathcal{S}_{A \cap B} = \{\mathcal{S}_{A \cap B} \setminus B, \ \mathcal{S}_{A \cap B} \cap B\}$.

As $\mathcal{S}_A \cap A = \emptyset$, the partition $\mathcal{S}_{A \cup B} = \{\mathcal{S}_{A \cup B} \setminus \mathcal{S}_A, \ \mathcal{S}_{A \cup B} \cap \mathcal{S}_A\}$ can be reduced to $\mathcal{S}_{A \cup B} = \{\mathcal{S}_{A \cup B} \setminus \mathcal{S}_A, \ \mathcal{S}_A \setminus (A \cup B)\}$. Similarly, $\mathcal{S}_B = \{\mathcal{S}_B \setminus \mathcal{S}_{A \cap B}, \ \mathcal{S}_{A \cap B} \setminus B\}$. Finally, $\mathcal{S}_A = \{\mathcal{S}_A \setminus B, \ (\mathcal{S}_A \cap B) \setminus \mathcal{S}_{A \cap B}, \ (\mathcal{S}_A \cap B) \cap \mathcal{S}_{A \cap B}\}$ can be reduced to $\mathcal{S}_A = \{\mathcal{S}_A \setminus (A \cup B), \ (\mathcal{S}_A \cap B) \setminus \mathcal{S}_{A \cap B}, \ \mathcal{S}_{A \cap B} \cap B\}$.. Hence,
$$|\mathcal{S}_A| + |\mathcal{S}_B| - |\mathcal{S}_{A \cup B}| - |\mathcal{S}_{A \cap B}| = |(\mathcal{S}_A \cap B) \setminus \mathcal{S}_{A \cap B}| + |\mathcal{S}_B \setminus \mathcal{S}_{A \cap B}| - |\mathcal{S}_{A \cup B} \setminus \mathcal{S}_A|.$$
To achieve proving the theorem, we prove that $\mathcal{S}_{A \cup B} \setminus \mathcal{S}_A \subseteq \mathcal{S}_B \setminus \mathcal{S}_{A \cap B}$. Indeed, let $z \in \mathcal{S}_{A \cup B} \setminus \mathcal{S}_A$. Then, $z \notin A \cup B$ and $H(z|xy)$ for all $x, y \in A$. Now, suppose that $z \notin \mathcal{S}_B$. Since z is not in $A \cup B$, we have $H(z|xy)$ for all $x, y \in B$. Furthermore, as A and B overlap and thanks to the transitivity of H, we have $z \notin A \cup B$ and $H(z|xy)$ for all $x, y \in A \cup B$, which is by definition $z \notin \mathcal{S}_{A \cup B}$. Contradiction. Finally, supposing $z \in \mathcal{S}_{A \cap B}$ would imply $z \in \mathcal{S}_A$. $\qquad \square$

Submodular functions are combinatorial objects with powerful potential (see e.g. [10]). Theorem 1 enables the application of this theory to homogeneous relations. In [17], T. Uno and M. Yagiura gave a (restricted) version of this theorem, and constructed a very nice algorithm computing the common intervals of a set of permutations. It would be interesting to generalise this approach to any homogeneous relation, as done in [3] for modular decomposition of graphs.

4   Strong Modules and Primality

In a family $\mathcal{F}$ of subsets of X, a subset is *strong* if it overlaps no other subset of $\mathcal{F}$. The other subsets are *weak*. If $\mathcal{F}$ contain X and the singletons $\{x\}$ for every element $x \in X$, then X and $\{x\}_{x \in X}$ form the *trivial* strong subsets. The set inclusion orders the strong subsets into a tree. This is a quick proof that there are at most $2|X| - 1$ strong subsets (and at most $|X| - 2$ nontrivial ones), as the tree has no internal node of degree 1.

The *parent* of a (possibly weak) subset M is the smallest strong subset M_P properly containing M, and M is said to be a *child* of M_P. If M is strong, M_P is by definition its parent in the inclusion tree.

An *overlap class* is an equivalence class of the transitive closure of the overlap relation $\oslash$ on $\mathcal{F}$. The *support* of an overlap class $\mathcal{C} = \{C_1, \ldots, C_k\}$ is $C_1 \cup \cdots \cup C_k$. A is an *atom* of the overlap class if it is included in at least one subset C_i, and it does not overlap any subset of the class, and is maximal for these properties. All the atoms of a class form a partition of its support, the coarsest partition compatible with the class. An overlap class is *trivial* if it contains only one subset; it is then clearly a strong one.

A strong subset is *prime* if all its children are strong, and *decomposable* otherwise. It is a classical result of set theory that

Lemma 1. *If $\mathcal{F}$ is a family closed under union of overlapping sets, then there is an one-to-one correspondence between the nontrivial overlap classes of $\mathcal{F}$ and the decomposable strong subsets of $\mathcal{F}$. More precisely, the overlap class $\mathcal{C}$ associated with a decomposable subset D is simply the set of weak children of D, and the support of $\mathcal{C}$ is D.*

Of course we apply all these notions to the family of modules of a homogeneous relation.

Theorem 2. *Let H be a homogeneous relation and $\mathcal{Z}$ be the family of modules containing x but not y, and maximal for this property, for all x and y. The strong modules of H are exactly the supports and atoms of all overlap classes of $\mathcal{Z}$.*

Proof. First, remark that, thanks to the closure under union of overlapping sets, the supports and atoms of every overlap class of $\mathcal{Z}$ are strong modules. Lemma 1 tells they cannot be overlapped by an element of $\mathcal{Z}$ and if one, say A, is overlapped by a module $B \notin \mathcal{Z}$ then for $x \in A \setminus B$, the maximal module containing y but not x overlaps A, a contradiction. So the family of supports and atoms is included in the family of strong modules. Conversely, let us prove that if M is a strong module then it is the support or an atom of some overlap class. We shall distinguish four cases.

Let M_P be the strong parent of M (for $M \neq X$). 1. M is trivial (X or $\{x\}$). There is no problem.

2. M is decomposable. It has k strong children $M_1, \ldots, M_k$. Let us pick an element x_i in each M_i. Then for all i and j we consider the maximal module containing x_i but not x_j. They form an overlap class of $\mathcal{Z}$. Its support is M, thanks to Lemma 1.

3. M is prime and M_P is prime. Then for all $x \in M$ and all $y \in M_P \setminus M$, M is the maximal module containing x but not y. As it is strong, it belongs to a trivial overlap class and is equals to its support.

4. M is prime and M_P is decomposable. Then for all $x \in M_P \setminus M$, M is included in some maximal module M_x not containing x (the one that contains the vertices of M). Let us consider the intersection I of all subsets of $\{M_x \mid x \in M_P \setminus M\}$. It is an atom of the overlap class associated with M_P and thus is strong. As M is a children of M_P, $I = M$.   □

5 Partitive Families of Homogeneous Sets

A generalisation of modular decomposition, known from [5], less general than homogeneous relations but more powerful, is the *partitives families*. The *symmetric difference* of two sets A and B, denoted $A\Delta B$, is $(A \setminus B) \cup (B \setminus A)$.

Definition 6. *A family* $\mathcal{F} \subseteq \mathcal{P}(X)$ *is* weakly partitive *if it contains X and the singletons $\{x\}$ for all $x \in X$, and is closed under union, intersection and difference of overlapping subsets, i.e.*
$$A \in \mathcal{F} \ \wedge \ B \in \mathcal{F} \ \wedge \ A \oslash B \ \Rightarrow \ A \cap B \in \mathcal{F} \ \wedge \ A \cup B \in \mathcal{F} \ \wedge \ A \setminus B \in \mathcal{F}$$
Furthermore a weakly partitive family $\mathcal{F}$ *is* partitive *if it is also closed under symmetric difference:* $A \in \mathcal{F} \ \wedge \ B \in \mathcal{F} \ \wedge \ A \oslash B \ \Rightarrow \ A \Delta B \in \mathcal{F}$

As mentionned before, strong subsets of a weakly partitive family $\mathcal{F}$ can be ordered by inclusion into a tree. Let us define three types of strong subsets, i.e. three types of nodes of the tree:

- *prime* nodes which have no weak children,
- *degenerate* nodes: any union of strong children of the node belongs to $\mathcal{F}$,
- *linear* nodes: there is an ordering of the strong children such that a union of children belongs to $\mathcal{F}$ if and only if the children follow consecutively in this ordering.

Theorem 3. *[5] In a partitive family, there are only prime and degenerate nodes. In a weakly partitive family, there are only prime and degenerate and linear nodes.*

The strong subsets are therefore an $O(|X|)$ space coding of the family: it is enough to type the nodes into complete, linear or prime, and to order the children of the linear nodes. All weak subsets can be outputted by making simple combinations of the strong children of decomposable (complete or linear) nodes. Now, the following properties state that modules of some homogeneous relations are a proper generalisation of (weakly) partitive families.

Proposition 5. *Let H be a homogeneous relation. If H is* Graphical *or is* Quotiental, *then H is* Digraphical.

Proposition 6. *The modules of a* Quotiental *(resp.* Digraphical*) relation form a weakly partitive family. The modules of a* Graphical *relation form a partitive family.*

Proof. Let us suppose $A \in \mathcal{F}_H$ and $B \in \mathcal{F}_H$ and $A \infty B$. Thanks to transitivity an element not in $A \cup B$ cannot distinguish $A \cup B$ (it would distinguish A or B). As an element not in A cannot distinguish A and an element not in B cannot distinguish B, then no element can distinguish $A \cap B$. For the same reason, only an element of $A \cap B$ can distinguish $A \setminus B$ or $A \Delta B$.

If $s \in A \cap B$ distinguishes $A \setminus B$, then this set contains x and y such that $\neg H(z|xy)$. As $B \setminus A$ is nonempty it contains t and we have $H(x|st)$ and $H(y|st)$ and $H(t|sx)$ and $H(t|sy)$ and $H(t|xy)$. Then H is neither *Quotiental* nor *Digraphical*.

Let us suppose H is *Graphical*. As it is also *Digraphical*, $A \setminus B$ and $B \setminus A$ are modules. If $z \in A \cap B$ distinguishes $A \Delta B$, then there exists $x \in A$ and $y \in B$ such that $\neg H(z|xy)$. Since $H(x|yz)$ and $H(y|xz)$, H is not *Graphical*, a contradiction. □

6   Modular Algorithmics for Homogeneous Relations

In the following, we consider a given ground set X and a homogeneous relation H on X, that are the input of all algorithms described here. The input H consists in $|X|$ partitions (the equivalence classes of H_x for each x) and thus can be stored in $O(|X|^2)$ space, instead of the naive $O(|X|^3)$ space representation storing all triples.

6.1   Smallest Module Containing a Subset

Let S be a nonempty subset of X. As $\mathcal{F}_H$ is closed under intersection, there is a unique smallest module containing S, the intersection of all modules containing S, denoted henceforth $SM(S)$.

Theorem 4. *Algorithm 1 computes $SM(S)$ in $O(|X|.|SM(S)|) = O(|X|^2)$ time.*

Proof. Time complexity is obvious as the **while** loop runs $|M| - 1$ times and the **for** loop $|X|$ times. The algorithm maintains the invariant that every splitter of M is in F. When M is replaced by $M \cup \{y\}$, every element that distinguishes $M \cup \{y\}$ distinguishes x from y, or already is in F. The algorithm ends therefore on a homogeneous set that contains S, and thus we have $SM(S) \subseteq M$. If $M \neq SM(S)$ let s be the first element of $M \setminus SM(S)$ added to F (eventually added to M). It distinguished two elements x and y from $SM(S)$, contradicting its homogeneity. So $SM(S) = M$. □

Algorithm 1. Smallest Module containing S

Let x be an element of S, $M := \{x\}$ and $F := S \setminus \{x\}$
while F *is not empty* **do**
   pick an element y in F ; $F := F \setminus \{y\}$; $M := M \cup \{y\}$
   for *every element* z **do**
      if $H(z|x,y)$ **then** $F := F \cup \{z\}$
   output M (now equals to $SM(S)$)

6.2 Maximal Modules Not Containing an Element

Proposition 7. *Let x be an element of X. As $\mathcal{F}_H$ is closed under union of intersecting subsets, there is a unique partition of $X \setminus \{x\}$ into $S_1, \dots, S_k$ such that every S_i is a module of $\mathcal{F}_H$ and is maximal w.r.t. inclusion in $\mathcal{F}_H$.*

We call $MaxM(x) \subset \mathcal{P}(V)$ this partition of maximal modules not containing x. We propose a partition refining algorithm [16]. It is obvious that

Lemma 2. *Every module (especially the maximal ones) not containing x is included in a x-class H_x^i of H.*

Therefore our algorithm starts with the partition $P = \{H_x^1, \dots, H_x^k\}$ of the x-classes of H, each part is an x-class. Then the partition is refined (parts are splitted) using the following rule. Let y be an element, called the *pivot*, and Y the part of P containing y.

Rule 1. *split every part A of P, except for Y, into $A \cap H_y^1, \dots, A \cap H_y^l$*

Notice that a part if broken in smaller ones iff it is distinguished by y.

Lemma 3. *Starting from the partition $P_0 = \{H_x^1, \dots, H_x^k\}$, the application of Rule 1 (for any pivot in any order) until no part can be actually splitted, produces $MaxM(x)$.*

Proof. The refining process ends when no pivot can split a part, i.e when every part is a module. Let us suppose one of these modules M is not maximal w.r.t. inclusion: it is included in a module M', itself included in a x-class H_x^i. Let us consider the pivot y that first broke M'. It cannot be out of M', as M' is module, nor within M', as a pivot does not break its own part. But M' was broken, contradiction. ☐

Let us now implement this lemma into an efficient algorithm. Let P_i be the partition after the ith application of Rule 1, y be a given vertex used as pivot, and Y_i the part of P_i containing y. We say that a part B of P_j descends from a part A of P_i if $i < j$ and $A \subset B$. Clearly, after y is chosen as pivot at step i, y does not distinguish any part of P_i excepted Y_i. If y is chosen as pivot after, at step $j > i$, y may only split the parts of P_{j-1} that descend from Y_i. Only these parts have to be examined for implementing Rule 1. But Y_j itself has not to be examined.

Let us suppose that, for a part A, we can split it in $O(|A|)$ time when applying Rule 1 with pivot y. Then the time spent at step j is $O(|Y_i| - |Y_j|)$, the sum of the size of the parts that descend from Y_i save Y_j. The time of all splittings with y as pivot is $O(|X|)$, leading to an $O(|X|^2)$ time complexity.

Let us suppose that the parts are implemented as a linked list [12], and the new parts created after splitting an old one replace it and follow consecutively in the list. Then for each pivot y two pointers, one on the first part that descend from Y_i and the second to the last part, are enough to tell the parts to be examinated. A simple sweep between the pointers, omitting Y_j, gives them.

Now let us see how a part A can be split in $O(|A|)$ time. It is a classical trick of partition refining [16,12]. If the y-classes are numbered from 1 to k, then A can be bucket sorted in $O(|A| + k)$ time, then each bucket gives a new part that descend from A. If $|A| < k$, we have to renumber the used y-classes from 1 to $k' \leq |A|$ before bucket sorting. A first sweep on A marks the used y-class numbers. A second sweep unmarks an used number the first time it is seen, and replace it by the new number (an incremented counter) which is less than $|A|$. The vector of y-classes numbers is initialized once in $O(k)$ time.

The last point is the order in which the pivot are taken. Using all elements as pivots, and repeating this $|X|$ time, i.e. $|X|^2$ applications of Rule 1, is enough. A clever choice is to use y only if Y_i has been split, keeping a queue of "active" pivots. We thus have:

Theorem 5. *$MaxM(x)$ can be computed in $O(|X|^2)$ time.*

6.3   Testing if a Homogeneity Relation Is Trivial

A homogeneous relation H on X is *trivial* if $\mathcal{F}_H$ contains only X and the singletons.

Theorem 6. *Let S be a nonempty subset of X. One can test in $O(|X|^2)$ time if H is trivial.*

Proof. If $|X| < 2$ the answer is yes. Otherwise let x and y be two elements of X. In $O(|X|^2)$ time, the algorithm of Section 6.2 outputs the maximal modules not containing x. If one of them is nontrivial the answer is no. Otherwise all nontrivial modules contain x. In $O(|X|^2)$ time, the algorithm of Section 6.2 outputs the maximal modules not containing y. If one of them is nontrivial the answer is no. Otherwise all nontrivial modules contain x and y. Then Algorithm 1 is used with $S = \{x, y\}$, in $O(|X|^2)$ time. The answer is yes iff $SM(\{x, y\}) = X$.   □

6.4   Strong Modules of a Homogeneous Relation

Theorem 2 straightforwardly leads to an algorithm:

Theorem 7. *The strong modules of a homogeneous relation H on X can be computed in $O(|X|^3)$ time.*

Proof. First compute $MaxM(x)$ for all $x \in X$. All these sets together exactly form the family $\mathcal{Z}$ defined in Theorem 2. It can be done in $O(|X|^3)$ time using the algorithm of Section 6.2 $|X|$ times. The size of this family (sum of the cardinals of every subsets) is $O(|X|^2)$ since they form $|X|$ partitions. Using Dahlhaus algorithm [7] the overlap components can be found in time linear on the size of the family, namely $O(|X|^2)$. According to Lemma 1 there are at most $|X|$ nontrivial overlap classes.

For each class it is easy to compute its support, and in $O(|X|^2)$ time easy to compute its atoms. For instance, consider the vector of parts of the overlap class containing a given element: the atoms are the elements with the same vector. Sorting the list of elements of the supports $O(|X|)$ times, one time per part, gives the elements with the same vector, thus the atoms.

And at least the $O(|X|^2)$ supports and atoms must be sorted by inclusion order into the inclusion tree of the strong modules. It can be done in $O(|X|^3)$ time using the same sorting technique. $\square$

7   Outcomes

Let us examine in the sequel some of the applications of this homogeneity theory to modular decomposition of graphs and 2-structures, and to other graph relations. The name of *Graphical, Quotiental,* and *Digraphical* relations are justified by the following proposition:

Proposition 8. *The standart homogeneous relation of a directed graph is* Quotiental *and* Digraphical. *If the graph is undirected, its standard relation also is* Graphical.

The notion of modules also extends to 2-structures [9]. A (symmetric) 2-structure is a complete edge-coloured (undirected) graph and $H(x|yz)$ is true when edges (xy) and (xz) have the same colour. We still have:

Proposition 9. *The standart homogeneous relation of a 2-structure* Quotiental *and* Digraphical. *If the 2-structure is symmetric, its standard relation also is* Graphical.

The modules of an undirected graph and of a symmetric 2-structure thus form a partitive family, while the modules of a directed graph just form a weakly partitive family. All know properties of modular decomposition [15] can be derived from this result. An $O(n^3)$ modular decomposition algorithm can also be derived from Section 6.4 algorithm, but it is less efficient than the existing algorithms [4,6,8,12,14].

In a graph we can consider different homogeneous relations, for instance the relation *"there exists a path from vertex x to vertex y avoiding the vertex s"*, or a more general relation *"there exists a path from x to y avoiding the neighbourhood of s"*. It is easy to see that these two relations fulfill the basic axioms (symmetry, reflexivity and transitivity). In the first case, the strong modules form a partition (into the 2-vertex-connected components, minus the articulation points). The second relation is related to decomposition into star cutsets.

Another interesting relation is $D_k(s|xy)$ if $d(s,x) \leq k$ and $d(s,y) \leq k$, where $d(x,y)$ denotes the distance between x and y. The case $k = 1$ corresponds to modular decomposition. It is worth investigating the general case.

8   Conclusion

We hope that this homogeneity theory will have many other applications and will be useful to decompose automata [1] and boolean functions [2]. Obviously, the algorithmic framework presented here can be optimised in each particular application, as it can be done for modular decomposition [4,6,8,12,14]. We think the homogeneity concept is a very general idea.

References

1. C. Allauzen and M. Mohri. Efficient algorithms for testing the twins property. *Journal of Automata, Languages and Combinatorics*, 8(2):117–144, 2003.
2. J. Bioch. The complexity of modular decomposition of boolean functions. *Discrete Applied Mathematics*, 149(1-3):1–13, 2005.
3. B.-M. Bui Xuan, M. Habib, and C. Paul. Revisiting T. Uno and M. Yagiura's Algorithm. In *16th International Symposium of Algorithms and Computation (ISAAC05)*, volume 3827 of *LNCS*, pages 146–155, 2006.
4. C. Capelle, M. Habib, and F. de Montgolfier. Graph decomposition and factorizing permutations. *D.M.T.C.S*, 5(1):55–70, 2002.
5. M. Chein, M. Habib, and M.C. Maurer. Partitive hypergraphs. *Discrete Mathematics*, 37(1):35–50, 1981.
6. A. Cournier and M. Habib. A new linear algorithm for modular decomposition. In *Trees in algebra and programming (CAAP 94)*, volume 787 of *LNCS*, 1994.
7. E. Dahlhaus. Parallel algorithms for hierarchical clustering, and applications to split decomposition and parity graph recognition. *Journal of Algorithms*, 36(2):205–240, 2000.
8. E. Dahlhaus, J. Gustedt, and R.M. McConnell. Efficient and practical algorithms for sequential modular decomposition. *Journal of Algorithms*, 41(2):360–387, 2001.
9. A. Ehrenfeucht and G. Rozenberg. Theory of 2-structures. *Theoretical Computer Science*, 3(70):277–342, 1990.
10. S. Fujishige. *Submodular Functions and Optimization*. North-Holland, 1991.
11. Tibor Gallai. Transitiv orientierbare Graphen. *Acta Math. Acad. Sci. Hungar.*, 18:25–66, 1967.
12. M. Habib, C. Paul, and L. Viennot. Partition refinement techniques: An interesting algorithmic tool kit. *I.J.F.C.S*, 10(2):147–170, 1999.
13. R.M. McConnell and F. de Montgolfier. Linear-time modular decomposition of directed graphs. *Discrete Applied Mathematics*, 145(2):189–209, 2005.
14. R.M. McConnell and J.P. Spinrad. Modular decomposition and transitive orientation. *Discrete Maths.*, 201:189–241, 1999. Extended abstract at SODA 94.

15. R.H. Möhring and F.J. Radermacher. Substitution decomposition for discrete structures and connections with combinatorial optimization. *Annals of Discrete Mathematics*, 19:257–356, 1984.
16. Robert Paige and Robert E. Tarjan. Three partition refinement algorithms. *SIAM J. Comput.*, 16(6):973–989, 1987.
17. T. Uno and M. Yagiura. Fast algorithms to enumerate all common intervals of two permutations. *Algorithmica*, 26(2):290–309, 2000.

Certifying Algorithms for Recognizing Proper Circular-Arc Graphs and Unit Circular-Arc Graphs

Haim Kaplan and Yahav Nussbaum

School of Computer Science, Tel Aviv University, Tel Aviv 69978, Israel
{haimk, nuss}@post.tau.ac.il

Abstract. We give two new algorithms for recognizing proper circular-arc graphs and unit circular-arc graphs. The algorithms either provide a model for the input graph, or a certificate that proves that such a model does not exist and can be authenticated in $O(n)$ time.

1 Introduction

A *certifying algorithm* for a decision problem is an algorithm that provides a *certificate* together with its answer. A certificate is an evidence that can be used to authenticate the correctness of the answer (cf. [7,11]). An *authentication algorithm* is an algorithm that validates the certificate. Certifying algorithms reduce the risk of erroneous answer, caused by bugs in the implementation.

For example, a recognition algorithm of bipartite graphs can provide as a certificates a 2-coloring when the graph is bipartite and an odd cycle when the graph is not bipartite. Graph classes that have certifying recognition algorithm include chordal graphs [15], planar graphs [11], interval graphs and permutation graphs [7], proper interval graphs [4,12], proper interval bigraphs [4], and more.

A *circular-arc graph* is an intersection graph of arcs on the circle. Every vertex is represented by an arc, such that two vertices are adjacent if and only if the corresponding arcs intersect. The arcs constitute a *circular-arc model* of the graph. Circular-arc graphs can be recognized in linear time [9,6].

A circular-arc model in which no arc contains another is a *proper circular-arc (PCA) model*. A circular-arc graph that admits a PCA model is a *proper circular-arc (PCA) graph*. Tucker gave characterizations of PCA graphs, in terms of the adjacency matrix [16] and forbidden subgraphs [17]. Skrien [13] and Deng, Hell and Huang [1] gave characterizations that use orientation of the edges. The characterization of [1] leads to a linear-time recognition algorithm for PCA graphs. Spinrad [14] showed that the characterization of [16] also leads to a linear-time recognition algorithm for PCA graphs. Both algorithms construct a PCA model if the graph is a circular-arc graph, but fail to provide a certificate otherwise.

A circular-arc model in which all arcs are closed (or all arcs are open) and of the same length is a *unit circular-arc (UCA) model*. A circular-arc graph that admits a UCA model is a *unit circular-arc (UCA) graph*. Every UCA graph is a PCA graph. Tucker [17] gave a characterization of PCA graphs which are not UCA graphs. Recently, Durán, Gravano, McConnell, Spinrad, and Tucker

F.V. Fomin (Ed.): WG 2006, LNCS 4271, pp. 289–300, 2006.

[2] presented a quadratic recognition algorithm for UCA graphs, based on this characterization. This algorithm does not provide a certificate for its answer. Even more recently, Lin and Szwarcfiter [8] gave new characterization for UCA graphs based on the length of the arcs in a PCA model. This gave a linear-time algorithm that constructs a UCA model if the input is a UCA graph, but fails to provide a certificate otherwise.

Circular-arc graphs generalize *interval graphs* which are the intersection graphs of intervals on the line. Kratsch, McConnell, Mehlhorn, and Spinrad [7] gave a linear-time certifying algorithm for interval graphs.

Another related graph class is *interval bigraphs*. A bipartite graph with the bipartition (X, Y) is an interval bigraph, if it can be represented by intervals on the line, such that the interval of $x \in X$ intersects the interval of $y \in Y$ if and only if x and y are adjacent in the graph. Two intervals corresponding to two vertices in X or to two vertices in Y, may or may not intersect. An interval bigraph that have a model in which no arc contains another, is a *proper interval bigraph*. Hell and Huang [5] showed that the class of proper interval bigraphs, is exactly the class of the complements of co-bipartite PCA graphs. These graph classes are known to be equivalent to many other well known graph classes including bipartite permutation graphs, bipartite AT-free graphs and bipartite trapezoid graphs. Hell and Huang [4] also gave a simple linear-time certifying algorithm for recognizing proper interval bigraphs, which we use in our algorithm.

In this paper, we present characterizations for PCA graphs and UCA graphs which are based on [16,17]. Those characterization leads to linear-time certifying algorithms for recognizing these classes of graphs. If the input graph is a member of the graph class, the algorithm provide the appropriate model for it. Otherwise, if the input graph is not a member of the graph class, we provide a certificate for this answer that can be authenticated in $O(n)$ time, where n is the number of vertices in the graph. This time bound is better than the time bound of the recognition algorithm, so the certificate is a *strong certificate* [7].

2   Preliminaries

Let $G(V, E)$ be a finite simple graph, and let $n = |V(G)|$ and $m = |E(G)|$. The *(closed) neighborhood* of v is $N[v] = \{v\} \cup \{u \mid vu \in E\}$. For $u, v \in V(G)$, if $uv \notin E(G)$ then we say that uv is a *non-edge*.

The sequence $P = (v_1, v_2, \ldots v_k)$ with $v_i v_{i+1} \in E(G)$ is a *path*. If $v_k v_1 \in E(G)$ then P is also a *cycle*. The sequence $P = (v_1, v_2, \ldots v_k)$ with $v_i v_{i+1} \notin E(G)$ is a *co-path*. If $v_k v_1 \notin E(G)$ then P is also a *co-cycle*. The length of a path, or a co-path, P is denoted by $|P|$. A path, cycle, co-path or co-cycle in which all the vertices are distinct is *simple*.

A graph that can be partitioned into two independent sets is a *bipartite graph*. The complement of a bipartite graph is a *co-bipartite graph*.

To simplify we refer to the clockwise direction of the circle as *right* and to the counterclockwise direction of the circle as *left*, as we view them if we stand at the center of the circle.

Table 1. Intersection types of two arcs in circular-arc model by order of their endpoints

Endpoints order (left to right)	Intersection type
$[\ell(x), r(y), \ell(y), r(x)]$	$\hat{x}$ and $\hat{y}$ cover the circle
$[\ell(x), \ell(y), r(x), r(y)]$	$\hat{y}$ overlaps the right side of $\hat{x}$
$[\ell(x), r(y), r(x), \ell(y)]$	$\hat{x}$ overlaps the right side of $\hat{y}$
$[\ell(x), r(x), \ell(y), r(y)]$	$\hat{x}$ and $\hat{y}$ are nonadjacent
$[\ell(x), \ell(y), r(y), r(x)]$	$\hat{x}$ contains $\hat{y}$
$[\ell(x), r(x), r(y), \ell(y)]$	$\hat{y}$ contains $\hat{x}$

For a PCA graph G with a PCA model ϱ, every vertex $v \in V(G)$ has an arc in ϱ with two endpoints. We denote the arc of v by $\hat{v}$, the left endpoint by $\ell(v)$ and the right endpoint by $r(v)$.

Every pair of arcs $\hat{x}$ and $\hat{y}$ in ϱ either *cover the circle*, *overlap*, or *nonadjacent*. Containment of arcs in a PCA model is impossible. If $\hat{x}$ overlaps $\hat{y}$ and covers $r(y)$ then $\hat{x}$ *overlaps the right side* of $\hat{y}$.

The *adjacency matrix* of a graph G, denoted by $M(G)$, has 1 in position (i, j) if $v_i v_j \in E$, and 0 otherwise. The *augmented adjacency matrix* of G is the adjacency matrix of G with 1's on the main diagonal, that is $M^*(G) = M(G) + I$, where I is the identity matrix. We refer to the row and column in $M^*(G)$ that correspond to the vertex v as row v and column v.

A $(0, 1)$-matrix has the *consecutive-ones property* if its columns can be ordered so that in every row the 1's are consecutive. McConnell [10] gave a linear-time certifying algorithm for this property. A $(0, 1)$-matrix has the *circular-ones property* if its columns can be ordered so that in every row the 1's are circularly consecutive.

2.1   Representation

The desired graph representation for certifying algorithms on graphs is discussed in [7]. We use, as [7], ordered adjacency list representation of graphs. This representation allow us to get the list of neighbors of a given vertex in constant time, and to certify adjacency of two vertices in constant time. An edge is certified by its location in the ordered adjacency list. A non-edge is certified by the location in the adjacency list of the edge that would be its predecessor, if the non-edge was an edge. To collect the locations of $O(n)$ edges and non-edges, we radix sort them, and scan the sorted list together with the adjacency lists of the graph. The running time for this sort and scan is $O(m + n)$.

We represent a PCA model by a cyclic order of the endpoints. The $2n$ endpoints in the model are indexed according to their ranks in the order, starting at any arbitrary endpoint and going right. Each arc has the indices of its two endpoints. The type of intersection between two arcs can be determined, in constant time, by the order of their endpoints [3] (see Table 1). A unit circular-model obey length constraint, so the exact location of the endpoints on the circle is required.

We represent $(0, 1)$-matrices in a sparse way, similar to the representation of graphs by ordered adjacency lists. This representation allows algorithms that process matrices to run in time proportional to the number of 1's in the matrix. For $M^*(G)$ the number of 1's is $O(m + n)$.

3   Characterization of Proper Circular-Arc Graphs

We define an *incompatibility graph* for PCA graphs, in a way similar to the definitions of incompatibility graphs for the consecutive-ones property [10] and for permutation graphs [7], as follows.

Let ϱ be a PCA model of G, and v_0 be a vertex in G. We start at $r(v_0)$ and traverse the circle to the right, we get a *traversal ordering* $(v_0, v_1, \ldots, v_{n-1})$ of the vertices according to the order in which we meet their right endpoints. This ordering defines a *traversal order relation* $R = \{(v_i, v_j) \mid i < j\}$.

The following holds for any PCA model of G. For every $x, y \in V(G)$, (x, y) and (y, x) cannot both appear in the same traversal order relation. We say that (x, y) is *incompatible* with (y, x). For every $w \in V(G)$, the right endpoints of all the vertices in $N[w]$ must be consecutive around the circle. Assume that $v_0 \notin N[w]$. Then, in a traversal ordering that starts with v_0 the vertices of $N[w]$ must be consecutive. Therefore, if $x, z \in N[w]$ and $y \notin N[w]$, the vertex y cannot be between x and z. So (x, y) and (y, z) are incompatible, with w as a *witness*. Assume that $v_0 \in N[w]$, now the vertices of $N[w]$ are not necessarily consecutive in a traversal ordering that starts with v_0, because it might be that v_{n-1} is also in $N[w]$. But $V(G) - N[w]$ must be consecutive in this ordering, so if $x, z \notin N[w]$ and $y \in N[w]$ then (x, y) and (y, z) are incompatible, with w as a witness.

The incompatible pairs define the *incompatibility graph* $IC(G; v_0)$ of G with starting vertex v_0. The vertex set of $IC(G; v_0)$ is $\{(x, y) \mid x, y \in V(G), x \neq y\}$. The edge set of $IC(G; v_0)$ are edges of the forms $(x, y)(y, x)$, $(x, y)(y, z)$ such that $x, z \in N[w]$, $y \notin N[w]$ for some $w \notin N[v_0]$ and $(x, y)(y, z)$ such that $y \in N[w]$, $x, z \notin N[w]$ for some $w \in N[v_0]$.

The definition of $IC(G; v_0)$ is analogous to the definition of the incompatibility graph for the consecutive-ones property $IC(M)$, presented by McConnell [10]. Since a consecutive-ones ordering is linear, we do not need a starting point to define $IC(M)$. The edge set of $IC(M)$ are $(x, y)(y, x)$, for every pair of columns x and y, and $(x, y)(y, z)$ such that there is a row w, with ones in the column of x, z but not in the column of y.

Theorem 1. *Let G be a PCA graph. For any $v_0 \in V(G)$, the incompatibility graph $IC(G; v_0)$ is bipartite.*

Proof. Let ϱ be a PCA model of G. Let $v_0 \in V(G)$ and let R be the traversal order relation defined by the traversal ordering of ϱ that starts with v_0. The relation R is made of vertices of $IC(G; v_0)$. The relation R cannot have any incompatible pairs, so the vertices of R are an independent set in $IC(G; v_0)$. Let ϱ' be a PCA model of G that is obtained by replacing the right and left directions of ϱ. Let R' be the traversal order relation defined by the traversal ordering of ϱ' starting with v_0. Any vertex in $IC(G; v_0)$ that is not in R is in R'. The vertices of R' are also an independent set, and therefore $IC(G; v_0)$ is bipartite.   □

To certify that G is not a PCA graph we can provide an odd cycle in one of its incompatibility graphs. We do so without explicitly constructing the entire incompatibility graph, since the size of this graph might be as large as $\Theta(n^4)$. Note that $IC(G; v_0)$ might be bipartite even when G is not a PCA graph.

4   Certifying Algorithm for Proper Circular-Arc Graphs

Our certifying algorithm for PCA graphs consists of two cases, depending on whether G is co-bipartite or not. We begin the algorithm by deciding whether G is co-bipartite. If G is co-bipartite, then it is covered by two cliques. At least one of those cliques should cover half of $V(G)$, so $m \geq \frac{n}{2}(\frac{n}{2} - 1)$. If this inequality does not hold then G is not co-bipartite. Otherwise, $m = \Theta(n^2)$, and we check if $\overline{G}$ is bipartite in $O(n^2) = O(m)$ time.

4.1   The Complement of G Is Not Bipartite

In the case where G is not co-bipartite, Tucker [16] showed that G is a PCA graph if and only if $M^*(G)$ has the circular-ones property.

To check if $M^*(G)$ has the circular-ones property, we use the following reduction of [16] from testing this property to testing the consecutive-ones property. Let M_1 be a $(0, 1)$-matrix. Fix a column j. Form the matrix M_2 by complementing those rows with 1 in column j of M_1. Then M_1 has the circular-ones property if and only if M_2 has the consecutive-ones property.

Let v_0 be a vertex of minimum degree in G. To perform the reduction stated above in linear time, we complement the rows of $M^*(G)$ which have one in the column of v_0. Since the degree of v_0 is $O(m/n)$, we complement $O(m/n)$ rows. It takes $O(n)$ time to complement a row so we perform the entire reduction in $O(m)$ time. We denote by M the new matrix that we obtain.

After the reduction we run the certifying algorithm of McConnell [10] to test if M has the consecutive-ones property. If M has a consecutive-ones ordering, we order the columns of $M^*(G)$ in the same way, to get a circular-ones ordering for $M^*(G)$. Tucker [17] showed an algorithm to produce a PCA model of G from a this ordering that can be implemented in $O(n + m)$ time.

If M does not have the consecutive-ones property, then the algorithm of [10] produces a certificate for this fact. This certificate is an odd cycle C of length at most $n + 2$ in the incompatibility graph $IC(M)$. Next, we show that all edges of C exist in $IC(G; v_0)$ so C is also an odd cycle in $IC(G; v_0)$.

Edges of C in $IC(M)$ have one of two forms. Edges of the form $(x, y)(y, x)$ always exist in $IC(G; v_0)$. Consider an edge $(x, y)(y, z)$ with a witness w, where w is a row in M such that the columns of x and z have 1 in this row, but the column of y has 0 in it. If $w \notin N[v_0]$ then the row of w in M is the same as in $M^*(G)$. So, $x, z \in N[w]$ while $y \notin N[w]$ and therefore $(x, y)(y, z)$ is an edge of $IC(G; v_0)$, with vertex $w \notin N[v_0]$ as a witness. Otherwise, if $w \in N[v_0]$, then the row of w in M is the complement of the row of w in $M^*(G)$. So, $y \in N[w]$ while $x, z \notin N[w]$ and therefore $(x, y)(y, z)$ is an edge of $IC(G; v_0)$, with vertex $w \in N[v_0]$ as a witness.

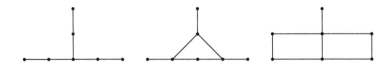

Fig. 1. Forbidden subgraphs

We provide the odd cycle C in $IC(G; v_0)$, together with v_0 as a certificate. To complete the certificate, we need to add a certificate for all edges and non-edges of G that are involved in it. For an edge $(x, y)(y, z)$ with a witness w in $IC(G; v_0)$, we need to provide a certificate for the edges or non-edges xw, yw, zw and wv_0 in G. The length of the cycle in $IC(G; v_0)$ is $O(n)$, and thus there are $O(n)$ edges or non-edges to certify.

4.2   The Complement of G Is Bipartite

In this case, when G is co-bipartite, we use the following forbidden subgraphs characterization of Tucker [17].

Theorem 2. *[17] Let G be a graph. If $\overline{G}$ contains an induced even cycle of length ≥ 6 or one of the graphs in Fig. 1, then G is not a PCA graph.*

A co-bipartite graph G is a PCA graph if and only if $\overline{G}$ is a proper interval bigraph [5]. So, we use the certifying algorithm for recognizing proper interval bigraph of Hell and Huang [4] on $\overline{G}$. Note that the graphs in Theorem 2 are exactly the graphs that the certifying algorithm for recognizing proper interval bigraphs [4] uses as certificates.

The graph G is covered by two cliques, one of those two cliques must cover at least $n/2$ of the vertices of G, therefore $m = \Theta(n^2)$. So we can produce $\overline{G}$ from G in $O(n^2) = O(m)$ time.

If $\overline{G}$ is an interval bigraph, we get a model for it, and we use an algorithm of Hell and Huang [5] to construct a PCA model of G, from this model. Otherwise, we have one of graphs in Theorem 2 as a certificate. For a graph of Fig. 1, we use its complement to certify that G is not a PCA graph.

If we have an induced even cycle of length ≥ 6 as a certificate that $\overline{G}$ is not a proper interval bigraph, we transform it into an odd cycle in an incompatibility graph of G. We do so for two reasons. First, a straightforward authentication of an even length cycle takes $O(m + n)$ time, while authentication of an odd cycle in an incompatibility graph takes $O(n)$ time. Second, we reduce the number of cases that the authentication algorithm has to deal with, since it has to verify an odd cycle in an incompatibility graph in the case where G is not co-bipartite.

Let $C = (x_0, x_1, \ldots, x_{2r-1})$ be an even induced cycle in $\overline{G}$. For every $i = 0, \ldots, 2r - 1$, and for every $j \neq i \pm 1$, we have that $(x_i, x_j) \in E(G)$, where the subscripts of the vertices are modulo $2r$. We find an odd cycle in the incompatibility graph $IC(G; x_1)$.

If r is even, we start the cycle in $IC(G; x_1)$ with (x_0, x_r). From the vertex (x_i, x_j) in the cycle, we continue to (x_j, x_{i+2}). We can use x_{i+1} as a witness for the edge $(x_i, x_j)(x_j, x_{i+2})$, since if we start with (x_0, x_r), we always have $x_{i+1} \in N[x_1]$ and $x_i, x_{i+2} \notin N[x_{i+1}]$ while $x_j \in N[x_{i+1}]$. After r edges we get to (x_r, x_0), we add an edge $(x_r, x_0)(x_0, x_r)$ to complete an odd cycle of length $r + 1$.

If r is odd, we start the cycle in the incompatibility graph in (x_0, x_{r+1}) and continue in the same way. After r edges we get back to (x_0, x_{r+1}), and we have odd cycle of length r.

Building the cycle in the incompatibility graph $IC(G; x_1)$, together with certificate for all the edges in it takes $O(n^2) = O(m)$ time.

4.3 The Certificate and the Authentication Algorithm

The certificate of the recognition algorithm is either a PCA model of G, an odd cycle in an incompatibility graph $IC(G; v_0)$, or one of the graphs of Fig. 1 as an induced subgraph of $\overline{G}$. If we got a PCA model for G, then G is a PCA graph. For the other certificates, G is not a PCA graph by Theorem 1 or Theorem 2.

To authenticate a PCA model we first authenticate that the model is a circular-arc model of G, this step is described by McConnell [9]. The model is a PCA model only if no arc is contained in another, we check this for every pair of adjacent vertices by checking the order of their endpoints (see Table 1). The size of this certificate is $O(n)$ and the time to authenticate it is $O(n + m)$.

To authenticate an odd cycle in an incompatibility graph $IC(G; v_0)$, we first verify that it has an odd length not larger then $n + 2$. Then, we verify that the certificate is indeed a cycle. We also verify that each edge of the cycle belongs to $IC(G; v_0)$, by checking that every edge is either of the form $(x, y)(y, x)$ or has a valid witness. The size of the cycle is $O(n)$ and validating it takes $O(n)$ time.

If the certificate is one of the graphs of Fig. 1, we verify that every edge exists in the certificate if and only if it exists in the graph. The size of each of these graphs is $O(1)$, and hence the authentication time is also $O(1)$.

When the algorithm found that G is not a PCA graph, both possible certificates can be authenticated in $O(n)$ time and therefore are strong certificates.

5 Characterization of Unit Circular-Arc Graphs

In this section we present the structure theorem of Tucker [17] for UCA graphs together with some relaxations of it that we use. Let G be a PCA graph with a PCA model ϱ. Every PCA graph has a PCA model in which no pair of arcs covers the circle [17,3]. In fact, the algorithm of Sect. 4 never constructs a model with a pair of arcs that covers the circle. Thus, in this section we assume that ϱ does not contain a pair of arcs that covers the circle.

Let $C = (x_0, \ldots, x_{p-1})$ be a simple cycle in G, such that for $i = 1, \ldots, p - 1$, the arc $\hat{x}_i$ overlaps the right side of $\hat{x}_{i-1}$, and the arc $\hat{x}_0$ overlaps the right side of $\hat{x}_{p-1}$, in ϱ. We call such a cycle C, a *bounding cycle*. Assume that we traverse the cyclic list of endpoints of ϱ, starting immediately after $\ell(x_0)$, going right to $r(x_0)$

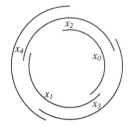   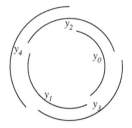

Fig. 2. Bounding cycle with ratio 5/2   **Fig. 3.** Bounding co-cycle with ratio 5/2

and continuing from $r(x_i)$ to $r(x_{i+1})$. We call the list of endpoints obtained this way the *walk* of C. The number of times that C goes around the circle is the number of times that the walk of C hits $\ell(x_0)$, we denote this number by $t(C)$. The *ratio* of C, denoted by $p(C)$, is $|C|/t(C)$. See Fig. 2. We call C a *minimum bounding cycle* if there is no other bounding cycle C' with $p(C') < p(C)$. We denote a minimum bounding cycle by C^m. If the union of the arcs in ϱ does not cover the circle, then there are no bounding cycles. In this case, we let $p(C^m) = \infty$.

Let $I = (y_0, \ldots, y_{p-1})$ be a simple co-cycle in G. We call I a *bounding co-cycle*. We define the *walk* of I as we define it for bounding cycles. To compute $t(I)$, the number of times that I goes around the circle, we add 1 to the number of times that the walk of I hit $\ell(y_0)$, to count also the last partial turn. The *ratio* of I, denoted by $p(I)$, is $|I|/t(I)$. See Fig. 3. We call I a *maximum bounding co-cycle* if there is no other bounding co-cycle I' with $p(I') > p(I)$. We denote a maximum bounding co-cycle by I^M.

The circumference of a UCA model of closed arcs with a bounding cycle C can be at most $p(C)$. On the other hand, the circumference of a UCA model with a bounding co-cycle I must be strictly greater than $p(I)$. So for any UCA model $p(I^M) < p(C^m)$. The following theorem shows that this condition is also sufficient. Furthermore, the bounds do not depend on the specific model.

Theorem 3. *[17] A PCA graph G is also a UCA graph if and only if for any PCA model of G with no pair of arcs that cover the circle, $p(I^M) < p(C^m)$.*

Given a PCA model ϱ, the algorithm of Durán et al. [2] finds a minimum bounding cycle and a maximum bounding co-cycle that start from a certain arc in the model. To do so in linear time they use complicated data structures.

We relax the definitions of a bounding cycle and a bounding co-cycle, in two ways, in order to get a simple implementation. First, we allow repetitions of vertices. Second, we use paths instead of cycles.

Let $P = (x_0, \ldots, x_{p-1})$ be a path of vertices in G, not necessarily simple, such that for $i = 1, \ldots, p-1$, the arc $\hat{x}_i$ overlaps the right side of $\hat{x}_{i-1}$, in ϱ. The path P is a *bounding path*. We define the *walk* of P and count the number of times that P goes around the circle, denoted by $t(P)$, in the same way as we do it for bounding cycles. The *ratio* of P is $p(P) = |P|/t(P)$. Note that since a bounding path is not necessarily a cycle, it might be that $t(P) = 0$, in this case we assume that $p(P) = \infty$.

Let $Q = (y_0, \ldots, y_{p-1})$ be a (not necessarily simple) co-path of vertices in G. We call Q a *bounding co-path*. We define the *walk* of Q and count the number of times that Q goes around the circle, denoted by $t(Q)$, in the same way as we do it for bounding co-cycles. The *ratio* of Q is $p(Q) = |Q|/t(Q)$.

6   Certifying Algorithm for Unit Circular-Arc Graphs

Every UCA graph is a PCA graph, so we start by testing whether G is a PCA graph using the algorithm of Sect. 4. If G is not a PCA graph then it is also not a UCA graph, and the algorithm of Sect. 4 certifies that. Otherwise, if G is a PCA graph then we have a PCA model of it which we denote by ϱ. As in Sect. 5, we assume that there is no pair of arcs in ϱ that covers the circle.

We generate bounding paths and bounding co-paths in ϱ, which are simple to find. Then, we show that we can find from this set of bounding paths and co-paths a minimum bounding cycle and a maximum bounding co-cycle. We compare the ratios of the minimum bounding cycle and maximum bounding co-cycle, and use Theorem 3 to decide whether G is a UCA graph.

It can be shown that a PCA graph with a dominating vertex or a PCA graph that has a PCA model in which the union of the arcs does not cover the circle, is a UCA graph. Therefore, we assume that G does not contain any dominating vertex and that the union of arcs in ϱ covers the circle.

To obtain a certificate when G is a UCA graph, we use the algorithm of [8] to find a UCA model for G. Note, that the first two steps of [8] are to find a PCA model of G and to eliminate pairs of arcs that cover the circle. Thus, the implementation of [8] can be simplified by using the algorithm in Sect. 4.

The algorithm of [2] iteratively looks for the minimal bounding cycle and maximal bounding co-cycle that starts with every vertex in the graph. This is not necessary since the bounding cycles and bounding co-cycles are cyclic, so we only need to start from one of their vertices, not from all of them. Furthermore, we show that for every vertex $v \in V(G)$, we can start only from vertices in $N[v]$.

Let $C^m = (x_0, \ldots, x_{p-1})$ be a minimum bounding cycle in a PCA model, and let $I^M = (y_0, \ldots, y_{q-1})$ be a maximum bounding co-cycle. Let v be any fixed vertex in $V(G)$. The arcs of C^m covers the circle at least once, so there must be an arc $\hat{x}_i$ that overlaps $\hat{v}$. Because C^m is a cycle, we may assume that $\hat{x}_i$ is $\hat{x}_0$, hence, $x_0 \in N[v]$. Assume that v is not adjacent to any vertex of I^M, we can add v to I^M and get a bounding co-cycle I, with $t(I) = t(I^M)$ and $|I| = |I^M|+1$, and therefore $p(I) > p(I^M)$, contradicting the fact that I^M is maximum. Because I^M is a co-cycle, we may assume that $y_0 \in N[v]$.

For a particular vertex u, we find n bounding paths, of lengths 1 to n, each starting with u, using the following greedy algorithm. Let $u_0 = u$. We start with $P_1 = (u_0)$ as a bounding path with $t(P_1) = 0$. For $i = 1, \ldots, n-1$, we generate $P_{i+1} = (u_0, \ldots, u_i)$ by adding to P_i the vertex u_i, where $\ell(u_i)$ is the rightmost left endpoint covered by the arc $\hat{u}_{i-1}$. Such a vertex u_i exists since the circle is covered by the union of all arcs. We let $t(C_i) = t(C_{i-1}) + 1$, if the arc $\hat{u}_i$ covers $\ell(u_0)$, and otherwise $t(C_i) = t(C_{i-1})$. We stop after generating P_n. We represent the n bounding paths by the list of vertices in P_n and the list of the values $t(P_i)$ for $i = 1, \ldots, n$.

Similarly, we find n bounding co-paths, of lengths 1 to n, each starting with u, using the following greedy algorithm. Let $u_0 = u$. We start with $Q_1 = (u_0)$ as a bounding co-path with $t(Q_1) = 1$. For $i = 1, \ldots, n-1$, we generate $Q_{i+1} = (u_0, \ldots, u_i)$ by adding to Q_i the vertex u_i, where $\ell(u_i)$ is the leftmost left endpoint not covered by the arc $\hat{u}_{i-1}$. Since u_{i-1} is not a dominating vertex, $u_{i-1}u_i \notin E(G)$. We let $t(Q_i) = t(Q_{i-1}) + 1$, if the arc $\hat{u}_i$ covers $\ell(u_0)$, and otherwise $t(Q_i) = t(Q_{i-1})$. We stop after generating Q_n. We represent the n bounding co-paths by the list of vertices in Q_n and the list of the values $t(Q_i)$ for $i = 1, \ldots, n$.

To implement these algorithms in $O(n)$ time, we identify in advance, for every arc, the rightmost left endpoint it covers, and the leftmost left endpoint not covered by it. We do so by going around the circle from left to right, starting at some left endpoint, and maintaining $\ell(x)$, the last left endpoint we encountered. When we encounter a right endpoint $r(y)$, the last left endpoint that the arc $\hat{y}$ covers is $\ell(x)$. We can find the leftmost left endpoint following each arc in the same way, by going around the circle from right to left.

Let $v_0 \in V(G)$ be a vertex with a minimum degree, so $|N[v_0]| = O(m/n)$. We find n bounding paths and n bounding co-paths that start with each of the arc of vertices in $N[v_0]$, by the greedy algorithms described above. This takes $O(n|N[v_0]|) = O(n \cdot m/n) = O(m)$ time and $O(m)$ space. We then find among the $O(m)$ bounding paths, the bounding path P^m for which $p(P^m)$ is the smallest and the bounding co-path Q^M for which $p(Q^M)$ is the largest. If there are more than one bounding paths or co-paths with the same ratio we take the shortest.

Lemma 1. *For the bounding path P^m and the bounding co-path Q^M that we have found, we have $p(P^m) \leq p(C^m)$ and $p(I^M) \leq p(Q^M)$.*

Proof. Let C^m be a minimum bounding cycle, starting with $x_0 \in N[v_0]$. Let $k = |C^m|$, since C^m is a simple cycle, $k \leq n$. Let C_i be the prefix of C^m with $|C_i| = i$. The path C_i is a bounding path. Let P_i be the bounding path of length i stating at x_0 that our greedy algorithm has found. We prove by induction on i, that the walk of C_i is a prefix of the walk of P_i. It follows that for every $i = 1, \ldots, k$, we have $t(C_i) \leq t(P_i)$, and in particular $p(C^m) = p(C_k) \geq p(P_k) \geq p(P^m)$, as required.

For $i = 1$, $C_1 = P_1$, and thus the walks of C_1 and P_1 are identical. Assume that the walk of C_i is prefix of the walk of P_i. To get C_{i+1}, we add to C_i a vertex x_i such that the last occurrence of $\ell(x_i)$ in the walk of C_i is not followed by an occurrence of $r(x_i)$. The walk of C_{i+1} starts with the walk of C_i and continues until $r(x_i)$. To get P_{i+1}, we add to P_i the vertex u_i such that $\ell(u_i)$ is the last left endpoint in the walk of P_i. The walk of P_{i+1} starts with the walk of P_i and continues until $r(u_i)$. Since by induction the walk of C_i is a prefix of the walk of P_i, the last occurrence of $\ell(x_i)$ in the walk of C_i corresponds to an occurrence of $\ell(x_i)$ in the walk of P_i preceding or equal to the last occurrence of $\ell(u_i)$ in the walk of P_i. Therefore the last occurrence of $r(x_i)$ in C_{i+1} corresponds to an occurrence of $r(x_i)$ in the walk of P_{i+1} preceding or equal to the last occurrence of $r(u_i)$ in the walk of P_{i+1}. Thus, the walk of C_{i+1} is a prefix of the walk of P_{i+1}.

The claim $p(I^M) \leq p(Q^M)$ is proved similarly. $\qquad\qquad\qquad\qquad$ □

Lemma 2. *The bounding path P^m and the bounding co-path Q^M that we have found, are a minimum bounding cycle and a maximum bounding co-cycle.*

Proof. Let $P^m = (x_0, \ldots x_{p-1})$ and let $Q^M = (y_0, \ldots y_{q-1})$. By Lemma 1 it suffices to prove that P^m is a simple cycle and that Q^M is a simple co-cycle.

Assume that P^m is not a simple path. So, there are i, j such that $x_i = x_j$ but $i \neq j$. Let i be the minimal index for which there exists $\ell > i$ such that $x_i = x_\ell$, let j be the minimal possible value of ℓ. The way that our greedy algorithm chooses the successor of each vertex does not depend on its location on the path, so $x_{i+k} = x_{j+k}$ for every $k = 1, \ldots, (p-1) - j$. Let $C = (x_{p-(j-i)}, \ldots x_{p-1})$. The path C is simple by the way we choose x_i and x_j, C is a cycle since $x_{p-1} = x_{p-(j-i)-1}$ and since C is a suffix of P^m, it is a bounding cycle. We have a bounding cycle with a ratio $p(C) \geq p(C^m) \geq p(P^m)$. So we can truncate P^m after $x_{p-(j-i)-1}$ and get a new bounding path P with $p(P) \leq p(P^m)$ which is a prefix of P^m. This contradicts the definition of P^m, to be the shortest path with the maximal ratio that the greedy algorithm found. Similarly, we can show that Q^M is a simple co-path.

Now, assume that P^m is not a cycle. The last arc $\hat{x}_{p-1}$ does not cover $\ell(x_0)$, so it does not start a new turn around the circle. Let P be the prefix of P^m of length $p - 1$. We have $|P| < |P^m|$ but $t(P) = t(P^m)$ therefore $p(P) < p(P^m)$. That is a contradiction to the way we found P^m.

Assume that Q^M is not a co-cycle. It follows that the arc $\hat{y}_{q-1}$ overlaps $\hat{y}_0$. If the arc $\hat{y}_{q-1}$ covers $r(y_0)$ then $\hat{y}_{q-2}$ covers $\ell(y_0)$, because otherwise the greedy algorithm would have chosen y_0 to be y_{q-1}. Since there is no dominating vertex in G, there is a pair of nonadjacent vertices, this pair is a bounding co-cycle with ratio 2, so we have $2 \leq p(Q^M)$. Let Q be the prefix of Q^M of length $q - 2$. Since there is one arc less in Q that covers $\ell(y_0)$, we have $t(Q) = t(Q^M) - 1$. And since $p(Q^M) \geq 2$ we have $p(Q^M) \leq p(Q)$, which contradict the way we define Q^M. $\quad$ □

Therefore, by Theorem 3, G is a UCA graph if and only if $p(Q^M) < p(P^m)$. If G is not a UCA graph we use P^m and Q^M as a certificate.

6.1   The Certificate and the Authentication Algorithm

If G is a UCA graph then the certificate is a UCA model. This certificate can be authenticated by authenticating that it is a PCA model as in Sect. 4.3 and comparing the length of all the arcs. This can be done in $O(n + m)$ time.

If G is not a PCA graph, the certificate and its authentication algorithm are as in Sect. 4.3. The size and the authentication time of this certificate are $O(n)$.

If we decided that G is a PCA graph but not a UCA graph then we have a bounding cycle P^m, and a bounding co-cycle Q^M with $p(Q^M) \geq p(P^m)$ in a PCA model ϱ. Authenticating that ϱ is a valid PCA model without pairs of arcs that cover the circle, takes $O(m+n)$ time. We can verify in the same time bound that P^m and Q^M are bounding cycle and bounding co-cycle respectively, and compute $t(P^m)$ and $t(Q^M)$ from the model, by following the walks of P^m and Q^M.

It is also possible to construct a strong certificate that can be authenticated in $O(n)$ time. This certificate proves, using edges and non-edges between vertices of P^m and Q^M, that in any PCA model of G, P^m and Q^M are a bounding cycle and a bounding co-cycle with $p(Q^M) \geq p(P^m)$. Due to space constraint we omit the details of this certificate.

References

1. X. Deng, P. Hell, and J. Huang. Linear-time representation algorithms for proper circular-arc graphs and proper interval graphs. *SIAM J. Comput.*, 25(2):390–403, 1996.

2. G. Durán, A. Gravano, R. M. McConnell, J. P. Spinrad, and A. Tucker. Polynomial time recognition of unit circular-arc graphs. *J. Algorithms*, 58(1):67–78, 2006.

3. M. C. Golumbic. *Algorithmic Graph Theory and Perfect Graphs*. Academic Press, 1980.

4. P. Hell and J. Huang. Certifying LexBFS recognition algorithms for proper interval graphs and proper interval bigraphs. *SIAM J. Discrete Math.*, 18(3):554–570, 2004.

5. P. Hell and J. Huang. Interval bigraphs and circular arc graphs. *J. Graph Theory*, 46(4):313–327, 2004.

6. H. Kaplan and Y. Nussbaum. A simpler linear-time recognition of circular-arc graphs. In *10th Scandinavian Workshop on Algorithm Theory (SWAT)*, Lecture Notes in Computer Science 4059, pages 41–52, 2006.

7. D. Kratsch, R. M. McConnell, K. Mehlhorn, and J. P. Spinrad. Certifying algorithms for recognizing interval graphs and permutation graphs. *SIAM J. Comput.*, 36(2):326–353, 2006.

8. M. C. Lin and J. L. Szwarcfiter. Efficient construction of unit circular-arc models. In *SODA '06: Proceedings of the seventeenth annual ACM-SIAM symposium on Discrete algorithm*, pages 309–315, 2006.

9. R. M. McConnell. Linear-time recognition of circular-arc graphs. *Algorithmica*, 37(2):93–147, 2003.

10. R. M. McConnell. A certifying algorithm for the consecutive-ones property. In *SODA '04: Proceedings of the fifteenth annual ACM-SIAM symposium on Discrete algorithms*, pages 768–777, 2004.

11. K. Mehlhorn and S. Näeher. *The LEDA Platform for combinatorial and geometric computing*. Cambridge University Press, 1999.

12. D. Meister. Recognition and computation of minimal triangulations for AT-free claw-free and co-comparability graphs. *Discrete Applied Math.*, 146(3):193–218, 2005.

13. D. J. Skrien. A relationship between triangulated graphs, comparability graphs, proper interval graphs, proper circular-arc graphs, and nested interval graphs. *J. Graph Theory*, 6:309–316, 1982.

14. J. P. Spinrad. *Efficient Graph Representations*. Fields Institute Monographs 19. American Mathematical Society, 2003.

15. R. E. Tarjan and M. Yannakakis. Addendum: Simple linear-time algorithms to test chordality of graphs, test acyclicity of hypergraphs, and selectively reduce acyclic hypergraphs. *SIAM J. Comput.*, 14(1):254–255, 1985.

16. A. Tucker. Matrix characterizations of circular-arc graphs. *Pacific J. Math.*, 39(2):535–545, 1971.

17. A. Tucker. Structure theorems for some classes of circular-arc graphs. *Discrete Math.*, 7:167–195, 1974.

Graph Labelings Derived from Models in Distributed Computing

Jérémie Chalopin[1] and Daniël Paulusma[2]

[1] LaBRI Université Bordeaux 1,
351 cours de la Libération, 33405 Talence, France
chalopin@labri.fr
[2] Department of Computer Science, Durham University,
Science Laboratories, South Road, Durham DH1 3LE, England
daniel.paulusma@durham.ac.uk

Abstract. We discuss eleven well-known basic models of distributed computing: four message-passing models that differ by the (non-)existence of port-numbers and a hierarchy of seven local computations models. In each of these models, we study the computational complexity of the decision problem whether the leader election and/or naming problem can be solved on a given network. It is already known that this problem is solvable in polynomial time for two models and co-NP-complete for another one. Here, we settle the computational complexity for the remaining eight problems by showing co-NP-completeness. The results for six models and the already known co-NP-completeness result follow from a more general result on graph labelings.

1 Introduction

In distributed computing, one can find a wide variety of models of communication. These models reflect different system architectures, different levels of synchronization and different levels of abstraction. In this paper we consider eleven well-known basic models that satisfy the following two underlying assumptions. Firstly, a distributed system is represented by a simple (i.e., without loops or multiple edges), connected, undirected graph. Its vertices represent the processors, and its edges represent direct communication links. Secondly, the distributed systems we consider are anonymous, i.e., all the processors execute the same code to solve some problem and they do not have initial identifiers.

The eleven basic models can be divided into four *message-passing* models [6,14,16] and seven *local computations* models [1,4,5,11,12]. In a message-passing model, processors communicate by sending and receiving messages. In a local computations model, a computation step (encoded by a local relabeling rule) involves neighboring processors that synchronize, exchange information, and modify their states.

Understanding the computational power of various models, the role of structural network properties and the role of the initial knowledge enhances our understanding of distributed algorithms. For this purpose a number of standard

F.V. Fomin (Ed.): WG 2006, LNCS 4271, pp. 301–312, 2006.
© Springer-Verlag Berlin Heidelberg 2006

problems in distributed computing are studied. The election problem is one of the paradigms of the theory of distributed computing. In our setting, a distributed algorithm solves the election problem if it always terminates and in the final configuration exactly one processor is marked as *elected* and all the other processors are *non-elected*. Elections constitute a building block of many other distributed algorithms, since the elected vertex can be subsequently used to make centralized decisions. A second important problem in distributed computing is the naming problem. Here, the aim is to arrive at a final configuration where all processors have been assigned unique identities. Again this is an essential prerequisite to many other distributed algorithms that only work correctly under the assumption that all processors can be unambiguously identified. For a reference book on distributed algorithms we refer to [13].

Our Results. Whether the naming or election problem can be solved on a given graph depends on the properties of the considered model. If it is possible to solve the election (naming) problem we call the graph a *solution graph* for the election (naming) problem. It is a natural question to ask how hard it is to check whether a given graph is a solution graph in a certain model. For two models this problem is known to be polynomially solvable [2] and for one model it is co-NP-complete [15]. What about the computational complexity of this problem for the other models? In this paper we solve this question by showing that this decision problem is co-NP-complete for all remaining models.

The paper is organized as follows. In Section 2 we define the necessary graph terminology. To obtain our results we translate known characterizations [1,4,5,6,7,11,12,14,16] of solution graphs in terms of graph labelings. This is shown in Section 3 for the message-passing models and in Section 4 for the local computations models. In Section 5 we introduce a new kind of labeling that does not correspond to any model of distributed computing but that enables us to present a simpler co-NP-completeness proof for seven basic models including the already known model in [15]. In Section 6 we give the results for the remaining two models.

2   Preliminaries

For graph terminology not defined below we refer to [3]. A *labeling* of a graph $G = (V_G, E_G)$ is a mapping $\ell : V_G \rightarrow \{1, 2, 3, \ldots, \}$. For a set $S \subseteq V_G$ we use the shorthand notation $\ell(S)$ to denote the image set of S under ℓ, i.e., $\ell(S) = \{\ell(u) \mid u \in S\}$. A labeling ℓ of G is called *proper* if $|\ell(V_G)| < |V_G|$. For any *label* $i \geq 1$, the set $\ell^{-1}(i)$ is equal to $\{u \in V_G \mid \ell(u) = i\}$. The subgraph of G induced by a subset $S \subseteq V_G$ is denoted by $G[S]$. For a label $i \geq 1$ we write $G[i] = G[\ell^{-1}(i)]$. For two labels i, j, we let $G[i, j]$ be the bipartite graph obtained from $G[\ell^{-1}(i) \cup \ell^{-1}(j)]$ by deleting all edges $[u, v]$ with $\ell(u) = \ell(v) = i$ or with $\ell(u) = \ell(v) = j$.

For a vertex $u \in V_G$ in a graph $G = (V_G, E_G)$, we denote its *neighborhood* by $N_G(u) = \{v \mid [u, v] \in E_G\}$. A graph is *regular*, if all its vertices have the same number k of neighbors (i.e. are of *degree* $\deg_G(u) = k$), in that case we also say that the graph is k-regular. A graph is *regular bipartite* if it is regular and bipartite. A graph is *semi-regular bipartite* if it is bipartite and the vertices of one class of the bipartition are of degree k and all others are of degree l, in that case we also say that the graph is (k, l)-*regular bipartite*. In our context a *perfect matching* is a $(1, 1)$-regular bipartite graph.

3   Message-Passing Models

In [14,15,16], Yamashita and Kameda study four message-passing models. In the *port-to-port* model, each processor can send different messages to different neighbors (by having access to unique *port-numbers* that distinguish between neighbors), and each processor knows the neighbor each receiving message is coming from (again by using the port-numbers). In the *broadcast-to-mailbox* model, port-numbers do not exist. A processor can only send a message to all of its neighbors and all receiving messages arrive in a *mailbox*, so it never knows their senders. The two mixed models are called the *broadcast-to-port* model and the *port-to-mailbox* model. There exists an election (or naming) algorithm for a graph G if and only if the algorithm solves the problem on G whatever the port-numbers are.

In [16], Yamashita and Kameda characterize these four models: a graph G is a solution graph for the election and naming problem in the port-to-port model if and only if G does not have a proper *symmetric regular labeling*, i.e., a proper labeling ℓ such that

(i) for all $i \in \ell(V_G)$, $G[i]$ is regular and contains a perfect matching if its vertices have odd degree, and
(ii) for all $i, j \in \ell(V_G)$ with $i \neq j$, $G[i, j]$ is regular bipartite.

A graph G is a solution graph for the election and naming problem in the port-to-mailbox model if and only if G does not have a proper *regular labeling*, i.e., a proper labeling ℓ such that

(i) for all $i \in \ell(V_G)$, $G[i]$ is regular, and
(ii) for all $i, j \in \ell(V_G)$ with $i \neq j$, $G[i, j]$ is regular bipartite.

A graph G is a solution graph for the naming problem in the broadcast-to-mailbox and the broadcast-to-port model if and only if G does not have a proper *semi-regular labeling*, i.e., a proper labeling ℓ such that

(i) for all $i \in \ell(V_G)$, $G[i]$ is regular, and
(ii) for all $i, j \in \ell(V_G)$ with $i \neq j$, $G[i, j]$ is semi-regular bipartite.

In these two models, a graph G is a solution graph for the election problem if and only if there does not exist any semi-regular labeling ℓ of G such that for all $i \in \ell(V_G)$, $|\ell^{-1}(i)| > 1$.

In [1,6], different characterizations for these models are obtained (based on fibrations and coverings of directed graphs). The problem of deciding whether a graph G is a solution graph for the election and naming problem in the port-to-port model is co-NP-complete [15]. On the other hand, in [2], it is shown that the problem of deciding whether a graph G is a solution problem for the election and naming problem is polynomially solvable in the broadcast-to-mailbox and the broadcast-to-port model (by computing the degree refinement of G).

4   Local Computations Models

In the local computations models, a computation step can be described by the application of some *local relabeling rule* that enables the modification of the states of the different vertices involved in the synchronization. Two local computation models are different in the types of relabeling rules that they allow, see Figure 1. In models (5), (6) and (7) of Figure 1, a computation step occurs on a star, i.e., it involves some synchronization between one vertex and all its neighbors, whereas in models (1), (2), (3) and (4), a computation step occurs on an edge, i.e., it involves some synchronization between two neighbors. All these models are asynchronous, in the sense that not all processors have to been involved in each computation step. In models (5), (6) and (7) (resp. (1), (2), (3) and (4)), two computations steps can occur concurrently if they occur on stars (resp. on edges) that do not share any vertex.

Mazurkiewicz [11] consider the model (7) of Figure 1, where in one computation step, a processor can modify its state and the states of its neighbors, according to the previous states of itself and its neighbors. In this model, a graph G is a solution graph for the election and naming problem if and only if G does not have a proper *perfect-regular coloring*, i.e., a proper labeling ℓ such that

 (i) for all $i \in \ell(V_G)$, $G[i]$ is edgeless, and
 (ii) for all $i, j \in \ell(V_G)$ with $i \neq j$, $G[i, j]$ is edgeless or else is a perfect matching.

Boldi et al. [1] consider the model (5) of Figure 1, where in one computation step one processor can modify its state, according to its previous state and to the states of its neighbors. In this model, a graph G is a solution graph for the naming problem if and only if G does not have a proper *semi-regular coloring*, i.e., a proper labeling ℓ of G such that

 (i) for all $i \in \ell(V_G)$, $G[i]$ is edgeless, and
 (ii) for all $i, j \in \ell(V_G)$ with $i \neq j$, $G[i, j]$ is semi-regular bipartite.

In this model, a graph G is a solution graph for the election problem if and only if there does not exist any semi-regular coloring ℓ of G such that for all $i \in \ell(V_G)$, $|\ell^{-1}(i)| > 1$.

In [5] the models (3), (4) and (6) of Figure 1 are considered; in these models, vertices and edges can be labelled. The model (6) differs from the model (5) by the fact that the processor that modifies its state can also modify the labels of the edges incident to it. In the models (3) and (4), a computation step occurs on an edge whose label can be modified. These two models differs by the fact

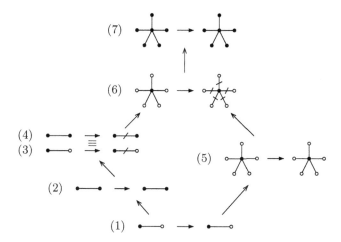

Fig. 1. A hierarchy of local computations models described by the different kinds of relabelling rules they use. Labels of black vertices can change when the local relabeling rule is applied. Labels of white vertices only enable to apply the rule but do not change. A rule can modify edge labels only in models $(3), (4)$ and (6). If $r_i \rightarrow r_j$ for models r_i and r_j then r_j can simulate r_i but not vice versa, i.e., r_j has a greater computational power than r_i; this relation is transitive. If $r_i \equiv r_j$ then r_i and r_j have the same computational power. The computational power of the model (5) is incomparable with the power of the models $(2), (3)$ and (4).

that in model (3), only one endvertex of the edge can modify its state, whereas in model (4), the two endvertices of the edge can modify their states. In each of these models, a graph G is a solution graph for the election and the naming problem if and only if G does not have a proper *regular coloring*, i.e., a proper labeling ℓ such that

(i) for all $i \in \ell(V_G)$, $G[i]$ is edgeless, and
(ii) for all $i, j \in \ell(V_G)$ with $i \neq j$, $G[i, j]$ is regular bipartite.

We note that Mazurkiewicz [12] has given an equivalent characterization of model (4) in terms of equivalence relations over vertices and edges. The characterizations for model (6) can also be obtained from [1].

In [4], the model (2) of Figure 1 is considered: in one computation step, two neighbors modify simultaneously their states according only to their previous states (the edges are not labeled). In this model, a graph G is a solution graph for the election and naming problem if and only if G does not have a proper *pseudo-regular coloring*, i.e., a proper labeling ℓ such that

(i) for all $i \in \ell(V_G)$, $G[i]$ is edgeless, and
(ii) for all $i, j \in \ell(V_G)$ with $i \neq j$, $G[i, j]$ is edgeless or else contains a perfect matching.

In [7], the model (1) of Figure 1 is considered: in one computation step, one processor modify its state according to both its previous state and the state of one of its neighbor (the edges are not labeled). In this model, a graph G is a solution graph for the naming problem if and only if G does not admit any proper *connected coloring*, i.e., a proper labeling ℓ such that

(i) for all $i \in \ell(V_G)$, $G[i]$ is edgeless, and
(ii) for all $i, j \in \ell(V_G)$ with $i \neq j$, $G[i,j]$ is edgeless or else has minimum degree at least one.

We note that the hierarchy in Figure 1 is also reflected by the labelings, e.g., a perfect-regular coloring is also a regular coloring, and so on.

5   Pseudo-regular Labelings

We call a labeling ℓ of a graph G a *pseudo-regular labeling* if

(i) for all $i \in \ell(V_G)$, $G[i]$ is regular, and
(ii) for all $i, j \in \ell(V_G)$ with $i \neq j$, $G[i,j]$ is edgeless or else contains a perfect matching.

In this section we prove that the problem whether a given graph G has a proper pseudo-regular labeling is NP-complete. This gives us a number of co-NP-completeness results for various models of distributed computing discussed in the previous two sections. The following observation is useful.

Observation 1. *Let ℓ be a pseudo-regular labeling of a connected graph G. Then* $|\ell^{-1}(i)| = \frac{|V_G|}{|\ell(V_G)|}$ *for all* $i \in \ell(V_G)$.

Let $G = (V_G, E_G)$ and $H = (V_H, V_G)$ be two graphs. We write $V_H = \{1, 2, \ldots, |V_H|\}$. For a mapping $f : V_G \to V_H$ and a set $S \subseteq V_G$, we write $f(S) = \{f(u) \mid u \in S\}$. A *graph homomorphism* from G to H is a vertex mapping $f : V_G \to V_H$ satisfying the property that for any edge $[u, v]$ in E_G, we have $[f(u), f(v)]$ in E_H, in other words, $f(N_G(u)) \subseteq N_H(f(u))$ for all $u \in V_G$. A homomorphism f from G to H that induces a one-to-one mapping on the neighborhood of every vertex is called *locally bijective*, i.e., for all $u \in V_G$ it satisfies $f(N_G(u)) = N_H(f(u))$ and $|N_G(u)| = |N_H(f(u))|$. In that case we write $G \xrightarrow{B} H$, and call the vertices of H *colors* of G. Sometimes, we also say that the labels $\ell(i)$ of a labeling ℓ of G are *colors* of G.

The H-COVER problem asks whether there exists a locally bijective homomorphism from an instance graph G to a fixed graph H. In our NP-completeness proof we use reduction from the K-COVER problem, where K is the graph obtained after deleting an edge in the complete graph K_5 on five vertices. The K-COVER problem is NP-complete [10]. Note that the two non-adjacent vertices have degree three. The other three vertices are adjacent to two vertices of degree three and two vertices of degree four. Then the following observation immediately follows from the definition of a locally bijective homomorphism.

Observation 2. *Let G be a graph with $G \xrightarrow{B} K$. Then the following holds:*

($*$) $V_G = B_1 \cup B_2$ *for two blocks B_1 and B_2 with $|B_1| = 2k$ and $|B_2| = 3k$ for some $k \geq 1$ such that*
- *for all $u \in B_1$, $|N_G(u) \cap B_1| = 0$ and $|N_G(u) \cap B_2| = 3$*
- *for all $u \in B_2$, $|N_G(u) \cap B_1| = 2$ and $|N_G(u) \cap B_2| = 2$.*

We call a graph satisfying the ($*$) condition in Observation 2 a K-*candidate*. Since ($*$) obviously can be checked in polynomial time, we may assume without loss of generality that any instance graph G of the K-Cover problem is a K-candidate.

For our NP-completeness structure we modify an instance graph G of the K-Cover as follows. Let u and v be vertices of G with $\deg_G(u) = 3$ and $\deg_G(v) = 4$. We replace the edge $[u, v]$ by a chain of $q \geq 1$ "diamonds" as described in Figure 2. We call the resulting graph G' a *diamond graph* of G with respect to the edge $[u, v]$. For $i = 1, \ldots, q$, the subgraph $D_i = G[\{a_i, b_i, c_i, d_i, e_i\}]$ is called a *diamond* of G'. The next lemma shows among others that a pseudo-regular

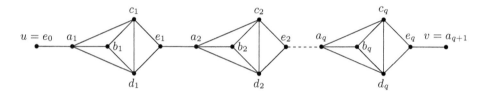

Fig. 2. The chain of q diamonds that replace the edge $[u, v]$

labeling is injective on the neighborhood of any vertex in a diamond. Its proof involves a case analysis and will be presented in the journal version of our paper.

Lemma 3. *Let G be a K-candidate that contains adjacent vertices u, v with $\deg_G(u) = 3$ and $\deg_G(v) = 4$. Let G' be a diamond graph of G with respect to $[u, v]$ that has diamonds $D_1, \ldots D_q$, where $q > k + 2$ and $q + k$ is a prime number. If ℓ is a proper pseudo-regular labeling of G', then $|\ell(V_{D_i})| = 5$ and $\ell(e_{i-1}) \notin \ell(V_{D_i} \setminus \{e_i\})$ for all $1 \leq i \leq q$.*

The following lemma is a key result.

Lemma 4. *Let G be a K-candidate that contains adjacent vertices u, v with $\deg_G(u) = 3$ and $\deg_G(v) = 4$. Let G' be a diamond graph of G with respect to $[u, v]$ that has diamonds $D_1, \ldots D_q$, where $q > k+2$ and $q+k$ is a prime number. If ℓ is a proper pseudo-regular labeling of G' then $|\ell(V_{G'})| = 5$.*

Proof. We write $p = q + k$. Then $|V_{G'}| = 5p$ and p is a prime number. Hence we find that $|\ell(V_{G'})| = 5$ or $|\ell(V_{G'})| = p$, due to Observation 1.

Suppose $|\ell(V_{G'})| = p > 5$. By our choice of q, there exist a vertex u in a diamond D_i with the same color as a vertex v in a diamond D_j. By Lemma 3, we may assume that $i < j$. We choose u and v such that there do not exist two vertices in $G[D_i \cup \ldots \cup D_{j-1}]$ having the same color. By Lemma 3, we can write $\ell(a_i) = 1$, $\ell(b_i) = 2$, $\ell(c_i) = 3$, $\ell(d_i) = 4$ and $\ell(e_i) = 5$, and we find that $\ell(e_{i-1}) \notin \{1, 2, 3, 4\}$. If $\ell(e_{i-1}) = 5$, then $\ell(a_{i+1}) = 1$ and consequently $|\ell(V_{G'})| = 5 < p$, so we write $\ell(e_{i-1}) = 6$.

By Observation 2 and the construction of G', every vertex of G has either degree 3 or 4. Note that, for each x in G' with $\ell(x) = 1$ (respectively $\ell(x) = 3$, $\ell(x) = 4$), we have that $\{2, 3, 4, 6\} \subseteq \ell(N_{G'}(x))$ (respectively $\{1, 2, 4, 5\} \subseteq \ell(N_{G'}(x))$, $\{1, 2, 3, 5\} \subseteq \ell(N_{G'}(x))$). Consequently, each vertex x with $\ell(x) \in \{1, 3, 4\}$ has $\deg_{G'}(x) = 4$.

By our choice of D_i and D_j, vertex a_{i+1} belongs to some diamond. By Lemma 3, we know that $|\ell(N_{G'}(a_{i+1}))| = 4$. Then each vertex x with $\ell(x) = \ell(a_{i+1})$ has $\deg_{G'}(x) = 4$. Suppose now that there exists a vertex y such that $\deg_{G'}(y) = 4$ and $\ell(y) = 2$ (respectively $\ell(y) = 5$). Then $\ell(N_{G'}(y)) = \{1, 3, 4\}$ (respectively $\ell(N_{G'}(y)) = \{3, 4, \ell(a_{i+1})\}$). Then y has three neighbors of degree four and this is not possible due to Observation 2. Consequently, each vertex y with $\ell(y) \in \{2, 5\}$ has $\deg_{G'}(y) = 3$.

We show that $1 \notin \ell(D_j)$. Suppose $\ell(a_j) = 1$. From our choice of D_i and D_j, we know that $\ell(e_{j-1}) \notin \{2, 3, 4\}$. Then $\ell(\{b_j, c_j, d_j\}) = \{2, 3, 4\}$ and $\ell(e_{j-1}) = 6$. Then $\ell(V_G) = \ell(D_i \cup \ldots \cup D_{j-1})$ and since all colors are different on diamonds $D_i, D_{i+1}, \ldots, D_{j-1}$, we find that $p = |\ell(V_G)| = 5(j - i)$. Since p is a prime number not equal to 5, this is not possible. We already know that $1 \notin \ell(\{b_j, e_j\})$ since $\deg_{G'}(b_j) = \deg_{G'}(e_j) = 3$. Suppose $\ell(c_j) = 1$ (respectively $\ell(d_j) = 1$). Then $\ell(d_j) \in \{3, 4\}$ (respectively $\ell(c_j) \in \{3, 4\}$) and $\ell(\{b_j, e_j\}) = \{2, 6\}$. Then a vertex with color in $\{3, 4\}$ is adjacent to a vertex with color 6. This is not possible.

We show that $2 \notin \ell(D_j)$. We already know that the only vertices in D_j that can be mapped to 2 are b_j and e_j in D_j. If $\ell(b_j) = 2$, then $1 \in \ell(\{a_j, c_j, d_j\})$. If $\ell(e_j) = 2$, then either $1 \in \ell(\{c_j, d_j\})$ or $\ell(\{c_j, d_j\}) = \{3, 4\}$ and in the second case $\ell(a_j) = 1$.

We show that $3 \notin \ell(D_j)$. We already know that only vertices a_j, c_j, d_j can be mapped to 3. If $\ell(a_j) = 3$ then 1, which does not occur on D_j, must be the color of $\ell(e_{j-1})$. This is not possible due to our choice of D_i and D_j. In the other two cases we find that $1 \in \ell(D_j)$. By symmetry, we deduce that $4 \notin \ell(D_j)$.

Finally, we show that $5 \notin \ell(D_j)$. We already know that only vertices b_j and e_j can be mapped to 5. In both cases, at least one of the colors 3, 4 is a color of a vertex in D_j. This finishes the proof of the lemma.   □

Lemma 5. *Let G be a graph that contains adjacent vertices u, v with $\deg_G(u) = 3$ and $\deg_G(v) = 4$. Let G' be a diamond graph of G with respect to (u, v). Then $G \xrightarrow{B} K$ if and only if $G' \xrightarrow{B} K$.*

Proof. We denote the vertices of K by $1, 2, 3, 4, 5$ and its edges by $[1, 2]$, $[1, 3]$, $[1, 4]$, $[1, 5]$, $[2, 3]$, $[2, 4]$, $[3, 4]$, $[3, 5]$, $[4, 5]$. Suppose $G \xrightarrow{B} K$. Without loss of generality we assume that u has color 5 and v has color 1. Then we assign color 1 to all a_i, color 2 all b_i, color 3 to all c_i, color 4 to all d_i and color 5 to all e_i.

Suppose $G' \xrightarrow{B} K$. The restriction of any locally bijective homomorphism $f' : V_{G'} \to V_K$ to V_G is a witness for $G \xrightarrow{B} K$.   □

Theorem 1. *The problems that ask whether a given graph G allows a proper pseudo-regular coloring, a proper pseudo-regular labeling, a proper regular coloring, a proper regular labeling, a proper symmetric regular labeling, or a proper perfect-regular coloring, respectively, are* NP-*complete.*

Proof. Obviously, all problems are in NP. We use reduction from the NP-complete problem K-COVER [10]. Let G be an instance graph of this problem. Since we may assume that G is a K-candidate, graph G has $5k$ vertices for some $k \geq 1$ and contains adjacent vertices u of degree three and v of degree four. We construct the diamond graph G' with respect to $[u, v]$ that has q diamonds $D_1, \ldots, D_q$, where we chose q such that $q > k + 2$ and $p = q + k$ is a prime number. By Lemma 5 we can consider G' as our instance graph for the K-COVER problem.

Any locally bijective homomorphism is a proper perfect-regular coloring, which is a regular coloring, which is a symmetric regular labeling, which is a regular labeling, which is a pseudo-regular labeling, and any regular coloring is a pseudo-regular coloring, which is a pseudo-regular labeling.

So we are left to show that a proper pseudo-regular labeling of G' implies that $G' \xrightarrow{B} K$. Suppose G' allows a proper pseudo-regular labeling ℓ. By Lemma 3, $|\ell(D_1)| = 5$. Let $\ell(a_1) = 1$, $\ell(b_1) = 2$, $\ell(c_1) = 3$, $\ell(d_1) = 4$ and $\ell(e_1) = 5$. By Lemma 3, $\ell(e_0) \notin \{1, 2, 3, 4\}$. Since $|\ell(V_G)| = 5$ due to Lemma 4, we then find that $\ell(e_0) = 5$. This means that ℓ defines a locally bijective homomorphism from G to K.   □

6   Connected Colorings and Semi-regular Colorings

A *hypergraph* $(Q, \mathcal{S})$ is a set $Q = \{q_1, \ldots, q_m\}$ together with a set $\mathcal{S} = \{S_1, \ldots, S_n\}$ of subsets of Q. A 2-*coloring* of a hypergraph $(Q, \mathcal{S})$ is a partition of Q into $Q_1 \cup Q_2$ such that $Q_1 \cap S_j \neq \emptyset$ and $Q_2 \cap S_j \neq \emptyset$ for $1 \leq j \leq n$. In our proofs we use reduction from the following, well-known NP-complete problem (cf. [9]).

HYPERGRAPH 2-COLORABILITY
Instance: A hypergraph $(Q, \mathcal{S})$.
Question: Does $(Q, \mathcal{S})$ have a 2-coloring?

With a hypergraph $(Q, \mathcal{S})$ we associate its *incidence graph* I, which is a bipartite graph on $Q \cup \mathcal{S}$, where $[q, S]$ forms an edge if and only if $q \in S$. From the incidence graph I we act as follows. Let C_k denote a cycle on k vertices. First we make a copy S' for each $S \in \mathcal{S}$. We add edges (S', q) if and only if $q \in S$. Let $\mathcal{S}' = \{S'_1, \ldots, S'_n\}$. Then we glue a cycle C_{q_i} isomorphic to a C_{6i-3} in I by vertex q_i for $1 \leq i \leq m$. We add a new vertex v and edges from v to all vertices in $\mathcal{S}$. Finally we glue a cycle C_v isomorphic to C_{6m+3} in I by v. We call the resulting graph I^* the C_3-*minimizer* of $(Q, \mathcal{S})$. See Figure 3 for an example. The proof of the following lemma will be included in the journal version.

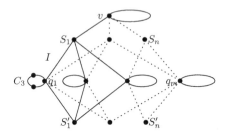

Fig. 3. Example of a C_3-minimizer I^* of a hypergraph $(Q, \mathcal{S})$

Lemma 6. *Let I^* be the C_3-minimizer of a hypergraph $(Q, \mathcal{S})$ with $S_j \neq S_k$ for all j, k. If ℓ is a proper connected coloring of I^* then $|\ell(V_{I^*})| = 3$.*

Theorem 2. *The problem that asks whether a given graph G has a proper connected coloring is NP-complete.*

Proof. Obviously, this problem is in NP. We prove NP-completeness by reduction from the HYPERGRAPH 2-COLORABILITY problem. Let $(Q, \mathcal{S})$ be a hypergraph. We assume without loss of generality that $S_j \neq S_k$ for $j \neq k$. We claim that $(Q, \mathcal{S})$ has a 2-coloring if and only if its C_3-minimizer I^* admits a proper connected coloring.

Suppose $(Q, \mathcal{S})$ has a 2-coloring $Q_1 \cup Q_2$. Define $\ell(v) = 1$, $\ell(S) = 2$ for all $S \in \mathcal{S} \cup \mathcal{S}'$, $\ell(q) = 1$ for all $q \in Q_1$ and $\ell(q) = 3$ for all $q \in Q_2$. Finish the coloring in the obvious way.

Suppose I^* has a proper connected coloring ℓ. By Lemma 6 we find $|\ell(V_{I^*})| = 3$. Let $\ell(v) = 1$. Then $\ell(S_j) \in \{2, 3\}$ for all j. If $\ell(S'_j) = 1$ for some j, then S'_j needs a neighbor of color 2 and a neighbor of color 3, both are adjacent to S_j. Hence, $\ell(S'_j) \in \{2, 3\}$ for all j. We define $Q_1 = \{q \in Q \mid \ell(q) = 1\}$ and $Q_2 = Q \backslash Q_1$. Since each S'_j needs at least two neighbors with different colors and at least one neighbor with color 1, the partition $Q_1 \cup Q_2$ is a 2-coloring of $(Q, \mathcal{S})$. □

The proof of Theorem 3 uses arguments of the proofs of Theorem 1 and Theorem 2 but the NP-completeness construction is more involved. We postpone it to the journal version.

Theorem 3. *The problem that asks whether a given graph G has a proper semi-regular coloring is NP-complete.*

7   Conclusions

By Theorems 1, 2 and 3 we have determined the computational complexity of the question whether the election and/or naming problem can be solved on a given graph in eleven different models of distributed computing that all have been studied in the literature.

Corollary 1. *It is co-NP-complete to decide if on a given graph G we can solve*

(a) the election problem in the models described in Sections 3 and 4 except for the broadcast-to-port model, the broadcast-to-mailbox model and models (1), (5) of Figure 1;

(b) the naming problem in the models described in Sections 3 and 4 except for the broadcast-to-port and broadcast-to-mailbox model.

As a matter of fact the above decision problem is co-NP-complete for the election problem in models (1) and (5) of Figure 1 as well. In both cases, the NP-hardness of the decision problem can be derived from the proofs of corresponding theorems for the naming problem (i.e., Theorems 2 and 3). From the characterization of Boldi et al. [1] of solution graphs for the election problem in the model (5) of Figure 1, one can see that the corresponding decision problem is in co-NP. We do not know any characterization of graphs that admits an election algorithm in the model (1) that can be expressed in terms of graphs labelings, and then the proof that the corresponding decision problem is in co-NP is not so obvious. Both proofs are postponed to the journal version.

We note that the problem that asks whether a given connected graph G has a proper perfect-regular coloring is equivalent to the problem that asks whether $G \xrightarrow{B} H$ for some connected graph H with $|V_H| < |V_G|$. A graph homomorphism f from G to H satisfying $f(N_G(u)) = N_H(f(u))$ for all $u \in V_G$ is called *locally surjective*. If such a homomorphism exists, we write $G \xrightarrow{S} H$. The problem that asks whether a connected graph G has a proper connected coloring is equivalent to the problem that asks whether $G \xrightarrow{S} H$ for some connected graph H with $|V_H| < |V_G|$. Let $\mathcal{C}$ denote the set of connected graphs (up to isomorphism). In [8] it has been proven that $(\mathcal{C}, \xrightarrow{B})$ and $(\mathcal{C}, \xrightarrow{S})$ are partial orders. Theorem 1 and 2 imply that it is co-NP-complete to check whether a graph is minimal in $(\mathcal{C}, \xrightarrow{B})$ and $(\mathcal{C}, \xrightarrow{S})$, respectively. Also the other studied graph labeling problems can easily be formulated as problems that ask whether there exist a homomorphism f, that satisfies a few extra constraints, from a given graph G to a smaller graph H. In future research we will study the relations between these *constrained* homomorphisms more carefully.

Acknowledgements. The authors thank Jiří Fiala for the idea on prime numbers in Lemma 4.

References

1. P. Boldi, B. Codenotti, P. Gemmell, S. Shammah, J. Simon, and S. Vigna, Symmetry breaking in anonymous networks: Characterizations. *Proceedings of the 4th Israeli Symposium on Theory of Computing and Systems* (ISTCS 1996), IEEE Press, 16–26.

2. P. Boldi and S. Vigna, Fibrations of graphs. *Discrete Mathematics 243* (2002), 21-66.

3. J.A. Bondy and U.S.R. Murty, *Graph Theory with Applications.* Macmillan, London and Elsevier, New York, 1976.

4. J. CHALOPIN, Election and Local Computations on Closed Unlabelled Edges. *Proceedings of the 31st Annual Conference on Current Trends in Theory and Practice of Informatics* (SOFSEM 2005), Lecture Notes in Computer Science 3381, Springer Verlag, 81-90.

5. J. CHALOPIN AND Y. MÉTIVIER, Election and Local Computations on Edges. *Proceedings of Foundations of Software Science and Computation Structures: 7th International Conference* (FOSSACS 2004), Lecture Notes in Computer Science 2987, Springer Verlag, 90–104.

6. J. CHALOPIN AND Y. MÉTIVIER, A Bridge between the Asynchronous Message Passing Model and Local Computations in Graphs. *Proceedings of the 30th International Symposium on Mathematical Foundations of Computer Science* (MFCS 2005), Lecture Notes in Computer Science 3618, Springer Verlag, 212–223.

7. J. CHALOPIN AND Y. MÉTIVIER AND W. ZIELONKA, Election, naming and cellular edge local computations. *Proceedings of the 2nd International Conference on Graph Transformation* (ICGT 2004), Lecture Notes in Computer Science 3256, Springer Verlag, 242–256.

8. J. FIALA, D. PAULUSMA, AND J.A. TELLE, Matrix and Graph Orders derived from locally constrained homomorphisms. *Proceedings of the 30th International Symposium on Mathematical Foundations of Computer Science* (MFCS 2005), Lecture Notes in Computer Science 3618, Springer Verlag, 340–351

9. M.R. GAREY AND D.S. JOHNSON, *Computers and Intractability*. W.H. Freeman and Co., New York, 1979.

10. J. KRATOCHVÍL, A. PROSKUROWSKI, AND J.A. TELLE, Complexity of graph covering problems. *Nordic Journal of Computing 5* (1998), 173–195.

11. A. MAZURKIEWICZ, Distributed enumeration. *Information Processing Letters 61* (1997), 233–239.

12. A. MAZURKIEWICZ, Bilateral ranking negotiations. *Fundamenta Informaticae 60* (2004), 1–16.

13. G. TEL, *Introduction to distributed algorithms*. Cambridge University Press, 2000.

14. M. YAMASHITA AND T. KAMEDA, Computing on anonymous networks: Part I - Characterizing the solvable cases. *IEEE Transactions on parallel and distributed systems 7* (1996), 69–89.

15. M. YAMASHITA AND T. KAMEDA, Computing on anonymous networks: Part II - decision and membership problems. *IEEE Transactions on parallel and distributed systems 7* (1996), 90–96.

16. M. YAMASHITA AND T. KAMEDA, Leader election problem on networks in which processor identity numbers are not distinct. *IEEE Transactions on parallel and distributed systems 10* (1999), 878–887.

Flexible Matchings

Miklós Bartha[1],[*] and Miklós Krész[2],[**]

[1] Memorial University of Newfoundland, St. John's, Canada
[2] University of Szeged, Szeged, Hungary

Abstract. A matching M is called flexible if there exists an alternating cycle with respect to M. Given a graph $G = (V, E)$ and $S \subseteq V$, a flexible matching $M \subseteq E$ is sought which covers a maximum number of vertices belonging to S. It is proved that the existence of such a matching is decidable in $\mathcal{O}(|V| \cdot |E|)$ time, and a concrete flexible maximum S-matching can also be found in the same amount of time.

1 Introduction

Flexibility of matchings is an interesting phenomenon emerging in a number of graph theory applications. In short, the problem is that, given a graph G with a maximum matching M, find another matching M' of G that covers the same vertices as M. The question is particularly interesting if G is allowed a so-called grace, i.e., a subset of vertices which need not necessarily be covered by maximum matchings.

As a motivating example, consider a project on which employees of a company work in couples. The company employs both full-time and part-time workers, but its preference is to assign as many full-time employees to the project as possible. In addition, a known compatibility relationship among the employees must be respected, which determines the possible couplings for the job. Furthermore, it is also desirable that the assignment be flexible in the sense that, while keeping the group of participating employees fixed, the coupling of them could be altered. Graph G in this case consists of the employees as vertices and the compatibility relation among them as edges. Vertices corresponding to full-time employees are considered relevant, whereas the group of part-time employees constitutes the grace set of G. The goal is to find a matching M of G that covers a maximum number of relevant vertices, so that M is also flexible in the above sense.

Flexibility of matchings without the grace feature has been studied in the papers [9,11,13,15]. The authors of [11] and [15] investigate the complexity of finding an alternating-cycle-free matching with a maximum cardinality, while [9] provides a linear time algorithm to decide if a given matching is free from

---

[*] Partially supported by Natural Science and Engineering Research Council of Canada, Discovery Grant #170493-03.
[**] Supported by a grant from the Hungarian Ministry of Economy and Transport (project No. GVOP-3.1.1-2004-05-0119/3.0) and by the MÖB-DAAD Hungarian-German Researcher Exchange Program (project No. 21).

F.V. Fomin (Ed.): WG 2006, LNCS 4271, pp. 313–324, 2006.

alternating cycles. The latter result suggests a naive method for finding out if a graph $G = (V, E)$ has a flexible maximum matching: list all the maximum matchings, and see if they contain an alternating cycle. Compared to the exponential time complexity of this method, our algorithm presented in Section 5 takes $\mathcal{O}(|V| \cdot |E|)$ time only.

Another application of flexible matchings with a grace concerns the study of soliton automata [2,3,6]. In this model the topological structure of a hydrocarbon molecule is represented by a graph G, and an alternating pattern of single and double chemical bonds between the atoms determines a matching in G. Some vertices are distinguished as external, and they are exempt from the general requirement that each vertex (atom) be connected to exactly one of its neighbors by a double bond. The existence of a flexible matching covering all of the internal (i.e., non-external) vertices then implies that the soliton automaton determined by G is non-deterministic [5].

Our concern in this paper is to design an algorithm that decides for a given graph $G = (V, E)$ with a grace set if G has a flexible maximum matching, and if so, find one. We shall provide an algorithm which solves this problem in $\mathcal{O}(|V| \cdot |E|)$ time.

2 Basic Concepts

By a graph, throughout the paper, we mean a finite undirected graph in which loops and multiple edges are allowed. Our notation and terminology will follow [14].

Let $G = (V, E)$ be a graph, fixed for the rest of the paper. A *matching* in/of G is a set $M \subseteq E$ such that no vertex in V occurs more than once as an endpoint of some edge in M. By this definition, loops are not allowed to participate in matchings. For a fixed set $S \subseteq V$, a *maximum (perfect) S-matching* of G is a matching M that covers a maximum number of (respectively, all) vertices in S. If $S = V$, then we speak of an ordinary maximum (perfect) matching of G. Vertices in $V \setminus S$ constitute the *grace* set of G in the sense that, although they may, they need not be covered by M.

Definition 2.1. *A matching M in G is flexible if there exists a matching $M' \neq M$ covering the same vertices as M. Graph G with a grace set $V \setminus S$ is flexible (rigid) if it has (respectively, does not have) a flexible maximum S-matching.*

It is clear by the above definition that matching M is flexible iff there exists an M-alternating cycle in G. Indeed, if an appropriate M' exists, then the symmetric difference of M and M' consists of pairwise disjoint even-length cycles alternating on edges belonging to M and M'. Thus, G is rigid iff it does not contain an alternating cycle with respect to any of its maximum S-matchings. Note that the edges connecting two grace vertices do not influence the cardinality of maximum S-matchings, therefore we can assume that such edges do not exist in G, and each grace vertex u is adjacent to at least one vertex $v_u \in S$. As it will be pointed out below, we can also assume without loss of generality that the vertex v_u is unique. If this is the case, and all vertices in S have a degree greater than 1, then $S = Int(G)$ and $V \setminus S = Ext(G)$ are called the set of *internal* and *external* vertices, respectively. (Notice that the degree of the external vertices, and only of these, is 1.) Edges incident with external vertices are called external themselves, and their internal endpoints will be referred to as *base* vertices. Graph G is called *open* if $Ext(G) \neq \emptyset$ and *closed* if $Ext(G) = \emptyset$.

The advantage of having an internal-external distinction between vertices in S and those in $V \setminus S$ is twofold. Technically, it simplifies the reasoning about grace vertices, but more importantly, it allows us to adopt the results obtained in [1,2,3] on soliton graphs and automata with no change. Following this philosophy, a maximum (perfect) S-matching will be called a maximum (respectively, perfect) *internal* matching, abbreviated as m.i.m. (respectively, p.i.m.).

To see how an arbitrary grace set $V \setminus S$ can be replaced by an external one, attach a new (external) edge to each vertex in $V \setminus S$, and add a loop around each vertex in S having degree 1. Let G_{ext} denote the resulting graph with its grace set being the set of newly introduced external vertices. Then, for every matching M of G there exists a unique matching M_{ext} of G_{ext} such that $M_{ext} \cap E = M$ and an external edge e is contained in M_{ext} iff the base vertex incident with e is not covered by M. Clearly, this correspondence between matchings of G and those of G_{ext} is one-to-one, which preserves the maximum and perfect properties. Moreover, flexibility of matchings is preserved as well.

3   Flexibility in Terms of the Gallai-Edmonds Decomposition

Recall from [14] that graph G is *factor-critical* if for every $v \in V$, the graph $G - v$ has a perfect matching. A *near-perfect* matching of G is a matching that covers all vertices but one. The factor-critical property has been generalized in [4] for open graphs with p.i.m.'s, and factor critical open graphs have been characterized as ones consisting of a single external family of elementary components [3].

One of the most useful results in matching theory is the so called Gallai-Edmonds Structure Theorem found independently by Gallai [10] and Edmonds [7]. The reader is referred to Section 3.2 in [14] for a detailed discussion of this theorem, which characterizes the structure of maximum matchings in graphs. A counterpart of the Gallai-Edmonds Structure Theorem for open graphs and m.i.m.'s, Theorem 3.1 below, has first been stated and proved in [1].

Let $D(G)$ denote the set of all internal vertices in G that are not covered by at least one m.i.m. Furthermore, let $A(G)$ be the set of vertices (internal or external) in $V \setminus D(G)$ adjacent to at least one vertex in $D(G)$. Finally, let $C(G) = V \setminus A(G) \setminus D(G)$.

Theorem 3.1. *The following four statements hold for the decomposition $D(G)$, $A(G)$, $C(G)$:*

(i) the components of the subgraph induced by $D(G)$ are factor-critical,

(ii) the subgraph induced by $C(G)$, augmented by a loop around each vertex that was originally internal in G but has become external after subtracting $A(G)$, contains all the external vertices of G and has a p.i.m.,

(iii) the bipartite graph $G[A, D]$ obtained from G by deleting the vertices of $C(G)$ and the edges spanned by $A(G)$ and by contracting each component of $D(G)$ to a single vertex has positive surplus (as viewed from $A(G)$),

(iv) if M is any m.i.m. of G, M contains a near-perfect matching of each component of $D(G)$, a p.i.m. of $C(G)$ and matches all vertices of $A(G)$ with vertices in distinct components of $D(G)$.

Recall that the *surplus* of a bipartite graph with bipartition (A, D) viewed from A is the number $\min\{(|\Gamma(X)| - |X|) \mid X \subseteq A, X \neq \emptyset\}$, where $\Gamma(X)$ is the set of vertices in D adjacent to some vertex in X.

In the light of Theorem 3.1 it is easy to prove the following necessary and sufficient condition for the existence of a flexible m.i.m. in G.

Proposition 3.2. *Graph G has a flexible m.i.m. iff one of the following three conditions is satisfied by the Gallai-Edmonds decomposition.*

(i) $C(G)$ *(with the augmentation as above) possesses a flexible p.i.m.,*

(ii) $G[A, D]$ *contains a flexible maximum matching that will necessarily cover all vertices in $A(G)$,*

(iii) *one of the components of the subgraph induced by $D(G)$ has a flexible near-perfect matching.*

By Proposition 3.2, the original problem of finding a flexible m.i.m. in G can be broken down to three subproblems concerning graphs having a p.i.m., bipartite graphs with a positive surplus, and factor-critical graphs. We shall deal with these subproblems separately in the following two sections.

4   Flexibility in Elementary Graphs

In this section we provide a characterization of rigid elementary open graphs. This characterization is the first main result of the paper, which will help us to solve the first and second subproblems specified above. For the rest of this section, assume that our graph G is open and has a p.i.m.

Recall from [3,14] that an edge $e \in E$ is *allowed (mandatory)* if e is contained in some (respectively, all) p.i.m.(s) of G. *Forbidden* edges are those that are not allowed. Graph G is called *elementary* if its allowed edges form a connected subgraph covering all the external vertices. Observe that if G is elementary, then it cannot contain a mandatory edge, unless G is a mandatory edge by itself with a number of loops incident with one of its endpoints.

Let M be a matching of G. An edge $e \in E$ is said to be *M-positive (M-negative)* if $e \in M$ (respectively, $e \notin M$). An *M-alternating path (cycle)* in G is a path (respectively, even-length cycle) stepping on M-positive and M-negative edges in an alternating fashion. Let us agree that, if the matching M is understood or irrelevant in a particular context, then it will not be explicitly indicated in these terms. Notice that an odd-length cycle can alternate on positive and negative edges only if it starts out and ends in a negative edge. Such cycles, however, will not be called alternating cycles. Rather, a *wrench* in G is a trail [14] consisiting of a (possibly empty) path with an odd-length cycle attached to one of its endpoints. An *alternating wrench* is a wrench alternating on positive and negative edges when followed as a trail from one endpoint to the other.

An *external alternating path* is one that has an external endpoint. If both endpoints of the path are external, then it is called a *crossing*. An alternating path is *positive* (*negative*) if it is such at its internal endpoints. An internal vertex $v \in V$ is called *accessible* with respect to matching M if there exists a positive external M-alternating path leading to v.

Let M be a p.i.m. of G. An M-*alternating unit* is either a crossing or an alternating cycle with respect to M. *Switching* on an alternating unit amounts to changing the sign of each edge along the unit. It is easy to see that the operation of switching on an M-alternating unit α creates a new p.i.m. $S(M, \alpha)$ for G. Moreover, it is easy to see that every p.i.m. M of G can be transformed into any other p.i.m. M' by switching on a collection of pairwise disjoint alternating units.

Now we introduce a concept which will be fundamental in the characterization of rigid open elementary graphs.

Definition 4.1. *A redex r in G consists of two adjacent edges $e = (u, z)$ and $f = (z, v)$ such that $u \neq v$ are both internal and the degree of z, $d(z) = 2$. The vertex z is called the center of r, while u and v (e and f) are the two focal vertices (respectively, focal edges) of r.*

Let r be a redex in G. *Contracting* r means creating a new graph G_r from G by deleting the center of r and merging the two focal vertices into one vertex s. The vertex s is called the *sink* of r in G_r. For a p.i.m. M of G, let M_r denote the restriction of M to edges in G_r. Clearly, M_r is a p.i.m. Notice that M can be reconstructed from M_r in a unique way. In other words, the connection $M \mapsto M_r$ is a one-to-one correspondence between p.i.m.'s of G and those of G_r. Graph G and p.i.m. M will be referred to as the *unfolding* of G_r and M_r, respectively.

For any walk [14] α in G, let $trace_r(\alpha)$ denote the restriction of α to edges in G_r. Obviously, $trace_r(\alpha)$ is a walk in G_r. It is also easy to see that if α is an alternating path (cycle, wrench) in G with respect to M, then $trace_r(\alpha)$ is such in G_r with respect to M_r. Moreover, α can uniquely be recovered from $trace_r(\alpha)$ by unfolding. Notice, furthermore, that if an alternating unit goes through both focal vertices of r, then it must do so along the center of r. As a consequence we have:

Proposition 4.2. *The function $trace_r$ establishes a one-to-one correspondence between alternating units of G and those of G_r. For any M-alternating unit α, $M' = S(M, \alpha)$ holds in G iff $(M')_r = S(M_r, trace_r(\alpha))$ holds in G_r.*

Corollary 4.3. *Every edge e of G_r is allowed in G_r iff e is allowed in G.*

Another natural simplifying operation on graphs is the removal of a loop from around a vertex v if this does not turn v into an external vertex. Such loops will be called *inner*. Let G_v denote the graph obtained from G by removing an inner loop at vertex v. Clearly, the p.i.m.'s of G_v are exactly the same as those of G.

Definition 4.4. *Graph G is reduced if it is free from redexes and inner loops.*

Lemma 4.5. *If G is reduced elementary and contains a cycle, then there exists an internal edge $h \in E$ such that $G - h$ is still elementary.*

Proof. For a fixed p.i.m. M of G, let us construct a sequence of subgraphs $G_0, \ldots, G_{n+1}$ of G in the following way. The sequence starts out with the empty graph G_0, and for every $0 \leq i \leq n$, G_{i+1} is obtained from G_i by adding an M-alternating unit α_i of G which covers at least one edge not already in G_i. The process stops when a connected graph G_{n+1} is reached that covers all vertices of G. Since G is elementary, the process is well-defined. During this process it is possible to cover the external vertices of G first, that is, to select for the unit α_i an appropriate crossing that covers at least one new external vertex, up to an index $i = k$ such that G_{k+1} already covers all the external vertices. Moreover, since each internal vertex of G_i is accessible (within G_i) with respect to M, we can assume that each graph G_i, $1 \leq i \leq k + 1$ is a forest.

If G contains edges that are not in G_{n+1}, then we are through, for any, or even all of these edges can be left out from G without losing the elementary property. Assume therefore that $G_{n+1} = G$. Then, clearly, $k < n$. (Remember that G contains a cycle and G_{k+1} is a forest.) On the other hand, observe that the unit α_n could not add a new vertex to G_n. Indeed, such a vertex would necessarily be internal in G, would also be different from the base vertices, and as such, it would have a degree greater than two. This contradicts the assumption that $G_{n+1} = G$. We conclude that G_n is not connected, covering all vertices of G though.

Let e be any edge of α_n that connects two different components of G_n. Since all internal vertices of G_n are accessible within their respective connected components, the edge e is allowed in $G_n + e$. Thus, if α_n contains an edge h that is neither in G_n, nor a cut-edge in G, then $G - h = G_n + \alpha_n - h$ is elementary. Otherwise G_n, too, contains a cycle in one of its connected components C. Augment C with the new edge(s) of α_n incident with C, considering them as external edges in the resulting graph. This graph satisfies the conditions of the lemma, and has fewer internal edges than G. The proof can now be finished by organizing the above argument as a proper induction on the number of internal edges in G.

For any path (wrench) α in G, *marking* α as an alternating path (respectively, wrench) amounts to specifying a matching M of G which consists of edges along α only, and for which α becomes an M-alternating path (wrench). A *generalized tree* is an open graph not containing even-length cycles.

Lemma 4.6. *Let G be a reduced generalized tree. Then the marking of any path or wrench α in an alternating way can be extended to a p.i.m. M_α in G.*

Proof. Without loss of generality we can assume that $\alpha = p$ is a path. Again, the proof is an induction argument on the number of internal edges in G. If G has no internal edges, then the statement is trivial. Let G therefore have at least one internal edge, and assume that the statement holds true for all reduced generalized trees having fewer internal edges than G. If G has an internal cut edge e, then cut G along e to obtain two generalized trees G_1 and G_2. In both G_1 and G_2, e appears as an external edge, ensuring that these graphs remain reduced. The statement of the lemma then follows easily from applying the induction hypothesis on G_1 and G_2 with suitable alternating paths p_1 and p_2.

If G does not have an internal cut edge, then every internal vertex of G, except possibly the base ones, is part of at least two odd-length cycles, and even the base vertices have a degree at least 3. Consequently, an arbitrary internal edge can be left out from G, still keeping it a reduced generalized tree. Since p is a path, the edge e to be left out can be chosen from outside p. Now the statement follows directly from the induction hypothesis by taking $e \notin M_p$.

Now we are ready to present our characterization of rigid open elementary graphs. The key to this result is Theorem 4.7 below. The underlying idea of the proof of this theorem is a simple induction argument, which is highlighted by the second paragraph of the proof. Yet, the complete proof is technically challenging, as one must cope with a number of cases and subcases.

Theorem 4.7. *If G is reduced elementary and contains an even-length cycle, then it also has an alternating cycle with respect to some p.i.m.of G.*

Proof. Induction on the number of internal edges of G. If G has no internal edges, then we have nothing to prove. Let G have at least one internal edge, and assume that the statement holds for all appropriate graphs having fewer internal edges than G. If G has a pair $\{h, g\}$ of parallel edges, then either both, or none of these edges is allowed. If they are allowed, then they form an alternating cycle with respect to every p.i.m. M in which one of them is present. If they are forbidden, then the graph $G - h$ is still elementary and reduced, otherwise some edge adjacent to h and g would be mandatory in G. Now the statement follows from the induction hypothesis.

We can therefore assume that G does not contain parallel edges. Let G, nevertheless, contain an even-length cycle. By Lemma 4.5 there exists an internal edge $h \in E$ such that $G - h$ is still elementary. Suppose first that $G - h$ is reduced. If it also contains an even-length cycle, then by the induction hypothesis we are through. If not, then $G - h$ is a reduced generalized tree, and G has an even-length cycle α containing h. Now the statement is obtained by applying Lemma 4.6 on $G - h$, marking $\alpha - h$ as a positive alternating path in that graph.

Suppose now that $G - h$ is not reduced, and let z_1, z_2 denote the two endpoints of h with z_1 being the center of an appropriate redex r in $G - h$. Since h is the only edge in G connecting z_1 and z_2, the focal vertices u_1, v_1 of the redex r are both different from z_2. We organize the rest of the proof according to whether both, only one, or neither of u_1 and v_1 is adjacent to z_2 in G. Let G' denote the graph obtained from G by contracting the redex r to a sink vertex s and removing a possible loop around s caused by the presence of the edge $g_1 = (u_1, v_1)$ in G. By Corollary 4.3 G' is elementary.

Case 1: Both u_1 and v_1 are adjacent to z_2.
In this case G' will contain a pair of parallel edges connecting z_2 with the sink s. Graph G' is also reduced, otherwise it would not be elementary. Now the statement follows from the induction hypothesis and Proposition 4.2.

Case 2: Only one of u_1 and v_1, say u_1, is adjacent to z_2.
To handle this case we need to further break it down into two subcases.

Subcase 2a: The edge $g_1 = (u_1, v_1)$ is present in G, as shown in Fig. 1.
If the edge $e_2 = (u_1, z_2)$ is allowed in $G - h$, then let M be any p.i.m. of G for which $e_2 \in M$. Clearly, the edges e_2, h, $f_1 = (z_1, v_1)$, and g_1 form an M-alternating cycle. If e_2 is forbidden in $G - h$, then G' must be reduced. For, if G' were not reduced, then either z_2 or the sink s would become the center of a new redex r' in G'. Either way, e_2 would be one of the focal edges of r', implying that the other focal edge is mandatory; a contradiction. The graph G' cannot be a generalized tree, because in that case Lemma 4.6 would force e_2 to be allowed. We conclude that G' is reduced and contains an even-length cycle. Now the statement follows again from the induction hypothesis and Proposition 4.2.

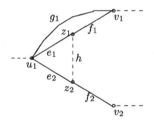

Fig. 1.

Subcase 2b: The edge $g_1 = (u_1, v_1)$ is not present in G.
First assume that $d(z_2) \geq 3$. Then G' is again reduced. If it also contains an even-length cycle, then we are done. Otherwise G' is a generalized tree, and G has an even-length cycle α passing through h. Concentrate on the composition of $\alpha' = trace_r(\alpha - h)$. It will consist of a possible odd-length cycle around the sink s and a path p connecting s with z_2. (See again Fig. 1.) If the odd-length cycle is present in α', then the length of p is odd, too, because α' contains two less edges than α. If $\alpha' = p$, then the length of p is even. In this case, however, α covers only one of u_1 and v_1. Should either of them be missed by α, there exists an edge $t \in E$ incident with that vertex such that $t \notin \{e_1, f_1, e_2\}$. (Remember that $g_1 = (u_1, v_1)$ is now not present in G.) Moreover, t is not incident with any vertex on α, otherwise t would give rise to an even-length cycle in G'.

   Mark the edges of α' in an alternating way, so that the endpoint z_2 of α' be positive. Furthermore, if $\alpha' = p$, then add the edge t specified above with a positive sign to the marking. By Lemma 4.6, this marking can be extended to a p.i.m. M' of G' by which α' is an M'-alternating path or wrench, positive at its z_2 end. Moreover, if $\alpha' = p$, then $t \in M'$. Finally, reconstruct the cycle α from α', and observe that α is M-alternating in G with respect to the unfolding M of M' to $G - h$, and further to G by taking $h \notin M$.

Now assume that $d(z_2) = 2$, so that G has a redex around both z_1 and z_2. By symmetry we can assume that $g_2 = (u_1, v_2)$ is not present in G. Then the graph G'' obtained from G by contracting its two redexes and removing a possible loop (caused by the edge (v_1, v_2)) around their common sink s is reduced. Again, if G'' contains an even-length cycle, then we are done. If not, then G'' is a generalized tree, and we proceed as in the second last paragraph. The trace α' of $\alpha - h$ is now a closed trail. This trail is either empty, when the even-length cycle α in G consists of the four edges f_1, h, f_2, (v_1, v_2), or it is a single odd-length cycle when (v_1, v_2) is not present in G. Indeed, if $(v_1, v_2) \notin E$, then α' has three less edges than α, so that if α' consisted of two cycles, then one of these would be even-length. Either way, α misses one of the vertices u_1, v_1, v_2. No matter which one of them is missed, there exists an edge $t \in E$ incident with that vertex, such that the other endpoint of t is not on α either.

Mark the edge t positive in G'', and continue the marking on α' in an alternating way if α' is not empty. Using Lemma 4.6, extend this marking to a p.i.m. M'' of G''. Reconstruct α from α', and observe that α becomes an M-alternating cycle with respect to the unfolding M of M'' to $G - h$, and on to G by taking $h \notin M$. The proof of Case 2 is now complete.

A similar proof works for Case 3, whereby neither u_1 nor v_1 is adjacent to z_2. This proof is left to the reader.

For an arbitrary graph G, contract all redexes and remove all inner loops in an iterative way to obtain a reduced graph $r(G)$. Observe that this reduction procedure has the so-called Church-Rosser property, that is, if G admits two different one-step reductions to graphs G_1 and G_2, then either G_1 is isomorphic to G_2, or G_1 and G_2 can further be reduced to a common graph G_3, actually by the very same reduction steps applied originally on G to obtain G_2 and G_1, respectively. In this context, one reduction step means contracting a redex or removing a single inner loop. As an immediate consequence of the Church-Rosser property, the graph $r(G)$ is unique up to graph isomorphism.

Corollary 4.8. *An open elementary graph G is rigid iff $r(G)$ is a generalized tree.*

Proof. Immediate by Proposition 4.2 and Theorem 4.7.

5 Algorithmic Issues

The earliest, and still one of the best matching algorithms is due to Edmonds [7]. See also [14] for a summary of different matching algorithms. Using the method of [8], the most efficient implementation of the Edmonds algorithm runs in $\mathcal{O}(|V| \cdot |E|)$ time. Not only does this algorithm find a maximum matching for G, but as a by-product it also constructs the Gallai-Edmonds decomposition of G. A straightforward modification of the Edmonds algorithm can be used to find a m.i.m. M and the Gallai-Edmonds decomposition of an open graph G, which algorithm still runs in $\mathcal{O}(|V| \cdot |E|)$ time.

Now we turn our attention to the three subproblems specified at the end of Section 3. To deal with the first of them, let G have a p.i.m. M and construct the graph G_M^* by augmenting G in the following way. Attach a new external edge to each external vertex not covered by M, and add these edges to M. This results in a new graph G_M' with a perfect matching M'. Then G_M^* is obtained from G_M' by adding two further vertices v_1 and v_2, and connecting them with each other and with all the external vertices of G_M'. Clearly, $M^* = M' + (v_1, v_2)$ is a perfect matching of G_M^*. Moreover, the following simple statement holds for the forbidden edges of G and those of G_M^*. (Recall that an edge is forbidden if it is not part of any p.i.m.)

Proposition 5.1. *For every edge $e \in E$, e is forbidden in G iff e is forbidden in G_M^*.*

Apply the method outlined in [16] to identify the forbidden edges in G_M^* and thus in G. As shown in [16], this can be done in $\mathcal{O}(|V| \cdot |E|)$ time, in terms of the graph G_M^*, though. Since G_M^* has $\mathcal{O}(|Ext(G)|)$ new vertices and edges compared to G, the complexity does not increase in the parameters of the original graph G. Deleting all forbidden edges splits G into its elementary components [3]. Obviously, G is flexible iff at least one of these components is such. Checking the flexibility of an internal elementary component is straightforward: see if the component consists of a single mandatory edge. If so, the component is rigid, otherwise it is flexible. Concerning open elementary components, we use Corollary 4.8 to check their flexibility in the way described below.

Let G be open elementary. Construct $r(G)$ and check if it contains an even-length cycle. Both steps require at most $\mathcal{O}(|V|^2)$ time. Indeed, the direct way of constructing $r(G)$ relies on the adjacency matrix of G and implements the contraction of a redex in $\mathcal{O}(|V|)$ time. Since the number of redexes is at most $\mathcal{O}(|V|)$, the total complexity of the reduction algorithm is $\mathcal{O}(|V|^2) \leq \mathcal{O}(|V| \cdot |E|)$. (Note that G is elementary, so at least connected.) Regarding the check for an even-length cycle, this can be done in $\mathcal{O}(|E|)$ time, e.g. by the help of the depth-first search algorithm.

If the elementary graph G turns out to be flexible, then a concrete p.i.m. M' with an M'-alternating cycle in G can also be constructed in $\mathcal{O}(|V| \cdot |E|)$ time. This is not an issue when G is closed, since one perfect matching M of G has already been created, and finding an M-alternating cycle in G only takes $\mathcal{O}(|E|)$ time according to [9]. Assume therefore that G is open. Observe that Theorem 4.7 is constructive in the sense that, starting from an arbitrary p.i.m. M, it does find an alternating cycle in a reduced graph G with respect to an appropriate p.i.m. M'.

The method first builds up the sequence $G_0, \ldots, G_{n+1} = G$ of elementary graphs according to Lemma 4.5, then successively eliminates edges and reduces the remainder elementary graph in a suitable way. The edge h selected for elimination always lies on the alternating unit α_i for some $i \geq k + 1$, so that the graphs G_i are gradually demolished in the reverse order of their creation. Remember that G_{k+1} is a forest, which definitely stops the demolition procedure. (See the proof of Lemma 4.5.) The selection of h itself does not require more than $\mathcal{O}(|V|)$ time, neither does the reduction of the remainder graph. Moreover, the sequence $G_0, \ldots, G_i$ need not be reconstructed after each reduction as it will consist of the traces of the alternating units α_i, which remain alternating after the reduction. Therefore this sequence only requires an adjustment, which can be accomplished in $\mathcal{O}(|V|)$ time. Thus, narrowing down the original graph G to the absolute minimum G_{k+1} costs at most $\mathcal{O}(|V| \cdot |E|)$ in time. At the end, the marking procedure that effectively constructs a p.i.m. M' and an M'-alternating cycle c using Lemma 4.6 takes at most $\mathcal{O}(|V|)$ time. After this, reconstructing the original graph G and blowing up the matching M' together with the cycle c costs at most as much as the demolition of G did, that is, $\mathcal{O}(|V| \cdot |E|)$ time.

Regarding the second subproblem, let G be bipartite with a bipartition $V = A \cup D$ as in (iii) of Theorem 3.1, and assume that the surplus of G is positive when viewed from A. Construct the graph G_D by deleting the isolated vertices in D and attaching a new external edge to each remaining vertex in D. Let D_{ext} denote the set of resulting external vertices in G_D.

Proposition 5.2. *The open graph G_D is elementary, and it has a flexible p.i.m. iff G has a flexible maximum matching covering all vertices of A.*

Proof. Notice that, due to the surplus condition, the degree of each vertex in A is at least 2. Therefore $Ext(G_D) = D_{ext}$. For each $v \in D$, the surplus of the bipartite graph $G - v$ is still non-negative when viewed from A. Consequently, $G - v$ has a matching from A to $D - v$ (cf. [14]), which implies that the external edge adjacent to v is allowed in G_D. The very same argument shows that each edge of G is allowed in G_D. Thus, G_D does not have forbidden edges, meaning that it is elementary (in fact 1-extendable [14]). The rest of the proof is obvious.

Proposition 5.2 immediately implies that our second subproblem can be reduced to the first one, hence it is solvable in $\mathcal{O}(|V| \cdot |E|)$ time.

As to the third subproblem, if G is factor-critical and M is a near-perfect matching of G missing vertex v, then we proceed as follows. Using the method of [9], check if $G - v$ contains an M-alternating cycle. If not, then switch to another vertex v' and a perfect matching M' of $G - v'$. The switch takes $\mathcal{O}(|E|)$ time to implement, as well as the testing of M' for flexibility. In total, the search for an alternating cycle costs $\mathcal{O}(|V| \cdot |E|)$ in time.

With the previous discussion we have proved the second main result of the paper.

Theorem 5.3. *It is decidable in $\mathcal{O}(|V| \cdot |E|)$ time if $G = (V, E)$ contains a flexible m.i.m. If it does, a concrete flexible m.i.m. can also be constructed in the same amount of time.*

6   Conclusion

We have solved the problem of finding a flexible maximum matching in graphs. Our graphs were equipped with a so-called grace set, i.e., a set of vertices that need not necessarily be covered by maximum matchings. The problem was first reformulated in terms of open graphs and maximum internal matchings, then it was broken down into three subproblems using the Gallai-Edmonds decomposition. Open elementary graphs have been studied in the light of a simple reduction procedure, and it was proved that an open elementary graph is flexible iff it reduces to a graph containing an even-length cycle. This characterization gave rise to an algorithm that decides if a graph $G = (V, E)$ having a p.i.m. is flexible, and if so, finds a concrete p.i.m. M with an M-alternating cycle in G in $\mathcal{O}(|V| \cdot |E|)$ time. An algorithm with the same time complexity has been worked out to solve the same problem for appropriate bipartite graphs and factor-critical graphs. Combining these algorithms, a $\mathcal{O}(|V| \cdot |E|)$ time algorithm was obtained which solves the question of flexibility for general graphs.

References

1. M. Bartha, E. Gombás, A structure theorem for maximum internal matchings in graphs, *Information Processing Letters* **40** (1991), 289-294.
2. M. Bartha, M. Krész, Isolating the families of soliton graphs, *Pure Mathematics and Applications* **13** (2002), 49–62.
3. M. Bartha, M. Krész, Structuring the elementary components of graphs having a perfect internal matching, *Theoretical Computer Science* **299** (2003), 179–210.
4. M. Bartha, M. Krész, Tutte type theorems in graphs having a perfect internal matching, *Information Processing Letters* **91** (2004), 277–284.
5. M. Bartha, M. Krész, Deterministic soliton graphs, *Informatica* to appear.
6. J. Dassow, H. Jürgensen, Soliton automata, *J. Comput. System Sci.* **40** (1990), 154–181.
7. J. Edmonds, Paths, trees, and flowers, *Canad. J. Math.* **17** (1965), 449–467.
8. H. N. Gabow, R. E. Tarjan, A linear-time algorithm for a special case of disjoint set union, *Journal of Computer and System Sciences* **30** (1985), 209–221.
9. H. N. Gabow, H. Kaplan, R. E. Tarjan, Unique maximum matching algorithms, *Journal of Algorithms* **40** (2001), 159–183.
10. T. Gallai, Kritische Graphen II, *Magyar Tud. Akad. Mat. Kutató Int. Közl.* **8** (1963), 373–395.
11. M. C. Golumbic, T. Hirst, M. Lewenstein, Uniquely restricted matchings, *Algorithmica* **31** (2001), 139–154.
12. M. P. Groves, C. F. Carvalho, and R. H. Prager, Switching the polyacetylene soliton, *Materials Science and Engineering* **C3** (1995) pp. 181–185.
13. V. Levit, E. Mandrescu, Uniquely restricted maximum matchings in unicycle graphs, in *Fifth Haifa Workshop on Interdisciplinary Applications of Graph theory, Combinatorics, and Algorithms*, May 16-19, 2005
14. L. Lovász, M. D. Plummer, *Matching Theory*, North Holland, Amsterdam, 1986.
15. H. Müller, Alternating cycle free matchings in chordal bipartite graph, *Order* **7** (1990) 11–21.
16. J. C. Regin, The symmetric alldiff constraint, In *Proc. of the 16th Int. Joint Conf. on Artificial Intelligence*, pp. 420–425, Morgan Kaufmann, 1999.

Simultaneous Graph Embeddings with Fixed Edges

Elisabeth Gassner[2], Michael Jünger[3,*], Merijam Percan[3,*],
Marcus Schaefer[1], and Michael Schulz[3,*]

[1] DePaul University, School of CTI, 243 South Wabash, Ste 401,
60604 Chicago, IL, USA
mschaefer@cs.depaul.edu
[2] Technische Universität Graz, Institut für Mathematik B, Steyrergasse 30/II,
8010 Graz, Austria
gassner@opt.math.tu-graz.ac.at
[3] Universität zu Köln, Institut für Informatik, Pohligstraße 1, 50969 Köln, Germany
{mjuenger, percan, schulz}@informatik.uni-koeln.de

Abstract. We study the problem of simultaneously embedding several
graphs on the same vertex set in such a way that edges common to two or
more graphs are represented by the same curve. This problem is known
as *simultaneously embedding graphs with fixed edges*. We show that this
problem is closely related to the *weak realizability* problem: Can a graph
be drawn such that all edge crossings occur in a given set of edge pairs?
By exploiting this relationship we can explain why the simultaneous
embedding problem is challenging, both from a computational and a
combinatorial point of view.

More precisely, we prove that simultaneously embedding graphs with
fixed edges is NP-complete even for three planar graphs. For two planar
graphs the complexity status is still open.

1 Introduction

Cologne offers various methods of public transport, including buses, trains, and
even a cable car to the zoo. Suppose your task is to design a system map dis-
playing all of the transport systems simultaneously.

We can easily model each transport system by itself as a graph: the bus system
has bus stops as vertices and edges correspond to direct connections between
two stops on a bus route. Similarly, for the train system we model train stations
as vertices and connections between train stations as edges. Let us make the
assumption that each system by itself (bus, train, cable car, etc.) is planar; for
instance, two bus routes do not intersect except at bus stops. We also assume
that all graphs share the same vertex set, so a bus stop can coincide with a train
stop or a cable car stop. To draw the full system map we need to embed vertices
and edges in the plane such that each of the systems—bus, train, cable car—is

---

* Partially supported by Marie-Curie Research Training Network (ADONET) and by
the German Science Foundation (JU204/10-1).

F.V. Fomin (Ed.): WG 2006, LNCS 4271, pp. 325–335, 2006.

planar by itself, but allowing, for instance, train lines and bus routes to cross outside stations.

More formally, a *simultaneous embedding* of a set of graphs sharing the same vertex set is a mapping of the vertices into the plane and a planar embedding of each graph on the vertex set [1]. Notice that edges belonging to different graphs are allowed to intersect. There are several natural ways to restrict the model; for example, we could require all edges to be drawn as straight-line segments. Such a drawing is called a *simultaneous geometric embedding* of the graphs [1].

Returning to the transport system example, it appears that asking for a simultaneous geometric embedding of the individual systems is too strict a requirement: we would be willing to accept Jordan curves to represent edges. However, requiring only a simultaneous embedding does not seem satisfactory either: if a train and bus route run in parallel, we would like to see them drawn in parallel; that is, the two routes should be represented by the same curve. This restriction leads to the notion of *simultaneous embeddings with fixed edges* introduced by Erten and Kobourov [2], and it is this notion that will mostly concern us in this paper: if several graphs share the same edge, that edge is represented by the same curve in the drawing. We see that every simultaneous geometric embedding is a simultaneous embedding with fixed edges, which, in turn, is a simultaneous embedding.

While different variants of simultaneous embeddings have been investigated from both theoretical and practical points of view [1,2,3,4,5,6,7], there is not much known about simultaneous embeddings with fixed edges. Erten and Kobourov [2] give an algorithm that constructs a simultaneous embedding with fixed edges of a tree and a path in linear time and uses at most one bend per edge. They point out that it is not known whether two trees can always be simultaneously embedded with fixed edges such that the number of bends is a small constant.

Studying the problem from a complexity-theoretic point of view gives a different challenge: it was not even clear whether the problem is decidable, let alone in NP or polynomial time. As we will see in Section 3 there is good reason for this: simultaneously embedding graphs with fixed edges is a different (and interesting) way to look at the weak realizability problem [9,10]. The complexity of the weak realizability problem is related to that of the string graph problem, whose complexity was only settled recently after being open for thirty years [8,11]. The relationship implies that the simultaneous embedding problem can be decided in NP. We settle the computational complexity question for three or more graphs in Section 3 by showing that it is NP-complete. The complexity of the case for two graphs remains open.

2   Preliminaries

Given a graph $G = (V, E)$ and a set $R \subseteq \binom{E}{2} = \{\{e, f\} \mid e, f \in E, e \neq f\}$, we call a drawing D of G in the plane in which all intersecting edge pairs belong to R a *weak realization* of (G, R).

Notice that not all edge pairs in R are required to intersect in D, R just defines the set of allowed crossings. (G, R) is *weakly realizable* if such a drawing D exists.

Problem: *Weak Realizability*
Instance: A graph $G = (V, E)$ and a set $R \subseteq \binom{E}{2}$.
Question: Is (G, R) weakly realizable?

The weak realizability problem is closely related to the string graph problem [8]. Both problems are now known to be NP-complete even though this question was open for a long time [11].

The simultaneous embedding with fixed edges problem was introduced by Brass et al. [1]. It has become a major theoretical problem in simultaneous graph drawing. We will present an example of two planar graphs which do not have a simultaneous embedding with fixed edges. Brass et al already gave an example of two outerplanar graphs as well as three paths with no simultaneous embedding with fixed edges.

Problem: *Simultaneous Embedding with Fixed Edges (SEFE)*
Instance: A set of planar graphs $G_i = (V, E_i)$ on the same vertex set V.
Question: Are there plane drawings D_i of G_i, such that each vertex is mapped to the same point in the plane in all D_i and every shared edge $e \in E_i \cap E_j$ is represented by the same simple open Jordan curve in D_i and D_j?

In Figure 1 two planar graphs G_1, G_2 are given for which no SEFE exist: the triangle induced by v_1, v_2, v_3 is equal to $G_1 \cap G_2$ (visualized by bold edges) and makes it impossible to include the edge $(v_4, v_5) \in G_1$ into the unique embedding (up to a homomorphism of the plane) of G_2 without crossing an edge of $G_1 \cap G_2$. This indicates that the SEFE problem is not trivial.

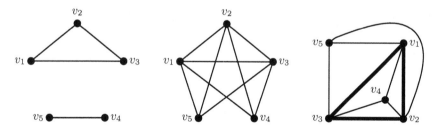

Fig. 1. Graph G_1 (left), graph G_2 (middle), unique planar embedding of G_2 (right)

3   Complexity Results on Simultaneous Embeddings

We start with a result dealing with the relationship between the two problems introduced in the previous section. This theorem implies the NP-completeness of SEFE in the general case. We will show later that the problem remains NP-complete in the case of three graphs.

Theorem 1. *The weak realizability problem is polynomially equivalent to the problem of simultaneous embedding with fixed edges.*

Proof. Let $((V, E), R)$ be an instance of the weak realizability problem. For every pair of edges $\{e, f\} \in \binom{E}{2} \setminus R$ we construct a graph $G_{e,f} = (V, \{e, f\})$. It is easy to verify that the weak realizability problem is solvable if and only if there exists a simultaneous embedding with fixed edges for the set of constructed graphs $G_{e,f}$.

Let now $G_i = (V, E_i)$, $i = 1, \ldots, n$, be a set of graphs on the same vertex set V. Let $E = \bigcup_{i=1}^{n} E_i$ be the set of edges. Define R as the set of those pairs of edges which are not contained in one graph G_i. Then the problem of simultaneously embedding the graphs G_i is equivalent to the weak realizability problem of the pair (G, R) with $G = (V, E)$.   □

Kratochvíl [8] and Schaefer, Sedgwick and Štefankovič [11] showed that weak realizability is NP-complete. Hence we get the following corollary.

Corollary 1. *Simultaneous embedding with fixed edges is NP-complete.*

In particular, SEFE lies in NP for every fixed number k of graphs. As we will show next, it is even NP-complete in that case, as long as $k \geq 3$.

Theorem 2. *Deciding whether three graphs have a simultaneous embedding with fixed edges is NP-complete.*

Proof. We have already seen that SEFE for a fixed number of graphs is in NP. We will now show that there exists a polynomial transformation from 3SAT (which is well-known to be NP-complete) to SEFE for three planar graphs $G_1 = (V, E_1)$, $G_2 = (V, E_2)$ and $G_3 = (V, E_3)$. Given an instance of 3SAT, we will construct an instance (G_1, G_2, G_3) of SEFE. Then we will prove that the instance of 3SAT is satisfiable if and only if there exists a simultaneous embedding with fixed edges of (G_1, G_2, G_3).

The decision problem 3SAT is given in the following way:

Problem:	*3SAT*
Instance:	A set U of boolean variables and a collection C of clauses in conjunctive normal form over U such that each clause $c \in C$ has exactly three literals.
Question:	Is there a truth assignment to U that satisfies C?

Construction: Let $C = \{c_1, \ldots, c_m\}$ be the set of clauses and let $U = \{u_1, \ldots, u_n\}$ be the variable set of a 3SAT-instance. Each clause is of the form $c_j = l_1^j \vee l_2^j \vee l_3^j$ with literals either $l_i^j = u_h$ or $l_i^j = \bar{u}_h$ for some $h = 1, \ldots, n$ and $i = 1, 2, 3$.

Our construction of a SEFE-instance is made up of several components. For our construction we assume an ordering of the clauses, say $(c_1, c_2, \ldots, c_m)$. Furthermore we choose an order of the three literals in each clause c_j and hence get an order of all literals in the following way $(l_1^1, l_2^1, l_3^1, l_1^2, \ldots, l_3^m)$.

For each clause c_j we define a *clause box* by introducing vertices $r_1^j, \ldots, r_6^j$, $y^{1,j}, y^{2,j}, y^{3,j}$. These vertices are connected by edges in E_2 and E_3 to form the cycle $(r_1^j, \ldots, r_6^j, y^{3,j}, y^{2,j}, y^{1,j})$.

Next, we introduce two global vertices R_1 and R_2. We add an edge (R_1, R_2) which is part of all three graphs G_1, G_2 and G_3. Furthermore, R_1 is connected to each clause box by four edges to the vertices r_2^j, r_3^j, r_4^j and r_5^j. These edges are part of all three graphs. R_2 is connected to each clause gadget by two edges to vertices r_1^j and r_6^j in E_3.

To make the graphs more rigid we glue together neighboring clause boxes. This is done by identifying r_2^{j+1} with r_5^j and r_1^{j+1} with r_6^j for $j = 1, \ldots, m-1$. Figure 2 shows the construction so far. We remark that all edges that have been constructed so far belong (among others) to E_3.

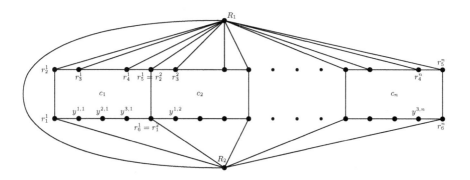

Fig. 2. The figure shows all clause boxes, the global vertices R_1 and R_2 and all connecting edges of E_3

For every literal l_i^j in clause c_j we define a *literal gadget* consisting of the vertices $V^{i,j} = \{x_k^{i,j}, z_k^{i,j} \mid k = 1, \ldots, 6\} \cup \{y^{i,j}\}$. The edge set of the gadget is shown in Figure 3. From now on edges of G_1 are illustrated by solid lines, edges of G_2 are dashed and edges of G_3 are dotted.

Furthermore, let $l_{i(1)}^{j(1)}, \ldots, l_{i(\omega_h)}^{j(\omega_h)}$ be all literals that belong to a variable u_h, that is $l_{i(\alpha)}^{j(\alpha)} = u_h$ or $l_{i(\alpha)}^{j(\alpha)} = \bar{u}_h$ for $\alpha = 1, \ldots, \omega_h$. Assume that the literals are given in the order defined above. We connect each pair of adjacent literals in this ordered list by the following edges of E_2 (see Figure 4):

$$\left(z_1^{i(k),j(k)}, z_6^{i(k+1),j(k+1)}\right), \left(z_2^{i(k),j(k)}, z_5^{i(k+1),j(k+1)}\right), \left(z_3^{i(k),j(k)}, z_4^{i(k+1),j(k+1)}\right)$$

with $k = 1, \ldots, \omega_h - 1$.

For each clause we define a *clause gadget* consisting of three literal gadgets, the clause box and some additional vertices and edges. Let c_j be a clause with literals l_1^j, l_2^j and l_3^j. Notice that the three literal gadgets are already connected to the clause box via the vertices $y^{i,j}$ with $i = 1, 2, 3$. Further connections are given by the additional edges $(r_3^j, x_2^{1,j}) \in E_3$ and $(r_4^j, x_2^{3,j}) \in E_2$. We also add two vertices s^j, t^j.

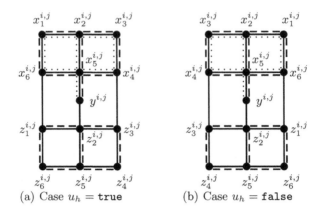

Fig. 3. Literal gadget for literal l_i^j (with corresponding variable u_h) of clause c_j. The edges in E_1 are solid, those in E_2 are dashed and those in E_3 are dotted. The two different drawings (a) and (b) will become important later.

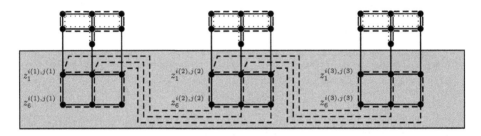

Fig. 4. Three literal gadgets that belong to variable u_h are linked with edges in E_2

In order to distinguish between negated and non-negated variables we define connections between the literal gadgets. The gadget of l_1^j is connected along s^j with the gadget of l_2^j. And the gadget of l_2^j is connected along t^j with the gadget of l_3^j. More precisely, we add edges $(p_1^j, s^j) \in E_3$ (where p_1^j is a vertex of the gadget of l_1^j), $(s^j, p_2^j) \in E_1 \cap E_2$ and $(q_2^j, t^j) \in E_1 \cap E_3$ (where p_2^j and q_2^j are vertices of the gadget of u_{i_2}) and $(t^j, q_3^j) \in E_2$ (where q_3^j is a vertex of the gadget of l_3^j). If l_1^j is not negated, we set $p_1^j = x_3^{1,j}$ (i.e., $x_3^{1,j}$ is connected with s^j), otherwise, if l_1^j is negated, we set $p_1^j = x_1^{1,j}$. Analogously we set $p_2^j = x_1^{2,j}$ and $q_2^j = x_3^{2,j}$ if l_2^j is not negated and $p_2^j = x_3^{2,j}$ and $q_2^j = x_1^{2,j}$ if l_2^j is negated. And finally, we set $q_3^j = x_1^{3,j}$ if l_3^j is not negated and $q_3^j = x_3^{3,j}$ if l_3^j is negated. See Figure 5 for a clause gadget.

We complete the construction by adding an edge from R_2 to $z_5^{i,j}$ to graph G_3 for each literal l_i^j.

1. Assume that the 3SAT-instance is satisfiable. We will prove that there exists a simultaneous embedding with fixed edges of the constructed instance.

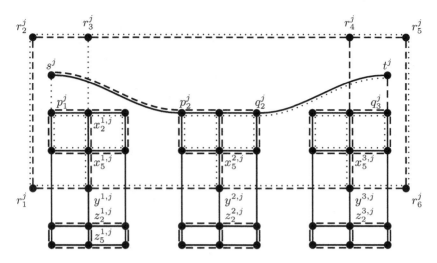

Fig. 5. Clause gadget for the case where $u_{k(2)}$ is the only variable that makes c_j true

If variable $u_h = $ **true** we draw each of its literal gadgets as shown in Figure 3(a). Otherwise we draw the gadgets as shown in Figure 3(b).

The clause gadgets are drawn side by side as illustrated in Figure 2 for the clause boxes. Furthermore, the x-vertices of each literal gadget lie inside the cycle of the corresponding clause boxes and the z-vertices lie outside (see Figure 5). Moreover, every variable gets its own horizontal level for the z-variables of its literal gadgets (in Figure 4 the horizontal level is marked gray). Since all literal gadgets that belong to one variable u_h are drawn in the same way we can draw the edges that connect literal gadgets which belong to the same variable without crossings (see Figure 4).

It remains to show that we can draw the edges inside the clause gadgets without crossings of edges of the same graph. We say that a variable u makes a clause c true if either $u \in c$ and $u = $ **true** or if $\bar{u} \in c$ and $u = $ **false**. Since the instance of 3SAT is satisfiable there exists at least one variable u in each clause c that makes c true.

Consider clause c_j with literals l_1^j, l_2^j, and l_3^j. Let $u_{k(i)}$ be the corresponding variable to literal l_i^j, thus either $l_i^j = u_{k(i)}$ or $l_i^j = \bar{u}_{k(i)}$. Assume first that $u_{k(1)}$ makes c_j true. Then there are two possibilities. Either $u_{k(1)} \in c_j$ and $u_{k(1)}$ is set to **true** or $\bar{u}_{k(1)} \in c_j$ and $u_{k(1)} = $ **false**. In the first case the literal gadget is drawn like Figure 3(a) and $p_1^j = x_3^{1,j}$ and in the other case we have the situation of Figure 3(b) and $p_1^j = x_1^{1,j}$. But in both of these cases the vertex p_1^j is the upper right vertex of literal gadget l_1^j (see Figure 6). Figure 6 shows that in the case where $u_{k(1)}$ is the only variable that makes c_j true there is a simultaneous embedding with fixed edges of the corresponding clause gadget. Simple modifications yield a simultaneous embedding for the case where $u_{k(1)}$ is not the only variable that makes c_j true. Due to symmetry an analogue drawing can be found for the case where $u_{k(3)}$ makes c_j true.

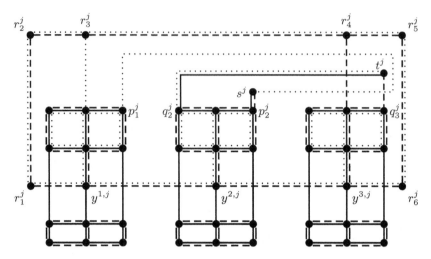

Fig. 6. Simultaneous embedding of the clause gadget when $u_{k(1)}$ is the only variable that makes c_j true

Finally, if $u_{k(2)}$ makes c_j true we can find a simultaneous embedding as shown in Figure 5. Hence, we have found a simultaneous embedding with fixed edges of the constructed instance.

2. Now assume that we are given a simultaneous embedding with fixed edges. We will show that we can find a satisfying truth assignment for the 3SAT-instance.

Notice that the subgraph of G_3 shown in Figure 2 is a triconnected subdivision. Consequently, it has a unique combinatorial embedding. We choose the planar embedding with the edge (R_1, R_2) on the boundary of the outer face such that the cycle $(R_1, r_2^1, r_1^1, R_2, r_6^n, r_5^n)$ has the same order as visualized in Figure 2.

Due to the edge $(R_2, z_5^{i,j})$ the vertex $z_5^{i,j}$ lies outside the clause box of clause c_j for each literal l_i^j. Hence, all z-vertices of every literal lie outside all clause boxes.

Moreover, all x-vertices of a literal l_i^j lie inside the clause box of clause c_j. Notice first that both paths $(r_3^j, x_2^{1,j}, x_5^{1,j}, y^{1,j})$ and $(r_4^j, x_2^{3,j}, x_5^{3,j}, y^{3,j})$ have to be completely within the clause box of clause c_j due to vertex R_1 and its incident edges. Then all $x_k^{1,j}$ and $x_k^{3,j}$, $k = 1, \ldots, 6$, have to lie within the clause box. Thus, s^j is also forced to lie within the clause box. Otherwise, the edge $(p_1^j, s^j) \in E_3$ crosses an edge of the clause box boundary which is also in E_3. Hence, the edge (s^j, p_2^j) and then all $x_k^{2,j}$, $k = 1, \ldots, 6$, have to lie within the clause box, otherwise there would be a crossing of two edges of graph G_2.

Observe further, that each literal gadget contains (among others) two cycles, $C_1^{i,j} = \{x_k^{i,j} \mid k = 1, \ldots, 6\}$ and $C_2^{i,j} = \{z_k^{i,j} \mid k = 1, \ldots, 6\}$. Since the literal gadgets in G_1 are planar graphs, that are a triconnected subdivision, these gadgets have a unique planar embedding. Therefore, the clockwise ordering of the vertices of $C_1^{i,j}$ and $C_2^{i,j}$ in a simultaneous embedding is either $(x_1^{i,j}, x_2^{i,j}, \ldots, x_6^{i,j})$ and $(z_1^{i,j}, z_2^{i,j}, \ldots, z_6^{i,j})$ or $(x_6^{i,j}, x_5^{i,j}, \ldots, x_1^{i,j})$ and $(z_6^{i,j}, z_5^{i,j}, \ldots z_1^{i,j})$. In the first case, we say the literal gadget has a *positive* ordering and in the second case it has a *negative* ordering. Thus, the clockwise ordering of $C_1^{i,j}$ implies the clockwise ordering of $C_2^{i,j}$ and vice versa.

Again, let $l_{i(1)}^{j(1)}, \ldots, l_{i(\omega_h)}^{j(\omega_h)}$ be all literals that belong to the same variable u_h. Notice that the subgraph of G_2 induced by the vertices $z_q^{i(k),j(k)}$ for $q = 1, \ldots, 6$ and $k = 1, \ldots, \omega_h$ is also a triconnected subdivision. Since G_2 is drawn planar, the ordering of all cycles $C_2^{i(k),j(k)}$ has to be the same. We conclude that all cycles C_1 and C_2 that belong to one variable u_h have the same ordering. Therefore, we can let

$$u_h = \begin{cases} \texttt{true} & \text{if the literal gadgets of } u_h \text{ have a positive ordering,} \\ \texttt{false} & \text{if the literal gadgets of } u_h \text{ have a negative ordering.} \end{cases}$$

We define the interior of cycle $C' = (r_1^j, r_2^j, r_3^j, x_2^{1,j}, x_5^{1,j}, y^{1,j})$ as region F_1^j, the interior of cycle $(r_3^j, r_4^j, x_2^{3,j}, x_5^{3,j}, y^{3,j}, y^{2,j}, y^{1,j}, x_5^{1,j}, x_2^{1,j})$ as region F_2^j and the interior of cycle $C'' = (r_4^j, r_5^j, r_6^j, y^{3,j}, x_5^{3,j}, x_2^{3,j})$ as region F_3^j (see Figure 7).

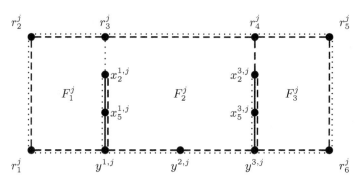

Fig. 7. p_1^j and s^j have to be embedded in region F_1^j, p_2^j and q_2^j have to be embedded in region F_2^j and q_3^j and t^j have to be embedded in region F_3^j

Assume that there exists a clause c_j that is not true. This implies that none of the three corresponding clause variables $u_{k(1)}, u_{k(2)}, u_{k(3)}$ makes c_j true. If $u_{k(1)} \in c_j$ then the literal gadget of l_1^j has a negative ordering and $p_1^j = x_3^{1,j}$ is connected with s^j (see Figure 8). Otherwise, if $\bar{u}_{k(1)} \in c_j$ then the variable gadgets of $u_{k(1)}$ have a positive ordering and $p_1^j = x_1^{1,j}$ is connected with s^j. But in both cases the vertex p_1^j has to be embedded in region F_1^j. With similar arguments for $u_{k(2)}$ and $u_{k(3)}$ we conclude that the clause gadget of c_j is of the form illustrated in Figure 8. To be more specific, vertex q_3^j has to lie within region F_3^j and both p_2^j and q_2^j have to be embedded within region F_2^j. For the last two vertices we have an additional restriction: running through cycle $C_1^{2,j}$ in clockwise order starting from vertex $x_5^{2,j}$ the vertex p_2^j comes first compared to vertex q_2^j (see Figure 8).

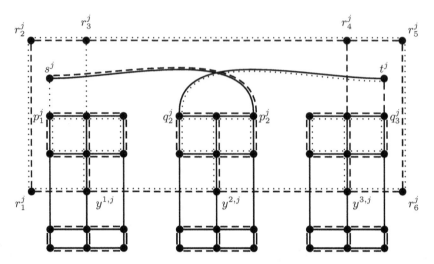

Fig. 8. If a clause is false then there exists a crossing of two edges of the same graph in the corresponding clause gadget

In Figure 8 we observe that vertex s^j has to be placed within region F_1^j. Otherwise edge $(p_1^j, s^j) \in E_3$ would cross at least one edge of the cycle $C' \subset E_3$. Moreover, t^j has to be placed within the cycle C'' that is region F_3^j. Otherwise we have a crossing of two edges of E_2, since $C'' \subseteq E_2$. If (q_2^j, t^j) is drawn without crossing an edge of the same graph there exists a cycle consisting of edges in $E_1 \cup E_2$ that separates p_2^j from s^j. Since $(s^j, p_2^j) \in E_1 \cap E_2$ we conclude that there exists a crossing of at least two edges of the same graph. If (s^j, p_2^j) is drawn, analogue implications lead to a non planar embedding of one of the graphs G_1, G_2 or G_3. This leads to a contradiction to the assumption that clause c_j is not true. Hence, all clauses are satisfied and thus, we have found a truth assignment. □

4   Summary and Future Work

We believe that in spite of their unwieldy name, simultaneous graph embeddings with fixed edges are a natural problem to study. Erten and Kobourov [2] supply an important reason why the fixed-edge model is more natural for many applications in comparison to the unrestricted version: Building a mental model of a complex graph drawing is easier if the underlying structure resembles the graph drawing, or, as Charles Peirce would have phrased it, if the drawing is *iconic*. In other words, edges belonging to several graphs should not have multiple representations, since this complicates forming a mental model: we have to identify the different curves as representations of the same edge. And while straight-line embeddings resolve this issue, they exclude the possibility of simultaneously embedding even very simple planar graphs, such as trees [5].

It has been known that the weak realizability problem is related to several famous problems, such as string graphs, topological inference, and Euler diagrams [9,11], but the connection to fixed edge embeddings is the most purely combinatorial connection we have seen so far.

The main open question of complexity is the question whether two planar graphs allow a simultaneous embedding with fixed edges.

Acknowledgments

We would like to thank Dániel Marx and Stephen G. Kobourov for helpful discussions. Moreover, we are grateful to Stefan Hachul and Alejandro Estrella Balderrama for proof-reading this paper.

References

1. Peter Brass, Eowyn Cenek, Christian A. Duncan, Alon Efrat, Cesim Erten, Dan Ismailescu, Stephen G. Kobourov, Anna Lubiw, and Joseph S. B. Mitchell. On simultaneous planar graph embeddings. In *Workshop on Algorithms and Data Structures. Lecture Notes in Computer Science*, volume 2748, pages 243–255. Springer-Verlag, 2003.
2. Cesim Erten and Stephen G. Kobourov. Simultaneous embedding of planar graphs with few bends. In *12th Intl. Symp. on Graph Drawing. Lecture Notes in Computer Science*, volume 3383, pages 195–205. Springer-Verlag, 2004.
3. Cesim Erten and Stephen G. Kobourov. Simultaneous embedding of a planar graph and its dual on the grid. *Theory of Computing Systems*, 38:313–327, 2005.
4. Cesim Erten, Stephen G. Kobourov, Vu Le, and Armand Navabi. Simultaneous graph drawing: Layout algorithms and visualization schemes. In *11th Intl. Symp. on Graph Drawing. Lecture Notes in Computer Science*, volume 2912, pages 437–449. Springer-Verlag, 2003.
5. Markus Geyer, Michael Kaufmann, and Imrich Vrt'o. Two trees which are self-intersecting when drawn simultaneously. In *13th Intl. Symp. on Graph Drawing. Lecture Notes in Computer Science*, volume 3843, pages 201–210. Springer-Verlag, 2005.
6. Emilio Di Giacomo and Giuseppe Liotta. A note on simultaneous embedding of planar graphs. In *21st European Workshop on Comp.Geometry*, pages 207–210, 2005.
7. Stephen G. Kobourov and Chandan Pitta. An interactive multi-user system for simultaneous graph drawing. In *12th Intl. Symp. on Graph Drawing. Lecture Notes in Computer Science*, volume 3383, pages 492–501. Springer-Verlag, 2004.
8. Jan Kratochvíl. String graphs II Recognizing string graphs is NP-hard. *Journal of Combinatorial Theory*, B 52:67–78, 1991.
9. Jan Kratochvíl. Crossing number of abstract topological graphs. In *6th Intl. Symp. on Graph Drawing. Lecture Notes in Computer Science*, volume 1547, pages 238–245. Springer-Verlag, 1998.
10. Jan Kratochvíl, Anna Lubiw, and Jaroslav Nešetřil. Noncrossing subgraphs in topological layouts. *SIAM Journal on Discrete Mathematics*, 4(2):223–244, May 1991.
11. Marcus Schaefer, Eric Sedgwick, and Daniel Štefankovič. Recognizing string graphs in *NP*. *Journal Comput. Syst. Sci.*, 67(2):365–380, 2003.

Approximation Algorithms for Restricted Cycle Covers Based on Cycle Decompositions[*]

Bodo Manthey[**]

Universität des Saarlandes, Informatik
Postfach 151150, 66041 Saarbrücken, Germany
manthey@cs.uni-sb.de

Abstract. A cycle cover of a graph is a set of cycles such that every vertex is part of exactly one cycle. An L-cycle cover is a cycle cover in which the length of every cycle is in the set $L \subseteq \mathbb{N}$. For most sets L, the problem of computing L-cycle covers of maximum weight is NP-hard and APX-hard.

We devise polynomial-time approximation algorithms for L-cycle covers. More precisely, we present a factor 2 approximation algorithm for computing L-cycle covers of maximum weight in undirected graphs and a factor 20/7 approximation algorithm for the same problem in directed graphs. Both algorithms work for arbitrary sets L. To do this, we develop a general decomposition technique for cycle covers.

Finally, we show tight lower bounds for the approximation ratios achievable by algorithms based on such decomposition techniques.

1 Introduction

A cycle cover of a graph is a spanning subgraph that consists solely of cycles such that every vertex is part of exactly one cycle. Cycle covers play an important role in the design of approximation algorithms for several variants of the travelling salesman problem [3,5,6,9,10,11,12,17], for the shortest common superstring problem [8,21], and for vehicle routing problems [14].

We consider cycle covers in (directed or undirected) edge-weighted complete graphs. Given such a graph, the aim is to find a cycle cover of maximum weight. In contrast to Hamiltonian cycles, which are special cases of cycle covers, cycle covers of maximum weight can be computed efficiently. This is exploited in the aforementioned approximation algorithms, which usually start by computing an initial cycle cover and then join cycles to obtain a Hamiltonian cycle.

Short cycles in a cycle cover limit the approximation ratios achieved by such algorithms. In general, the longer the cycles in the initial cover, the better the approximation ratio. Thus, we are interested in computing cycle covers that do not contain short cycles. Moreover, there are approximation algorithms that

---

[*] A full version of this work is available at http://arxiv.org/abs/cs/0604020.
[**] Work done in part at the Institut für Theoretische Informatik of the Universität zu Lübeck and supported by DFG research grant RE 672/3.

F.V. Fomin (Ed.): WG 2006, LNCS 4271, pp. 336–347, 2006.
© Springer-Verlag Berlin Heidelberg 2006

perform particularly well if the cycle covers computed do not contain cycles of odd length [5]. Finally, some vehicle routing problems [14] require covering vertices with cycles of bounded length.

Therefore, we consider *restricted cycle covers*, where cycles of certain lengths are ruled out a priori: For $L \subseteq \mathbb{N}$, an *L-cycle cover* is a cycle cover in which the length of each cycle is in L.

Unfortunately, computing L-cycle covers of maximum weight is hard in general [16,19]. Thus, in order to fathom the possibility of designing approximation algorithms based on computing cycle covers, our aim is to find out how well L-cycle covers can be approximated.

Beyond being a basic tool for approximation algorithms, cycle covers are interesting in their own right. Matching theory and graph factorisation are important topics in graph theory. Cycle covers of undirected graphs are also known as *two-factors* since every vertex is incident to exactly two edges in a cycle cover. A considerable amount of research has been done on structural properties of graph factors and on the complexity of finding graph factors (cf. Lovász and Plummer [18] and Schrijver [20]). In particular, the complexity of finding restricted two-factors, i.e. L-cycle covers in undirected graphs, has been investigated, and Hell et al. [16] and Manthey [19] showed that finding L-cycle covers in graphs is NP-hard for almost all L.

1.1 Preliminaries

Let $G = (V, E)$ be a graph with vertex set V and edge set E. If G is undirected, then a *cycle cover* of G is a subset $C \subseteq E$ of the edges of G such that all vertices in V are incident to exactly two edges in C. If G is a directed graph, then a cycle cover of G is a subset $C \subseteq E$ such that all vertices are incident to exactly one incoming and one outgoing edge in C. Thus, the graph (V, C) consists solely of vertex-disjoint cycles. The length of a cycle is the number of edges it consists of. We are concerned with simple graphs, i.e. the graphs do not contain multiple edges or loops. Thus, the shortest cycles of undirected and directed graphs are of length three and two, respectively. We will refer to a cycle of length ℓ as an ℓ-cycle for short. Furthermore, cycles of odd or even length will simply be called odd or even cycles, respectively.

An *L-cycle cover* of an undirected graph is a cycle cover in which the length of every cycle is in $L \subseteq \mathcal{U} = \{3, 4, 5, \ldots\}$. An L-cycle cover of a directed graph is analogously defined except that $L \subseteq \mathcal{D} = \{2, 3, 4, \ldots\}$. A *k-cycle cover* is a $\{k, k + 1, \ldots\}$-cycle cover. In the following, let $\overline{L} = \mathcal{U} \setminus L$ in the case of undirected graphs and $\overline{L} = \mathcal{D} \setminus L$ in the case of directed graphs. (This will be clear from the context.)

Given a weight function $w : E \to \mathbb{N}$, the *weight* $w(C)$ of a subset $C \subseteq E$ of the edges of G is $w(C) = \sum_{e \in C} w(e)$. In particular, this defines the weight of a cycle cover since we view cycle covers as sets of edges.

Max-L-UCC is the following optimisation problem: Given an undirected complete graph with non-negative edge weights, find an L-cycle cover of maximum weight. Max-k-UCC is defined for $k \in \mathcal{U}$ like Max-L-UCC except that k-cycle covers are sought instead of L-cycle covers.

Max-L-DCC and Max-k-DCC are defined for directed graphs like Max-L-UCC and Max-k-UCC for undirected graphs except that $L \subseteq \mathcal{D}$ and $k \in \mathcal{D}$.

A *single* is a single edge (or a path of length one) in a graph, while a *double* is a path of length two.

1.2   Previous Results

Undirected Cycle Covers. Max-$\mathcal{U}$-UCC, i.e. the undirected cycle cover problem without any restrictions, can be solved in polynomial time via Tutte's reduction (cf. Lovász and Plummer [18, Sect. 10.1]) to the perfect matching problem, which can be solved in polynomial time [1, Chap. 12]. By a modification of an algorithm of Hartvigsen [13], it is possible to show that 4-cycle covers of maximum weight in graphs with edge weights zero and one can be computed efficiently [19].

For the problem of computing k-cycle covers of minimum weight in graphs with edge weights one and two, there exists a factor 7/6 approximation algorithm for all k [7]. Hassin and Rubinstein [15] devised a randomised approximation algorithm for Max-$\{3\}$-UCC that achieves an approximation ratio of $169/89 + \epsilon$. Max-L-UCC can be approximated within a factor of 2.5 for arbitrary sets L [19].

Testing whether an undirected graph contains an L-cycle cover as a spanning subgraph is NP-hard if $\overline{L} \not\subseteq \{3,4\}$, i.e. for almost all L [16]. Vornberger showed that Max-5-UCC is NP-hard [22]. Max-L-UCC is APX-hard if $\overline{L} \not\subseteq \{3\}$ [19], i.e. for almost all L, Max-L-UCC is unlikely to possess a polynomial-time approximation scheme. (We refer to Ausiello et al. [2] for a survey on optimisation problems and their approximability.) Even a restriction of Max-L-UCC where only edge weights zero and one are allowed is APX-hard for all L with $\overline{L} \not\subseteq \{3,4\}$ [19].

Directed Cycle Covers. Max-$\mathcal{D}$-DCC, which is also known as the *assignment problem*, can be solved in polynomial time by a reduction to the maximum weight perfect matching problem in bipartite graphs [1, Chap. 12]. The only other L for which Max-L-DCC can be solved in polynomial time is $L = \{2\}$. For all $L \subseteq \mathcal{D}$ with $L \neq \{2\}$ and $L \neq \mathcal{D}$, Max-L-DCC is APX-hard and NP-hard, even if only edge weights zero and one are allowed [19].

There are a factor 4/3 approximation algorithm for Max-3-DCC [6] and a factor 3/2 approximation algorithm for computing k-cycle covers of maximum weight for $k \geq 3$ with the restriction that the only edge weights allowed are zero and one [4]. Max-L-DCC admits a factor 3 approximation for arbitrary L [19].

1.3   New Results

In this paper, we present approximation algorithms for Max-L-UCC and Max-L-DCC that work for arbitrary sets L. Our algorithms achieve an approximation ratio of 2 for Max-L-UCC (Section 3.1) and an approximation ratio of 20/7 for Max-L-DCC (Section 3.2). The best approximation algorithms previously known for these problems achieve ratios of 2.5 and 3, respectively [19].

As a main ingredient of the algorithms, we prove a decomposition lemma that shows how an arbitrary cycle cover can be decomposed while preserving as much of its weight as possible (Section 2).

Finally, we show the limits of decomposition-based approximation algorithms (Section 4): For approximating undirected L-cycle covers, a ratio of 2 is the best one can achieve using decomposition techniques. Thus, our algorithm for Max-L-UCC is an optimal decomposition-based algorithm. For directed L-cycle covers, only a ratio of 3 can be achieved. Our approximation algorithm for Max-L-DCC achieves the ratio of $20/7 < 3$ by a combination of the decomposition technique and a matching-based algorithm.

2   Decomposing Cycle Covers

In this section, we present a general decomposition technique for cycle covers. The technique can be applied to all cycle covers that do not contain 2-cycles, thus in particular to cycle covers of undirected graphs. But it can also be applied to directed cycle covers without 2-cycles.

We decompose cycle covers into a collection of vertex-disjoint singles, doubles, and isolated vertices. Our aim is to decompose a cycle cover C on n vertices into roughly $n/6$ singles and $n/6$ doubles. Thus, we retain half of the edges of C. We aim at decomposing the cycle covers such that at least half of the weight of the cycle cover is preserved.

The reason why we decompose cycle covers into singles and doubles is the following: If we decomposed them into longer paths, then we would run into trouble when trying to decompose a 3-cycle. If we restricted ourselves to decomposing the cycle covers into singles only, then 3-cycles would limit the weight preserved: We would get only one third of the edges of the 3-cycles, thus at most one third of their weight in general. Finally, if we restricted ourselves to doubles, then 5-cycles would limit the weight we could obtain since we would get only a fraction of $2/5$ of their edges.

In our approximation algorithms, we exploit the following observation: If every cycle cover on n vertices can be decomposed into α singles and β doubles, then, for every L, every L-cycle cover on n vertices can be decomposed in the same way. This implies that we can build cycle covers from such a decomposition: Given α singles and β doubles, and $n - 2\alpha - 3\beta$ isolated vertices, we can join them to form an L-cycle cover. (The only restriction is that there must exist L-cycle covers on n vertices. We refer to Section 3 for more details.)

If n is not divisible by six, we replace $n/6$ by $\lfloor n/6 \rfloor$ or $\lceil n/6 \rceil$: Assume that $n = 6k + \ell$ for $k, \ell \in \mathbb{N}$ and $\ell \leq 5$. Then we take $k + \alpha_\ell$ singles and $k + \beta_\ell$ doubles, where α_ℓ and β_ℓ are given in Table 1.

Since we want a decomposition weighing at least half of the weight of the cycle cover, we need to take at least half of the edges of the cycle cover. Otherwise, we would get a decomposition of less weight. It can be checked easily that by taking $k + \alpha_\ell$ singles and $k + \beta_\ell$ doubles, we obtain $\lceil n/2 \rceil$ edges of C.

Lemma 1 (Decomposition Lemma). *Let $C = (V, E)$ be a cycle cover on $n = 6k + \ell$ vertices such that the length of each cycle is at least three. Let $w : E \to \mathbb{N}$ be an edge weight function.*

Then there exists a decomposition $D \subseteq E$ of C with the following properties:

Table 1. A cycle cover on $n = 6k + \ell$ vertices will be decomposed into $k + \alpha_\ell$ singles and $k + \beta_\ell$ doubles

ℓ	0	1	2	3	4	5
α_ℓ	0	1	1	0	0	1
β_ℓ	0	0	0	1	1	1

- (V, D) consists of $k + \alpha_\ell$ singles, $k + \beta_\ell$ doubles, and $n - 5k - 3\beta_\ell - 2\alpha_\ell$ isolated vertices, such that all these subgraphs are pairwise vertex-disjoint, and
- $w(D) \geq \frac{1}{2} \cdot w(E)$.

The following two lemmas will simplify the proof of the decomposition lemma.

Lemma 2. *Let $\lambda, \alpha, \beta \in \mathbb{N}$ with $\alpha + 2\beta \geq \lambda/2$ and $2\alpha + 3\beta \leq \lambda$. Then every cycle c of length λ can be decomposed into α singles and β doubles such that the weight of the decomposition is at least $w(c)/2$.*

Lemma 3. *Let $\lambda \in \mathbb{N}$. Suppose that every cycle c of length λ can be decomposed into α singles and β doubles of weight at least $w(c)/2$. Then every cycle c' of length $\lambda + 6$ can be decomposed into $\alpha + 1$ singles and $\beta + 1$ doubles of weight at least $w(c')/2$.*

Lemma 3 also holds if we consider more than one cycle: Assume that every collection of k cycles of lengths $\lambda_1, \ldots, \lambda_k$ can be decomposed into α singles and β doubles such that the weight of the decomposition is at least half the weight of the cycles. Then k cycles of lengths $\lambda_1 + 6, \lambda_2, \ldots, \lambda_k$ can be decomposed into $\alpha + 1$ singles and $\beta + 1$ doubles such that also at least half of the weight of the cycles is preserved.

Due to Lemma 3, we can restrict ourselves to cycles of length at most eight in the following. The reason for this is the following: If we know how to decompose cycles of length λ, then we also know how to decompose cycles of length $\lambda + 6, \lambda + 12, \ldots$ from Lemma 3.

We now come to the proof of the decomposition lemma. The decomposition described can clearly be done in polynomial time.

Proof (Decomposition Lemma). We prove the lemma by induction on the number of cycles. As induction basis, we consider two cases:

One cycle. Due to Lemma 3, we can restrict ourselves to considering cycles of length at most eight.

3-cycles and 4-cycles have to be decomposed into one double. 5-cycles and 6-cycles have to be decomposed into one double and one single. Finally, 7-cycles and 8-cycles have to be decomposed into two singles and one double. According to Lemma 2, these decompositions can be made such that at least half of the weight of the cycle is preserved.

Two odd cycles. The two cycles can be of length three, five, or seven. Thus, there are six cases to be considered. We describe exemplarily how to decompose two 3-cycles. The other five cases are treated similarly.

Six vertices are involved, thus we need one single and one double. A single can be chosen such that at least one third of the weight of the cycle is preserved. Analogously, a double can be chosen such that at least two thirds of the weight of the cycle are preserved. We take the double of the heavier cycle and the single of the lighter cycle. Both the single and the double are chosen such that their weight is maximised. Then their total weight is at least one half of the sum of the weight of the two cycles.

As induction hypothesis, we assume that the lemma holds if the number of cycles is less than r. Assume that we have a cycle cover C consisting of r cycles. Let $n = 6k + \ell$ for the number of its vertices for $k, \ell \in \mathbb{N}$ and $\ell \leq 5$. This means that C has to be decomposed into $k + \alpha_\ell$ singles and $k + \beta_\ell$ doubles. In the following, let C' be the new cycle cover obtained by removing one or two cycles of C.

We proceed as follows: First, we show how to remove an even cycle from C. Second, we show how to remove a pair of odd cycles from C. Special care is needed when removing a pair of one 3-cycle and one 5-cycle.

Let us start by considering the removal an even cycle. It suffices to consider cycles of length four, six, or eight. The easiest case is removing a 6-cycle: We decompose it into one single and one double preserving at least half of its weight. The new cycle cover C' consists of $n - 6$ vertices. Consequently, C' can be decomposed into $k + \alpha_\ell - 1$ singles and $k + \beta_\ell - 1$ doubles by the induction hypothesis. In addition, we have one single and one double from the 6-cycle. Thus, C can be decomposed into $k + \alpha_\ell$ singles and $k + \beta_\ell$ doubles such that at least half of its weight is preserved.

If we want to remove a 4-cycle or an 8-cycle, several cases have to be distinguished. A 4-cycle has to be decomposed either into one double or into two singles. This depends on the value of ℓ: If, for instance, $\ell = 4$, then C' consists of $6k$ vertices. Thus, C' has to be decomposed into k singles and k doubles. Since $\alpha_4 = 0$ and $\beta_4 = 1$, the 4-cycle has to be decomposed into one double. On the other hand, if $\ell = 2$, then C' consists of $6(k - 1) + 4$ vertices. Thus, C' has to be decomposed into k doubles and $k - 1$ singles, while C has to be decomposed into k doubles and $k + 1$ singles. In this case, we have to decompose the 4-cycle into two singles.

If $\ell \in \{0, 3, 4, 5\}$, the 4-cycle has to be decomposed into one double. Otherwise, i.e. if $\ell \in \{1, 2\}$, it has to be decomposed into two singles. By the induction hypothesis and Lemma 2, C' and the 4-cycle can be decomposed appropriately such that at least half of the weight of both is preserved.

Analogously, an 8-cycle has to be decomposed into two doubles if $\ell \in \{3, 4\}$ or into two singles and one double if $\ell \in \{0, 1, 2, 5\}$.

Now we consider removing a pair of odd cycles. We have to distinguish six cases as we already did in the proof of the induction basis. As an example, we consider the cases of two 3-cycles and of a 3-cycle and a 7-cycle. Furthermore, we consider the case of a 3-cycle and a 5-cycle since this case needs special attention.

Twice length three. We decompose the heavier of the 3-cycles into a double and the lighter one into a single. The new cycle cover C' has $n - 6$ vertices. It thus has to be decomposed into $k + \alpha_\ell - 1$ singles and $k + \beta_\ell - 1$ doubles. Plus one single and one double from the two 3-cycles, we obtain a feasible decomposition of C.

Length three and seven. If $\ell \in \{0, 3, 4, 5\}$, then we decompose the two cycles into one single and two doubles. We take either a double of the 3-cycle and a single and a double of the 7-cycle or a single of the 3-cycle and two doubles of the 7-cycle. This depends on which alternative yields more weight. In this way, we preserve at least half of the weight of the two cycles.

If $\ell \in \{1, 2\}$, then we take three singles and one double. Either we decompose the 3-cycle into a double and the 7-cycle into three singles or the 3-cycle into a single and the 7-cycle into two singles and one double. Again, we choose the alternative that yields more weight, and we preserve at least half of the weight of the cycle cover.

Length three and five. The case of a 3-cycle and 5-cycle is a bit more complicated than the other cases. We run into troubles if, for instance, $\ell = 3$. In this case, we have to decompose the two cycles into two doubles. If the 5-cycle is much heavier than the 3-cycle, then it is impossible to preserve half of the weight of the two cycles.

However, we can avoid the problem as follows: As long as there is an even cycle, we decompose this one. After that, as long as their are at least three odd cycles, we can choose two of them such that we do not have a pair of one $(3 + 6\xi)$-cycle and one $(5 + 6\xi')$-cycle for some $\xi, \xi' \in \mathbb{N}$.

Thus, the only situation in which we cannot avoid to decompose a $(3 + 6\xi)$-cycle and a $(5 + 6\xi')$-cycle is when there are only two cycles left. In this case, we have $\ell = 2$, and we have treated this case already in the induction basis.

If there is only one odd cycle, then either $r = 1$ or all other cycles are of even length. We have already dealt with the former case in the induction basis. In the latter case, we proceed by removing even cycles as described above. □

3   Approximation Algorithms

In this section, we apply the decomposition lemma to devise approximation algorithms for restricted cycle covers both in undirected and directed graphs. The catch is that for most L it is impossible to decide whether some cycle length is in L since there are uncountably many sets L: If, for instance, L corresponds to the halting problem, then deciding whether a cycle cover is an L-cycle cover is impossible. One option would be to restrict ourselves to sets L such that the unary language $\{1^\lambda \mid \lambda \in L\}$ is in P. For such L, Max-L-UCC and Max-L-DCC are NP optimisation problems. Another possibility for circumventing the problem is to include the permitted cycle lengths in the input. While such restrictions are mandatory when we want to compute optimum solutions, they are not needed for our approximation algorithms.

A necessary and sufficient condition for a complete graph with n vertices to have an L-cycle cover is that there exist (not necessarily distinct) lengths $\lambda_1, \ldots, \lambda_k \in L$ for some $k \in \mathbb{N}$ with $\sum_{i=1}^{k} \lambda_i = n$. We call such an n L-*admissible* and define $\langle L \rangle = \{n \mid n$ is L-admissible$\}$. Although L can be arbitrarily complicated, $\langle L \rangle$ always allows efficient membership testing. In fact, it has been proved that for all $L \subseteq \mathbb{N}$, there exists a finite set $L' \subseteq L$ with $\langle L' \rangle = \langle L \rangle$ [19].

For every fixed L, we can not only test in time polynomial in n whether n is L-admissible, but we can, provided that $n \in \langle L \rangle$, also find numbers $\lambda_1, \ldots, \lambda_k \in L'$ that add up to n, where $L' \subseteq L$ denotes a finite set with $\langle L \rangle = \langle L' \rangle$. This can be done via dynamic programming in time $O(n \cdot |L'|)$, which is $O(n)$ for fixed L.

Instead of computing L'-cycle covers in the following two sections, we assume without loss of generality that already L is a finite set. This does not affect the approximation ratios achieved by our algorithms since the L-cycle covers computed are compared to optimal cycle covers without restrictions.

In general, our algorithms work as follows: They start by computing an initial cycle cover C^{init}. Then C^{init} is decomposed according to Lemma 1. Finally, the singles, doubles, and isolated vertices are joined to form an L-cycle cover C^{apx}. Since no weight is lost during the final merging, the weight of the decomposition is a lower bound for the weight of C^{apx}. With this approach, we can achieve approximation ratios of 2 and 3 for undirected and directed L-cycle covers, respectively (see Section 4, where a decomposition technique for directed graphs is sketched). We improve on the factor of 3 for directed graphs by using a more sophisticated algorithm.

3.1   Undirected Cycle Covers

Theorem 1. *Algorithm 1. is a factor 2 approximation algorithm for* Max-L-UCC *for all* $L \subseteq \mathcal{U}$.

Proof. Algorithm 1. returns $\bot$ if and only if $n \notin \langle L \rangle$. Otherwise, an L-cycle cover C^{apx} is returned. Let $C^\star$ denote an L-cycle cover of maximum weight of G. We have $w(C^\star) \leq w(C^{\mathrm{init}}) \leq 2 \cdot w(D) \leq 2 \cdot w(C^{\mathrm{apx}})$. $\square$

The running-time of the algorithm is dominated by the time needed to compute the initial cycle cover, which is $O(n^3)$ according to Ahuja et al. [1, Chap. 12].

3.2   Directed Cycle Covers

For directed graphs, we cannot apply the decomposition lemma directly. The reason is that we have to cope with 2-cycles. Therefore, we balance two approaches. The first approach is a simple matching-based algorithm: We compute a maximum-weight matching of a certain cardinality and join the edges of the matching to obtain an L-cycle cover. This works particularly well if an optimum cycle cover has much of its weight in cycles of even length. Since 2-cycles are even cycles, this works well if an optimum L-cycle cover has much of its weight in 2-cycles. The cardinality of the maximum is chosen such that an

Input: undirected complete graph $G = (V, E)$, $|V| = n$; edge weights $w : E \to \mathbb{N}$
Output: an L-cycle cover C^{apx} of G if n is L-admissible, $\perp$ otherwise
 1: **if** $n \notin \langle L \rangle$ **then**
 2:       **return** $\perp$
 3: compute a cycle cover C^{init} in G of maximum weight
 4: decompose C^{init} into a set $D \subseteq C^{\text{init}}$ of edges according to Lemma 1
 5: join the singles and doubles in D to obtain an L-cycle cover C^{apx}
 6: **return** C^{apx}

Algorithm 1. A 2-approximation algorithm for Max-L-UCC

L-cycle cover can be built from such a matching. A cycle of length λ yields a matching of cardinality $\lfloor \lambda/2 \rfloor$. Thus, a matching of cardinality d in a graph of n vertices can be extended to form an L-cycle cover if and only if $d \leq D(n, L) = \max\{\sum_{i=1}^{k} \lfloor \lambda_i/2 \rfloor \mid k \in \mathbb{N}, \sum_{i=1}^{k} \lambda_i = n, \text{ and } \lambda_i \in L \text{ for } 1 \leq i \leq k\}$.

If an optimum L-cycle cover has much of its weight in cycles of length at least three, then we compute an approximate 3-cycle cover using an approximation algorithm by Bläser et al. [6]. This 3-cycle cover is then decomposed according to the decomposition lemma. A problem with this approach is that an optimum L-cycle cover may contain 2-cycles if $2 \in L$. But a collection of τ cycles of length two can be rejoined to form two τ-cycles for some $\tau \in L$. In this way, we lose at most two thirds of their weight. We still might have to cope with $\xi < \tau$ cycles of length two. Since L is fixed, τ is a constant. Thus, we can simply try all subsets of vertices of even cardinality $2\xi \leq 2\tau - 2$, join them to form ξ cycles of length two, and remove them to proceed with the approximation algorithm on the remaining graph.

Theorem 2. *Algorithm 2. is a factor* 20/7 *approximation algorithm for* Max-L-DCC *for all* $L \subseteq \mathcal{D}$.

Proof. First, we consider the case that $2 \notin L$. The algorithm starts by computing a 4/3-approximation C_3^{init} to an optimal 3-cycle cover by using an algorithm of Bläser et al. [6]. This cycle cover, which does not contain 2-cycles, is then decomposed into singles and doubles according to the decomposition lemma. The resulting set D of edges has a weight of at least 3/8 of an optimal L-cycle cover. By joining the singles and doubles of D to an L-cycle cover, we obtain a factor 8/3 approximation. Note that $8/3 < 20/7$.

If $L = \{2\}$, then we can solve the problem optimally in polynomial time.

What remains to be considered is the case that $2 \in L$ and $L \neq \{2\}$. The main idea in this case is that we balance two approaches for computing L-cycle covers: One is that we form an L-cycle cover from a (not necessarily perfect) matching, which works particularly well if much of the weight of an optimum L-cycle cover is contained in even cycles. The other approach is using the 4/3-approximation algorithm by Bläser et al. [6] and the decomposition lemma. This works well if an optimum L-cycle cover does not contain too much weight in 2-cycles. We omit the details of the proof due to space constraints. $\square$

Input: directed complete graph $G = (V, E)$, $|V| = n$; edge weights $w : E \to \mathbb{N}$
Output: an L-cycle cover C^{apx} of G if n is L-admissible, $\bot$ otherwise
1: **if** $n \notin \langle L \rangle$ **then**
2:      **return** $\bot$
3: **if** $2 \notin L$ **then**
4:      compute a 4/3-approximation C_3^{init} to an optimal 3-cycle cover
5:      decompose C_3^{init} into a set $D \subseteq C_3^{\mathrm{init}}$ of edges according to Lemma 1
6:      join the singles and doubles in D to obtain an L-cycle C^{apx}
7: **else if** $L = \{2\}$ **then**
8:      compute a $\{2\}$-cycle cover C^{apx} of maximum weight
9: **else**
10:      compute a matching M of cardinality at most $D(n, L)$ that has maximum
           weight among all such matchings
11:      construct an L-cycle cover $C_{\mathrm{match}}^{\mathrm{apx}} \supseteq M$
12:      $\tau \leftarrow \min\{L \setminus \{2\}\}$
13:      **for all** $\xi \leftarrow 0, 2, 4, \ldots, 2\tau - 2$ **do**
14:           **for all** $V' \subseteq V$ of cardinality ξ **do**
15:                compute a maximum weight $\{2\}$-cycle cover $C_{2,V'}$ on V'
16:                remove V' from G to obtain G'
17:                compute a 4/3-approximation C_3^{init} to an optimal 3-cycle cover of G'
18:                decompose C_3^{init} into a set $D \subseteq C_3^{\mathrm{init}}$ of edges according to Lemma 1
19:                join the singles and doubles in D to obtain an L-cycle cover $C_{V'}^{\mathrm{apx}}$
20:                add $C_{2,V'}$ to $C_{V'}^{\mathrm{apx}}$
21:      let C_3^{apx} be the cycle cover of maximum weight among all $C_{V'}^{\mathrm{apx}}$
22:      let C^{apx} be the heavier cycle cover of $C_{\mathrm{match}}^{\mathrm{apx}}$ and C_3^{apx}
23: **return** C^{apx}

Algorithm 2. A 20/7-approximation algorithm for Max-L-DCC

4   Limits for Decomposition-Based Algorithms

The aim of this section is to fathom the possibilities of designing approximation algorithms for L-cycle covers that base on cycle decompositions as described in Section 2.

An approximation ratio of 2 is the best possible for undirected L-cycle covers. Hence, the algorithm presented in Section 3.1 is an optimal decomposition-based algorithm. For further improvements of the approximation ratio, we thus need more sophisticated techniques that in particular take the set L into account.

For directed L-cycle covers, already the previously known factor 3 approximation algorithm [19] can be viewed as a decomposition algorithm: Every directed cycle cover on n vertices can be decomposed into $\lceil n/3 \rceil$ singles such that at least one third of the weight of the cycle cover is preserved.

We have presented an algorithm for directed cycle covers that exploits properties of the set L: The bottleneck for the decomposition-based approach are cycles of length two since they can only be decomposed into paths of length one. By taking special care of 2-cycles, and by applying an approximation algorithm for 3-cycle covers, we were able to achieve the improved approximation ratio of 20/7.

Overall, every approximation algorithm for Max-L-UCC that works for arbitrary sets L and is purely decomposition-based achieves at best an approximation ratio of 2. For Max-L-DCC, such algorithms achieve achieve at best a ratio of 3.

5   Conclusions

One way to get better approximation algorithms is balancing several approximation algorithms as we did to achieve the ratio of 20/7 for directed L-cycle covers. One option to do this for undirected graphs might be to start with a 4-cycle cover instead of 3-cycle cover. This is possible for approximating Max-L-UCC restricted to edge weights zero and one since the corresponding 4-cycle cover problem can be solved in polynomial time [13,19]. Another option is to use approximation algorithms for 4-cycle covers. In either case, we need a decomposition lemma that preserves more than half of the weight of the cycle cover.

Finally, from a more abstract point of view, we are interested in structural properties of restricted cycle covers: Let u_L and d_L denote the best approximation ratios for undirected and directed L-cycle covers, respectively, that can be achieved by polynomial-time algorithms. What is the minimum number $u^\star$ such that all L-cycle cover problems can be approximated with a ratio of $u^\star$, i.e. what is $\sup_{L \subseteq \mathcal{U}, L \neq \emptyset} u_L$? Analogously, what is the minimum number $d^\star$ such that all L-cycle cover problems can be approximated with a ratio of $d^\star$, i.e. what is $\sup_{L \subseteq \mathcal{D}, L \neq \emptyset} d_L$? For the moment, we know $u^\star \leq 2$ and $d^\star \leq 20/7$. On the other hand, does there exist an $r > 1$ as a general lower bound for the approximability of L-cycle cover problems? What we mean by a general lower bound r is the following: If an L-cycle cover problem is NP-hard, then it cannot be approximated with a ratio of less than r unless P = NP. The reductions known so far do not yield such a general lower bound.

References

1. Ravindra K. Ahuja, Thomas L. Magnanti, James B. Orlin. *Network Flows: Theory, Algorithms, and Applications*. Prentice-Hall, 1993.
2. Giorgio Ausiello, Pierluigi Crescenzi, Giorgio Gambosi, Viggo Kann, Alberto Marchetti-Spaccamela, Marco Protasi. *Complexity and Approximation: Combinatorial Optimization Problems and Their Approximability Properties*. Springer, 1999.
3. Markus Bläser. A 3/4-approximation algorithm for maximum ATSP with weights zero and one. In *Proc. of the 7th Int. Workshop on Approximation Algorithms for Combinatorial Optimization Problems (APPROX)*, vol. 3122 of *Lecture Notes in Comput. Sci.*, pp. 61–71. Springer, 2004.
4. Markus Bläser, Bodo Manthey. Approximating maximum weight cycle covers in directed graphs with weights zero and one. *Algorithmica*, 42(2):121–139, 2005.
5. Markus Bläser, Bodo Manthey, Jiří Sgall. An improved approximation algorithm for the asymmetric TSP with strengthened triangle inequality. *J. Discrete Algorithms*, to appear.

6. Markus Bläser, L. Shankar Ram, Maxim I. Sviridenko. Improved approximation algorithms for metric maximum ATSP and maximum 3-cycle cover problems. In *Proc. of the 9th Workshop on Algorithms and Data Structures (WADS)*, vol. 3608 of *Lecture Notes in Comput. Sci.*, pp. 350–359. Springer, 2005.

7. Markus Bläser, Bodo Siebert. Computing cycle covers without short cycles. In *Proc. of the 9th Ann. European Symp. on Algorithms (ESA)*, vol. 2161 of *Lecture Notes in Comput. Sci.*, pp. 368–379. Springer, 2001. Bodo Siebert is the birth name of Bodo Manthey.

8. Avrim L. Blum, Tao Jiang, Ming Li, John Tromp, Mihalis Yannakakis. Linear approximation of shortest superstrings. *J. ACM*, 41(4):630–647, 1994.

9. Hans-Joachim Böckenhauer, Juraj Hromkovič, Ralf Klasing, Sebastian Seibert, Walter Unger. Approximation algorithms for the TSP with sharpened triangle inequality. *Inform. Process. Lett.*, 75(3):133–138, 2000.

10. L. Sunil Chandran, L. Shankar Ram. Approximations for ATSP with parameterized triangle inequality. In *Proc. of the 19th Int. Symp. on Theoretical Aspects of Computer Science (STACS)*, vol. 2285 of *Lecture Notes in Comput. Sci.*, pp. 227–237. Springer, 2002.

11. Zhi-Zhong Chen, Takayuki Nagoya. Improved approximation algorithms for metric Max TSP. In *Proc. of the 13th Ann. European Symp. on Algorithms (ESA)*, vol. 3669 of *Lecture Notes in Comput. Sci.*, pp. 179–190. Springer, 2005.

12. Zhi-Zhong Chen, Yuusuke Okamoto, Lusheng Wang. Improved deterministic approximation algorithms for Max TSP. *Inform. Process. Lett.*, 95(2):333–342, 2005.

13. David Hartvigsen. *An Extension of Matching Theory*. PhD thesis, Carnegie Mellon University, Pittsburgh, USA, 1984.

14. Refael Hassin, Shlomi Rubinstein. On the complexity of the k-customer vehicle routing problem. *Oper. Res. Lett.*, 33(1):71–76, 2005.

15. Refael Hassin, Shlomi Rubinstein. An approximation algorithm for maximum triangle packing. *Discrete Appl. Math.*, 154(6):971–979, 2006.

16. Pavol Hell, David G. Kirkpatrick, Jan Kratochvíl, Igor Kríz. On restricted two-factors. *SIAM J. Discrete Math.*, 1(4):472–484, 1988.

17. Haim Kaplan, Moshe Lewenstein, Nira Shafrir, Maxim Sviridenko. Approximation algorithms for asymmetric TSP by decomposing directed regular multigraphs. *J. ACM*, 52(4):602–626, 2005.

18. László Lovász, Michael D. Plummer. *Matching Theory*, vol. 121 of *North-Holland Mathematics Studies*. Elsevier, 1986.

19. Bodo Manthey. On approximating restricted cycle covers. In *Proc. of the 3rd Workshop on Approximation and Online Algorithms (WAOA 2005)*, vol. 3879 of *Lecture Notes in Comput. Sci.*, pp. 282–295. Springer, 2006.

20. Alexander Schrijver. *Combinatorial Optimization: Polyhedra and Efficiency*, vol. 24 of *Algorithms and Combinatorics*. Springer, 2003.

21. Z. Sweedyk. A $2\frac{1}{2}$-approximation algorithm for shortest superstring. *SIAM J. Comput.*, 29(3):954–986, 1999.

22. Oliver Vornberger. Easy and hard cycle covers. Technical report, Universität/Gesamthochschule Paderborn, 1980.

Circular-Perfect Concave-Round Graphs

Sylvain Coulonges

Laboratoire Bordelais de Recherche Informatique (LaBRI)
Talence, France

Abstract. For $1 \leq d \leq k - d$, $K_{k/d}$ denotes the graph with vertices $0, 1, \ldots, k-1$, in which i is adjacent to j if and only if $d \leq |i - j| \leq k - d$. A graph G is circular-perfect if, for every induced subgraph H of G, the infimum k/d for which H admits a homomorphism to $K_{k/d}$ is equal to the supremum k/d for which $K_{k/d}$ admits a homomorphism to H. We answer a question af Bang-Jensen and Huang by giving a complete characterization of circular-perfect concave-round graphs.

1 Introduction

Coloring the vertices of a graph is an important concept with a large variety of applications. Let $G = (V, E)$ be a graph with vertex set V and edge set E, a *k-coloring* of G is a mapping $f : V \rightarrow \{1, \ldots, k\}$ with $f(u) \neq f(v)$ if $uv \in E$, i.e., adjacent vertices receive different colors. The minimum k for which G admits a k-coloring is called the *chromatic number* $\chi(G)$; calculating $\chi(G)$ is NP-hard in general. In a set of k pairwise adjacent vertices, called *clique* K_k, all k vertices have to be colored differently. Thus the size of a largest clique in G, the *clique number* $\omega(G)$, is a trivial lower bound on $\chi(G)$; this bound is hard to evaluate as well.

Berge [2] proposed to call a graph G *perfect* if each induced subgraph $G' \subseteq G$ admits an $\omega(G')$-coloring. Perfect graphs have been recently characterized as those graphs without chordless odd cycles C_{2k+1} with $k \geq 2$, termed *odd holes*, and their complements $\overline{C}_{2k+1}$, the *odd antiholes*, as induced subgraphs (Strong Perfect Graph Theorem [4]). (The *complement* $\overline{G}$ of a graph G has the same vertex set as G and two vertices are adjacent in $\overline{G}$ if and only if they are non-adjacent in G.) In particular, the class of perfect graphs is stable under complementation [8]. Perfect graphs turned out to be an interesting and important class with a rich structure, see [9] for a recent survey. For instance, both parameters $\omega(G)$ and $\chi(G)$ can be determined in polynomial time if G is perfect [6].

1.1 Circular-Perfect and Strongly Circular-Perfect Graphs

As a generalization of perfect graphs, Zhu [14] introduced recently the class of circular-perfect graphs based on the following more general coloring concept. Define a (k, d)-*circular coloring* of a graph $G = (V, E)$ as a mapping $f : V \rightarrow \{0, \ldots, k-1\}$ with $|f(u) - f(v)| \geq d \bmod k$ if $uv \in E$. The *circular chromatic number* $\chi_c(G)$ is the minimum $\frac{k}{d}$ taken over all (k, d)-circular colorings of G; we have $\chi_c(G) \leq \chi(G)$ since every $(k, 1)$-circular coloring is a usual k-coloring of G. (Note that $\chi_c(G)$ is sometimes called the star chromatic number [3,12].)

In order to obtain a lower bound on $\chi_c(G)$, we generalize cliques as follows: Let $K_{k/d}$ with $k \geq 2d$ denote the graph with the k vertices $0, \ldots, k-1$ and edges ij if and only if $d \leq |i - j| \leq k - d$. Such graphs $K_{k/d}$ are called *circular cliques* (or sometimes antiwebs [10,13] as complements of *webs*) and are said to be *prime* if $\gcd(k, d) = 1$. Circular cliques include all cliques by $K_k = K_{k/1}$, all odd antiholes by $\overline{C}_{2k+1} = K_{2k+1/2}$, and all odd holes by $C_{2k+1} = K_{2k+1/k}$, see Figure 1. The *circular clique number* is defined as $\omega_c(G) = \max\{\frac{k}{d} : K_{k/d} \subseteq G, \gcd(k, d) = 1\}$, and we immediately obtain that $\omega(G) \leq \omega_c(G)$.

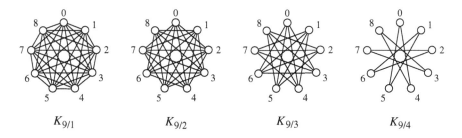

$$K_{9/1} \qquad\qquad K_{9/2} \qquad\qquad K_{9/3} \qquad\qquad K_{9/4}$$

Fig. 1. The circular cliques on nine vertices

Every circular clique $K_{k/d}$ clearly admits a (k, d)-circular coloring (simply take the vertex numbers as colors, as in Figure 1), but no (k', d')-circular coloring with $\frac{k'}{d'} < \frac{k}{d}$ by [3]. Thus we obtain, for any graph G, the following chain of inequalities:

$$\omega(G) \leq \omega_c(G) \leq \chi_c(G) \leq \chi(G). \tag{1}$$

A graph G is called *circular-perfect* if, for each induced subgraph $G' \subseteq G$, circular clique number $\omega_c(G')$ and circular chromatic number $\chi_c(G')$ coincide. Obviously, every perfect graph has this property by (1) as $\omega(G')$ equals $\chi(G')$. Moreover, any circular clique is circular-perfect as well [14,1]. Thus circular-perfect graphs constitute a proper superclass of perfect graphs.

Gyárfás introduced in [7] an other natural extension of perfect graphs. A family $\mathcal{G}$ of graphs is called χ-bound with χ-binding function b if $\chi(G') \leq b(\omega(G'))$ holds for all induced subgraphs G' of $G \in \mathcal{G}$. It is known for any graph G that $\omega(G) = \lfloor \omega_c(G) \rfloor$ by [14] and $\chi(G) = \lceil \chi_c(G) \rceil$ by [12], so, we obtain that circular-perfect graphs satisfy the following property:

$$\omega(G) \leq \chi(G) \leq \omega(G) + 1. \tag{2}$$

Thus, circular-perfect graphs are a class of χ-bound graphs with the smallest non-trivial χ-binding function. In particular, this χ-binding function is best possible for a *proper* superclass of perfect graphs implying that circular-perfect graphs admit coloring properties almost as nice as perfect graphs.

In contrary to perfect graphs, circular-perfect graphs are not stable under complementation and the list of forbidden subgraphs is unknown. In [5], *strongly circular-perfect graphs* are introduced as the subclass of circular-perfect graphs that is closed under complementation (it entails perfect graphs, odd holes, and odd antiholes) and the problem of finding the minimal not strongly circular-perfect graphs is addressed.

1.2   Convex-Round and Concave-Round Graphs

A graph G is called *convex-round* if the vertices of G can be circularly ordered $\mathcal{L} = v_1, v_2, \ldots, v_n$ such that the neighbourhood $N(v_i)$ of each vertex v_i forms an 'interval', i.e., is of the form $\{v_{l_i}, v_{l_i+1}, \ldots, v_{r_i}\}$ where additions are modulo n. We shall refer to the circular ordering $\mathcal{L}$ as a *convex-round enumeration* of G. The complement of a convex-round graph is called concave-round. For concave-round graphs, we shall refer to the circular ordering as a *concave-round enumeration*.

It is easy to see that every induced graph of a circular clique is convex-round. In [1], Bang-Jensen and Huang showed that convex-round graphs are circular-perfect. They also state the following problem (Problem 5.3): *characterize circular-perfect concave-round graphs*. This is equivalent to characterize which convex-round graphs are strongly circular-perfect.

In [5], a complete characterization of strongly circular-perfect circular cliques is given:

Theorem 1. *The only prime circular cliques induced subgraphs of a strongly circular-perfect graph are cliques, odd holes and odd antiholes*

2   Results

In this paper, we solve the problem given by Bang-Jensen and Huang by giving a complete characterization of minimal circular-imperfect concave-round graphs. By Theorem 1, we know that we have 2 cases to study: concave-round graphs containing an odd hole, and concave-round graphs containing an odd antihole. The problem is solved by the 2 following theorems.

Theorem 2. *Let $p \geq 3$. Let G be a concave-round graph containing $\overline{C_{2p+1}}$. G is minimal circular-imperfect if and only if G has only 2 vertices u and v not belonging to $\overline{C_{2p+1}}$ such that the non-neighborhood of u and v in $\overline{C_{2p+1}}$ is at most 2 vertices and*
(1) $uv \notin E(G)$ (see Figure 2 for an example), or
(2) there are $p - 1$ vertices between u and v in the ordering and u (resp. v) is not adjacent to the vertex following v (resp. preceding u) (see Figure 3 for an example).

Theorem 3. *Let G be a concave-round graph containing an induced odd hole with a concave-round enumeration $\mathcal{L} = v_1, v_2, \ldots, v_n$. Let $A_k = \{i \mid v_i v_{i+k} \notin E(G)\}$. G is minimal circular-imperfect if and only if*

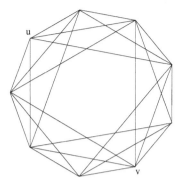

Fig. 2. A minimal circular-imperfect graph containing a $\overline{C_7}$

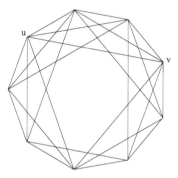

Fig. 3. Minimal circular-imperfect graphs containing an odd antihole

  – $A_1 = \emptyset$,
  – $A_2 = \{3\lambda_1, 3\lambda_2, \ldots, 3\lambda_l\}$ *with* $\lambda_1 < \lambda_2 < \ldots < \lambda_l$ *integers,* $\lambda_l - \lambda_1 < n/3$
  – $A_k = \{1, 2, \ldots, n\}$ *for every* $2 < k < n - 2$,
  – $n = 1 \ (mod\ 3)$.

(see Figure 4 for an example)

Theorem 2 solve the problem for concave-round graphs containing an odd anti-hole of size at least 7, and Theorem 3 for concave-round graphs containing an odd hole. Remark that the graphs of Theorem 3 satisfy $\omega(G) = 3$.

The proofs of those 2 theorems can be found in the next two sections.

3   Proof of Theorem 2

In this section, we give the proof of the characterization of minimal circular-imperfect concave-round graphs containing an odd antihole of size at least 7.

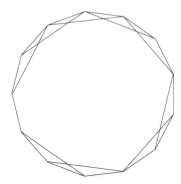

Fig. 4. A minimal circular-imperfect graph containing C_9

Lemma 1 ([1]). *Let G be a connected concave-round graph with concave-round enumeration $\mathcal{L} = v_1, v_2, \ldots, v_n$. If $v_i v_{i+k} \in E(G)$ for some i, then either G can be partitioned into two cliques, or at least one of $[v_i, v_{i+k}]$, $[v_{i+k}, v_i]$ is a clique.*

Lemma 2. *Consider the odd antihole with $2p + 1$ vertices. This graph is a concave-round graph with the ordering $\mathcal{L} = c_1, \ldots, c_{2p+1}$ such that $c_i c_j$ is an edge if and only if $\min\{|i - j|, |j - i|\} < p$. $\mathcal{L}$ is unique up to translation and inversion.*

Proof. For every concave-round graph, the inversion or a translation of a concave-round enumeration is a concave-round enumeration. For every enumeration $\mathcal{L}'$ not equivalent to $\mathcal{L}$ by translation or inversion, there exists an integer i such that c_i and c_{i+1} are not successive in $\mathcal{L}'$. $\mathcal{L}'$ is not a concave-round enumeration since the non-neighborhood of c_{i-p} is $\{c_i, c_{i+1}\}$, and so, by definition, must be an interval.

In the next 3 lemmas, we will study the concave-round graphs formed by an odd antihole and an additional vertex.

Lemma 3. *Let G be a concave-round graph with a concave-round enumeration $\mathcal{L} = v_1, v_2, \ldots, v_{2p+2}$, such that $V(G) - \{v_{2p+2}\}$ induce the odd antihole of $2p + 1$ vertices. Then, v_{2p+2} is adjacent to $v_1, v_2, \ldots, v_{p-1}, v_{2p+1}, v_{2p}, \ldots, v_{p+3}$, and possibly to one or both of v_p and v_{p+2}.*

Proof. In $G - \{v_{2p+2}\}$, the non-neighborhood of v_{p+1} is $\{v_{2p+1}, v_1\}$, so, in G, v_{p+1} is not adjacent to v_{2p+2}. Since v_{2p+1} is adjacent to v_{p-1}, by lemma 1, either $[v_{2p+1}, v_{p-1}]$ or $[v_{p-1}, v_{2p+1}]$ is a clique. Since $v_{2p+1} v_p \notin E(G)$, $[v_{2p+1}, v_{p-1}]$ is a clique, and so, v_{2p+2} is adjacent to $v_1, v_2, \ldots, v_{p-1}$. With the same arguments, we obtain that v_{2p+2} is adjacent to $v_{2p+1}, v_{2p}, \ldots, v_{p+3}$.

Lemma 4. *Let $p \geq 3$. Let G be a concave-round graph with a concave-round enumeration $\mathcal{L} = v_1, v_2, \ldots, v_{2p+2}$ such that $V(G) - \{v_{2p+2}\}$ induce the odd antihole of $2p + 1$ vertices. Then, $\omega(G) = p + 1$ and $\alpha(G) = 2$.*

Proof. By Lemma 3, the non-neighborhood of v_{2p+2} is 1, 2 or 3 consecutive vertices in the ordering. So, since $p \geq 3$, the non-neighborhood of any vertex forms a clique, and so, every stable set containing u is of cardinality at most 2. We obtain $\alpha(G) = 2$.

Moreover, $\{v_{2p+1}, v_{2p+2}, v_1, \ldots, v_{p-1}\}$ form a clique of cardinality $p + 1$.

Lemma 5 ([5]). *Let $p \geq 3$. Let G be a concave-round graph with a concave-round enumeration $\mathcal{L} = v_1, v_2, \ldots, v_{2p+2}$ such that $V(G) - \{v_{2p+2}\}$ induce the odd antihole of $2p + 1$ vertices. Then, G is circular-imperfect if and only if $v_{2p+2}v_p$ and $v_{2p+2}v_{p+2}$ are not edges of G. Moreover, if G is not circular-perfect, then G is not minimal circular-imperfect neither.*

Proof of Theorem 2: First, let's prove that if (1) is satisfied, then G is minimal circular-imperfect. Since u and v are not adjacent, by Lemma 2, we know that there are at least $(p - 1)$ vertices between them in the ordering. Consider that there exists a vertex not adjacent to both u and v. This vertex must also be at distance at least p of u and v in the ordering. So G must contain at least $3p$ vertices. Moreover, since the non-neighborhood of u and v in $\overline{C_{2p+1}}$ is at most 2 vertices, G must contain at least $3p + 1$ vertices. So, since $|V(G)| = 2p + 3$ and $p \geq 3$, such a vertex does not exists and the only stable set containing u and v is $\{u, v\}$. By Lemma 4, we obtain $\alpha(G) = 2$. Since u and v are not adjacent, Lemma 4 tells us that $\omega(G) = p + 1$. It is now clear that we cannot partition G into $\omega(G)$ stable sets. So, G is not circular-perfect. Lemma 5 tells us that G is minimal circular-imperfect.

If G verifies the condition (2), it is also easy to see that $\alpha(G) = 2$, and $\omega(G) = p + 1$.

Now, consider the other graphs. Let $\{u_1, \ldots, u_k\}$ be the vertices of G not belonging to $\overline{C_{2p+1}}$, $\{u_1, \ldots, u_k\}$ form a clique. If u_1 and u_k have p vertices of the odd antihole between them in the ordering, then those p vertices and $\{u_1, \ldots, u_k\}$ form a clique of cardinality $p + k$. If u_1 and u_k have $(p - 1)$ vertices of the odd antihole between them in the ordering, then those p vertices and $\{u_1, \ldots, u_k\}$ form a clique of cardinality $p + k - 1$. Moreover, since we are not in the condition (2) , either the vertex preceding u_1 or the vertex following u_k is adjacent to all those vertices and so, $\omega(G) = p + k$. If u_1 and u_k have less than $p - 1$ vertices of the odd antihole between them in the ordering, then, we also have a clique of cardinality $p + k$. So, in any case, we have $\omega(G) = p + k$. We also have $\alpha(G) = 2$. We can easily cover those $2p + 1 + k$ vertices with $p + k$ stable sets.

4   Proof of Theorem 3

In this section, we are going to show which concave-round graphs containing an induced odd hole are minimal circular-imperfect. The only concave-round graphs with clique size 2 are the holes and so are circular-perfects. We are first going to study the case where $\omega(G) = 3$, and then show that no concave-round graphs containing an induced odd hole with clique size at least 4 are minimal circular-imperfect.

Lemma 6. *Let G be a K_4-free concave-round graph containing an induced odd hole with concave-round enumeration $\mathcal{L} = v_1, v_2, \ldots, v_n$. Let $A_k = \{i \mid v_i v_{i+k} \notin E(G)\}$. Then, $A_1 = \emptyset$, $A_l = \{1, 2, \ldots, n\}$, for every $3 \leq l \leq n - 3$.*

Proof. This is a direct application of Lemma 1.

Proof of Theorem 3 when $\omega(G) = 3$: If $n \equiv 0 \pmod 3$, then the 3 stable sets $S_i = \{v_{3p+i} \mid 0 \leq p < n/3\}$, $1 \leq i \leq 3$ form a partition of G, and, so G is not minimal circular-imperfect.

If $n \not\equiv 0 \pmod 3$, and if the first 3 conditions are not verified, we are going to show that G is 3−colorable, and so not minimal circular-imperfect. Let $I = \{i_1, i_2, \ldots, i_k\} \subseteq [n]$, such that $i \in I$ if and only if $v_{i-1} v_{i+1} \notin E(G)$, and $i_j < i_{j'}$ if and only if $j < j'$. Since the first 3 conditions of the theorem are not verified, there exist l and l' such that $i_{l+1} - i_l \not\equiv 0 \pmod 3$ and $i_{l'+1} - i_{l'} \not\equiv 0 \pmod 3$. Let's 3-color G. First assign $c(v_{i_l}) = 1$. Now, we are going to color the vertices of the graph according to the concave-round enumeration. Without lost of generality, suppose that for every $l < m < l'$, $i_{m+1} - i_m \equiv 0 \pmod 3$. Depending on the choice of $c(v_{i_l+1})$, we can color v_{i_l+1} with either color 2 or 3. The same property occurs for $v_{i_{l'}}$. Next, depending on the choice of $c(v_{i_l+1})$ and $c(v_{i_{l'}+1})$, we can color $v_{i_{l'+1}}$ with any of the 3 colors. Since the choice of $c(v_{i_{l'+1}+1})$ does not depend of $c(v_{i_{l'+1}-1})$, we can color the rest of the graph with 3 colors.

Now, we are going to show that if $n \not\equiv 0 \pmod 3$, and if the first 3 conditions are verified, then G is not circular-perfect. We can see in Theorem 2 that no minimal circular-imperfect graphs containing an induced odd antihole contains an induced odd hole. So, we just have to prove that $\chi(G) > 3$. Let's try to 3-color G. First let's assign $c(v_{3\lambda_1}) = 1$. Since $\{v_{3\lambda_1}, v_{3\lambda_1+1}, v_{3\lambda_1+2}\}$ and $\{v_{3\lambda_1+1}, v_{3\lambda_1+2}, v_{3\lambda_1+3}\}$ form cliques, we are forced to assign $c(v_{3\lambda_1+3}) = 1$. We can do this recursively, and since $n \not\equiv 0 \pmod 3$, one of $v_{3\lambda_1-1}$ and $v_{3\lambda_1-2}$ is colored with color 1 which is not possible, since $\{v_{3\lambda_1-2}, v_{3\lambda_1-1}, v_{3\lambda_1}\}$ form a clique.

Moreover, if $n \equiv 2 \pmod 3$, deleting $v_{3\lambda_1-1}$ leads to a circular-imperfect graph. Finally, if $n \equiv 1 \pmod 3$, deleting any vertex leads to a circular-perfect graph.

Proof of the Theorem 3 when $\omega(G) > 3$: We are first going to prove that if G is minimal circular-imperfect, then $n \equiv 1 \pmod{\omega(G)}$. Let $n = p\omega(G) + r$, $0 \leq r < \omega(G)$. If $r = 0$ then G is $\omega(G)$-colorable, and so not minimal circular-imperfect. We are going to prove that $\alpha(G) = p$. Clearly, $\{v_{\omega(G)i} \mid 0 \leq i < p\}$ form a stable set of size p. Suppose that there exists stable sets of size $p + 1$. Without lost of generality suppose that v_1 belongs to one of them. By Lemma 1, a natural way to find a maximal stable set containing v_1 is to take v_1, then the first one in the concave-round enumeration not adjacent to v_1, and so on. This gives us a stable set S with $|S| \geq p + 1$. If the last vertex taken is at distance from v_1 greater than $\omega(G)$ in the enumeration, delete this one from S and take $v_{1-\omega(G)}$ instead. This gives us a new stable set. Now, if the one before the last is at distance from $v_{1-\omega(G)}$ more than $\omega(G)$, delete it and take $v_{1-2\omega(G)}$. Let's go on until it stops (it is necessarily the case since $|S| \geq p + 1$). Finally we obtain $\omega(G - S) = \omega(G) - 1$, and so, if G is not $\omega(G)$-colorable, then $G - S$ is not

$\omega(G-S)$-colorable. This is a contradiction since G is minimal circular-imperfect. So, $\alpha(G) = p$. If $r > 1$, then, for any vertex u, $G - \{u\}$ cannot be covered with $\omega(G)$ stable sets and so, is not $\omega(G)$-colorable. We finally obtain that $r = 1$.

Now let's prove that if $n = p\omega(G) + 1$, then G is not minimal circular-imperfect. By theorem 1 and Trotter [11], we know that no webs of clique size greater than 3 are minimal circular-imperfect. So, there exists i such that $v_i v_{i+\omega(G)-1} \notin E(G)$. Let $S = \{v_{i-1}, v_{i+\omega(G)}\} \bigcup \{v_{i\omega(G)k} \mid 2 \leq k < p\}$. S is a stable set whose deletion decreases the maximal clique size. So, with same argument as before, G is not minimal circular-imperfect.

References

1. J. Bang-Jensen and J. Huang, *Convex-round graphs are circular-perfect*. J. Graph Theory 40 No.3 (2002) 182–194
2. C. Berge, *Färbungen von Graphen, deren sämtliche bzw. deren ungerade Kreise starr sind*, Wiss. Zeitschrift der Martin-Luther-Universität Halle-Wittenberg 10 (1961) 114–115.
3. J.A. Bondy and P. Hell, *A note on the star chromatic number*. Journal of Graph Theory 14 (1990) 479–482.
4. M. Chudnovsky, N. Robertson, P. Seymour, and R. Thomas, *The Strong Perfect Graph Theorem*. To appear in: Annals of Mathematics.
5. S. Coulonges, A. Pêcher, and A. Wagler, *Triangle-free strongly circular-perfect graphs*. submitted to Discrete Mathematics
6. M. Grötschel, L. Lovász, and A. Schrijver, *Geometric Algorithms and Combinatorial Optimization*. Springer-Verlag (1988).
7. A. Gyárfás, *Problems from the world surrounding perfect graphs*. Zastos. Mat. 19 (1987) 413–431.
8. L. Lovász, *Normal hypergraphs and the weak perfect graph conjecture*, Discrete Math. 2 (1972) 253–267.
9. B. Reed and J. Ramirez-Alfonsin, *Perfect Graphs*, Wiley, 2001.
10. F.B. Shepherd, *Applying Lehman's Theorem to Packing Problems*. Math. Programming 71 (1995) 353–367.
11. L.E. Trotter, jr., *A Class of Facet Producing Graphs for Vertex Packing Polyhedra*. Discrete Math. 12 (1975) 373–388
12. A. Vince, *Star chromatic number*, Journal of Graph Theory 12 (1988) 551–559.
13. A. Wagler, *Antiwebs are rank-perfect*. Quarterly Journal of the Belgian, French and Italian OR Societies 2 (2004) 149-152.
14. X. Zhu, *circular-perfect Graphs*. J. of Graph Th. 48 (2005), 186–209.

Author Index

Lecture Notes in Computer Science

For information about Vols. 1–4188

please contact your bookseller or Springer

Vol. 4231: J. F. Roddick, R. Benjamins, S.S.-S. Cherfi, R. Chiang, C. Claramunt, R. Elmasri, F. Grandi, H. Han, M. Hepp, M. Hepp, M. Lytras, V.B. Mišić, G. Poels, I.-Y. Song, J. Trujillo, C. Vangenot (Eds.), Advances in Conceptual Modeling - Theory and Practice. XXII, 456 pages. 2006.

Vol. 4229: E. Najm, J.F. Pradat-Peyre, V.V. Donzeau-Gouge (Eds.), Formal Techniques for Networked and Distributed Systems - FORTE 2006. X, 486 pages. 2006.

Vol. 4228: D.E. Lightfoot, C.A. Szyperski (Eds.), Modular Programming Languages. X, 415 pages. 2006.

Vol. 4227: W. Nejdl, K. Tochtermann (Eds.), Innovative Approaches for Learning and Knowledge Sharing. XVII, 721 pages. 2006.

Vol. 4225: J.F. Martínez-Trinidad, J.A. Carrasco Ochoa, J. Kittler (Eds.), Progress in Pattern Recognition, Image Analysis and Applications. XIX, 995 pages. 2006.

Vol. 4224: E. Corchado, H. Yin, V. Botti, C. Fyfe (Eds.), Intelligent Data Engineering and Automated Learning – IDEAL 2006. XXVII, 1447 pages. 2006.

Vol. 4223: L. Wang, L. Jiao, G. Shi, X. Li, J. Liu (Eds.), Fuzzy Systems and Knowledge Discovery. XXVIII, 1335 pages. 2006. (Sublibrary LNAI).

Vol. 4222: L. Jiao, L. Wang, X. Gao, J. Liu, F. Wu (Eds.), Advances in Natural Computation, Part II. XLII, 998 pages. 2006.

Vol. 4221: L. Jiao, L. Wang, X. Gao, J. Liu, F. Wu (Eds.), Advances in Natural Computation, Part I. XLI, 992 pages. 2006.

Vol. 4219: D. Zamboni, C. Kruegel (Eds.), Recent Advances in Intrusion Detection. XII, 331 pages. 2006.

Vol. 4218: S. Graf, W. Zhang (Eds.), Automated Technology for Verification and Analysis. XIV, 540 pages. 2006.

Vol. 4217: P. Cuenca, L. Orozco-Barbosa (Eds.), Personal Wireless Communications. XV, 532 pages. 2006.

Vol. 4216: M.R. Berthold, R. Glen, I. Fischer (Eds.), Computational Life Sciences II. XIII, 269 pages. 2006. (Sublibrary LNBI).

Vol. 4215: D.W. Embley, A. Olivé, S. Ram (Eds.), Conceptual Modeling - ER 2006. XVI, 590 pages. 2006.

Vol. 4213: J. Fürnkranz, T. Scheffer, M. Spiliopoulou (Eds.), Knowledge Discovery in Databases: PKDD 2006. XXII, 660 pages. 2006. (Sublibrary LNAI).

Vol. 4212: J. Fürnkranz, T. Scheffer, M. Spiliopoulou (Eds.), Machine Learning: ECML 2006. XXIII, 851 pages. 2006. (Sublibrary LNAI).

Vol. 4211: P. Vogt, Y. Sugita, E. Tuci, C. Nehaniv (Eds.), Symbol Grounding and Beyond. VIII, 237 pages. 2006. (Sublibrary LNAI).

Vol. 4210: C. Priami (Ed.), Computational Methods in Systems Biology. X, 323 pages. 2006. (Sublibrary LNBI).

Vol. 4209: F. Crestani, P. Ferragina, M. Sanderson (Eds.), String Processing and Information Retrieval. XIV, 367 pages. 2006.

Vol. 4208: M. Gerndt, D. Kranzlmüller (Eds.), High Performance Computing and Communications. XXII, 938 pages. 2006.

Vol. 4207: Z. Ésik (Ed.), Computer Science Logic. XII, 627 pages. 2006.

Vol. 4206: P. Dourish, A. Friday (Eds.), UbiComp 2006: Ubiquitous Computing. XIX, 526 pages. 2006.

Vol. 4205: G. Bourque, N. El-Mabrouk (Eds.), Comparative Genomics. X, 231 pages. 2006. (Sublibrary LNBI).

Vol. 4204: F. Benhamou (Ed.), Principles and Practice of Constraint Programming - CP 2006. XVIII, 774 pages. 2006.

Vol. 4203: F. Esposito, Z.W. Raś, D. Malerba, G. Semeraro (Eds.), Foundations of Intelligent Systems. XVIII, 767 pages. 2006. (Sublibrary LNAI).

Vol. 4202: E. Asarin, P. Bouyer (Eds.), Formal Modeling and Analysis of Timed Systems. XI, 369 pages. 2006.

Vol. 4201: Y. Sakakibara, S. Kobayashi, K. Sato, T. Nishino, E. Tomita (Eds.), Grammatical Inference: Algorithms and Applications. XII, 359 pages. 2006. (Sublibrary LNAI).

Vol. 4200: I.F.C. Smith (Ed.), Intelligent Computing in Engineering and Architecture. XIII, 692 pages. 2006. (Sublibrary LNAI).

Vol. 4199: O. Nierstrasz, J. Whittle, D. Harel, G. Reggio (Eds.), Model Driven Engineering Languages and Systems. XVI, 798 pages. 2006.

Vol. 4198: O. Nasraoui, O. Zaiane, M. Spiliopoulou, B. Mobasher, B. Masand, P. Yu (Eds.), Advances in Web Minding and Web Usage Analysis. IX, 177 pages. 2006. (Sublibrary LNAI).

Vol. 4197: M. Raubal, H.J. Miller, A.U. Frank, M.F. Goodchild (Eds.), Geographic, Information Science. XIII, 419 pages. 2006.

Vol. 4196: K. Fischer, I.J. Timm, E. André, N. Zhong (Eds.), Multiagent System Technologies. X, 185 pages. 2006. (Sublibrary LNAI).

Vol. 4195: D. Gaiti, G. Pujolle, E. Al-Shaer, K. Calvert, S. Dobson, G. Leduc, O. Martikainen (Eds.), Autonomic Networking. IX, 316 pages. 2006.

Vol. 4194: V.G. Ganzha, E.W. Mayr, E.V. Vorozhtsov (Eds.), Computer Algebra in Scientific Computing. XI, 313 pages. 2006.

Vol. 4193: T.P. Runarsson, H.-G. Beyer, E. Burke, J.J. Merelo-Guervós, L.D. Whitley, X. Yao (Eds.), Parallel Problem Solving from Nature - PPSN IX. XIX, 1061 pages. 2006.

Vol. 4192: B. Mohr, J.L. Träff, J. Worringen, J. Dongarra (Eds.), Recent Advances in Parallel Virtual Machine and Message Passing Interface. XVI, 414 pages. 2006.

Vol. 4191: R. Larsen, M. Nielsen, J. Sporring (Eds.), Medical Image Computing and Computer-Assisted Intervention – MICCAI 2006, Part II. XXXVIII, 981 pages. 2006.

Vol. 4190: R. Larsen, M. Nielsen, J. Sporring (Eds.), Medical Image Computing and Computer-Assisted Intervention – MICCAI 2006, Part I. XXXVVIII, 949 pages. 2006.

Vol. 4189: D. Gollmann, J. Meier, A. Sabelfeld (Eds.), Computer Security – ESORICS 2006. XI, 548 pages. 2006.